Recent Advances in Robotics

Edited by

GERARDO BENI

SUSAN HACKWOOD

University of California
Santa Barbara, California

A Wiley-Interscience Publication

JOHN WILEY & SONS

New York • Chichester • Brisbane • Toronto • Singapore

Published by John Wiley & Sons, Inc.

ISBN 0-471-88383-2

ISSN 0749-1603

Printed in the United States of America

10 9 8 7 6 5 4 3 2 1

Contributors

S. Desa
Design Division
Department of Mechanical Engineering
Stanford University
Stanford, California

George Devol
Devol Research Associates
Fort Lauderdale, Florida

Joseph Duffy
Center of Intelligent Machines and Robotics
University of Florida
Gainesville, Florida

Bernard Espiau
IRISA, Institut de Recherche en Informatique et Systemes Aleatoires
Campus Universitaire de Beaulieu
Rennes, France

Ernest L. Hall
Center for Robotics Research
University of Cincinnati
Cincinnati, Ohio

Leon D. Harmon
Department of Biomedical Engineering
Case Western Reserve University
Cleveland, Ohio

Ewald Heer
Jet Propulsion Laboratory
California Institute of Technology
Pasadena, California

J. J. Kessis
Laboratoire de Robotique et Intelligence Artificielle
Université Paris
Paris, France

Harvey Lipkin
Center of Intelligent Machines and Robotics
University of Florida
Gainesville, Florida

Charles A. McPherson
C. S. Draper Laboratory
Cambridge, Massachusetts

J. Penne
Laboratoire de Robotique et Intelligence Artificielle
Université Paris
Paris, France

J. P. Rambaut
Laboratoire de Robotique et

Intelligence Artificielle
Université Paris
Paris, France

B. Roth
Design Division
Department of Mechanical Engineering
Stanford University
Stanford, California

Carl F. Ruoff
Jet Propulsion Laboratory
California Institute of Technology
Pasadena, California

J. Kenneth Salisbury, Jr.
Artificial Intelligence Laboratory
Massachusetts Institute of Technology
Cambridge, Massachusetts

G. N. Saridis
School of Engineering
Electrical, Computer, and Systems Engineering Department
Rensselaer Polytechnic Institute
Troy, New York

R. Wood
Laboratoire de Robotique et Intelligence Artificielle
Université Paris
Paris, France

Series Preface

The knowledge of robotics science and technology has grown tremendously since the beginning of the 1980s. This growth has occurred on an international scale and perhaps has been more rapid than growth in most other major fields of engineering. Although this growth of robotics has not been characterized by radical changes in basic science, it has greatly extended the theory and the understanding of its implications. The viewpoints and activities in certain closely allied fields—particularly control, mechanics, AI systems, and sensor science—have been influenced markedly by developments in robotics.

As a result of this expansion in knowledge roboticists are finding that in order to make significant contributions they must concentrate their efforts in narrower fields than formerly. Because of this specialization it is desirable that a mechanism exist whereby investigators and students can readily obtain a balanced view of the whole field. Also, it would be very helpful to workers in allied fields if the results of robotics science pertinent to their activity were readily available. However, descriptions of robotics developments are widely dispersed in many journals so that it is difficult to obtain a broad and unified picture of the field. Although excellent texts have appeared recently, many scientists and engineers have come to recognize the need for an up-to-date treatment of robotics science and technology that reviews comprehensively all of the important facets of the subject. The purpose of the present series is to fulfill this need, at least in part, by publication of compact and authoritative reviews of the important areas of the field.

The plan of publication selected is similar to that used in the "Advances" series covering various other scientific fields. Well-qualified scientists and engineers, many of whom are in the early years of their professional lives, will prepare articles on parts of the field that seem ripe for presentation. Three general types of article are solicited: (1) broad elementary surveys that have particular value in orienting the advanced graduate student or an investigator having little previous knowledge of the subject; (2) broad surveys of fields of advanced research that inform and stimulate the more experienced investigators; and (3) more specialized articles describing important new techniques, both experimental and theoretical. It is planned that the authorship be international even though the articles will be written in English.

It would be desirable to publish the articles in a more highly organized sequence than that which actually will be followed. However, it appeared that an ideally organized presentation sequence could not be adhered to without serious delays in publication. Therefore, the actual selection of articles for a given volume results from compromise between considerations such as timeliness of the subjects, availability of authors, and interrelation of subject matter.

The volumes will be published annually for a period of time. It is hoped that this schedule will be sufficient to fulfill, in a reasonable period, the pressing need for a survey and to ensure that individual articles will not be subjected to long publication delays.

GERARDO BENI
SUSAN HACKWOOD

Santa Barbara, California
October 1984

Preface

We begin this first volume of Recent Advances in Robotics with a word of caution. This collection of chapters does not have, and was never intended to have, the level of homogeneity and completeness that is required by a treatise. The chapters should be regarded as articles of the type described in the Preface to the series. As explained there, the actual selection of articles for a given volume results from a compromise between timeliness of subject, availability of authors, and interrelation of topics. Within these limits, for this first volume we strove to choose topics that would present as broad a view as possible of the general field. Thus this volume can be read also as a general orientation to robotics.

The volume is divided in three parts. Part I is dedicated to applications, Part II to mechanics, and Part III to sensors.

The text begins with an interview with George Devol, inventor of the industrial robot. This interview is an exception to the publication criteria described in the Preface to the series. The interview is included as a special recognition to the pioneering contributions of George Devol to robotics.

In Chapter 1, Ewald Heer presents an overview of the general problems of robotics in modern industry, including future trends and arguments for and against robotics. In the second chapter, G. N. Saridis reviews robotic control for application to helping the disabled. A hierarchically intelligent control approach, developed to manage man–machine interactive systems, requiring advanced decision making, is used to develop "intelligent" artificial limbs and robotic all-purpose servants in a hospital environment. It is based on the principle of increasing intelligence with decreasing precision and it is meant to provide the means to assist disabled people with minimum training or other mental effort.

Part II comprises Chapters 3 to 6. In Chapter 3, S. Desa and B. Roth describe several recent developments in the kinematic and dynamic analysis of manipulators. This chapter is also an exceptionally clear introduction to the mechanics of open-loop manipulators. The fourth chapter, by J. Kenneth Salisbury, Jr., deals with issues central to extending our use and understanding of articulated hands as manipulators. Coupled with more intelligent robot systems, articulated hands promise to be of major importance in the future of robotics.

In Chapter 5, Harvey Lipkin and Joseph Duffy develop a novel method for generating the trigonometrical equations that determine the computation of joint displacements in a manipulator. The method is based on elementary vector operations, relating the geometry of the robot to the trigonometric formulation. In the sixth chapter, J. J. Kessis, J. P. Rambaut, J. Penne, and R. Wood analyze six-legged walking robots. Mobile robots can perform tasks such as spatial and terrestrial exploration, mineral prospecting, maintenance and repair of difficult environment (e.g., nuclear plants), and farming. Current research on mobile robots has the potential to open such tasks to robots if a wide range of robotics problems can be solved.

The third part of this book contains four chapters on different sensors. First and foremost among the robotic senses is vision. Ernest L. Hall presents in Chapter 7 a review of three-dimensional robot vision techniques with emphasis on curved surface measurement and representation. Techniques for obtaining surface shape from shading information are also described.

In Chapter 8 Bernard Espiau discusses the use of optical reflectance sensors for parts acquisition and pattern recognition. The author raises the issues of sensor integration and interpretation and describes a practical embodiment.

Chapter 9, by Carl F. Ruoff, covers several aspects of device organization in advanced robot systems. It describes hierarchical ways of collecting devices in task-oriented structures.

The last chapter is a detailed review of tactile sensing. It contains a lucid discussion of one of the most exciting new fields of robotics. Unfortunately it has also become an occasion of sadness for the passing of Leon D. Harmon, author of the chapter, during the publication stage of this volume.

We wish to thank our contributors, each an acknowledged expert in his own specialty, for their fine cooperation in the preparation of the camera-ready manuscript. We are also very grateful to Ms. Connie Santini for her excellent contribution to the completion of this volume. Finally we thank Ms. Darlene McGuire and her co-workers in the AT&T Bell Laboratories Text Processing Center for their technical and formatting skills.

GERARDO BENI
SUSAN HACKWOOD

Santa Barbara, California
October 1984

Contents

Recent Advances in Robotics

The Future of the Automated Factory: An Interview with GEORGE DEVOL

Question: Mr. Devol, the next ten years promise to bring exciting new developments in factory automation. Where do you think robotics is heading?

Answer: Well, I think the overall business will keep increasing, but one of the main problems with Robots (or as I call them numerically controlled manipulators) is that they are not isolated entities but part of a system. At the moment people's idea of automating a process is to install specialized robotic equipment in a production line. These so-called work cells perform a small part of the total production. Partial robotization really does not have the payoff that people are looking for. I believe this has held back the overall sale of robots. I am in favor of putting together a robot system to automate a whole production line.

Question: Do you think that you should restructure the factory from scratch?

Answer: With completely automatic manufacturing facilities it will be necessary to design the building from scratch. This is because the work carriers (such as mobile carts) that transport materials from one machining center to another have to be under computer control. One way to do this is to bury wires in the concrete floor on a matrix basis. You can then direct the work carriers to the correct loading, storage or work station. A further point is that the overall storage facilities necessary for work currently in progress or finished materials will be part of the building. The third consideration is concerned with tax breaks. The write-off for real estate in such a building is long term compared to the write-off for machinery. Therefore, building a special factory may make more sense as far as business is concerned.

Question: Will this not prevent small businesses from using robots because they cannot afford to remake a whole factory?

Answer: That is exactly why we are intending to assemble complete plants and then lease them to the users. This eliminates a considerable portion of the financial consideration.

Question: How does this leasing work?

Answer: For instance if we go to a certain industry, let's say, that manufactures centrifugal pumps. By analysis we find out we can build a facility that will make many different types of centrifugal pumps as well as small gas engines, gear boxes and compressors in the same plant with the same machinery. The only changes to be made to the plant is in reprogramming and changing the actual tooling on the machines, but other than that it is the same facility. In that way the economics will pay off fast. The overall savings are enormous. By running a plant in three shifts, 24 hours a day you automatically gain about 20 percent. That means a 20 percent reduction in costs even before taking into account the labor saved.

Question: Where would a factory of this sort be?

Answer: Due to the fact that few employees are required (mostly supervisory and maintenance personnel), the plant can be almost anywhere. The only consideration would be the availability of raw materials. It is likely the plant will be somewhere in the center part of the US, to minimize shipping costs.

Question: So you would lease to users whose main location is in a different part of the country but they would do their manufacturing in your centrally located factory?

Answer: Yes, that could be. This factory does not have to be directly connected to any of their own manufacturing facilities. It is a whole new facility. It, therefore, can be in almost any location.

Question: Does that mean that all manufacturing is going to be done in one location?

Answer: Only the part that is being done completely automatically. Their own normal business can be wherever they want. In other words, once we have reduced the labor content of a given product to a minimum there is nowhere else to go. All competition will have to do the same thing. We are talking about going so far that there is practically no longer any consideration for labor cost in manufacturing. This is quite a large jump.

Question: Would this apply only to new industries?

Answer: No, centrifugal pumps for instance is an old industry. To start with, though it will only apply to industries that are doing machining operations. We are not talking about sheet metal production or electronic assembly. We are mostly talking about high labor content, reasonable quantity machining operations and automatic assembly. For instance, we are intending to take a casting as the input of this plant and turn out a complete product, in a package ready to be shipped, at the other end of the plant.

Question: Why do you not consider the electronics industry? Could Robotics in high technology have a fast payback?

Answer: Well, that is another business and there is an enormous amount of work going on in that field now. Automated factories making high technology products are being built by a number of companies. I am interested in a different market. There is very little activity in the area about which I'm talking.

Question: For the person who already has a running plant, what makes it more convenient to shut down that plant and use your facilities?

Answer This is just a question of economics. A plant is leased to them just like they lease trucks, automobiles, and real estate. So they can do whatever they want, although we are operating the plant. The result is, obviously, if they save considerably more in operating the plant we have leased to them, they would naturally eventually close down or eliminate the facilities that are less efficient.

Question: How many people would be manning your automated facilities?

Answer: We ran through an exercise taking as an example a production line where there were 1200 people now required. There would be somewhere between 80 and 90 people required to operate the automated plant.

Question: If you lease space in your facility to a manufacturer, what happens when they stop producing that item or they change over and start producing something else?

Answer: That is where the flexibility comes in. We can quickly turn around and start making a completely different product. The tooling on the various machines would be changed but not the machines or the work carriers. In fact a very small percentage of the overall facility has to be changed to make different product.

Question: Can you give an example of the range and flexibility of production?

Answer: It would not make much sense to make a product for only one day. We try to pick products with a production run of at least a week.

Question: Can you give an example of the extremes of the type of the products that you could be making?

Answer: Pumps, air compressors, small gas engines and gear boxes. There are other examples but these four would fit into the same type of facility. We could also make many different types of pumps, compressors and engines as long as they were similar in size to allow handling by the same machinery.

Question: Who decides if a product can be manufactured in your automatic facility?

Answer: That is something I have not mentioned yet. We will not undertake making any kind of a product unless we are in on the design of it in the first place. That is very important. A small design modification can make a big difference in the degree in which it can be made automatically.

Question: Are the 80 or 90 people in the automatic facility involved in the design of the product?

Answer: No, that is a secondary consideration. The manufacturer's own people (with very little supervision) can decide on the type of changes that we require. Most people in the automated facility are concerned with supervision, maintenance and tool changing. Of course a lot of the tool changing is automatic but some still require human intervention.

Question: You are proposing a number of new ideas for manufacturing. Can you summarize the key points?

Answer: Yes, first you must have numerically controlled manipulators or robots in order to make this possible. Without them, you could not make an automatic manufacturing facility (unless you made one that was made very inflexible). Second, we intend to lease the facilities and not sell. That is new, no one is doing this at the present time. Third, no one that we know of is assembling complete, self contained, automatic manufacturing plants, either for sale or lease to anyone.

Question: Is this type of facility already in existence in Japan?

Answer: Not that I am aware of.

Question: Is the technology available for building factories like this?

Answer: I would say at least 80 percent of everything that we need to build this plant is available today. The other 20 percent is concerned with what I call "machine interface equipment." These are devices that will monitor the operation of the machine and will tell you whether a tool is getting dull, or broken, or if the coolant temperature in the cutting tools is too high. Information of that type would be monitored by microprocessors in each individual machine. The information is then given to an on-line computer that is controlling the entire operation.

Question: At what stage is the development of these "machine interfacing" techniques?

Answer: There are a number of companies working on it now and it is development rather than primary research.

Question: Can you give us an example?

Answer: There are a number of companies who make cutting tools.

Some are working on the design of equipment that will monitor when the tool is getting dull, how dull it is or if it is broken. That information has to be digitized so it can be given to the computer for monitoring the operation. The interface equipment is not commercially available but that is a matter of just making up the circuit.

Question: Do you need to have new sensors for this type of system?

Answer: Part of this monitoring equipment requires sensors. There are a number of ways of monitoring cutting tools. For example, you could do a complete operation with a particular cutting tool and a particular casting. The overall sonic input received while the cutting tool is operating should be normally within a certain range as far as noise is concerned. If this varies from the normal, there is something wrong with the tool and it has to be checked. If it breaks you would also get information immediately. Monitoring coolant temperature is not hard because there are plenty of temperature sensors around.

Question: What about the role of visual sensors and visual inspection?

Answer: There are a number of companies working on visual inspection equipment. We would use vision for inspection in the production process. We are not intending to use much vision in sorting out parts because it is a very inefficient way to do things if you deliberately mix things up and then sort them out again. That slows the production process down.

Question: If you are designing the whole process from scratch can you eliminate most of the processes that jumble up parts?

Answer: We are intending a continuous rather than discontinuous manufacturing process. If a process is totally automated you never have to pick up random parts from bins. The factory floor is designed as a grid matrix with parts and products in known locations. A mobile vehicle carries parts from station to station. There is no need to look for objects as you already know where they are. Some vision and tactile sensing systems are very interesting. However, they make the system more complex and obviously more expensive. It is one of those things that if you have to use, you do, but if not, you design around it.

Question: What computer problems do you see?

Answer: If we use a completely on-line, main frame computer without separate microcomputers at individual work stations the overall size would be enormous. We have to continuously monitor thousands of different things per second. The result is that we do a lot of the processing at each individual station with much simpler pieces of equipment. The balance we deliver to the on-line computer.

Question: So, is it a distributed system?

Answer: Yes. Each machine has its own intelligence. As an aside, on the subject of automation, we have found a new word for all of this. We call it the "Degree of Automaticity." There does not seem to be any word that fits when you are talking about how automated a process should be. You cannot just use a lot of words to say how automated something can be. You merely say, what is the degree of automaticity.

Question: Can you give an example of something that has a low degree of automaticity and of something that has a high degree of automaticity?

Answer: Today's manufacturing plants that have a number of machines working along-side laborers would have a reasonable to low degree of automaticity. On the other hand, the digital switching system in the telephone company would have a very high degree of automaticity.

Question: If you had to pick a subject of current or future academic relevance for the advancement of automation what would you think would be the most useful?

Answer: I am working with different universities right now along these lines. To establish a course, in let's call it, "Automatic Manufacturing Methods," we should take into consideration Robots and automatic machines, but there has to be an enormous increase of emphasis in the implementation of manufacturing methods. One of the reasons the Japanese are so far ahead as they are, is that they have concentrated on manufacturing methods, not developing new things. We have given up a large part of our manufacturing processes to the Japanese because they are more efficient in doing it. There is no reason we cannot do it too.

Question: Are you saying that we have many things available at our finger tips now; it is just that we do not put them together?

Answer: That is right. The thinking prevalent in US industries is a little strange. It seems it is a lot easier to go and buy somebody else out than it is to develop anything new. From a financial stand point it may make some sense in the short term. In the long run, it is going to be a disaster because at some point we are just running out of manufacturing facilities.

Question: Should we put effort and money into building new robots, new interfacing equipment or new sensors?

Answer: The reduction of cost of a robot is one of the considerations but it is not the primary one. Robots today are very slow. There is no such thing as what I would call a high speed robot. They are enormous pieces of machinery. For instance, some of them weigh 500 to 1000 pounds and they are picking up 5 or 10 pound weights. This is pretty silly. What you need is much lighter, stiffer, high speed robots. Now this is an area where further development has to be done. Another useful development would be a fast, mobile robot-platform. We really need speeds of 5-10 mph to efficiently move parts around on the factory floor.

Question: Are these more important than developing sensors for existing robots?

Answer: There is a large amount of work going on in sensors at the present time and obviously all of those things can be needed. Further development has to be done in order to make them practical. A fast robot if available now would increase the sale of robots more than the availability of sensors. However in some applications sensors are necessary. Let's take a robot and put it on a packaging line where you are making up broken lot shipments for a drug company. The normal operator can (by vision) pick out the various different products and package them many times faster than most robots can. However, even with sensors this does not open up the field to automation because the robot is too slow.

Question: Do you see fast robots coming from America or coming from Japan?

Answer: I hope it is going to come from America.

Question: Are you going to use Japanese robots?

Answer: We will not use anything Japanese unless we absolutely have to. I can see no reason why we should keep putting the Japanese in business and putting outselves out of business.

Question: What kind of training should a person have in Robotics?

Answer: One of the things that I found over the years is that robotics is a multi-disciplinary subject. You should have a good knowledge of mechanics, control systems, computers, various different input devices and sensors. A robot is a complete system in itself. In order to be a good robot engineer you should have a background of information available so that you have certain expertise in all of these various fields. This is presently something that is missing because there are mechanical engineers that do not understand anything about computers and computer engineers that do not understand mechanics. You have to think of all those things at once, to design effective numerically controlled manipulators.

Question: What sort of training do you need for the design of new production lines and automated new factories?

Answer: This is where you would need long years of experience in manufacturing methods. We have people that have many years of experience. They are coming into our company having a background of experience and knowledge, able to say what type of machine will stand up and operate continuously with very little trouble. A production engineering job is quite a bit different thing than designing Robots.

Question: How can you judge these people because it is unlikely that they have had experience in Robotics?

Answer: We will not need their experience in robotics. If they know that a robot will do a certain thing, that is all they need to know. If we put in a robot to move a piece of equipment in between a manufacturing process, all they need to know is the various technical characteristics of the reach and the speed and so forth and so on.

Question: How many people are you going to employ on your new adventure?

Answer: We have 6 or 8 of the primary people.

Question: When are you going to finish your first automated facility?

Answer: We will be operational within 3 to 5 years.

Question: What do you think are the most important aspects of your new manufacturing methods?

Answer: First of all we will be in on the original design of the product. We generally do not require large design modifications, just small touches that make a part easy to handle and recognize. For example putting a little dent in the mold for making plastic bottles so you can tell where to put the label on. Second, with our flexibility, we can offer small manufacturers efficient automation. A small manufacturer may not have the throughput to keep an automated facility going 24 hours a day, 7 days a week. We can let them sub lease to another company making similar items. In this way everyone gains.

PART I APPLICATIONS

1

Robots in Modern Industry

EWALD HEER

1. INTRODUCTION

Robots have come to symbolize high-level industrialization of a society. It appears that industrial robots are the wave of the future when it comes to manufacturing, and the perceived demand for greater industrial productivity. A flood of articles, magazines, and conferences on production automation, and, in particular robotics, attests to this. So do trade shows and meetings.

Improved productivity, reduced costs, and better manufacturing quality are the objectives on which most agree. How to achieve them is not so clear. Politicians and labor leaders want the government's intervention to rebuild conventional, old plants to put workers back on the job. Academics want new approaches to the manufacturing missions. Economic planners call for a shift to manufacturing industries of the future, like aerospace electronics and computers. Aerospace engineers see huge benefits to be derived from the utilization and industrialization of space, although not in the immediate future.

To increase industrial productivity and to alleviate some of the social problems, changes in methods have been and are expected to be introduced at all levels of industrial production through the use of computers. An overall trend stands out, namely, the development and implementation of computer automated manufacturing leading toward the realization of the computer integrated automatic factory as a complete entity in the future. This trend is impeded, however, by a loss of flexibility to change products quickly and maintain cost-effective operation. This is particularly true for small volume and mid-volume production. The production in companies manufacturing in the batch-type mode is therefore in most cases still done by human operated machines organized according to conventional workshop or assembly line principles. In contrast, high-volume producers employ highly automated transfer lines where possible. These usually become more cost effective with increasing production volume.

The advent of numerical control and the development of numerically controlled machining centers improve the situation somewhat for the low-volume producers of parts. Nevertheless, most mid-volume manufacturers continue to search for production systems that can significantly reduce per-piece part costs below those of the job-shop with traditional technologies. Computerized manufacturing systems, flexible manufacturing systems, higher level direct numerical control systems, etc. are some of the more promising answers offered by builders or machine tools to the mid-volume manufacturers. Although the approach varies with each machine-tool builder, the ultimate objectives are the same:

1. Minimize in-process inventory;
2. Minimize lead time;
3. Minimize direct labor;
4. Minimize indirect labor;
5. Minimize tool changing and setups;
6. Maximize equipment utilization, and
7. Maximize flexibility.

In working towards these objectives, robots quietly take their places alongside humans on the production line to raise productivity and to do the "dirty work". They can do most of the work still performed by humans, even in plants filled with automated heavy machinery. They can handle materials, load and unload, sort, stack, and do assembly operations. They can position workpieces on machines, weld, spray, rivet, rout, sand, and grind. They can do many of the monotonous, hot, disagreeable, dangerous tasks formerly assigned to humans, as well as new tasks that humans cannot do. They can work for thousands of hours with, typically, less than two percent downtime. Robots can be reprogrammed to do different tasks, and the digital electronics of their control systems places them squarely within the computer-aided manufacturing (CAM) arena.

Looking back over the last two decades in manufacturing, it is probably fair to categorize the 60's as the decade when numerical control came to maturity. The 70's established the computer and, in particular, computerized numerical control. Perhaps the 80's will be the decade of the robot, when robots will achieve maturity and full acceptance in industry and take over many boring and dangerous jobs heretofore done by humans. If so, productivity and product quality should indeed substantially increase.

2. HISTORICAL PERSPECTIVE

Robots had their origins with the creation of the first tools. The idea of the functional robot was born in classical Greece. In the fourth century B.C., Aristotle wrote: "If every instrument could accomplish its own work, obeying or anticipating the will of others...if the shuttle could weave, and the pick touch the lyre, without a hand to guide them, chief workmen would not need servants nor masters slaves." This sounds very familiar in the context of robotic systems. However, it took many centuries before the vision began to be possible.

By the eighteenth century, the industrial revolution was fully under way. With its proliferation of new power sources, new tools, new industry, and new mechanisms, it became possible to create machinery capable of controlling a whole sequence of actions. However, the true beginning of the machine age occurred towards the end of the nineteenth century. Machines were everywhere, and exposition after exposition was held to display the latest achievements in technology. The large machines were steam powered, the gasoline engine was developed, and the electric motor was being introduced. The machines where substituting for, or were tremendously amplifying, human physical capabilities. It was hard to imagine how people had ever been able to get along without them.

Except for the application of simple controllers, equivalent to Watts' governor, the machines required human involvement to detect changes in the environment or in the required objectives and, accordingly, to effect changed actions by the machine. Such machine capabilities were beyond the available technologies and had to wait for later years.

World War I brought on many changes. The raw power of the machine was shown in its most destructive form. Humans seemed to be expendable and replaceable in that war. Ideas along these lines carried over into the post war industrial society and spawned in 1921 the play R.U.R. (Rossum's Universal Robots) by the Czech dramatist Karel Capek. Thus, the word *robot* was coined. It is derived from the Slavik word *robota,* meaning *heavy work*.

Capek's robots were manufactured for profit, as a replacement for workers. But, these machines were becoming uncomfortably sophisticated and "smart". Their values had nothing to do with human values. Contrary to the purpose for which they had been created, they tore down what humans had taken centuries to build, and in the end destroyed man-kind itself.

Given the probability that one day there would be robots with intelligence, Issac Asimov formulated in 1940 the following three laws of robotic behavior [1].

1. A robot may not injure a human being, or, through inaction allow a human being to come to harm.
2. A robot must obey the orders given it by human beings, except, where such orders would conflict with the first law.
3. A robot must protect its own existence as long as such protection does not conflict with the first or second law.

In the real world, a robot that would perform useful work had to wait until several subsystems were perfected to a certain level of technical accomplishment. Articulations and linkages of joints had been possible for at least a century, but a sensitive means of controlling them was lacking. Also, it was required that motors, in particular electric motors, be miniaturized to be convenient for a working robot. Working robots became really possible through the development of controls to keep them performing, if not optimally, at least within acceptable limits, and to have them make corrections themselves if anything went wrong. Simple feedback devices had long existed, but they could not be used effectively for robots until they could be attached to a computer that could identify and prescribe the correction. Not until the 1950s did all the elements come together. The phenomenally rapid development of solid state electronics and computers since that time opened entirely new perspectives in many areas. In addition to the physical amplification provided by powered machines, the computer now provided amplification of the human mental capabilities. The computer made it possible to equip the robot with a "brain," however rudimentary at this time.

Remotely controlled manipulators which perform the motions of a master replica without a computer involvement are so-called *master-slave manipulators* (Figure 1). Such mechanisms had been developed by the early 1950s for use in the nuclear-energy field to avoid radiation hazards. Although not qualifying as robot devices, they were forerunners of present-day robot manipulator technology. Many mechanisms and control strategies developed at the time found their way into today's robotic technology.

In the late 1960s, George C. Devol and associates made a number of robot related inventions to the ultimate benefit of their company, Unimation Inc., and robotics in general [2]. As a result, the first Unimate robot was installed to tend a die casting machine in 1961. The development of industrial robots has been steadily advancing ever since, and robots have found applications in a multitude of production branches (Figure 2). In the late 1960s, Europe and Japan also started to adopt this technology, and by the early 1970s, industrial robots proved to be practical possibilities. They are now numbering many thousands in all kinds of industrial applications throughout the industrialized world.

To date, robots with manipulators have been applied in the space program as well [3]. In 1967/1968, U.S. Surveyor spacecrafts landed on the moon. Manipulators under a remote control mode from Earth, the so called *teleoperator* mode, were used to dig trenches and execute other simple tasks. In 1970 and 1976, the USSR used a drill on Luna 16 and Luna 24, respectively, to acquire a Moon sample and return it to Earth. The USSR Luna 17 (1970) and 21 (1973) were remotely controlled lunar surface vehicles (Lunachod) which performed soil analyses.

Perhaps the most complicated space robot built so far is the Viking Lander that was sent to Mars in 1976. It was a working laboratory, fully equipped to perform a series of experiments on anything it could reach with its manipulator. Such remote operations in space were possible because of telecommunication. In addition to physical amplification through powered machines, and mental amplification through computers, telecommunication provided now to humans also a tremendous amplification of the accessible space.

In 1982, the Space Shuttle has been using a 48-ft-long manipulator in space in the semiautomatic control mode, i.e., the *supervisory control mode,* to

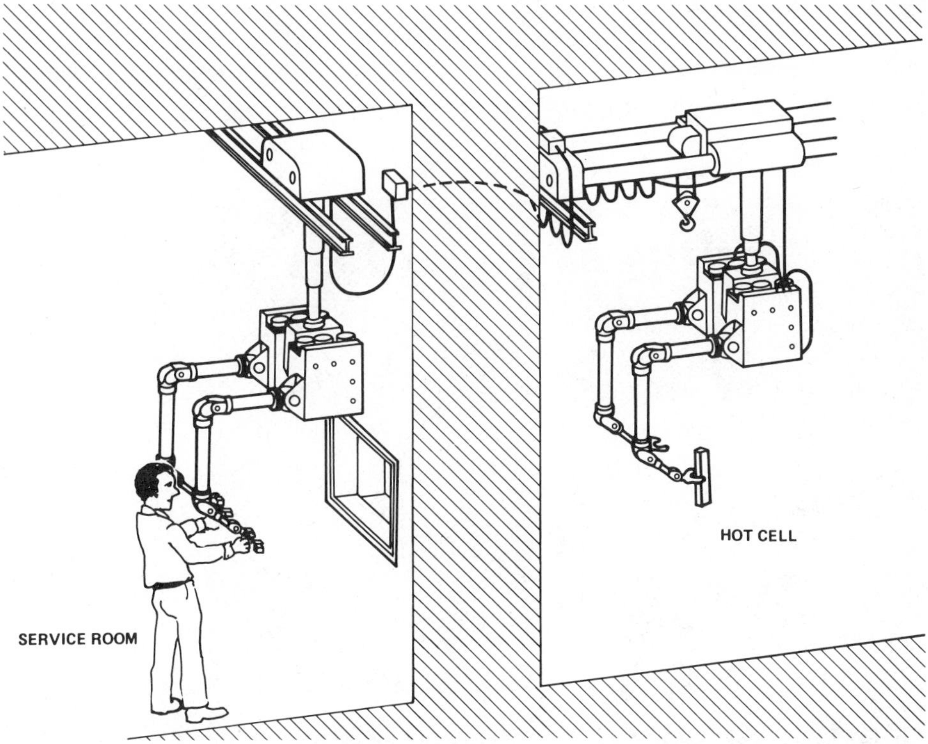

Figure 1 Master-slave manipulators for operation in nuclear hot cells. The manipulators in the hot cells are controlled by exact replicas in the control room. Forces can be amplified by a factor of about six, which is realized by asynchronously operating electrical motors and associated mechanisms. Direct and/or TV visual feedback and force feedback is available to the human operator.

unload payloads from its cargo bay and to perform other tasks in space (Figure 3). This initiated a new era of large space robot manipulators that will ultimately play a major role in the industrialization of space [4].

Undersea manipulators have found much more frequent application than manipulators in space. In a 1979 report [5], the U.S. Department of Commerce identified a third of over a hundred remotely operated undersea vehicles as having one or two manipulators (one had three). Some of these manipulators are simple grasping devices which rely upon the vehicle's maneuvering to place them correctly at the work site; others are quite sophisticated. However, none had significant computer control. Most were controlled in the master-slave control mode with man intimately involved in every detail of the control process.

Numerous other scattered instances of robotic device applications are in areas such as:

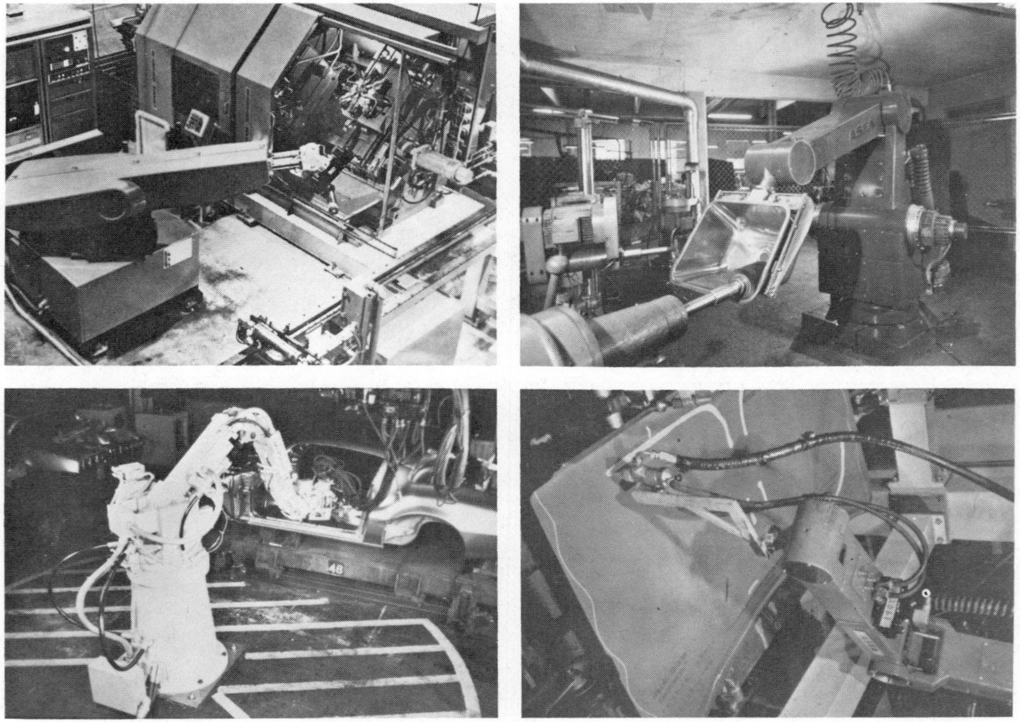

Figure 2 *Modern industrial robots perform typical tasks. (Courtesy: Unimation, Cincinnati Milacron, ASEA): Top left: A Unimate robot performs machine loading operations. Bottom left: A Cincinnati Milacron T-3 performs spot welding operations in an automobile plant. Top right: An ASEA robot performs kitchen sink polishing operations. Bottom right: An ASEA robot places glue in a precisely prescribed pattern.*

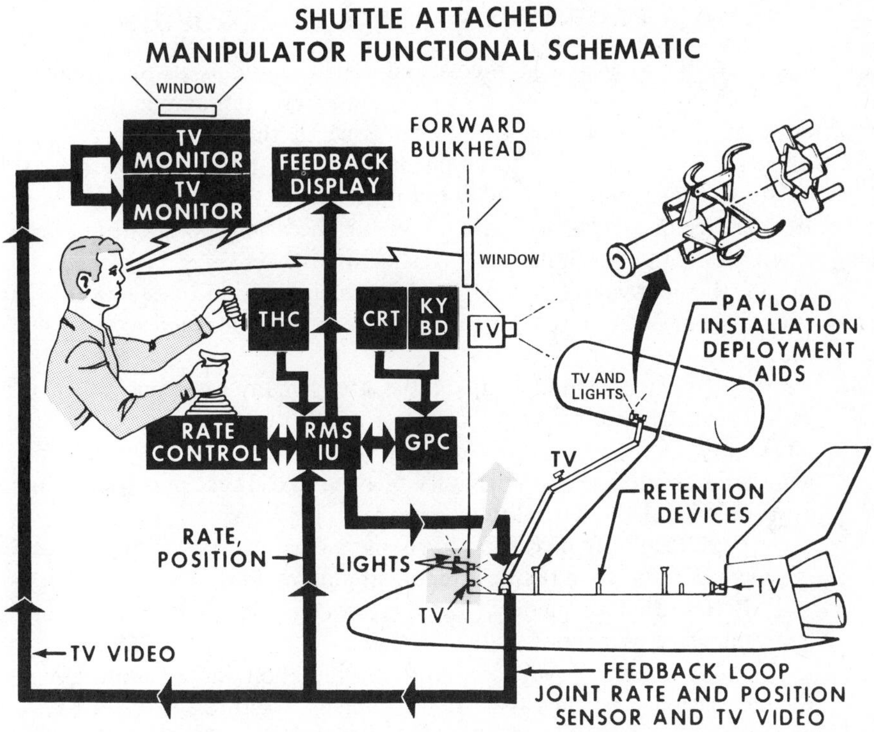

Figure 3 The Space Shuttle attached manipulator will play a major role to develop and test manipulation strategies at various levels of automation for the industrial utilization of the space environment in the future.

1. Medicine [6], e.g., organ replacement, bionic limbs, manipulator aids to quadriplegics.
2. Mining, e.g., automatic coal mining machines.
3. Human protection, e.g., fire fighting vehicle, bomb disposal vehicle.
4. Military, e.g., target seeking bombs, unmanned airplanes.

At the present time, the non-industrial applications in space, undersea, medicine, etc., satisfy specific needs usually for reasons other than economics. They help to solve important problems in each area, but the overall frequency of their use is too small to have appreciable economic impact in comparison to robotics use in the production industry. Nevertheless, the level and complexity of the technology required in these areas is often at the fringes of the possible, and the experience from its application can be of great value in other areas, especially in the production industry. The concern in the remainder of this article is primarily with the production industry because of its overriding economic importance.

3. ARGUMENTS FOR AND AGAINST ROBOTS

The arguments for and against robots and robotic systems in the production industry are closely related to those of industrial automation in general. Production automation became a national issue in the late 1950s and early 1960s, when labor leaders and government officials had debated the pros and cons of automation technology. Even business leaders, who see themselves as advocates of technological progress, have on occasion questioned whether automation was really worth its high investment cost.

Some of the motivating factors for introducing robotic systems into the production industry can be subdivided into technical, economical and social categories:

1. *Technical Factors:* Human capabilities are in many cases not sufficient to satisfy modern requirements of precision, speed, endurance, strength, uniformity, etc. Robotic systems offer such capabilities. They provide in addition a link between the rigidity of fixed and direct numerical control automation and the flexibility of humans. They offer:
   a. High flexibility of product type and variation, and smaller losses for preparation time than conventional automation;
   b. Better product quality, fewer rejects and less waste than human intensive production.
2. *Economical Factors:* An overriding consideration in the application of robotic systems is the associated economic picture. Ever increasing competition and, at the same time, increasing compensation costs for labor call for increased productivity. Robotic systems offer to contribute to productivity increases through items such as:
   a. Providing maximal utilization of capital intensive production facilities, possibly in three shifts around the clock;
   b. Reducing production losses due to interruptions, absenteeism, and labor shortages;
   c. Reducing in-process inventory;
   d. Reducing manufacturing lead time.
3. *Sociological Factors:* The impact on the human factors of production can be beneficial in various respects. Many low level or otherwise undesirable tasks can be done by robots. This includes work in dangerous or unhealthy environments, monotonous activities with short recurring cycles and heavy physical tasks. The benefits of robotic systems include:
   a. Reduction of accidents (safety);
   b. Removal of conditions dangerous to human health;
   c. Shorter working hours;
   d. Increased living standards.

To these should be added that the growth of the robotic systems industry will itself provide employment opportunities. This has been especially exemplified in the computer industry. As the companies in this industry have grown (IBM, Burroughs, Digital Equipment Corp., Honeywell, etc.) new jobs have been created. These new jobs include not only workers directly employed by these companies, but also computer programmers, systems engineers, and others needed to use and operate the computers. Similar arguments could be made for robotic systems.

The arguments against robotic systems are also closely related to those against automation in general. Such arguments include:

1. The subjugation of humanity by the machine;
2. Reduction of the labor force with resulting unemployment;
3. Reduction of the purchasing power as a consequence of (2).

A resolution of these questions will not be easily found. In fact, the impact of robotics on industrialized nations and society in general must be viewed in the light of new evidence each year.

Labor in America has come to give at least tacit support to advances in automation and industrial robotics. The United Auto Workers membership views robotics as just one of the technological advances facing the union today. In fact, the union membership has favored the introduction of such advances and has recognized them as essential in promoting economic progress through increased productivity. Employees recognize that automation and robotics have the potential of relieving humans of dangerous or undersirable and boring jobs. Employees generally also recognize that wages, working hours, fringe benefits, and safety will ultimately improve as productivity increases. However, automation also has the potential to cause economic hardships for workers whose jobs are directly affected and eventually for others through its effect on overall employment. This picture remains in the public consciousness. Labor does not yet feel secure about the long-term effects of automation and robotics on the industrial job market. Thus industrial robotics and other industrial automation can be characterized as *ambivalent* technology. This causes, at least temporarily, tensions between requirements for social equitability and demands for productivity through automation, and will (and should) largely determine the speed at which robotics diffuses through industry.

4. ROBOTS AND ROBOTIC SYSTEMS DEFINED

At this point we should perhaps ask what is a robot, and what is a robotic system? What attributes make a machine a robot irrespective of its application? Many definitions have been proposed and discarded. In the technical literature, robots are understood to be machines which have the capability of performing independently complex tasks of a physical and mental nature. To overcome problems of changing situations, they must interact, to some degree "intelligently," with their environment and must be able to store experience once it has been gained.

The Robot Institute of America (RIA) defines a robot as:

> *A robot is a programmable, multifunction manipulator designed to move material, parts, tools or specialized devices through variable programmed motions for the performance of a variety of tasks.*

The emphasis here is on the handling of physical objects in a flexible manner with the possibility of changing the job content through reprogramming. It implies foremost a set of powered actuators consisting of one or more versatile manipulators and a computer that can be programmed in such a manner that it will send the correct signal to the right activating motor in the required sequence and at the appropriate time for executing a prescribed task.

More complex robots may also have mobility units such as wheels, tracks, and legs to move from point to point. They may have sensors to sense unpredictably changing tasks and other external environments, thus enabling them to incorporate this information, store it, and effect automatic task reprogramming for the robot computer. Hence, advanced robots are machines which have the ability to solve problems that require, in addition to manipulative capabilities, also *autonomous decision-making* and *task planning* capabilities. Such "intelligence" resides in the robot's computer, i.e., in the computer software. It is primarily the computer programs, designed in accordance with *artificial intelligence* principles, that provide to the robot the required degree of autonomy.

These computer programs grow exponentially in complexity with the number of sensors and with the number of branch points in the program. If the data from more than a few sensors affects the selection of operational strategies, it has proven successful to design the structure of such command and control programs in a hierarchical fashion. The advantage of a hierarchical control system is that it can be partitioned to confine the complexity of any module to manageable limits regardless of the complexity of the entire structure.

Although the design of such program hierarchies is not unique, certain conventions have been established. For example, the U.S. Air Force adopted as an integral part of its Integrated Computer Aided Manufacturing (ICAM) program a five-level hierarchy concept originated in the National Bureau of Standards (Figure 4). The question is how high up the hierarchy must one go to select the control level for a stand-alone robot that may replace some of the simpler human roles in the factory. According to the indicated definitions, level three should be right for a robot that is essentially by itself. Level four would be indicative of a robot integrated into its *work station* or *work cell* forming a robotic system. We have thus made a vague distinction between *robot* and *robotic system,* where a robot incorporates control levels 1, 2, and 3, and a robotic system includes these lower levels plus those above level 3, i.e., it includes at least one robot plus the integrated peripheral equipment belonging to the robot work cell.

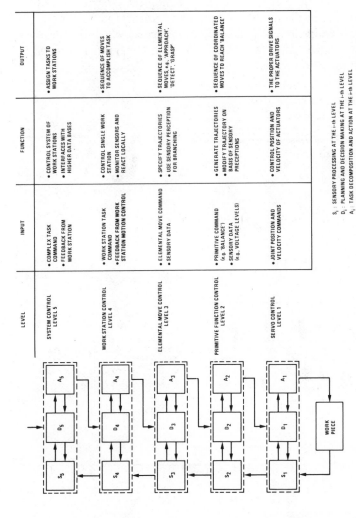

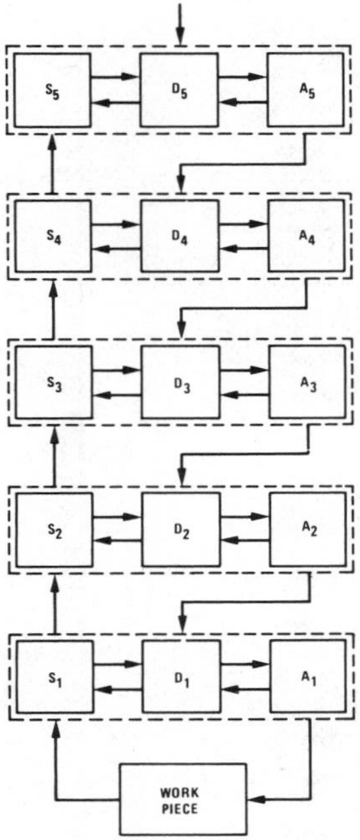

LEVEL	INPUT	FUNCTION	OUTPUT
SYSTEM CONTROL LEVEL 5	• COMPLEX TASK COMMAND • FEEDBACK FROM WORK STATION	• CONTROL SYSTEM OF WORK STATIONS • INTERFACES WITH HIGHER DATA BASES	• ASSIGN TASKS TO WORK STATIONS
WORK STATION CONTROL LEVEL 4	• WORK STATION TASK COMMAND • FEEDBACK FROM WORK STATION MOTION CONTROL	• CONTROL SINGLE WORK STATION • MONITOR SENSORS AND REACT LOCALLY	• SEQUENCE OF MOVES TO ACCOMPLISH TASK
ELEMENTAL MOVE CONTROL LEVEL 3	• ELEMENTAL MOVE COMMAND • SENSORY DATA	• SPECIFY TRAJECTORIES • USE SENSORY PERCEPTION FOR BRANCHING	• SEQUENCE OF ELEMENTAL MOVES, e.g., 'APPROACH', 'DETECT', 'GRASP'
PRIMITIVE FUNCTION CONTROL LEVEL 2	• PRIMITIVE COMMAND (e.g. 'BALANCE') • SENSORY DATA (e.g., VOLTAGE LEVELS)	• GENERATE TRAJECTORIES • MODIFY TRAJECTORY ON BASIS OF SENSORY PRECEPTIONS	• SEQUENCE OF COORDINATED MOVES TO REACH 'BALANCE'
SERVO CONTROL LEVEL 1	• JOINT POSITION AND VELOCITY COMMANDS	• CONTROL POSITION AND VELOCITY OF ACTUATORS	• THE PROPER DRIVE SIGNALS TO THE ACTUATORS

S_i : SENSORY PROCESSING AT THE i-th LEVEL
D_i : PLANNING AND DECISION MAKING AT THE i-th LEVEL
A_i : TASK DECOMPOSITION AND ACTION AT THE i-th LEVEL

Figure 4 Hierarchical control system. At each level in the hierarchy, the sensory-processing module S_i extracts information from the sensory data stream so that the planning and decision module D_i can make behavioral decisions at that level which can then be implemented by the task decomposition and action module A_i. Each S_i receives a continuous stream of expectations and predictions generated by a world model in D_i, and each D_i receives inputs describing the actions generated in A_i. The lowest level in the hierarchy corresponds to the observable output.

The robot work cell is thus the basic building block of a robotic system. Work cells may be linked together to form flexible, integrated robotic manufacturing systems. The nodal points between work cells usually incorporate a buffer storage unit of some type, allowing shutdown of a work cell for scheduled or unscheduled maintenance or set-up of a new process without shutting down the entire system. The work cell upstream can then continue to process parts and feed the storage unit serving the inoperative cell. The down stream work cell can continue to draw parts from its buffer storage unit. This entire process is monitored by a supervisory control which is at least at level 5. Figure 5 shows two examples of such integrated robotic manufacturing systems under supervisory control.

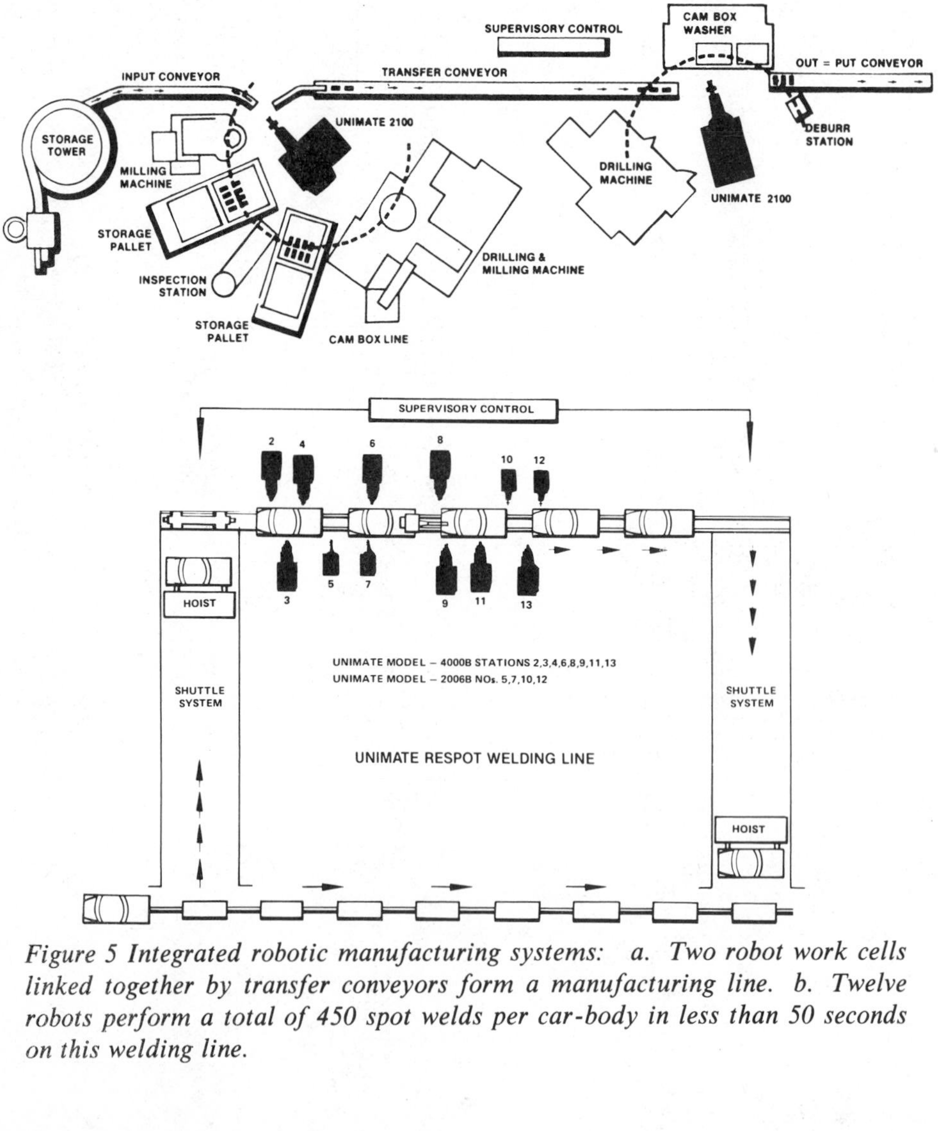

Figure 5 Integrated robotic manufacturing systems: a. Two robot work cells linked together by transfer conveyors form a manufacturing line. b. Twelve robots perform a total of 450 spot welds per car-body in less than 50 seconds on this welding line.

At a higher level of control, such integrated robotic manufacturing systems can be linked to centralized or distributed data bases that will coordinate and manage product handling within the manufacturing context. These manufacturing data bases are still rare, because they usually require long-term development and investment commitments. However, the trend is marked in that direction, and is expected to lead to the end-to-end computer integrated automated factory in which robotic systems play a vitally important role.

5. CHARACTERISTIC APPLICATIONS

Production robots fall roughly into two broad categories: those that handle workpieces in some fashion, and those that use and handle tools to perform tasks.

5.1 Workpiece Handling Robots

These robots are equipped with grippers appropriate for the workpieces to be handled. In certain cases, these grippers can be automatically exchanged, if required. The typical task for the robot is to bring workpieces from a defined initial position to a defined end position. The path between the individual positions can remain undefined. In most cases, the manipulators are point-to-point controlled and differ conceptually from ordinary feeding devices only in the variety of cycles of possible movements. These cycles can be programmed to fit particular requirements. Workpiece handling robots perform tasks for example in:

a. Die casting;
b. Investment casting;
c. Press loading;
d. Forging and heat treating;
e. Plastic molding;
f. Machine loading;
g. Palletizing.

More specific descriptions of the operational processes required in the performance of these tasks can be found in the literature, e.g., [2 and 7].

5.2 Tool Handling Robots

These robots use special tools in place of grippers or securely grasp the tools with their grippers. Such tools are, for example:

a. Paint sprayers for finishing;
b. Welding guns for spot or arc welding;
c. Cutters, grinders, drills, riveters, etc., for machining and fettling;
d. Measuring instruments for inspection;
e. Various tools for assembly operations.

The movement characteristics of tool handling robots are usually more complex than those of workpiece handling robots. Many tool handling robots must be

	REQUIREMENTS PROFILE / TASKS	TYPE OF TRAJECTORY				TYPE OF VELOCITY			VELOCITY (m/s)			POSITION DEVIATION (mm)			TYPE OF PROGRAM		
		POINT-TO-POINT	STREIGHT LINE	DEFINED TRAJECTORY	UNDEFINED TRAJECTORY	SINGLE STEP	MULTISTEP	CONTINUOUS	LOW < 0.2	INTERMEDIATE 0.2-1.0	HIGH > 1.0	SMALL < 0.2	INTERMEDIATE 0.2-1.0	LARGE > 1.0	CONTINUOUS	DISCONTINUOUS	PROBABILISTIC
WORK PIECE H.	LOADING/UNLOADING	●				●	●				●		●			●	
	PALLETIZING	●				●	●				●			●		●	
	TEMPERING/FORGING	●		●			●		●	●				●	●	●	
	CASTING/MOLDING				●			●						●			●
	ASSEMBLING		●				●		●			●				●	
TOOL HANDLING	SPOT WELDING	●									●			●		●	
	ARC WELDING			●				●	●	●			●		●		
	TORCH CUTTING			●				●	●					●	●		
	DEBURRING			●				●		●			●		●		
	GRINDING			●				●		●			●		●		
	POLISHING				●		●			●			●		●		
	WASHING			●		●				●				●	●		
	SAND-BLASTING			●		●				●				●	●		
	CAST CLEANING			●	●	●				●				●			●
	STEAM CLEANING			●	●	●				●				●			●
	COATING			●				●			●			●	●		
	X-RAY TESTING	●		●		●	●		●	●				●	●	●	

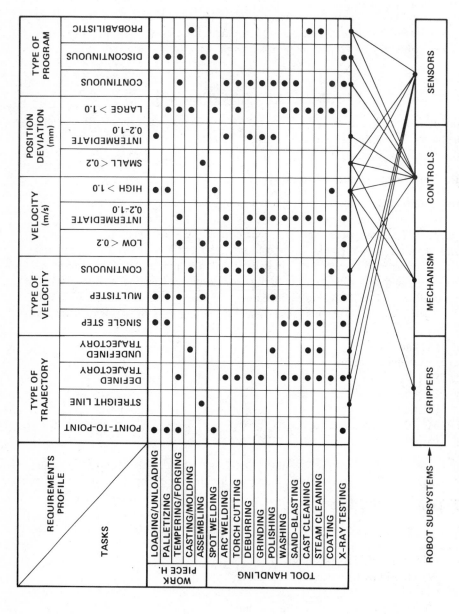

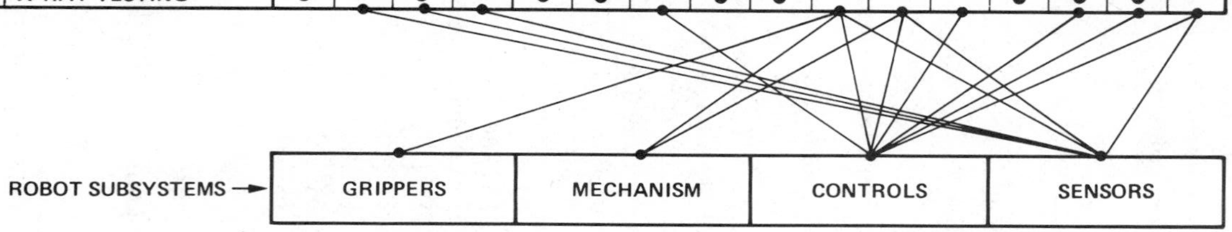

Figure 6 Influence of robot tasks on manipulation requirements and on the flexibility of robot subsystems.

able to follow prescribed curves in space (e.g., painting, spraying, welding), or they must perform delicate placement and joining operations which put a great demand on sensory feedback and control techniques.

Figure 6 gives a general overview of requirements imposed on robotic systems by different robot applications [8]. This matrix can be used to determine the necessary flexibility of robotic systems for a given task. It can be used to search systematically for application areas for robotic systems and to extend the application areas by changing the performance profile.

Figure 6 shows that in most cases the permitted positioning deviation of 1 mm is sufficient. The types of task implementation speeds are approximately evenly divided among single step, multi-step, and continuous operation. More than half of the cases require a speed of about 0.2 to 1.0 m/sec with more than a quarter above 1.0 m/sec. The majority of the listed cases requires that a defined trajectory be maintained for task implementation and hence continuous path control of the robot manipulator.

6. FUNCTIONAL AND PHYSICAL CHARACTERISTICS

Robotic systems are synthesized from a large number of interrelated and interacting subsystems which characterize the entire system with respect to its flexibility and applicability for a specific task. The subsystems perform subfunctions of the total process. Appropriately coordinated, they enable complex sequences of manipulation. The most important subsystems with their functional characteristics are [9]:

1. *Robotic Mechanisms:*
   * Establish the spacial relationships between end effectors (grippers, tools) and the workpieces.
   * Maintain relations in time and space between manipulator joints and end effector motions.
   * Assure the required position and orientation of workpieces.
2. *End Effectors:*
   * Grasp, hold and manipulate tools and work pieces.
   * Establish interactions between workpieces by assembling, fitting, joining, bolting, welding, spraying, etc.
3. *Power Sources:*
   * Provide motive power to all robot joints, end effectors and controls (electrical, hydraulic, pneumatic).
4. *Controls:*
   * Supervise and regulate all manipulation sequences.
   * Coordinate and synchronize the manipulators with the handling processes.
5. *Sensors:*
   * Measure and determine the internal state of the robot system and manipulators (positions, velocities, forces, moments, etc.)

* Measure and determine the state of tools, workpieces and environments.
* Identify and determine the position and orientation of workpieces in relation to their environment.

6. *Programming Systems:*
   * Develop the software for the control program (compiler, interpreter, test system, simulator, etc.).
7. *Computers:*
   * Implement the various control programs.
   * Support the development of the control program.
   * Process the sensor data and modify the control program accordingly.

The physical characteristics of industrial robots are determined during the design process to satisfy a broad spectrum of functional requirements at an affordable cost. A multitude of such properties must be traded off to arrive at an acceptable design. Some of the most important physical and operational characteristics which are important for robot selection and design are identified in the following list.

1. *Data for basic work units:*
   * Space requirements for robot operations.
   * Robot weight.
   * Required weight ranges of handled objects, including workpieces, grippers and tool weights.
   * Power supplies and their required values (electric, hydraulic, pneumatic).
2. *Data for control units:*
   * Space requirements.
   * Weight.
   * Power supplies and their required values.
3. *Robot mechanisms:*
   * Design and arrangement of manipulator linkages (degrees of freedom, rotational joints, translational joints).
   * Workspace (rectangular, cylindrical, spherical, torus-like or a combination of these).
   * Number of reachable positions within the work space.
   * Limits of reachability, velocity, and acceleration for each degree of freedom and for different weights of handled objects.
   * Precision of position repeatability.
4. *Control system:*
   * Type of control, point-to-point or continuous path.
   * Control flexibility, fixed manipulation sequences, sensor controlled sequences, adaptive manipulation sequences, etc.
   * Control stops (e.g., mechanical or drive controlled).
5. *Programming methods:*
   * Manual programming, where the motion of each axis is limited by fixed stoppers.

* Switching method, where every required manipulator position and orientation is approached with related drive brakes, is switched into memory, and is subsequently repeated with the help of position sensors.
* Lead-through programming where the robot is led by hand through the desired end points (position and orientation) and/or trajectory points which are automatically stored in memory for subsequent repetition.
* Master-slave programming where a master replica of the robot (usually smaller) is used to perform lead-through programming.
* Teach-in methods were a programming unit or teach pendant is used to drive the robot through the desired end points (position and orientation) and/or trajectory points which are automatically stored in memory together with additional typed-in data such as speed, interrupts, single step module, etc. for subsequent repetitive operations.
* Textual programming where a symbolic description of the operations (desired points, trajectories, etc.) is developed based on computer languages which range from low level machine languages and assembly languages to high level robot programming languages such as AL, AML, VAL, etc.

7. MECHANISM AND CONTROL CONSIDERATIONS

At this point, it is not yet clear whether the *universal robot* or the trend towards specialized robots will previal. Nevertheless, an increasing number of modularized robot systems is being offered with which the user can assemble his robot to satisfy specialized requirements. This may be an indicator that the trend toward specialization will continue for some time. Since the general purpose or universal robot is likely to exceed the technical requirements, it is in many cases an uneconomical solution.

A good concept to robot modularization has been developed, for example, by the Volkswagen Company (Figure 7). The basic unit of the robot is the same. It can be combined with a linear manipulator serving presses, a telescopic manipulator for spot welding, and an articulated manipulator for various handling and welding tasks, etc. The control structure and computer memory units are in each case the same, while the amplifying units are designed in accordance with the individual drive requirements. Such approaches to repetitive modular building blocks can often lead to economically optimal solutions.

While modularization has an immediate, visible influence on the physical structure (mechanisms) of the robotic system, its greatest economic impact will probably be through modular control, software and programming language structures. The effective modularization of current robotic systems requires a clear understanding of broadly based spectra of manipulation requirements. It also requires the application of group technology which categorizes products into families of similar size, configuration, and processing needs. This will

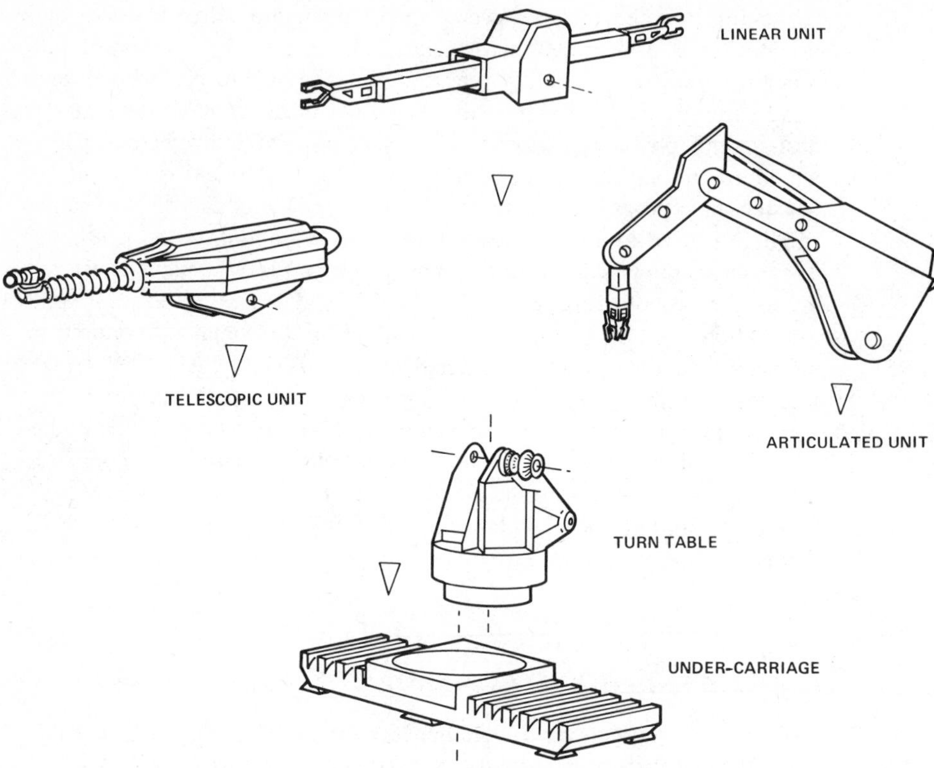

Figure 7 Example of modularized robot system.

make possible the design of modularized robotic systems with optimally located interfaces in the mechanical, electrical and software systems.

7.1 Kinematics and Dynamics of Industrial Robots

An arbitrarily located body (work piece) in space requires six quantities (three translational and three rotational coordinates) to describe its position and orientation uniquely. To grasp or approach such workpieces in a prescribed manner, the robot manipulator must be able to bring its end effector into an arbitrary position within its work space. This requires six degrees of freedom for the manipulator. If the workpieces and their motions are constrained to certain positions and orientations, it may be possible to have as few as two degrees of freedom, and if the workpieces are hidden, requiring to reach around an object, it may be necessary to have seven or more degrees of freedom.

A robot manipulator may have rotational, translational (sliding), or a combination of these degrees of freedom forming a kinematic chain. The last element of this chain is the end effector, which must trace a prescribed trajectory, or obtain a required position in relation to the workpiece (Figure 8). If the desired motion of the end effector, i.e., the coordinate system with the

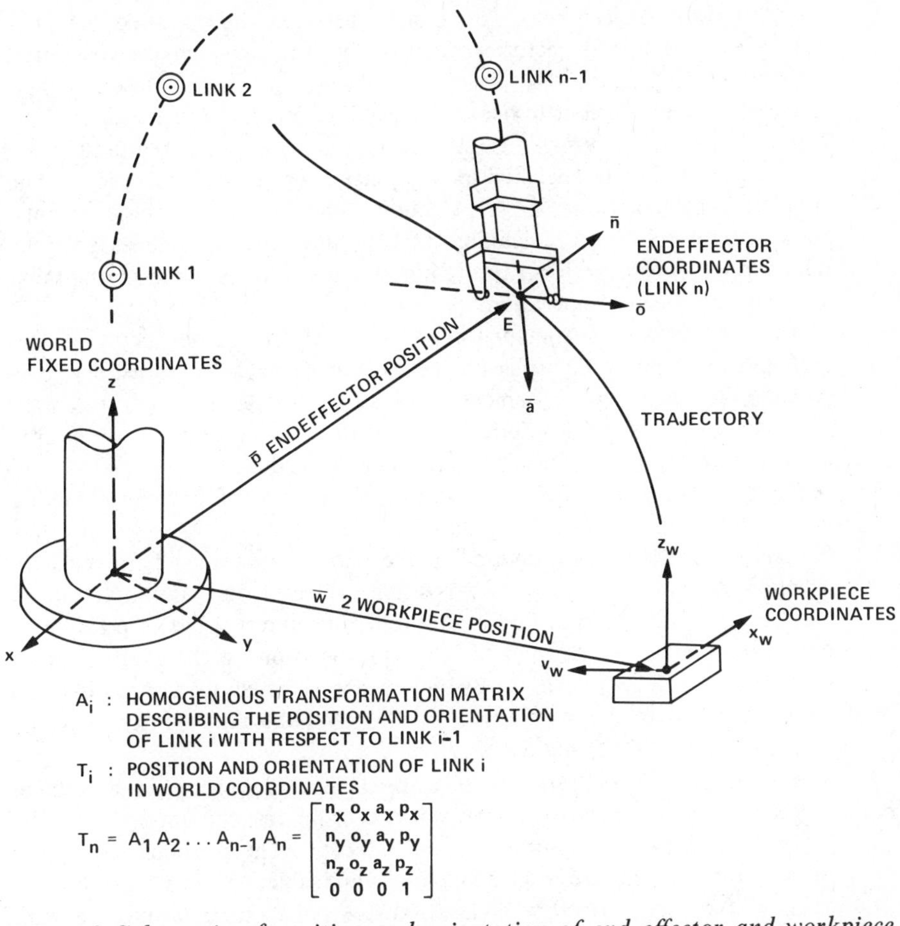

Figure 8 Schematic of position and orientation of end effector and workpiece with respect to the world coordinate system.

origin at E, is known as a function of time in world coordinates (XYZ), it is required to specify the manipulator coordinates (joint rotations and translations) as functions of time to effect the trajectory. The relationship between world coordinates and manipulator coordinates is, in general, characterized by a product of Danavit-Hartenberg transformation matrices A_i [11,12]. The transformation of the manipulator coordinates into world coordinates is unique, however, the inverse transformation is strongly dependent on the geometrical configuration and is generally not unique.

The trajectory of the end effector depends on the task to be implemented. The corresponding manipulator coordinates are continuously computed by the control computer and the set point functions are sent to the actuators for implementation. For robots with servo loop control, the actual manipulator coordinates are then measured, and corrective computations are made as required. Some of the more important types of trajectory planning are:

1. *Point-to-point trajectory* -- The motion between the specified points is not defined. These trajectories are usually used for pick-and-place and machine loading operations. When the specified points are close together, it becomes more efficient to use trajectory interpolation routines.
2. *Interpolated trajectories* -- Interpolation techniques are required when the motion between specified points has a prescribed precision. The simplest interpolation is a straight line between these points. Intermediate points are calculated at time intervals as small as possible, but long enough to perform the required computations while maintaining the required speed of operation.
3. *Transitions between trajectory segments* -- If the corner point of two joining trajectory segments is not necessarily part of the trajectory, it is possible to construct a smooth transition using a quadratic path interpolation by superposition of a constant acceleration for a suitably chosen time interval. This is a first step toward trajectory control and requires in addition to position control the control of linearly changing velocities during the transition phase.
4. *Trajectory control* -- The motion of the manipulator must be given as a function of time to control its position, velocity and acceleration. If trajectory points are specified at discrete time intervals, it is possible to determine intermediate points by interpolation with higher order polynominals. The relevant parameters are selected to assure that the computations can be performed in the available time interval while satisfying specified optimality conditions.
5. *Known relative trajectories* -- If a workpiece is moving in a known, prescribed manner relative to the world coordinates, this motion must be superimposed on the manipulator end effector. The previous techniques of trajectory control are in principle still applicable. A simple example of a known relative trajectory is a workpiece on a conveyor moving with known velocity where the workpiece must be picked up by a stationary robot manipulator.
6. *Random relative trajectories* -- If a workpiece moves in an unpredictable manner relative to the world coordinates, the robot system must be equipped with sensors (vision, range, etc.) to determine any changes in position and velocity. The new information is provided to the control computer, where the new trajectory parameters are computed and sent to the actuators. Examples are space and undersea robot manipulators which must capture floating objects.

The motion of a robot manipulator, with or without workpiece, is brought about by appropriately selected drives (electric, hydraulic, pneumatic) for each manipulator joint. These drives must be capable of delivering the torques and/or forces to achieve the required accelerations and velocities, while overcoming inertia, friction and externally applied forces and moments. These torques and forces must be continuously computed in a timely fashion to assure smooth and stable dynamic behavior by the robot.

There are two basic approaches available to model the dynamics of industrial robots. First, the Lagrangian approach is in principle the simpler one. However, for more than four manipulator joints it quickly becomes algebraically involved and obscure. Second, the Newton-Euler method is based on recursive relations between the elements of the kinematic chain. It lends itself to develop these recursive relations for the kinematic as well as the dynamic description of the individual elements, and is therefore especially suited for digital computers. Both techniques can be applied to compute the unknown joint torques and forces to move the end effector along a defined trajectory. This is known as the *inverse problem* [13].

7.2 Control of Industrial Robots

Within certain control constraints, the robot control system assures that the industrial robot performs preplanned motions and manipulations taking into account the kinematics and dynamics of the manipulator. It assures communication and synchronization between the operational process periphery and the human interface and/or a higher level control system. It thus satisfies specific requirements for the type of operation, programming, reliability, safety and fault detection. Figure 9 shows in principle the overall control structure and the interrelations of the elements of a robotic system.

In accordance with the above identified types of trajectories, robot controls can be subdivided into: (1) Point-to-point controls, (2) continuous trajectory controls, and (3) relative or adaptive controls. The type of control has a profound influence on the required sensor system and robot flexibility and basically determines the overall sophistication of the industrial robot.

A measure of robot flexibility is its adaptability to the required manipulation process. By this criteria one can characterize industrial robots as:

1. *Open-loop controlled* -- Such robots merely move their arms between fixed points without sensory feedback. They can achieve comparatively high speeds.
2. *Servo-controlled* -- These robots employ sensory feedback from each actuator joint (internal state sensors) to the servo units to maintain a predetermined trajectory which has been preprogrammed into the control computer and is provided as set point time functions to the servo units. In general, there are position sensors, tachometers, and accelerometers.
3. *Adaptive-controlled* -- In addition to the servo feedback sensors, adaptive control requires sufficient information about the surrounding world to take into account changes which influence the operational process. Based, in general, on visual, optical, tactile and force sensory information (external state sensors), the control computer program is then modified, e.g., through logical decisions in accordance with predetermined algorithms, to accommodate the changed situation.
4. *"Intelligent"-controlled* -- These robots use the same sensory systems as for adaptive control, but they are programmed to identify and recognize their environment in relation to the required task through a world model

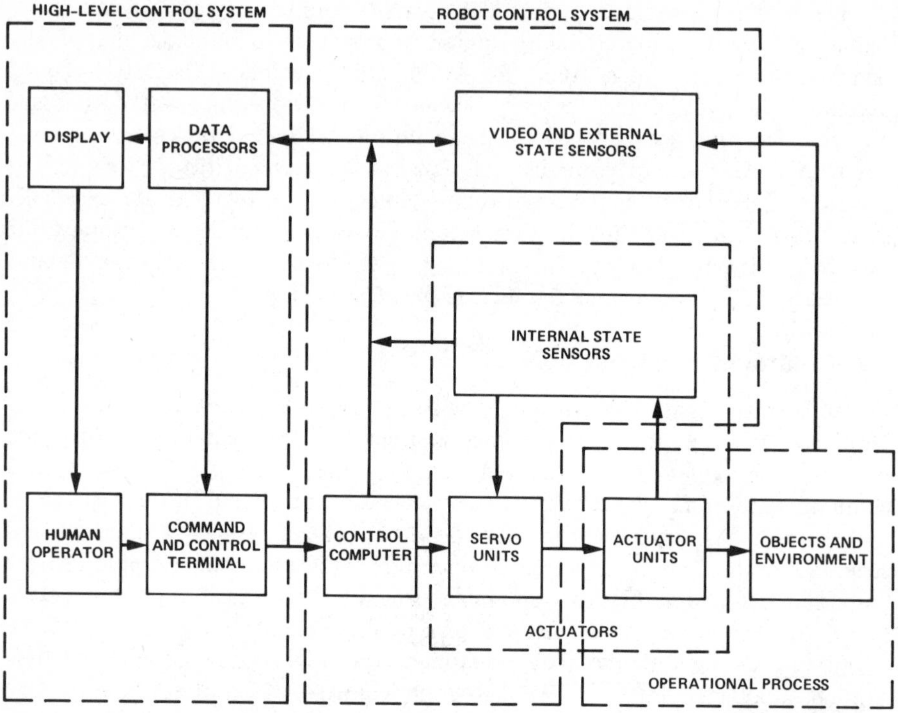

Figure 9 Control structure of robotic system. The operational process includes the actuator units (joint motors, grippers, tools, etc.) and the external environment and objects. The robot control system includes internal feedback loops, consisting of internal state sensors (position, velocity, acceleration) and corresponding servo units, and external feedback loops consisting of external sensor systems (video, range, force, touch), and a control computer. The control computer can be programmed on-line (lead-through, teach-in, etc.), or off-line using command and control terminals, displays and appropriate data processing equipment containing CAD/CAM and modeling data.

in the control computer. They plan and derive action decisions for task implementation and self maintenance; they are able to learn from experience, adapt to unanticipated events, and work towards optimal performance [14].

By far, most robots in industry today are in the open-loop and servo-controlled category. It is expected that these control systems will also in the future satisfy the greatest number of requirements. This is especially true for the relatively simple, but frequently occurring applications which can be programmed on-line by non-textual methods.

Adaptive and intelligent industrial robots generally require off-line textual programming. This calls for specially trained personnel, which may be one reason why adaptive and intelligent robots are in most cases still in research laboratories and not on the factory floor. Yet, numerous robot programming languages and associated software have appeared on the market since Unimation Inc. first introduced VAL in 1979. Some present commercial offerings are [15,16]:

AL by Robot Technology Inc.

AML by IBM Company

HELP by GE Company

MCL by McDonnel-Douglas Company

RAIL by Automatix Inc.

VAL by Unimation Inc.

The important design characteristics of robot programming languages are determined by a number of parameters, such as:

1. Language modalities (e.g., textual, menu, etc.);
2. Language type (e.g., subroutines, new language, etc.);
3. Geometric data types (e.g., frame, joint angles, vector transformation, rotation, path);
4. Display and specification of rotation matrix (e.g., rotation matrix, angle about a vector, quaternions, Euler angles, roll-pitch-yaw);
5. Ability to control multiple arms;
6. Control structures (e.g., statement labels, if-then, if-then-else, while-do, do-until, case, for, begin-end, cobegin-coend, procedure/function/subroutine);
7. Control modes (e.g., position, guarded moves, bias force, stiffness/compliance, visual servoing, conveyor tracking, object tracking);
8. Motion types (e.g., coordinated-joint between two points, straight-line between two points, splined through several points, continuous path, implicit geometry circles, implicit geometry patterns);
9. Signal lines (e.g., binary input, binary output, analog input, analog output);
10. Sensor interfaces (e.g., vision, force/torque, proximity, limit switch);
11. Support modules (e.g., text editor, file system, hot editor, interpreter, compiler, simulator, MACROs, INCLUDE statement, command files, logging of sessions, error logging, HELP functions, tutorial dialogue);
12. Debugging features (e.g., single stepping, break points, trace, dump).

The above mentioned languages and others still under development are strongly dependent on the robots with which they are marketed. They make the user think about the joints of the robot instead of the task to be performed. It is therefore difficult to use these languages in a practical setting.

A robot language should be task oriented, i.e., the programmer should be required to think about the task in his terms not the machine's terms. It should be robot independent, i.e., portable, and the associated software systems should be modular and should be interfaceable to many robots of many types from many vendors. A robot language should be designed so that it can be integrated into existing and future CAD/CAM facilities and should be flexible enough to describe parallel activities and their priority. It should contain facilities for handling descriptions of the available processes and interconnections, for estimating processor run time and for distributing processes among processors. It should be able to handle graceful degradation, process migration, fault tolerance, and optimization. Languages with these facilities are still in the research stages and are not expected to be available within the next five years.

Another related issue is for what user a new language should be written. Should it be for the assembly line worker, the factory foreman, CAD operators, NC part programmers, manufacturing engineers, or computer scientists? The user will ultimately determine what language characteristics will survive the test of practical applications.

8. TRENDS IN INDUSTRIAL ROBOTICS

Present development trends show that a process of change is taking place towards the integration of flexible automation, including robotics, and towards the automation of production related planning and design processes. From a planner and designer's point of view, it is important to recognize that robotics requires highly interdisciplinary technologies and associated developments. Mechanical, electrical, electronics, computer, information, industrial and systems engineering disciplines play a role in the development of robotic systems. The planner and designer must be conscious of the fact that during implementation numerous, complex interfaces must be coordinated. He must take into account the fact that automation will pervade increasing areas of industrial production. Not only the actual machining and assembly of workpieces will be automated, but also handling, transport, measuring, testing, etc. at various finishing stages will be included into the automation process.

Further development of robotics technology and its flexible adaptation into variable forms of production is an important characteristic in structuring and controlling production processes. The flexibility of production systems must increase with the demand for smaller batch manufacturing. Considering the respective spectrum of workpieces, transport systems, handling systems, and machine tools must be automated as flexibly as necessary but also as cost effective as possible. To achieve the required flexibility for small batch manufacturing, the trend is towards robotics and, in particular, towards the modularized construction of robotic systems.

An important step towards modern production systems is the automation of all forms of handling and manipulative functions which are related to the flow of material within the sphere of the work station and the manufacturing installation. The required flexibility for these operations can be provided by programmable, multiaxis industrial robots which are equipped with appropriate end effectors, grippers or hands.

To develop a high flexibility in overall production, it is required to rethink the concepts of material flow in relation to the serviced machines. Manipulation and handling operations can be accomplished by single robots or by integrated robotic systems. Analyses of the requirements for flexibility in the area of manipulation and handling leads to modular building blocks for robotic systems. The synthesis of such flexible modules with machine tools and other peripheral equipment results in new types of manufacturing concepts called robot work cell or flexible work cell.

A robot work cell is arranged to perform a series of manipulative, handling and transfer functions completely automatically. The robot work cell is the basic building block of a robotic line. The robot work cell is linked to other cells to form a flexible, integrated robotic manufacturing system. Such robotic manufacturing systems can be structured to produce prescribed groups or families of parts with small-to-medium batch sizes following the concepts of group technology.

Robotics manufacturing systems can be linked and coordinated with each other using computer automated control in a hierarchical fashion. It is then not difficult to visualize the combination of a collection of automated production system concepts into an integrated computer automated factory.

With these developments in mind, it is to be expected that automation will penetrate all production related operations beyond the application of robots and robot work cells. The planner and designer of robotic systems must take into account these developments and should consider, already during the planning phase, the impact of the entire manufacturing system. An advanced state of technology of robotic systems must be followed by a corresponding and compatible state of technology in the entire manufacturing process.

REFERENCES

[1] James S. Albus, *Brains, Behavior, and Robotics,* BYTE Publications, Inc., Peterborough, N.H., 1981.

[2] Joseph F. Engelberger, *Robotics in Practice,* American Management Association, 1981.

[3] Ewald Heer, *New Luster for Space Robots and Automation,* Astronautics and Aeronautics, September, 1978 Vol. 16, No. 8, p. 48-60.

[4] Ewald Heer, *Prospects for Robots in Space,* Robotics Age, Vol. 1, No. 2, 1979.

[5] R. F. Busby Associates, Inc., *Remotely Operated Vehicles,* NOAA

Office of Ocean Engineering, Rockville, Maryland, Contract No. 03-78-G03-0136, August, 1979.

[6] Larry Liefer, *Rehabilitative Robots,* Robotics Age, May/June, 1981, p. 4-15.

[7] H. J. Warnecke and R. D. Schraft, *Industrial Robots -- Applications Experience,* 1982, I.F.S. Publications, Ltd., U.K.

[8] G. Spur, et at., *Industrieroboter,* Carl Hanser Verlag, Munich, Vienna, 1979.

[9] U. Rembold, et al., *Technische Anforderungen an zukuenftige Montageroboter, Teil 2* (Technical Requirements on Future Assembly Robots, Part 2). VDI-Z, Bd. 123 (1982) No. 19, p. 790-796.

[10] F. Weissgerber, *Fuenf Jahre Industrieroboter-Einsatz bei Volkswagen* (Five years of Industrial Robots -- Applications at Volkswagen). Proceedings of the 8th International Symposium on Industrial Robots, Stuttgart, 1978 Vol. 1, p. 78-91.

[11] J. Denavit and R. S. Hartenberg, *A Kinematic Notation for Lower-Pair Mechanisms Based on Matrices,* ASME Journal of Applied Mechanics, June, 1955, Vol. 22 p. 215-221.

[12] R. P. Paul, *Robot Manipulators-Mathematics, Programming and Control,* MIT Press Cambridge, Massachusetts, 1981.

[13] J. Y. S. Luh, M. W. Walker, and R. P. Paul, *Resolved-Acceleration Control of Mechanical Manipulators,* IEEE Transactions of Automatic Control, Vol. AC25 No. 3, June, 1980, p. 468-474.

[14] G. N. Saridis, *Toward the Realization of Intelligent Controls,* Proceedings of the IEEE, Vol. 67, No. 8, August, 1979, pp. 1115-1131.

[15] B. I. Soroka, *What Can't Robot Languages Do?,* Proceedings 13th International Symposium on Industrial Robots, April 17-21, 1983, Chicago, Illinois.

[16] W. A. Gruver, et at., *Evaluation of Commercially Available Robot Programming Languages,* Proceedings 13th International Symposium on Industrial Robots, April 17-21, 1983, Chicago, Illinois.

2

Robotic Control To Help The Disabled

G. N. SARIDIS

1. INTRODUCTION

The discipline of rehabilitative engineering was created when humans used two pieces of wood to brace a broken leg, or attached a socket with a hook to replace a missing hand. However, for several centuries this was the exclusive practice of medical practitioners. It was not until after World War I, that some mobility and control of the function of the artificial limb or brace was required, that the engineer was invited to offer his services. Since then, and in particular, after World War II, a large number of artificial limbs, joints, orthotic devices, e.g., braces powered by the wearer or externally have been produced as bodily aids to the disabled [1]. With the recent developments in robotics and manipulators, their applications were extended to provide maid service to paralyzed people like quadroplegics, by creating electrically powered wheel chairs or electronically controlled manipulators [15].

Most of these devices were ingeniously designed, cosmetically functional and highly sophisticated, and inspite of their relatively high cost, well accepted. However, the introduction of the digital computer in engineering presented opportunities of developing devices adaptable to a patient with less mental and training effort, facilitating thus the rehabilitation process. On the other hand, research on structural design of artificial limbs and robots, produced lighter and more versatile structures capable of cosmetically reproducing accurate anthropomorphic movements [3,7,9,15,16].

Arm prostheses or orthoses and manipulative devices require more sophisticated intelligent functions to execute cosmetically anthropomorphic tasks. A systematic approach to generate, organize, coordinate and execute such sophisticated tasks has been developed at the Advanced Automation Research Laboratory at Purdue University [8,12]. This approach is briefly discussed in the next section while two applications are presented in the sequel:

1. A voice-controlled seven degree of freedom mechanical arm, with visual object recognition suitable for a hospital environment to aid people with disabled upper limbs [6,8].
2. An Electromyographic (EMG) signal controlled prosthetic arm for above the elbow amputees with four degrees of freedom [12,13].

Both projects have been sponsored by NSF and rely heavily in developing, decision making, learning and task coordination capabilities on mini and microcomputers, with specialized functions for machine intelligence. The hardware was also designed to be light and efficient for optimal mobility and comfort. In total, they represent the realization of intelligent interactive machines capable of executing complex tasks with minimum mental effort and training from the part of their operator. Experimental and simulation results of the application of the hierarchically intelligent control approach have been impressive and have encouraged the author and his colleagues to plan the design of the intelligent machines of the future based on specialized microcomputer architectures.

2. HIERARCHICALLY INTELLIGENT CONTROL

Cognitive Systems have been traditionally developed as part of the field of artificial intelligence to implement, on a computer, functions similar to the ones encountered in human behavior [11]. Such functions as speech recognition and analysis, image and scene analysis, data base organization and dissemination, learning and high level decision making, have been based on methodologies emanating from simple logic operations to advanced reasoning as a pattern recognition, linguistic and fuzzy set theory approaches. The results have been well documented in the literature [4,5,11,13,17].

In order to solve the modern technological problems that require control systems with intelligent functions such as simultaneous utilization of a memory, learning, or multilevel decision making in response to "fuzzy" or qualitative commands, *Intelligent Controls* have been developed. They utilize the results of cognitive systems' research effectively with various mathematical programming control techniques. Each cognitive system associated with the specific process under consideration may be considered as subtask of the process request by an original general qualitative command, programmed by a special high-level symbolic computer language, and sequentially executable along with decision making and control of the hardware part of the process.

Many systems have been designed to perform in the above manner. In the area of manipulators and robotics many such systems have been developed for object handling in an industrial assembly line, remote manipulation in hazardous environments, the planet-exploration Mars-vehicle, hospital aids to disabled, and autonomous robots [2]. In most cases the control process is remotely performed from the operator, its function is semi-autonomous and the system must utilize some cognitive systems to understand the task requested to execute, identify the environment and then decide for the best plan to execute the task.

Various pattern recognition, linguistic or even heuristic methods have been used to analyze and classify speech, images or other information coming in through sensory devices as part of the cognitive system. Decision making and motion control were performed by a dedicated digital computer using either kinematic methods, like trajectory tracking, or dynamic methods based on compliance, dynamic programming or even approximately optimal control.

A *Hierarchically Intelligent Control* approach has been proposed by Saridis [11] as a unified theoretic approach of cognitive and control systems methodologies. The control intelligence is hierarchically distributed according to the principle of *Decreasing Precision with Increasing Intelligence,* evident in all hierarchical management systems. They are composed of three basic levels of controls even though each level may contain more than one layer of tree-structure functions:

1. The Organization level
2. The Coordination level
3. The Hardware Control level.

The *organization level* is the master mind of such a system. It accepts and interprets the input commands and related feedback from the system, defines the task to be executed and segments it into subtasks in their appropriate order of execution. An appropriate subtask library and a learning scheme for continuous improvement provide additional intelligence to the organizer. Since the organization level takes place on a medium to large size computer appropriate "translation and decision making schemata" and linguistically implementing the desirable functions [6].

The *coordination level* receives instructions from the organizer and feedback information from the process for each subtask to be executed and coordinates the execution at the lowest level. The coordinator, composed usually of a decision making automaton representing a context free language, may assign both the performance index and end conditions as well as possible penalty functions designed to avoid inaccessible areas in the space of the motion. The decisions of the coordinator are obtained with the aid of a performance library and a learning decision scheme, recursively updated to minimize the cost of operation.

A *lowest-level control* process usually involves the execution of a certain motion and requires besides the knowledge of the mathematical model of the process the assignment of end conditions and a performance criterion or cost function defined by the coordinator. Optimal or approximately optimal control system theory may be used for the design of the lower level controls of decentralized subprocesses of the overall process to be controlled [33].

The method has been successfully applied to control a general purpose manipulator with visual and voice inputs [8] and an EMG controlled prosthetic

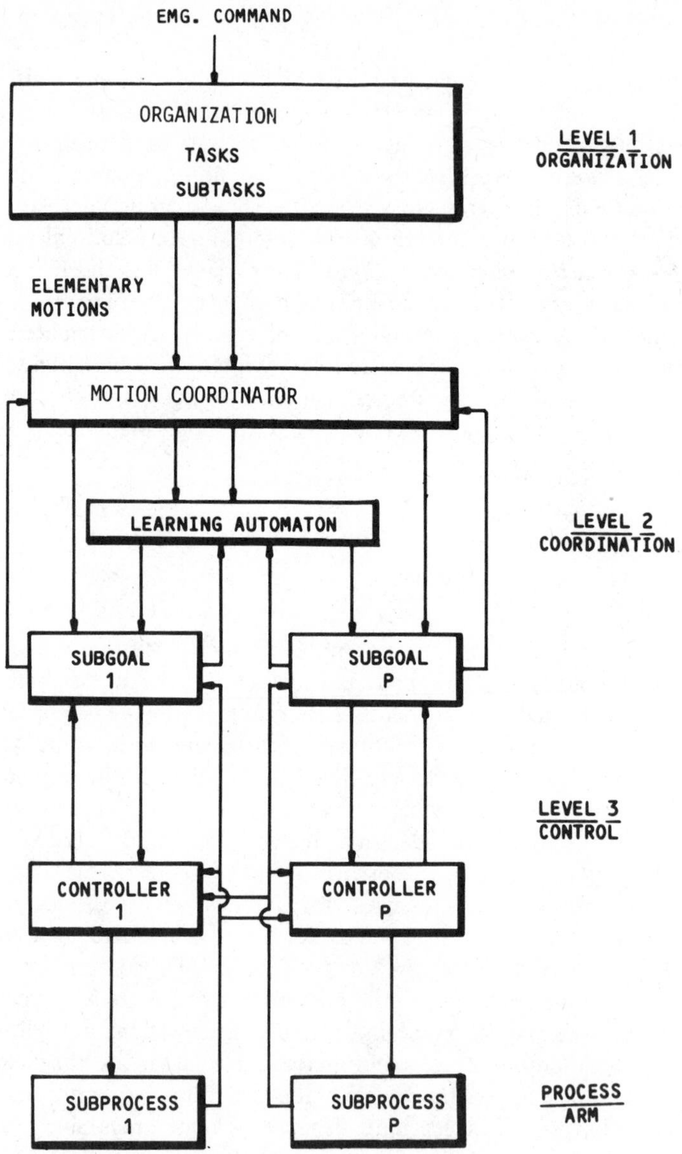

Figure 1 Hierarchical Intelligent Control of a Prosthetic Arm.

arm [12,13]. They are described in the next sections. A block diagram of the general hierarchical intelligent control system is given in Figure 1. Other applications of this approach may be found in space exploration, work in hazardous environments, like nuclear reactors, etc., urban traffic control systems, robotics and manufacturing systems and many more.

3. A VOICE-CONTROLLED MANIPULATOR WITH VISUAL INPUTS FOR A HOSPITAL ENVIRONMENT

A hierarchically intelligent manipulator system consists of a general purpose digital computer with appropriate peripheral devices, a seven degrees of freedom electric arm, three for positioning the wrist, three for the orientation of the hand, and one for opening and closing the gripper, a television digitizing camera, external sensors for the arm, and an interfacing device between the arm and the computer. A block diagram of the control system is given in Figure 2. Mechanical and other hardware specifications are given in Figures 3, 4 and 5 [6,8].

In order to accomplish a complex manipulative task, a computer-controlled manipulator system and its control algorithm must show the following properties:

1. Man-manipulator communication - to recognize the linguistic commands from the operator and interact with him.
2. Coordinated motion control - to possess some levels of autonomous coordinated position and rate control without the assistance of the operator. That is, the operator is taken out of the control loop.
3. Interaction with the environment - to ability to integrate the feedback signals from television camera and other external sensors into the system

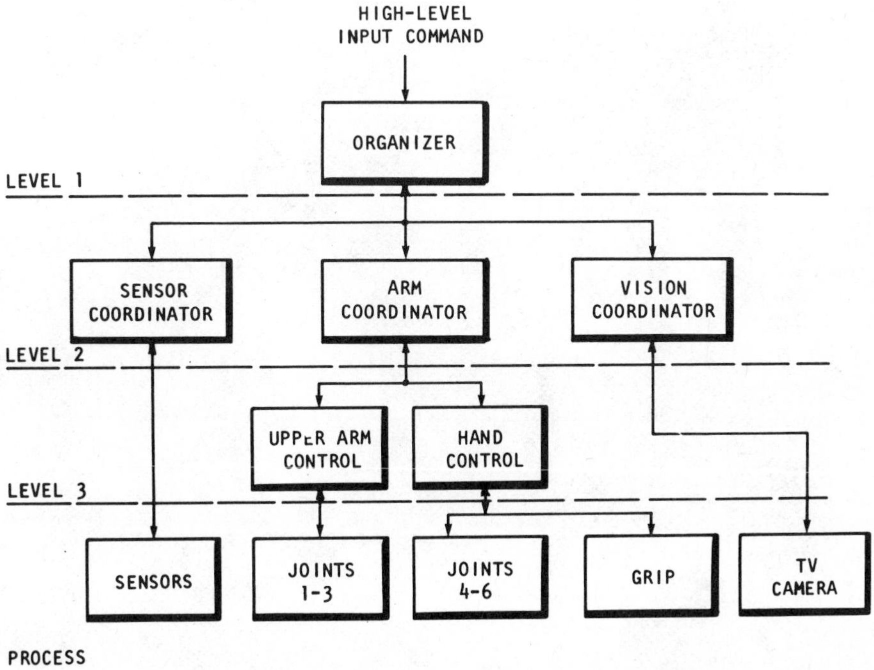

Figure 2 Hierarchically Intelligent Control of a Noise-Controlled Manipulator with Visual Feedback.

and update it strategies or sequences of control actions to accomplish the task.

Hierarchically Intelligent Control is mostly suitable to control such a general purpose manipulator. According to the principle of *Increasing Intelligence with Decreasing Precision,* the lowest level, the run-time control level, in the hierarchy must execute a local task with high precision by satisfying certain performance criterion, thus requiring a rather sophisticated and precise model. These models of the arm subsystems are formulated from the state space approach. The next higher level in the hierarchy is that of the coordinator, where the individual subtasks are put to work together by appropriately selecting their performance evaluation requiring less precision and modes of operations while the need of higher level decision-making capabilities is apparent to improve the overall performance of the system. In addition to coordinating and supervising the decision-making units in the lower level, the coordinator should be endowed with learning capability to improve the system performance under reappearing control situations. An automaton capable of executing fuzzy or stochastic inputs can be implemented to perform such learning function. Finally, the organization level which serves as a linguistic organizer at the top of the hierarchy also possesses certain learning capabilities

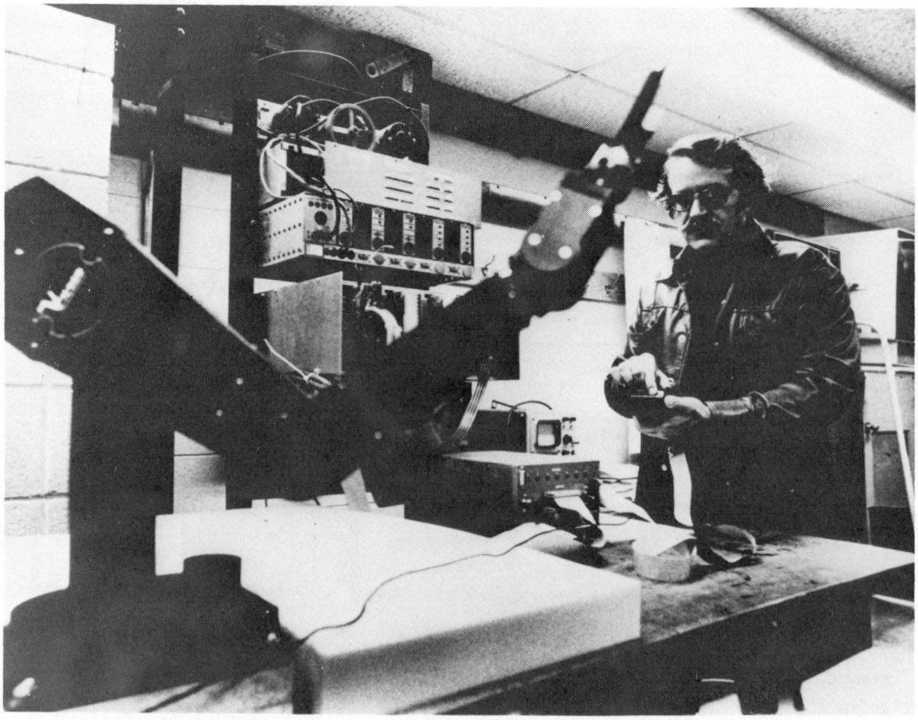

Figure 3 The AARL Manipulator.

and decision-making. This highest level decision-making involves parsing the stochastic linguistic input strings, organizing the task, identifying the control situation, and assigning the appropriate control pattern without much knowledge of the detailed execution of the task.

3.1 System Hardware

The MIT Scheinman arm, shown in Figure 3, is connected through an interfacing device, built by Perceptronics, which is in turn connected to the direct memory access peripheral device DR-11B of the computer, Figure 5. The arm has six degrees of freedom plus the opening and closing of the gripper, Figure 4. The hand can be placed anywhere in the work space within 50 cm

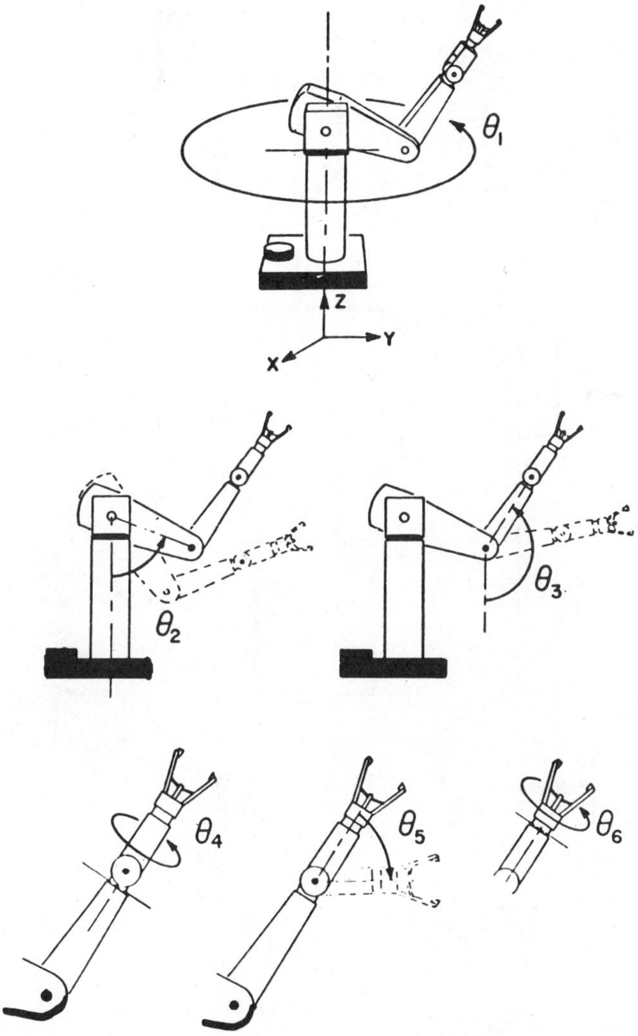

Figure 4 Joint Angles of the MIT Arm.

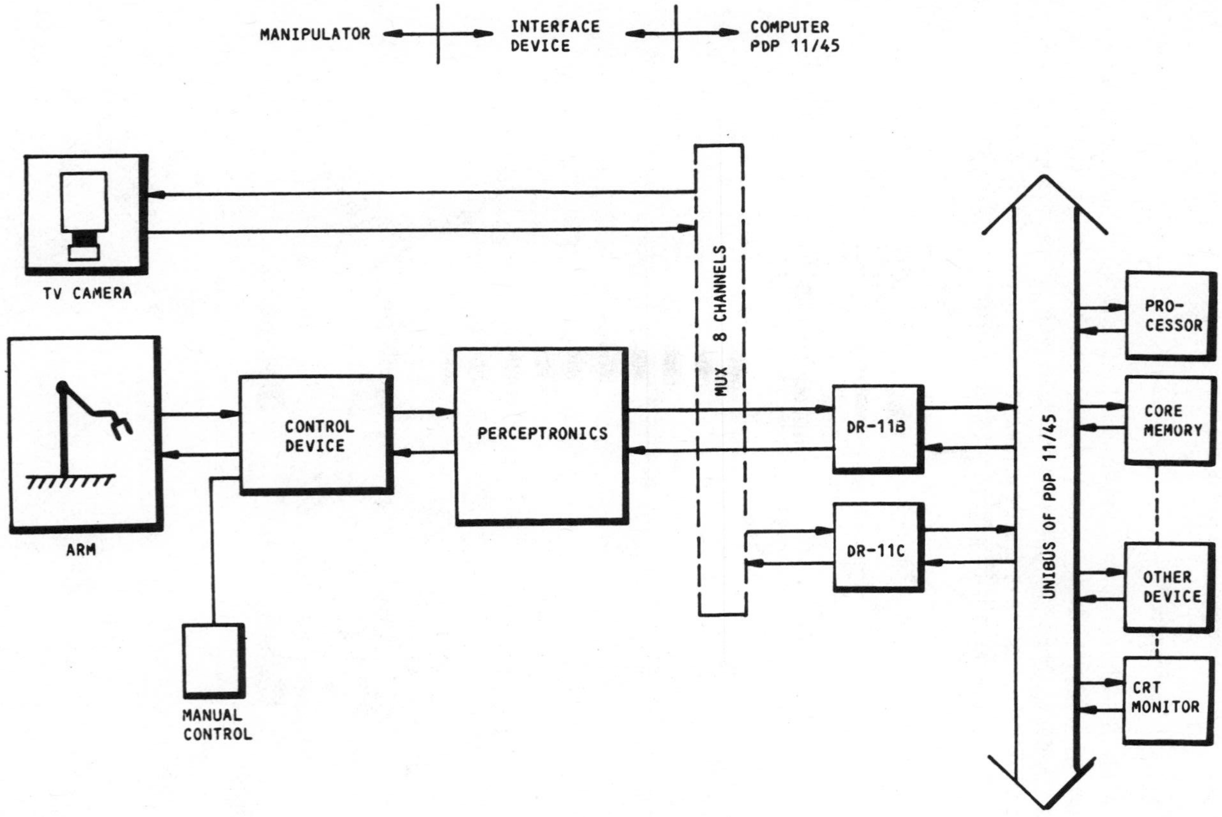

Figure 5 Purdue University's AARL Manipulator System Configuration.

radius from the shoulder of the arm, and can pick up a maximum load of 1.4 Kg. All the joints are rotary and are driven by DC torque motors. Each joint is equipped with a potentiometer and a tachometer, which provide feedback signals of position and velocity, respectively, except joint 6 and the "gripper". Moreover, the upper 5 joints are equipped with electromechanical brakes which hold the arm in position when activated. The hardware is designed to shut off the voltage supply to the joint motors automatically when the brakes are on.

The computer, a PDP 11/45 with Disk Operating System has a core memory of 128 K words (124 KW addressable memory). Besides the usual input-output peripheral devices, it has two television cameras for scanning the work space of the arm. Both cameras can provide pictures with gray level up to 128. Using a Ramtek digitizer, the picture can be displayed on a black and white or a color monitor. At the present time all the picture files are automatically stored in an auxiliary memory of 96 K words.

3.2 The Task Organization Level and Voice Command Input

The organization level serves two major purposes; interface with the operator and to organize the various tasks for different control and environmental situations.

External supervision of the manipulator has been minimized and replaced by the hierarchically intelligent controller, but it should still be capable of accepting commands by a user to execute a certain job involving all the tasks pertinent to the process. Therefore the organizer is designed to accept voice inputs from the user, decode them and then organize the sequence of tasks necessary for its execution by providing appropriate inputs to the coordinators.

Implementation of the organizer is obtained by a syntax directed translation schema [6] which generate a speech recognition algorithm and then another translation schema to organize the required tasks. The process was assumed to take place in a hospital environment as an aid to disabled patients and therefore no learning was required at the organization level.

An English like input language was assigned for the speech recognition and the schema. Analogous translation schemata may be assigned for the motion organization, vision and sensory tasks.

The discrete-word speech recognition programs developed for this project operate on the digitized speech input without any pre-filtering or hardware feature extraction. Speech is digitized for one second at a sampling rate of 10 kHz, which allows the speaker to utter one word. The endpoints of the utterance are detected by comparing the magnitude of the signal sampled with that of "silence." Once the endpoints of the word are found, the word is divided into 30 time-segments, and four features are extracted for each time segment.

The features chosen are pole frequency and normalized error, zero crossing rate, and absolute magnitude. These features have been shown to characterize a word accurately enough to allow discrimination between dissimilar words, but are still general enough to be computationally efficient.

Therefore a spoken word is represented by a 4 × 30 matrix, or "template:" four feature vectors whose components correspond to the 30 time segments. Prior to use, the computer is "trained" by forming average templates for each word in the vocabulary as spoken by the user, which become reference templates used in the actual recognition. This recognition is done by comparing the unknown word with each of the reference templates, and the closest match is selected as the word recognized. The distance function used for matching is a weighted sum of the absolute difference between corresponding feature vectors. The weighting coefficient have been selected empirically by an off-line program to optimize recognition sources.

A total of seven seconds is required to recognize a single word, including the one second sampling time. An additional three seconds is needed to complete a verification sequence which allows the user to correct an error in recognition. However, these programs are written in Fortran, and it is expected that significantly faster recognition can be achieved when the programs are converted to assembly language. This system has attained a recognition accuracy of 88$ correct for 24 repetitions of each of the 16 words in the vocabulary.

A 16 word vocabulary has been selected which allows easy formulation of commands which correspond to specific tasks for the arm to perform. These spoken commands are recognized one word at a time according to a pre-programmed syntax of Table 1, and each command obeying the syntax is associated with a task. The use of a syntax also allows the program to match the unknown word against a subset of the total vocabulary. Table 2 lists the user commands and their meaning.

3.3 Task Coordination Level

The function of an intelligently controlled mechanical arm may be subdivided into three major tasks as shown in Figure 2.

1. Sensory
2. Vision
3. Mechanical Motion.

The first task deals with the collection of information from proximity sensors, pressure gauges and other sensory devices. The selection of the proper device and the processing of the data to provide needed feedback information should be performed by an appropriate coordinator at the second level of the hierarchically intelligent control system.

The second task deals with processing visual information provided from one or more TV cameras fixed or moving with the arm. This involves object recognition and classification for appropriate selection and end point coordinate evaluation, object tracking using three dimensional vision, object avoidance and vision feedback for motion control purposes. A coordinating device may

TABLE 1.

VOICE COMMAND SYNTAX

---

1. ⟨ Command ⟩ → TV ⟨ arg 1 ⟩ |
Radio ⟨ arg 1 ⟩ |
Drink ⟨ arg 1 ⟩ |
Lights ⟨ arg 2 ⟩ |
Books ⟨ arg 2 ⟩ |
Fan ⟨ arg 2 ⟩ |
Window ⟨ arg 2 ⟩ |
Bed ⟨ arg 3 ⟩ |
Eat ⟨ arg 3 ⟩ |
Nurse

11. ⟨ arg 1 ⟩ → Channel ⟨ arg 3 ⟩ |Volume ⟨ arg 3 ⟩ | ⟨ arg 2 ⟩

14. ⟨ arg 2 ⟩ → Please|Thanks

16. ⟨ arg 3 ⟩ → More ⟨ arg 3 ⟩ |Less ⟨ arg 3 ⟩ |Thanks

---

generate all close subtasks in the proper order at the second level of the hierarchically intelligent controls system.

Finally the third task deals with the selection of the proper control gains based on information about the end points, and the type of motion of the upper three joints requested e.g., fast, slow, etc., the orientation of the hand for object handling and finally the coordination of the wrist motion with the hand orientation during special motions, e.g., the transportation of a cup full of water.

A fuzzy or stochastic automaton was originally suggested to implement the coordinator. This is a finite state machine designed to select one particular subtask from a library, using a learning (optimization) procedure to avoid external supervision in a unfamiliar environment. Since the equivalence of such a automaton to a formal language has been established in the literature [5], the coordinator was implemented by *linguistic decision schemata* in the form of software implementable on the PDP 11/45 minicomputer [6].

An example of the syntax for coordination of the wrist motion (upper three joints), with the hand orientation is given in Table 3.

TABLE 2

INSTRUCTION SET FOR LP 4.5

Instruction	Description
ALLU n,t	Select gait n and execute that gait for t e.t.'s.
ALLC t	Execute the current gait for t e.t.'s.
CHBA a	Switch to the ath gait bank.
REPO t,n	Initialization command. The robot retracts its legs (the robot rests on a pedestal), resets the gait cycle counter, and initializes for gait n, and wait t e.t.'s.
DEBT	Stand up command (executed after a REPO)
TMPO n	Sets the value of e.t. to n
CNTR p,v	Equilibrium adjustment. Sets leg p's vertical centering to v.
BRAN 1	Unconditional branch to label 1.
BEVE c, 1	Branch on condition code c to label 1.
BARI p, 11, 12, 13	Arithmetic branch on contents of register p.
CONS vl,vd	Set sensor register to range vl $\pm$ vd.
RADF p	Set telemeter to angular position p.
RADM p	Set telemeter to p and rotate 360 degrees.
EXAM p,n	Call routine p, passing argument n.
MONI n	Branch to monitor entry point n.
EXIT	Return.

3.4 Coordinated Suboptimal Control of the Wrist Motion

Due to the nonlinearities and dimensionality of the dynamic equations of the arm, an optimal control solution is very difficult to obtain. Even if one could find the optimal solution, the real-time implementation of the optimal controller may be very difficult if not impossible. As a result, a suboptimal feedback

TABLE 3.

SYNTAX OF THE MOTION COORDINATOR

1.	⟨ Statement ⟩ → ⟨ Query ⟩ | ⟨ Commands ⟩ | ⟨ Relation ⟩
4.	⟨ Query ⟩ → Report present state
5.	⟨ Commands ⟩ → Int ⟨ Argument list 1 ⟩ |
	Goto ⟨ Argument list 2 ⟩ |
	Move ⟨ Argument list 3 ⟩ |
	Pickup ⟨ Argument list 4 ⟩ |
	Rlse ⟨ Argument list 4 ⟩ |
	Scan ⟨ Parameters ⟩ |
	Twist ⟨ Rotation ⟩ ⟨ Angle ⟩ |
	End
13.	⟨ Relation ⟩ → If ⟨ Query ⟩ then ⟨ Commands ⟩ |
	If ⟨ Commands ⟩ then ⟨ Commands ⟩ |
	If ⟨ Commands ⟩ then ⟨ Query ⟩
16.	⟨ Argument list 1 ⟩ → ⟨ POSX ⟩ , ⟨ POSY ⟩ , ⟨ POSZ ⟩ , ⟨ Orientation ⟩ |
	⟨ POSX ⟩ , ⟨ POSY ⟩ , ⟨ POSZ ⟩ |
	⟨ Orientation ⟩
19.	⟨ Argument list 2 ⟩ → ⟨ Final position ⟩ | ⟨ Object ⟩
21.	⟨ Argument list 3 ⟩ → ⟨ Object ⟩ to ⟨ Final position ⟩
22.	⟨ Argument list 4 ⟩ → ⟨ Object ⟩ at ⟨ Final position ⟩ with ⟨ Orientation ⟩
24.	⟨ Parameters ⟩ → 128 | 256
25.	⟨ Orientation ⟩ → ⟨ Approach vector ⟩ , ⟨ Orientation vector ⟩
26.	⟨ Approach vector ⟩ → ⟨ POSX ⟩ , ⟨ PSOY ⟩ , ⟨ POSZ ⟩
27.	⟨ Orientation vector ⟩ → ⟨ POSX ⟩ , ⟨ POSY ⟩ , ⟨ POSZ ⟩
28.	⟨ Final position ⟩ → ⟨ POSX ⟩ , ⟨ POSY ⟩ , ⟨ POSZ ⟩
29.	⟨ POSX ⟩ → ⟨ Limit ⟩ ⟨ Integer ⟩ | ⟨ Integer ⟩
31.	⟨ POSY ⟩ → ⟨ Limit ⟩ ⟨ Integer ⟩ | ⟨ Integer ⟩
33.	⟨ POSZ> ⟩ → ⟨ Limit> ⟩ ⟨ Integer> ⟩ | ⟨ Integer> ⟩
35.	⟨ Limit ⟩ → 1 | 2 | 3 | 4
39.	⟨ Integer ⟩ → 0 | 1 | 2 | 3 | 4 | 5 | 6 | 7 | 8 | 9
49.	⟨ Object ⟩ - ⟩ Red/Blue
51.	⟨ Rotation ⟩ → CLKW|CCLKW
53.	⟨ Angle ⟩ → 0 | 1... | 180

controller, which is simpler in structure and easier to implement, has been developed. The structure of the controller consists of nonlinear and linear control efforts;

$$u(x) = u_{NL}(x) + u_L(x).$$

An approximation theory of nonlinear optimal control has been developed to provide a procedure for selecting an efficient suboptimal feedback controller for the motion of the wrist, defined as the first control subtask. The remaining four subtasks listed here in their order of execution - Orientation, Searching, Sensing, and Forcing - are further broken down into combinations of six primitive movements which govern the positioning/orientation of the hand. A television camera digitizing system is available for providing visual input to the manipulator. The location of the object to be manipulated by the arm is treated as the terminal state for the feedback controller.

The dynamic equations for the upper three joints of the arm are obtained from the lagrange equation of motion and expressed in matrix differential equation as

$$x(t) = \begin{bmatrix} 0 & 1 \\ 0 & 0 \end{bmatrix} x(t) + \begin{bmatrix} 0 \\ J^{-1}(x)N(x) \end{bmatrix} + \begin{bmatrix} 0 \\ J^{-1}(x) \end{bmatrix} u \qquad (1)$$

where $x(t)$ = state of the system = $(\theta_1,\theta_2,\theta_3,\dot{\theta}_1,\dot{\theta}_2,\dot{\theta}_3)^T$

$J^{-1}(x)N(x)$ = coupled gravitational, Coriolis and centrifugal torques and gear friction of the motor

$u(x)$ = external applied torques

and its associated performance index is given by

$$IP(u) = \int_O^T [(x-x^d)^T Q(x-x^d) + u^T u]\, dt. \qquad (2)$$

The problem is formulated as a infinite time problem for the obvious reason that the feedback gains of the controller will be constants. This is justified because the time constants of the motors are small as compared to the time constants of the Matrix Riccatti Equation and will be verified experimentally. The performance index physically represents some functions of the energy expenditure of the physical system. It is desired to find an admissible feedback control law $u(x)$ that causes the system Eq. (1) starting from an initial state $x(O)$ to follow an admissible trajectory while minimizing the performance index Eq. (2). The gain matrix Q in the performance index is obtained from the off-line training of the time samples taken from the Goniometer curves.

Mathematically, Q is found from the inverse of the integral square error of $x(t)$

$$Q = \left[\int_{O}^{T} (x - x^d)\ (x - x^d)^{T} dt \right]^{-1} . \tag{3}$$

The resulting suboptimal control is stable and it should drive the wrist of the arm to a smooth well-coordinated motion referencing the Cartesian coordinate system only once to transform the final end point of the trajectory x^d to that coordinate system.

Experimental investigation performed at AARL of Purdue University has conclusively indicated that the Approximately Optimal Controller was very sensitive to parameter variations like moments of inertia and especially friction. Therefore the model (1) of the system that assumes constant coefficients is not suitable for such a solution.

Instead of improving and thus complicating the system's model the following hierarchical design has been proposed. An analog minor loop with a compensator built in to increase the robustness of the system is designed around each of the DC motors, obtaining measurements through torque sensors. Such a compensator may be designed using frequency domain methods. The resulting compensated system which will be insensitive to parametered variations now could be globally controlled using the Approximately Optimal Design procedure to retain the advantages of feedback and single command control mentioned earlier. A block diagram in Figure 6 explains the procedure.

3.5 The Vision System

An integral part of the manipulative system is the vision coordinating system for the arm. The vision system that is being developed at the AARL is quite similar to the vision system developed at SRI for industrial automation [11]. A fixed thresholding technique is used for the segmentation of the image data and the recognition algorithm processes the clear binary images, and uses area and perimeter of the object as the features for recognition. The objective is to identify the objects and their locations surrounding the arm from its environmental library or model. The library is then updated to initiate the arm to complete its execution of task. Current complex recognition algorithms would be inappropriate as recognition time is a crucial factor for real time on-line implementation. Because of this time constraint, the algorithm that is implemented on the MIT arm is simple and efficient and takes less than one minute to recognize all the objects surrounding that arm, and supplies enough information about the environment to the arm to initiate the execution of the task.

As stated earlier, the arm is intended to be used as a hospital aid to the immobilized people. It is assumed that the objects are distinct and separable

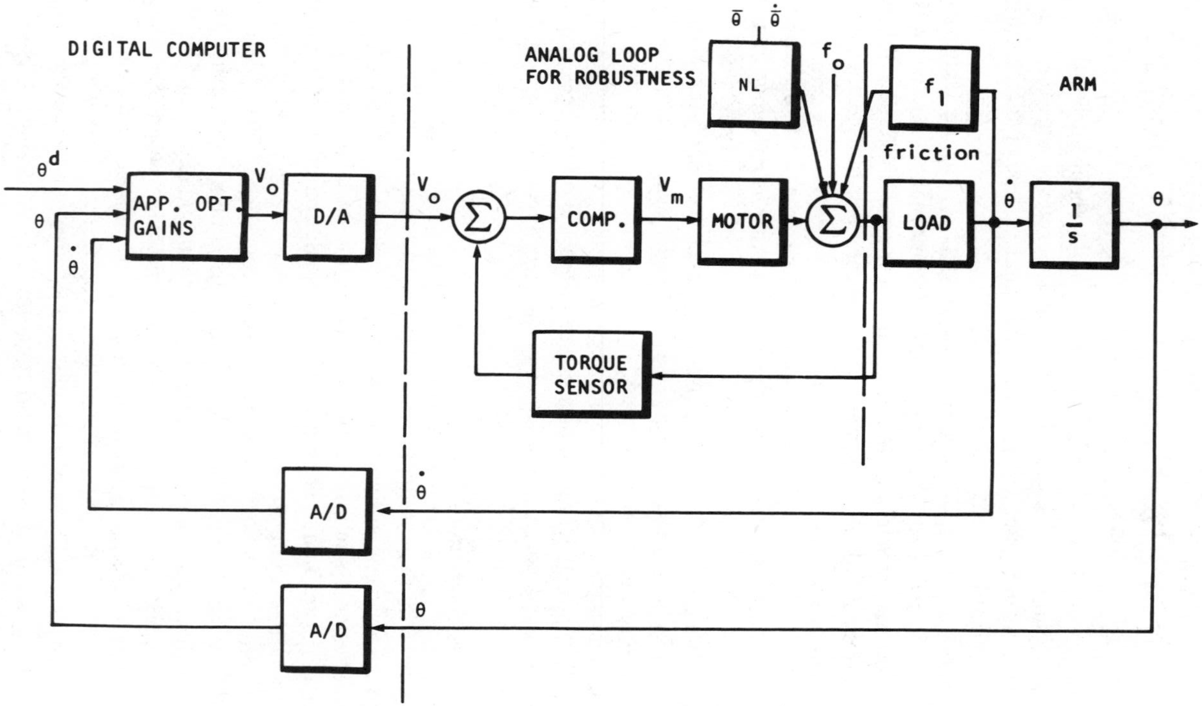

Figure 6 Integrated Frequency Domain and Suboptimal Control of Manipulator.

from the background which is usually white. The camera is tentatively fixed in position and placed above the arm and scans the work space when it is told to do so.

The arm system has a tree-like environmental library which contains the information about the area, the perimeter, the position, and the orientation of the objects surrounding the arm. However, the position of the objects may be altered by the nurse, the doctor or patient's visitors. Thus, it is the objective of the recognition algorithm to identify the objects and their locations and update or modify the environmental library which is used by the arm to initiate the execution of the task. The recognition algorithm assumes that no two objects are stacked up or aggregated into one object. That is, a minimum spatial distance exists between the neighboring objects.

The television camera digitizes the work space into 256×256 frame picture file with 128 grey levels and stores it in the auxiliary memory in byte format. A histogram curve is immediately calculated and displayed on the monitor. The first global minimum reached from the left of the histogram curve is used as a threshold value to separate the objects from the background. The fixed threshold segmentation provides a fast and efficient method to separate the objects from the background. However, the lighting surrounding the arm's work space has to be carefully controlled in order to produce a good image data for segmentation. Such a control on the lighting can be easily done in a hospital without any difficulty.

After segmentation, the image data has changed from gray-scale image data to binary image data that consists only of white (background) and black (objects and noises) segments. Undoubtedly, noise is always present in every scanning. To eliminate these undesirable noises, a clean-up algorithm which labels the black segments with numbers and counts their areas and perimeters is implemented by a push down stack. If the areas and perimeters of the segments are below the smallest area and perimeter of the object present in the environmental library, they are regarded as noises and will then be filtered out.

Two features from each scanned object are taken to form the feature vector $y^T = [y_1, y_2]$, namely its area (y_1) and its perimeter (y_2). To recognize an object, a linear discriminate function $g(y)$ based on the selected feature vector is used. That is,

$$g(y) = w_1 y_1 + w_2 y_2 = w^T y \tag{4}$$

where the w's are the weighting coefficients. The value $g(y)$ that is closest to the $g(y^*)$, which is the discriminate function value when the true pattern feature vector y^* from the environmental library is used, will automatically classify the scanned object to the pattern object from which y^* is selected.

The weighting coefficients (w_1, w_2) are trained initially with all the pattern objects in the environmental library. By holding w_2 constant, w_1 is changed by Δw_1 such that the set of weights $(w_1 + \Delta w_1, w_2)$ will produce least recognition error among all the pattern objects. Similar, the procedure is repeated for w_2 by holding w_1 constant. This training procedure to obtain the weighting coefficient has to be performed when new objects are introduced to the arm and incorporated into the environmental library.

The positions of the recognized objects are located from the grip coordinates of the picture file. The offset of the picture grip coordinates from the arm's work space is added to the locations of the objects. Thus the positions of the objects, expressed in the Cartesian coordinates, are obtained from the binary image data. The position of the object is then considered as the desired position and orientation for the end effector of the arm.

3.6 Experimental Results

Several experiments with voice commands, vision object recognition and suboptimal control have been successfully performed at the Advanced Automation Research Laboratory at Purdue University. They involved:

1. Picking up a cup of water and taking it to a patient to drink.
2. Turning on a television set and changing channels.
3. Pushing the nurse and light buttons.

Other experiments like picking up a spoon and fetching a pill are also planned.

4. EMG CONTROLLED UPPER EXTREMITY PROSTHESIS

The author and his colleagues have been involved in the design of a p-degree of freedom hierarchically intelligently controlled prosthetic arm for above the elbow amputees. Such a arm would receive commands directly from the intact masculature (e.g. biceps and triceps) of the amputee through synergistic Electromyographic signals (EMG) without extra mental effort. It is conjectured that such an anthropomorphic function of the arm, in response to crude qualitative commands, can be obtained only by a hierarchical, say, three-level control, with hierarchically increasing order of intelligence and hierarchically decreasing order of precision of control signals as one moves from the electromechanical actuators of the arm to the interface with brain. A block diagram in Figure 7 illustrates the concept of such a hierarchically intelligent control system. The three levels of control are

1. A linguistic organizer,
2. A fuzzy automaton as a coordinator, and
3. A bank of self-organizing controls.

They are described briefly below along with their function in the control system.

An arm of four degrees of freedom is considered (e.g. $p=4$). It was built by parts of the N.Y. Prosthetic Center of the Veteran's Administration and is depicted in Figure 8. It should be driven to perform all possible simultaneous combinations in the following six primitive motions.

1. Humeral Rotation in
2. Humeral Rotation out
3. Elbow Flexion

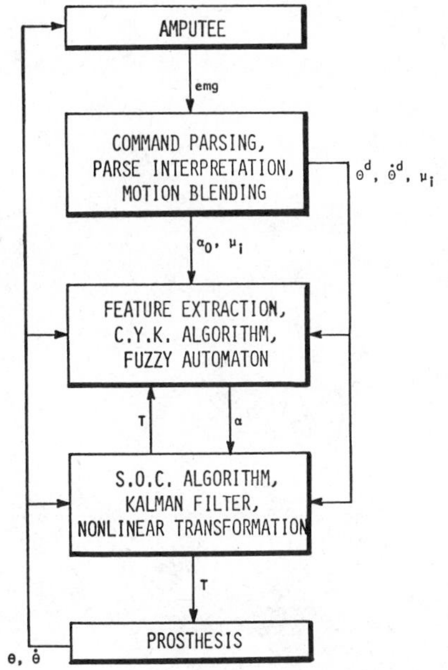

Figure 7 Block diagram of EMG Controlled Prosthetic Arm.

4. Elbow Extension
5. Wrist Pronation
6. Wrist Supination.

The hand grasp was not included in the final classification because no synergistic signals were obtained experimentally on the biceps and triceps. The grasp could be generated through other muscles of the body.

4.1 EMG Signal Analysis and Classification

The EMG signals are easily gathered from skin electrodes [13]. In this study, two electrode sites were used. At each site, two differential silver-silver chloride electrodes are separated 1.75 inches by a center ground electrode. They are separated from the skin by gel impregnated foam, and attached by an adhesive foam pad. The electrodes are placed perpendicular to the humeral axis, just lateral of the center bulge of each of the biceps and triceps. The locations are the dorsal most placements that are undamaged in most above-the-elbow amputees for whom this work will be applicable. An on-board microcomputer should sample the two signals and sequentially update its decision as to which motion is meant to be in progress. The stream of decisions is fed into a coordinator of the motions of the arm. This is done at intervals of several msecs. The amputee's visual feedback of the arm's position and velocity should thus provide accurate enough correction of the motion. We note that the coordinator is designed with a memory to provide trend characteristics of the decision stream to eliminate inadvertant errors and improve classification accuracy.

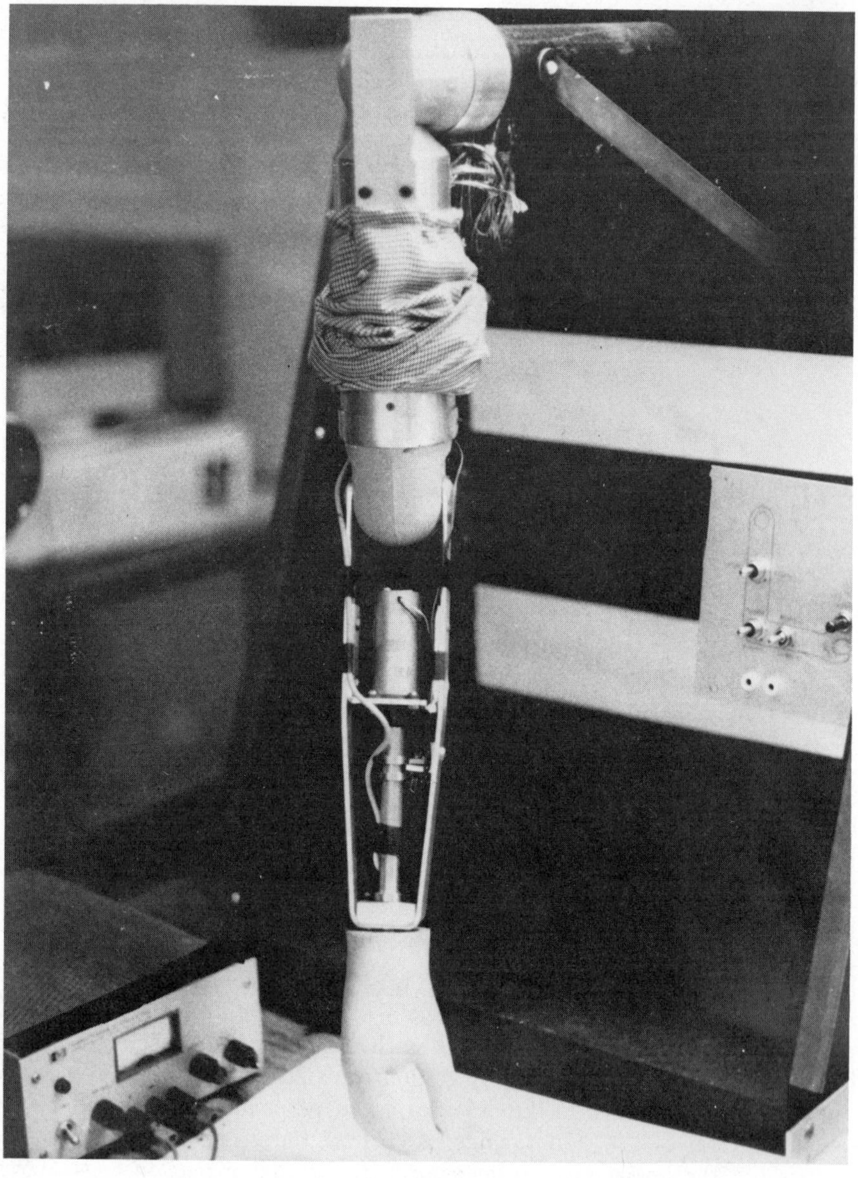

Figure 8 The AARL Prosthetic Arm.

Two high gain (5000) high input impedance (22 MΩ) EMGG amplifiers are used to amplify the signals collected by the electrodes, Figure 9.

Twenty-nine motions and one control (zero) motion are defined, each consisting of either grasp open-close, or one to three of the wrist, elbow, and humeral primitives. For each motion, twenty trials were performed. At each trial, 0.17 second of both the electrodes' waveforms were sampled at 3000 Hz, through 5 Hz high-pass filters.

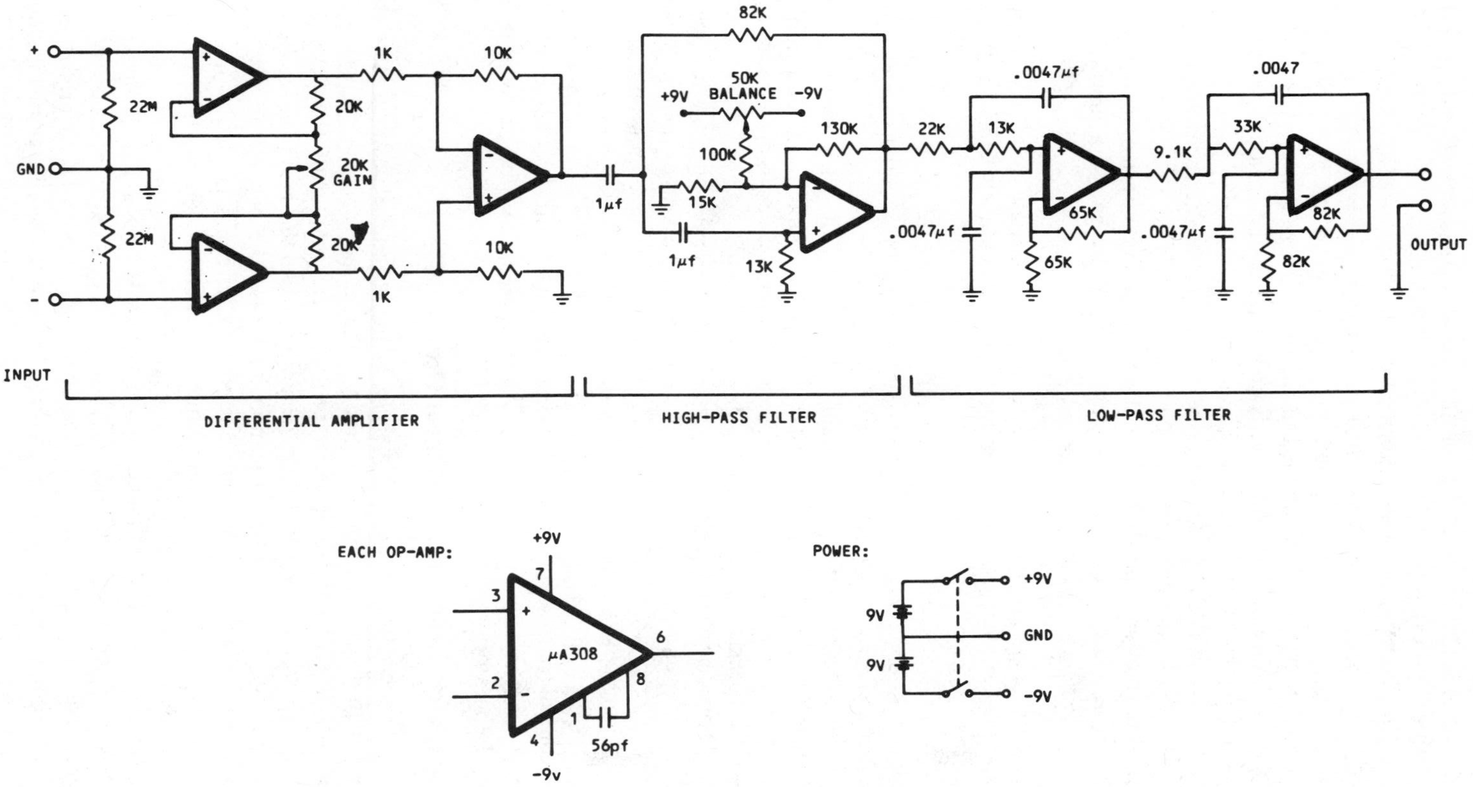

Figure 9 EMG Amplifier.

A thorough statistical analysis of the EMG signals corresponding to the above motions, indicated that

1. The EMG signals are nonstationary.
2. The EMG signals are bandlimited to virtually 1200 Hz.
3. No synergistic signals are generated from the hand grasp.
4. Pattern information is contained in the time moments of the signals.
5. Loading information of the arm is contained in the zero crossings which depends on the neutral firing frequency.
6. Motion information is concentrated mostly in the variances and less and less on the higher order moments.
7. In most motions the variance and zero crossing information groups in separable clusters that may be approximated by gaussian densities and their moments represent the classes.

Such signals are unreliable for time series analysis and physiological interpretation but their true averages may serve as feature vectors for pattern classification.

Systematic statistical pattern recognition algorithms were then used to classify separately each of the combined single motions and their misclassification error was studied.

Using the variances and zero crossings as features for pattern recognition, 85-90 of the pairs of motions can be separated. A simple linear discriminate function approach was taken for classification and the separable classes with less than 10$ misclassification error.

Finally, by discovering certain superposition properties on the statistics of the features, it was possible to decompose composite motions to their primitives and simplify the recognition problem, Figure 10. Load information was obtained through the count of "zero crossings" which are proportional to the frequency of nerve firing.

The control hardware as well as the computer software based on the classification algorithms are presently developed at the Advanced Automation Research Laboratory at Purdue University.

4.2 Linguistic Methods for Intelligent Organization

Based on the preceding discussion, given a command and a terminal state, the fuzzy automaton can be trained to produce the proper compound motion for the arm, which represents a level of control more intelligent than the self-organizing. However, the commands generated from the brain of the human operator are more in the form of compound tasks, like picking up a glass of water to drink. Therefore an intelligent control system is needed to interface the EMG signals with the fuzzy automaton, the motion coordinator of the next level and translate the above qualitative command to a sequence of compound motions of the area that will accomplish the task. Such a control will be required to produce a segmentation of the task providing appropriate $x^d(T)$ and qualitative information of the compound motion of the arm for each segment. It should produce *on-line* information about the change of direction,

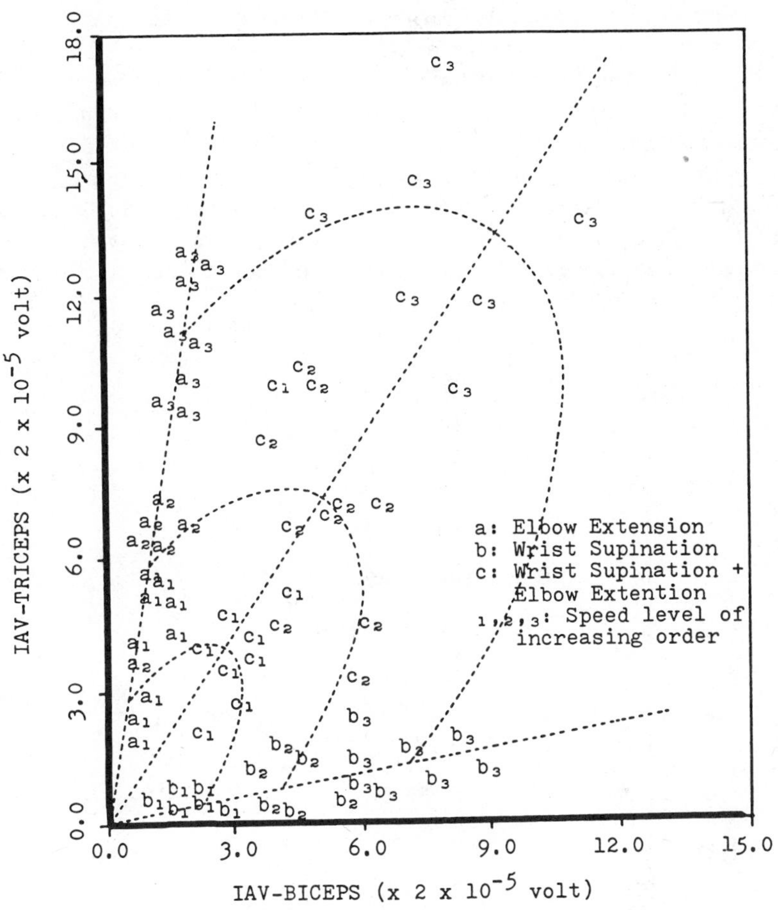

Figure 10 A Typical Pattern Trajectory.

combination, or expansion of segments, evaluation of the accomplishment of the task and processing of sensory feedback information from the brain, etc., without burdening the operator with unnecessary details about the function of the arm.

A machine producing decisions and functions of such a high level of intelligence must be an advanced digital computer, capable of processing qualitative information of high content, but also a fuzzy nature in the sense that high precision in execution is not required. A natural system for this type of information processing is the linguistic methods approach which has been developed in the model literature for artificial intelligence, pattern-recognition, scene analysis, and other functions [6]. Some methods process strings of words with logic instructions to accomplish the task according to certain predetermined grammar and syntax in manner similar to natural languages. In particular, *Linguistic Decision Schemata and Translators,* developed as high level decision making devices by Saridis and Graham are suitable to generate strings appropriate to organize the motions of the artificial arm. The reader is referred to Reference [6] for detailed information on this project. Figure 11 shows a block diagram of a linguistic organized commanded by Electromyographic (EMG) signals. However, through the example of the control by the prosthetic arm on the concept of a hierarchical intelligent man-machine interactive control system has been proposed and its feasibility established. Generalization to other man-machine interactive systems or even autonomous robots should be straightforward and would be one of the areas of future research in control systems.

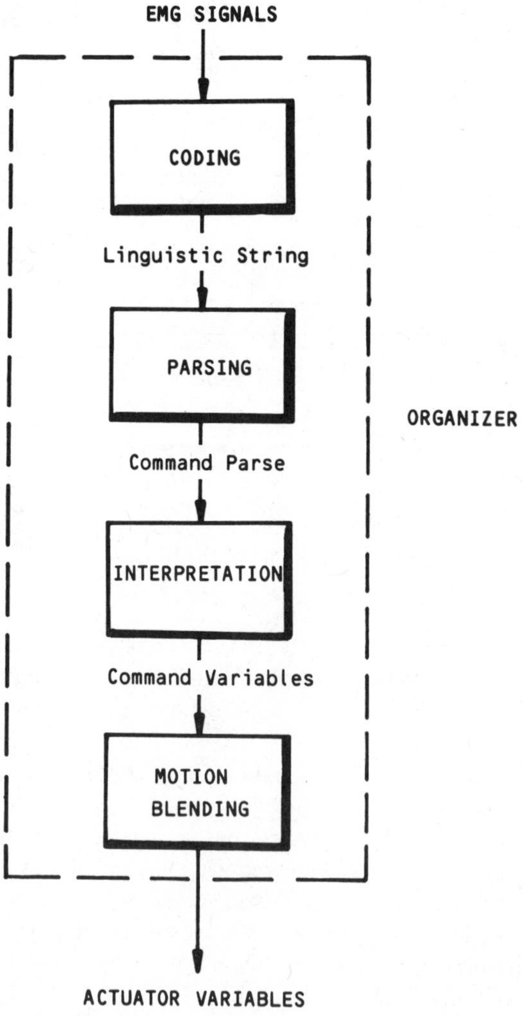

Figure 11 Block diagram of the organization level.

4.3 A Fuzzy Automaton as a Control Coordinator

A *fuzzy automaton* has been developed as an extension of the variable structure stochastic automata to accommodate inputs of a "fuzzy" or uncertain nature [5].

A fuzzy quantity, belongs to a set of values describing an "object" with undefined boundaries that are characterized individually by a membership coefficient to the set.

A fuzzy automaton is therefore defined as as sixtuple $[Z,Q,U,F,H,\zeta]$, where, in addition to the finite set of "fuzzy" inputs $Z = \{z\}$, finite set of states $Q = \{q\}$, finite set of outputs $U = \{u\}$, state transition function F, and output function H, a fuzzy membership vector ζ is assigned to the states of the automaton. A membership transition matrix $\xi^k(n)$ for each fuzzy input z^k may be assigned to update the membership functions ζ.

Such a fuzzy automaton may be used to coordinate the p primitive motions of the subprocess to form a desired compound movement of the arm from an initial state $x_0(t_0)$ to a final predefined state $x^d(T)$. Such a coordination is needed to put together the right amount of primitive motions and their proper velocities in order to accomplish the proper compound motion in response to a fuzzy command of the higher-level intelligent controller which is not to be bothered with the details of coordination. A *fuzzy automaton* $[C,Q,Q,F,I,\zeta]$ has a structure natural for such a coordinator, if $C = \{c\}$ is the set of fuzzy command inputs transmitted from the organizer, $Q = \{q\}$ is the set of the states as well as outputs of the automaton representing the appropriate performance criteria of each of the subprocesses assigned to generate the appropriate motion. Figure 12 depicts such a fuzzy automaton with learning capabilities.

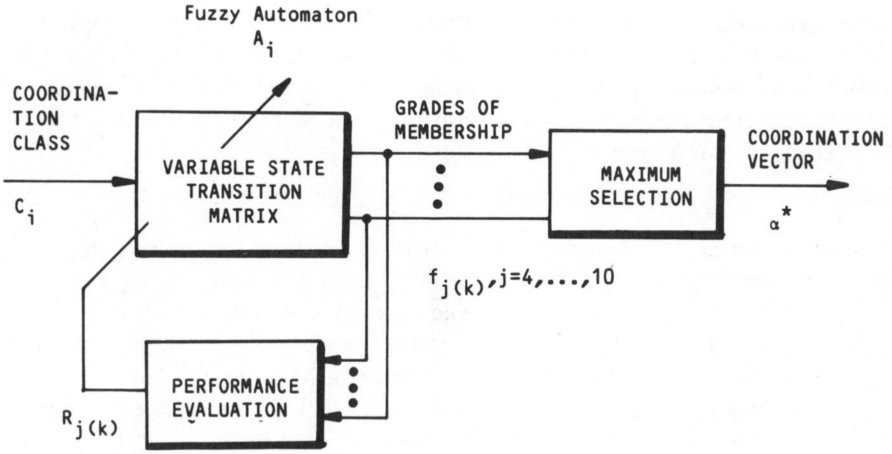

Figure 12 Structure of a Learning Fuzzy Automaton.

4.4 The Self-Organizing Control Level

The arm as whole process may in general be subdivided into p subprocesses, one per degree of freedom, described by the following set of generalized differential equations:

$$\dot{x}_i = F_i x_i + B_i(\bar{x}_i)u_i + f_i(x,\dot{\bar{x}},w_i)$$

$$z_i = H_i x_i + v_i \qquad i = 1,2,...,p \tag{5}$$

where x_i is the n_i-dimensional state vector z_i is the r_i-dimensional output vector, and u_i is the m_i-dimensional control vector of the ith subprocess; $F_i(B_i\bar{x}_i)$, and H_i are matrices of appropriate dimension, w_i and v_i appropriate noise vectors, and $f_i(\cdot)$ nonlinear functions representing the gravity influence and the coupling terms from other subsystems through various reaction forces [12]. If the subsystem was isolated from those force fields, these terms would be zero and the system would be linear in x_i. Furthermore, the state of the overall system would be

$$x^{\mathrm{T}} = [x_1^{\mathrm{T}}, \ldots, x_p^{\mathrm{T}}]\ \bar{x}_i^{\mathrm{T}} = [x^{\mathrm{T}}, \ldots, x_{i-1}^{\mathrm{T}}, x_{i+1}^{\mathrm{T}}, \ldots, x_p^{\mathrm{T}}]$$

$$z^{\mathrm{T}} = [z_1^{\mathrm{T}}, \ldots, z_p^{\mathrm{T}}]. \tag{6}$$

A performance criteria for the proper mechanical function of the system may be defined as

$$J = \sum_{i=1}^{p} \mu_i J_i(\alpha_i) \tag{7}$$

where $j_i(\alpha_i) = 1,...,p$ are appropriate performance criterion, α_i are adjustable coefficients relative to the speed of response of the subprocesses, and μ_i are adjustable coefficients $\mu_i = 0,1,\ i = 1,...,p$, relative to the appropriate blending of primitive motions defined by the individual subprocesses to generate an appropriate compound motion for the arm. Obviously $J_i(\alpha_i)$ is the performance criterion for each subprocess, assumed to be infinite duration for simplicity of implementation, a conjecture verified experimentally. A feedback control may be structured per subsystem as

$$u_i(z) = K_i\phi(z_i) + C_i\psi(z) \tag{8}$$

where the first part is the optimal control (O.C.) for the uncoupled subsystem, while the second term represents a nonlinear term depending on the nonlinear coupling with the other subsystems. The expanding subinterval algorithm may be applied to yield the asymptotically optimal coefficients K_i and C_i for each subprocess, thus creating performance-adaptive S.O.C.s for the lowest level of the hierarchically intelligent control system. A block diagram of such a device is shown in Figure 13. Self-organization is necessary to avoid parameter identification of the system dynamics. In the case of the above the elbow VA arm of Figure 8, p=4.

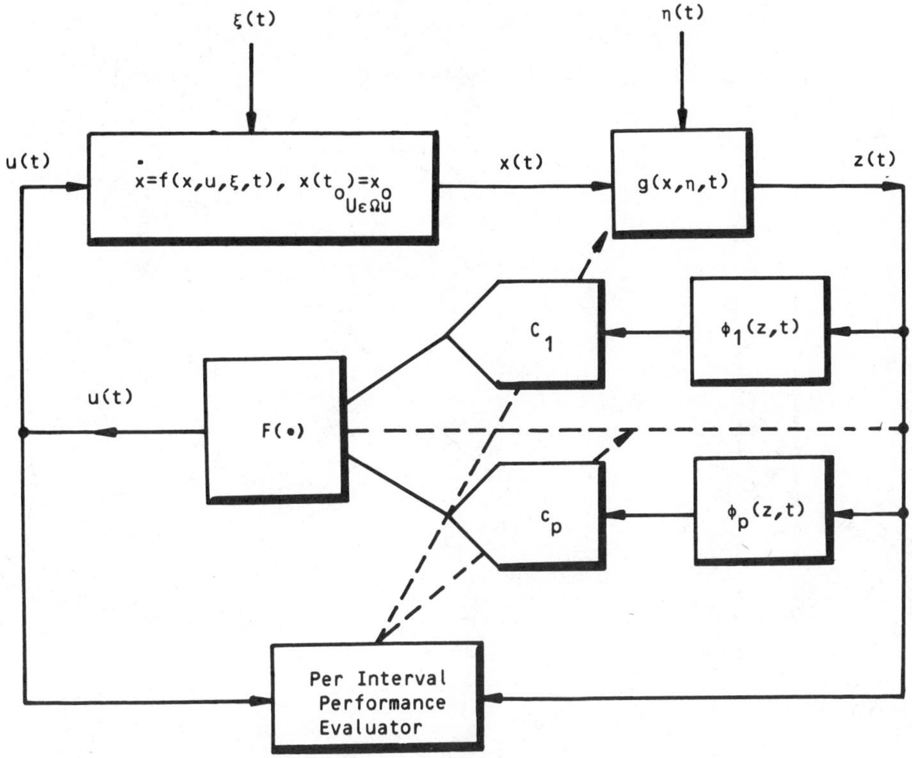

Figure 13 Block Diagram of the Expanding Subinterval Self-Organizing Control Algorithm.

The higher levels are interfaced with the individual self-organizing controls through the adjustable relative speed coefficients α_i, the blending coefficients μ_i, and the desired final states $x_i^d(T)$.

From the preceding discussion, it is obvious that the third-level S.O.C.s are designed for precision control of the mechanical subprocesses at hand but do not exhibit higher-quality intelligent functions, such as intelligent decision making for motion coordination, direction, and goal accomplishment. With their limited capabilities, they resemble more the reflexes of a biological system. The higher intelligent functions are hierarchically distributed to the higher levels of the controls which are described next.

4.5 Simulation Results

Since the construction of hierarchically intelligent control hardware for the VA arm have not yet been completed at the present time, simulations results on the Purdue University CDC 6500 computer will be given. A motion which humeral rotation in, θ_2 and elbow extension was simulated and compared to the output of a "goniometer" a device that records actual motions of a human arm.

The learning properties of the coordinating automaton and self-organizing controls were demonstrated in Figures 14 and 15. Showing the trajectories

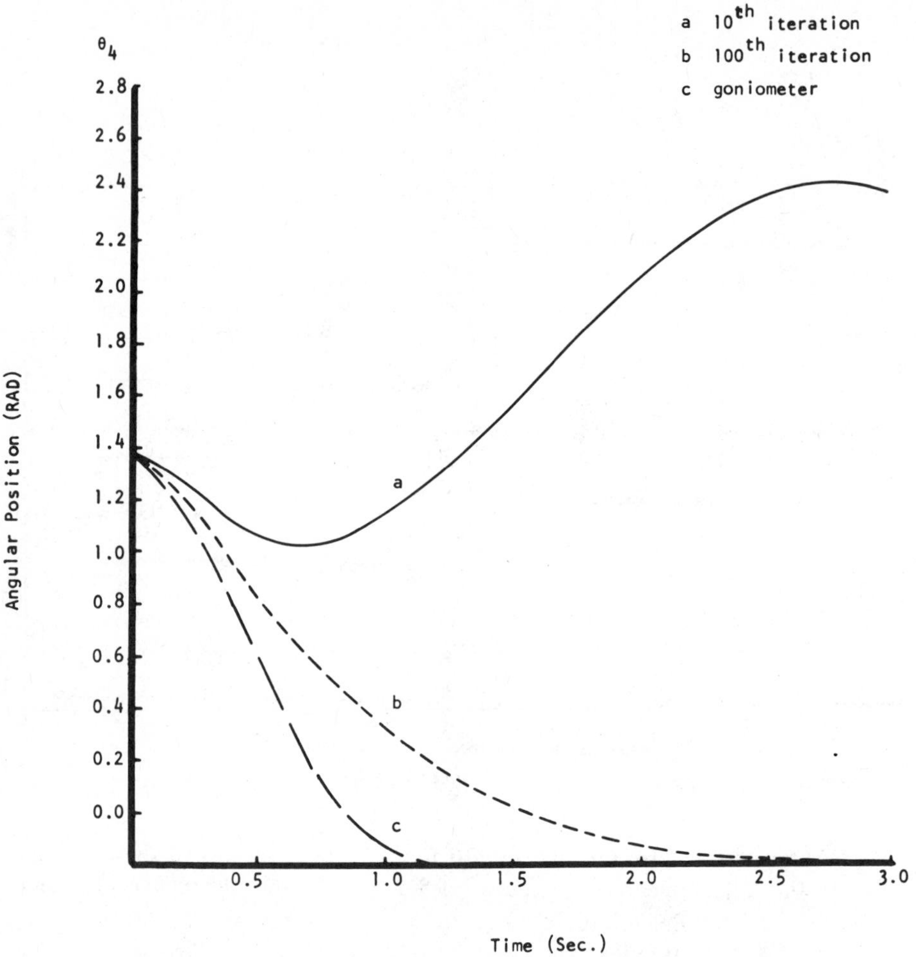

Figure 14 Example of ARM Movement θ_2 vs. time.

after and 10 and 100 training iterations as well as the trajectory of the "goniometer." Figure 16 shows the phase plane trajectory of the learned motion. The effect of learning and its desirability for generations of anthropomorphic motions cosmetically acceptable is made obvious from these plots [12].

Actual implementation on the VA arm involves hierarchically structured microprocessors for the implementation of the intelligent control and low power actuators for efficient use of arm. It should prove that intelligent control may reduce the long and tedious rehabilitation period required to train an upper extremity amputee.

5. REMARKS AND CONCLUSIONS

The purpose of writing this chapter was twofold. One is to familiarize the reader with the Hierarchically Intelligent Control Approach of designing

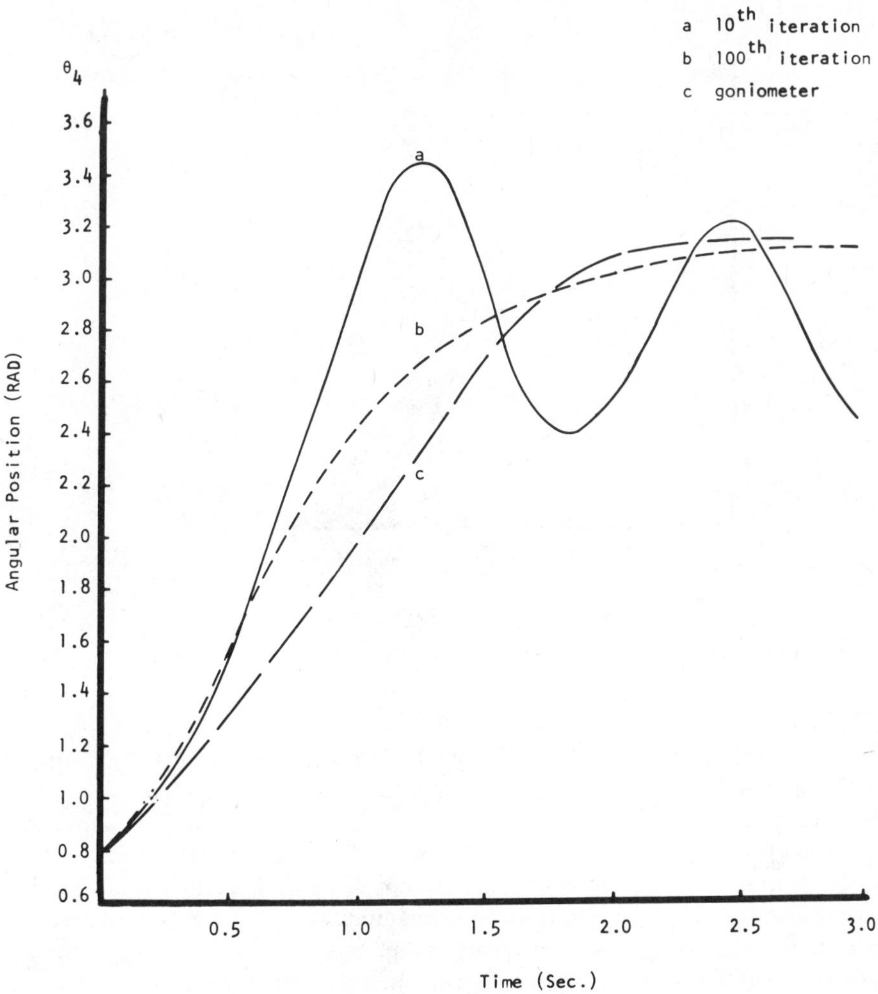

Figure 15 Example of Arm Movement θ_4 vs. time.

machines that can operate almost autonomously in an unfamiliar environment with minimum interaction with a human operator. The second one is to demonstrate the application of the above approach as a systematic approach to develop aids to disabled people. The voice-controlled manipulator with visual inputs and the EMG-controlled prosthetic arm were successfully developed at the Advanced Automation Research Laboratory of Purdue University to prove the applicability and efficiency of the Hierarchically Intelligent Control Approach. However, it is applicable in many other processes requiring autonomous machine intelligence like, traffic control systems, nuclear reactor maintenance robots, space exploration, and even automated manufacturing [11]. Inspite of the fact that specialized computer architectures and dedicated hardware were used, in contrast to the other "Artificial Intelligence" generated machines, the cost of this design is a limitation for its

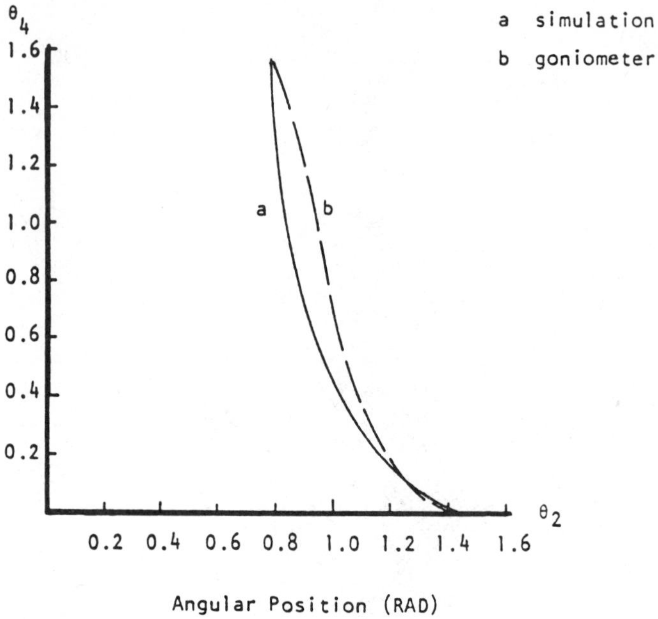

Figure 16 Phase Plane Trajectory for a Coordinated Motion.

immediate production. It is like Joseph Engelbergers mobile machine, quoted in his interview with Robotics Age, which he can afford to build as his own butler but he is probably not going to put in production in the immediate future [14].

In contrast the cost of production is not a limiting factor for a hospital environment were the voice-controlled manipulator may be time-shared by more than one patient in an emergency room, or a wheelchair or the intelligent prosthetic arm of an amputee. The major requirements, in such an environment, are versatility, minimum training from the past of the patient, small size, and average object manipulation and/or recognition in a standard environment. Therefore, I believe that Engelberger's butler could be commercially realized for a hospital environment first. Similar experimental machines have already been developed, constructed and tested in many hospitals and rehabilitation centers in this country and abroad [15]. The proposed approach will hopefully systematize and optimize their design.

ACKNOWLEDGMENTS

The author is grateful to Drs. H. Stephanou, G. Lee, J. Graham and S. Lee for their assistance in developing these projects. These projects were supported by NSF Grants ENG7802400 and ECS7916802.

REFERENCES

[1] Samuel Alderson, The Electric Arm, Chapter 13 in Klopsteg and Wilson's "Human Limbs and Their Substitutes," McGraw-Hill, 1954, reprinted by Hafner Press, 1969.

[2] A. K. Bejczy, "Remote Manipulator System Technology Review," Tech. Rep. No. 760-77, Jet Propulsion Lab., Pasadena, CA, July, 1972.

[3] A. Freedy, F. C. Hull, L. F. Lucaccini, and J. Lyman, "A Computer-Based Learning System for Remote Manipulator Control," *IEEE Trans. Systems, Man, Cybern., SMC-1,* (4), 356-364, 1971.

[4] K. S. Fu, "Learning Control Systems and Intelligent Control Systems: An Intersection of Artificial Intelligence and Automatic Control," *IEEE Trans. Automatic Control, AC-16,* (1), 70-72, 1971.

[5] K. S. Fu and L. W. Fung, "Decision Making in a Fuzzy Environment," Tech. Rept. TR-EE 73-22, Purdue University, West Lafayette, IN, May, 1973.

[6] J. H. Graham and G. N. Saridis "Linguistic Decision Structures for Hierarchical Systems," IEEE Trans. on SMC, Vol SMC-12, No. 3, May/June 1982, pp. 325-333.

[7] D. Graupe, et al., "A Microprocessor System for Multifunctional Control of Upper Limb Prostheses via EMG Identification," *IEEE Trans. on Aut. Control,* Vol. AC-23, pp. 538-544, August, 1978.

[8] C. S. G. Lee and G. N. Saridis, "Computer control of a Trainable Manipulator," Technical Report TR-33 78-42, Purdue University, December, 1978.

[9] R. B. McGee, Control of Legged Locomotion Systems, Proceedings 1978, JACC, San Francisco, June, 1977.

[10] C. A. Rosen and N. J. Nilsson, "An Intelligent Automaton," 1967 IEEE International Convention Record, Part 9, New York, March, 1967.

[11] G. N. Saridis, "Toward the Realization of Intelligent Controls," IEEE Proceedings, Vol. 67, No. 8, August, 1979, pp. 1115-1133.

[12] G. N. Saridis and H. E. Stephanou, "A Hierarchical Approach to the Control of a Prosthetic Arm," *IEEE Trans. on SMC,* Vol. SMC-7, No. 6, pp. 407-420, June, 1977.

[13] G. N. Saridis and T. Gootee, "EMG Pattern Analysis and Classification for a Prosthetic Arm," IEEE Trans. on EMBS, Vol. BME-29, No. 6, pp. 403-412, June, 1982.

[14] J. W. Saveriano, "An Interview with Joseph Engleberger," Robotics Age, January/February, 1981, pp. 10-23.

[15] Veterans Administration "Bulletin of Prosthetic Research," Fall 1978.

[16] D. W. Whitney, "Resolved Motion Control of Manipulators and Human Prosthesis," *IEEE Trans. Man-Machine Systems,* MMX-10, (2), 1969.

[17] R. W. Wirta, D. R. Taylor and F. R. Finley, "Pattern Recognition Arm Prosthesis," Bulletin of Prosthesis Research BPR 10-30, Fall 1978, Veterans Administration, Washington, D.C.

PART II MECHANICS

3

Mechanics: Kinematics and Dynamics

S. DESA
AND
B. ROTH

1. INTRODUCTION

This chapter describes several recent developments in the kinematic and dynamic analysis of manipulators. No attempt has been made at completeness (in regard to either topics or authors); we deal with what we feel are the highlights and what in our opinion will prove to be the most enduring and valuable developments in this rapidly evolving field. We address ourselves exclusively to open-loop manipulators, and model these as idealized systems composed solely of rigid body members. We do not treat what are generally considered as second order effects (such as link flexibility, drive-train backlash and non-linearities, and joint clearances), even though we are aware they can, at times, be very important.

In dealing with an idealized open-loop manipulator, it is convenient to represent the system by a skeleton diagram, such as Figure 1. From a mechanics point of view manipulators are composed of links jointed together. We are generally concerned with only two types of joints: i) the simple hinge or turning joint, usually called a *revolute joint,* and denoted by the letter R; ii) the sliding or rectilinear joint, usually called a *prismatic joint,* and denoted by the letter P.

The function of the skeleton diagram is to show the type of each joint and its axis. The line connecting each pair of adjacent axes, usually taken as the common normal between the axes, is used to represent the actual link and to indicate its function of maintaining the link's axes at a fixed distance and angle from one another.

As short-hand notation it is customary to describe manipulators by an ordered sequence of letters, each denoting the joint type as one passes from the base to the free end. Thus, Figure 1 would be described as a RRRPR (5-bar)

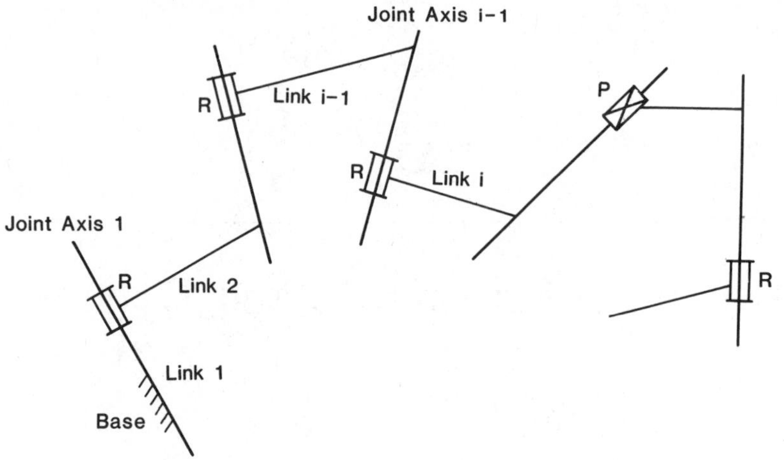

Figure 1 Skeleton diagram of a manipulator.

manipulator or, for short, a 3RPR. A six degree-of-freedom manipulator with six revolute joints would be denoted as a 6R, whereas if it had 3 prismatic joints followed by 3 revolutes we would call it a 3P3R.

From a kinematic point of view, once quantified, the information implied by the skeleton diagram is sufficient to determine all the position, velocity, acceleration, and static force information for the manipulator. So, all that is required, in addition to the diagram itself, is a convenient labeling of parameters with which to describe these concepts analytically. In the next section we review the notation which is commonly used for this purpose.

Matrix transformations are fundamental tools for the kinematic and dynamic analysis, as well as the control and design of manipulators (and also associated robotic elements such as vision systems, walking machines and dextrous end-effectors). We begin our next section by introducing the commonly used matrix transformations and then show how to use them in conjunction with the skeleton diagram description to determine the position of the free-end of a manipulator relative to its base.

In Section 3 we review the state of the art in solving the so-called inverse kinematic problem; viz, given a desired position for the free end of the manipulator, what are the required corresponding displacements in each of the joints. This problem is of fundamental importance in using computer controlled manipulators in "advanced", unstructured and variable environment applications. (It is of little or no interest for systems which are programmed by manual lead through or pendant control techniques.)

In Section 4 we treat several problems associated with determining the workspace and the extreme positions of a manipulator. The information presented in this section is useful in the design as well as geometric analysis of manipulators. Although some of the problems discussed in this section still remain unsolved, we are able to point to several extremely interesting and fundamental concepts and theorems which have recently been developed.

In this same section the concept of "hand size" becomes of some interest. To do useful work a manipulator needs to carry a tool or some holding or gripping device. We will refer to this generic category of devices as end-effectors or hands. For most of the material in this Chapter the end-effector can be regarded as simply another link. So unless otherwise stated the last link of the manipulator is the end-effector, and the fact that it can open and close is ignored. In dealing with workspace we deal only with the size of what we take to be a simple gripping device.

In Sections 5 and 6 we treat, respectively, the related concepts of small motions and statics, thereby reviewing the important concept of a manipulator Jacobian, and also setting the stage for our section on dynamics (Section 7).

In the section on dynamics we explain a method based on Kane's dynamical equations, which we feel has certain advantages over the more widely known Lagrangian and Newton-Euler methods. We remark in passing that, although the final word has probably not yet been written, it now seems that most of the techniques of dynamic analysis, *if properly applied,* are approximately equivalent in terms of computer time.

We end our article with a short list of references. For other materials on the mechanics of manipulators the reader is referred to the bibliographies contained in the following: [5,17,21,22,29].

2. KINEMATICS

The fundamental kinematic problem for manipulators is that of determining the relationship between the joint variables and the position and orientation of the free end of the manipulator. In the next few sections we describe the following kinematic/geometric tools which are useful in treating this problem.

1. Matrix representation of the relative position and orientation of two rigid bodies with respect to each other (Section 2.1).
2. Homogenous coordinates for the compact representation of the relative position and orientation of several rigid bodies (Section 2.2).
3. Kinematic representation of links and joints of a manipulator (Section 2.3).
4. Kinematic representation of the relative position and orientation of two links of a manipulator (Section 2.4).

We conclude Section 2 with the general formulation (Section 2.5) and some examples (Section 2.6).

2.1 Representation of the Relative Position and Orientation of Two Rigid Bodies

Given two rigid bodies, $\sum$ and E, the (relative) position of points of E with respect to $\sum$ (Figure 2) can be expressed in the following form:

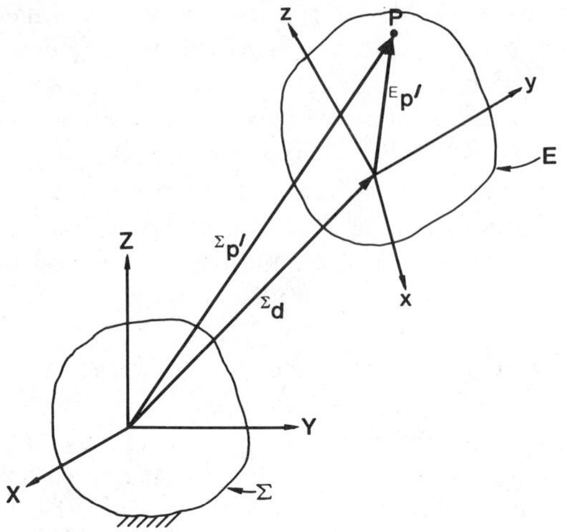

Figure 2 Relative position and orientation of two rigid bodies.

$$^{\Sigma}\boldsymbol{p}' = (^{\Sigma}\boldsymbol{B}^E)\ (^E\boldsymbol{p}') + {}^{\Sigma}\boldsymbol{d}, \tag{1}$$

where $^{\Sigma}\boldsymbol{p}' = (X,Y,Z)^{\mathrm{T}}$ and $^E\boldsymbol{p}' = (x,y,z)^{\mathrm{T}}$ are column matrices denoting the position, measured with respect to coordinate systems fixed in Σ and E respectively, of a generic point P of E (Figure 2); $^{\Sigma}\boldsymbol{d} = (d_1,d_2,d_3)^{\mathrm{T}}$ is a column matrix representing the position vector in Σ of the origin of the xyz coordinate system of E; $^{\Sigma}\boldsymbol{B}^E$ is a (3×3) matrix representing the orientation of E in Σ. (Superscripts are used to clearly identify the coordinate frames in question.)

2.2 Homogenous Coordinates and the Relative Motion of Several Rigid Bodies

Homogenous coordinates are useful when we wish to analytically express the relative motion of several rigid bodies - as in the case of a manipulator - in compact form. We now develop a rationale for using homogenous coordinates in studying the kinematics of several rigid bodies.

Given $^{\Sigma}\boldsymbol{B}^E$ and $^{\Sigma}\boldsymbol{d}$ (Section 2.1), if we define a matrix $^{\Sigma}\boldsymbol{A}^E$,as follows,

$$^{\Sigma}\boldsymbol{A}^E = \begin{bmatrix} ^{\Sigma}\boldsymbol{B}^E & ^{\Sigma}\boldsymbol{d} \\ 0 & 1 \end{bmatrix} = \begin{bmatrix} B_{11} & B_{12} & B_{13} & d_1 \\ B_{21} & B_{22} & B_{23} & d_2 \\ B_{31} & B_{32} & B_{33} & d_3 \\ 0 & 0 & 0 & 1 \end{bmatrix}, \tag{2}$$

then equation (1) can be written as follows,

$$\begin{bmatrix} X \\ Y \\ Z \\ 1 \end{bmatrix} = {}^{\Sigma}\boldsymbol{A}^{E} \begin{bmatrix} x \\ y \\ z \\ 1 \end{bmatrix}, \tag{3}$$

or symbolically as

$${}^{\Sigma}\boldsymbol{p} = ({}^{\Sigma}\boldsymbol{A}^{E})\ ({}^{E}\boldsymbol{p}), \tag{4}$$

where ${}^{E}\boldsymbol{p} = (x,y,z,1)^{\mathrm{T}}$ and ${}^{\Sigma}\boldsymbol{p} = (X,Y,Z,1)^{\mathrm{T}}$.

Here, $(X,Y,Z,1)$ and $(x,y,z,1)$ are a special set of *homogenous coordinates* which, for our purposes, are simply new sets of coordinates describing the positions of points in E and Σ respectively. Homogenous coordinates are useful in geometry and theoretical kinematics. They are used extensively in computer graphics, but in manipulator theory they are used mainly as a device to augment the $\boldsymbol{B}$ and $\boldsymbol{d}$ matrices in equation (1).

The advantage of homogenous coordinates is that they yield a compact representation of relative positions and orientations when several rigid bodies are involved. Given three rigid bodies E_1, E_2, and Σ such that the position and orientation of E_2 relative to E_1, and that of E_1 relative to Σ are specified, then the position and orientation of E_2 relative to Σ can be obtained, using equation (4), as follows,

$${}^{E_1}\boldsymbol{p} \underset{(4)}{=} ({}^{E_1}\boldsymbol{A}^{E_2})\ {}^{E_2}\boldsymbol{p}, \tag{5}$$

$${}^{\Sigma}\boldsymbol{p} \underset{(4)}{=} ({}^{\Sigma}\boldsymbol{A}^{E_1})\ {}^{E_1}\boldsymbol{p}, \tag{6}$$

$${}^{\Sigma}\boldsymbol{p} \underset{(5,6)}{=} ({}^{\Sigma}\boldsymbol{A}^{E_1})\ ({}^{E_1}\boldsymbol{A}^{E_2})\ {}^{E_2}\boldsymbol{p}. \tag{7}$$

If the positions of points of E_2 measured in Σ are expressed as follows,

$${}^{\Sigma}\boldsymbol{p} = ({}^{\Sigma}\boldsymbol{A}^{E_2})\ ({}^{E_2}\boldsymbol{p}), \tag{8}$$

then,

$${}^{\Sigma}\boldsymbol{A}^{E_2} \underset{(7,8)}{=} ({}^{\Sigma}\boldsymbol{A}^{E_1})\ ({}^{E_1}\boldsymbol{A}^{E_2}). \tag{9}$$

The reader can easily verify that the above form is much more compact than the corresponding one obtained by using (1).

Given $(n+1)$ rigid bodies $(E_1,\ldots,E_i,\ldots,E_n)$ and Σ, such that the positions of E_i relative to E_{i-1}, $(i=2,\ldots,n)$ and the position of E_1 relative to Σ are specified, the following equations can be written:

$$^{\Sigma}\boldsymbol{p} \underset{(4)}{=} ({}^{\Sigma}\boldsymbol{A}^{E_1})\ ({}^{E_1}\boldsymbol{p}), \tag{10}$$

$$^{E_{i-1}}\boldsymbol{p} \underset{(4)}{=} ({}^{E_{i-1}}\boldsymbol{A}^{E_i})\ {}^{E_i}\boldsymbol{p}; \quad (i=2,...n). \tag{11}$$

The above n equations can be combined (as in equation (7)) to yield

$$^{\Sigma}\boldsymbol{p} \underset{(10,11)}{=} ({}^{\Sigma}\boldsymbol{A}^{E_1}) \cdots ({}^{E_{i-1}}\boldsymbol{A}^{E_i}) \cdots ({}^{E_{n-1}}\boldsymbol{A}^{E_n})\ {}^{E_n}\boldsymbol{p}, \tag{12}$$

or,

$$^{\Sigma}\boldsymbol{p} \underset{(12)}{=} ({}^{\Sigma}\boldsymbol{A}^{E_1}) \left[\prod_{i=2}^{n} ({}^{E_{i-1}}\boldsymbol{A}^{E_i}) \right] ({}^{E_n}\boldsymbol{p}). \tag{13}$$

If the positions of points of E_n measured in Σ are expressed as follows:

$$^{\Sigma}\boldsymbol{p} = ({}^{\Sigma}\boldsymbol{A}^{E_n})\ ({}^{E_n}\boldsymbol{p}), \tag{14}$$

then,

$$^{\Sigma}\boldsymbol{A}^{E_n} \underset{(13,14)}{=} ({}^{\Sigma}\boldsymbol{A}^{E_1}) \prod_{i=2}^{n} ({}^{E_{i-1}}\boldsymbol{A}^{E_i}). \tag{15}$$

2.3 Kinematic Representation of the Links and Joints of a Manipulator

A generic link i of a manipulator is a rigid body connecting the axes of joints $i-1$ and i (see Figure 3). We set up a dextral, orthogonal coordinate system $X_i Y_i Z_i$ - fixed in link i - as follows (Figure 3):

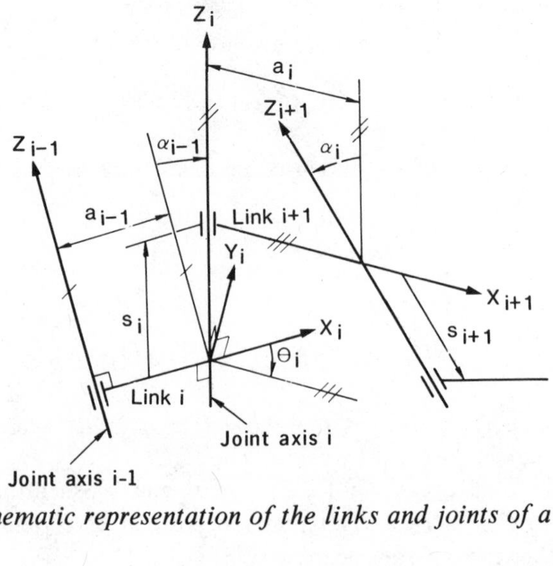

Figure 3 Kinematic representation of the links and joints of a manipulator.

Z_i is along the joint axis i, X_i is the common normal between Z_{i-1} (along joint axis $i-1$) and Z_i; Y_i is then determined by the right-hand rule. The positive sense of Z_i is chosen arbitrarily, while the (positive) X_i axis points from Z_{i-1} towards Z_i.

Link i is characterized (kinematically) by two parameters (a_{i-1}, α_{i-1}) defined as follows (Figure 3):

a_{i-1} is the distance, measured along X_i, between the joint axes Z_{i-1} and Z_i and α_{i-1} is the (twist) angle, measured in the Y_{i-1} Z_{i-1} plane, which Z_i makes with the Z_{i-1} axis. (a_{i-1} is always positive; the sense of α_{i-1} is determined by the right-hand rule with the thumb pointing in the direction of the positive X_i axis).

Joint i is characterized by two parameters, (s_i, θ_i) defined as follows: s_i is the directed distance, measured along Z_i, from X_i to X_{i+1} and θ_i is the angle, in the $X_i Y_i$ plane, which X_{i+1} makes with the X_i axis. (s_i is positive if it is in the direction of the positive Z_i axis; the sense of θ_i is determined by the right-hand rule with the thumb pointing in the direction of the positive Z_i axis).

Remark 1: If joint i is revolute, then Z_i is (along) the axis of rotation of the joint, s_i is constant and θ_i is the joint variable; if joint i is prismatic, then Z_i is (along) the axis of sliding of the joint, θ_i is fixed and s_i is the joint variable.

Remark 2: If joint axes Z_{i-1} and Z_i intersect, then $a_{i-1} = 0$; if Z_{i-1} and Z_i are parallel, then $\alpha_{i-1} = 0$.

Remark 3: For the base - denoted as Link 1 - the Z_1 axis is along the axis of joint 1, and the X_1 axis can be chosen arbitrarily in a plane perpendicular to Z_1.

Remark 4: For the last link of a manipulator - denoted as link $(n+1)$ - the only constraint on the choice of the X_{n+1} Y_{n+1} Z_{n+1} axes is that the X_{n+1} axis be perpendicular to the Z_n axis.

2.4 Kinematic Representation of the Relative Position and Orientation of Two Successive Links of a Manipulator

Let $(X_{i+1}, Y_{i+1}, Z_{i+1})$ be the coordinates of a generic point P, of link $i+1$, in the X_{i+1} Y_{i+1} Z_{i+1} coordinate system fixed to link $i+1$, and let (X_i, Y_i, Z_i) be the coordinates of P in the coordinate system X_i Y_i Z_i fixed to link i. The position (and hence the motion) of points of link $i+1$ relative to link i can then be written in the form required by equation (4) as follows:

$$\begin{bmatrix} X_i \\ Y_i \\ Z_i \\ 1 \end{bmatrix} \underset{(4)}{=} {}^iA^{i+1} \begin{bmatrix} X_{i+1} \\ Y_{i+1} \\ Z_{i+1} \\ 1 \end{bmatrix} \tag{16}$$

where,

$${}^iA^{i+1} = \begin{bmatrix} c\theta_i & -s\theta_i c\alpha_i & s\theta_i s\alpha_i & a_i c\theta_i \\ s\theta_i & c\theta_i c\alpha_i & -c\theta_i s\alpha_i & a_i s\theta_i \\ 0 & s\alpha_i & c\alpha_i & s_i \\ 0 & 0 & 0 & 1 \end{bmatrix} \tag{17}$$

($s\theta_i = \sin\theta_i$, $c\theta_i = \cos\theta_i$ ...)

The quantities α_i, θ_i, a_i, and s_i have been defined in the previous section (also see Figure 3).

2.5 Equation Relating Joint Angles to Position and Orientation of the End-Effector

We now have all the tools necessary to study an entire manipulator. Given an n degree-of-freedom manipulator, i.e., one with n moving links and n one degree-of-freedom joints (Figure 4), we can use equations (14), (15) to write:

$$^1\boldsymbol{p} \underset{(14)}{=} ({}^1\boldsymbol{A}^{n+1})\ ({}^{n+1}\boldsymbol{p}) \tag{18}$$

and

$${}^1\boldsymbol{A}^{n+1} \underset{(15)}{=} \prod_{i=1}^{n} {}^i\boldsymbol{A}^{i+1} \tag{19}$$

where ${}^i\boldsymbol{p} = (X_i, Y_i, Z_i, 1)$, $(i=1,\ldots,n+1)$, denotes the coordinates of a point measured in the $X_i\ Y_i\ Z_i$ system and ${}^i\boldsymbol{A}^{i+1}$ is given by (17).

(Equations (18,19) are obtained by replacing superscripts Σ and E_i in equations (14,15), by superscript 1 and $i+1$ respectively.)

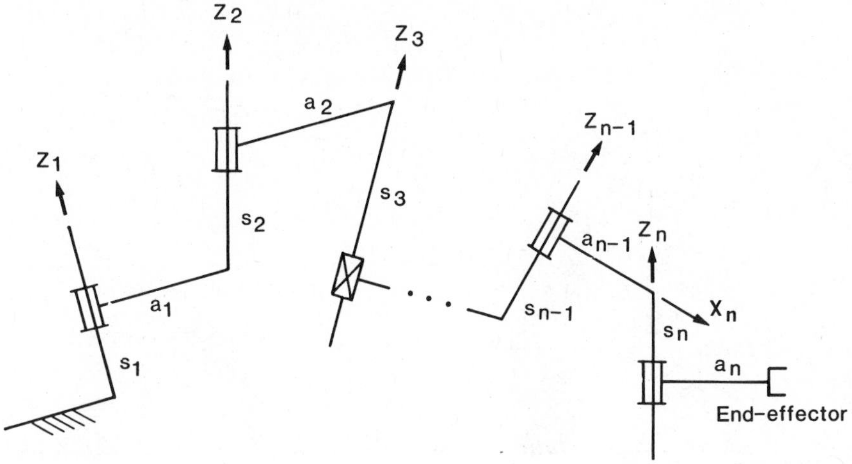

Figure 4 Kinematic representation of an n degree-of-freedom manipulator.

For most manipulators ${}^{i}\boldsymbol{A}^{i+1}$ is a function of a single joint variable: θ_i in the case of a revolute joint and s_i in the case of a prismatic joint. If we use q_i to denote the joint variable, equations (18,19) can be rewritten as

$$ {}^{1}\boldsymbol{p} \underset{(18)}{=} \left[{}^{1}\boldsymbol{A}^{n+1} \; (q_1,q_2,\ldots,q_n) \right] \left({}^{n+1}\boldsymbol{p}. \right) \tag{20} $$

and

$$ {}^{1}\boldsymbol{A}^{n+1} \; (q_1,\ldots,q_n) \underset{(19)}{=} \prod_{i=1}^{n} {}^{i}\boldsymbol{A}^{i+1} \; (q_i) \tag{21} $$

Equation (20) gives us the position ${}^{1}\boldsymbol{p}$ in the base (link 1), of a point ${}^{n+1}p$, fixed with respect to the end-effector (link $n+1$) of the manipulator.

Let $(\boldsymbol{h}_1, \boldsymbol{h}_2, \boldsymbol{h}_3)$ be unit vectors fixed in the end-effector (hand) and parallel to the X_{n+1}, Y_{n+1}, Z_{n+1} axes respectively, $(\boldsymbol{n}_1, \boldsymbol{n}_2, \boldsymbol{n}_3)$ be unit vectors fixed in the base and parallel to X_1, Y_1, Z_1 respectively and let ${}^{1}\boldsymbol{p}$ $(p_1, p_2, p_3)^{\dagger}$ be the position vector of the origin of the coordinate system of link $n+1$ with respect to the origin of the base coordinate system, X_1,Y_1,Z_1, (Figure 5). If $\boldsymbol{C}$ is a direction cosine matrix with (i,j)th element $C_{ij} = \boldsymbol{n}_i \cdot \boldsymbol{h}_j$, then applying (20) successively to points (1,0,0,1), (0,1,0,1), (0,0,1,1), and (0,0,0,1) of the end-effector we obtain (after some manipulation)

† $p_i = {}^{1}\boldsymbol{p} \cdot \boldsymbol{n}_i$; $(i=1,2,3)$

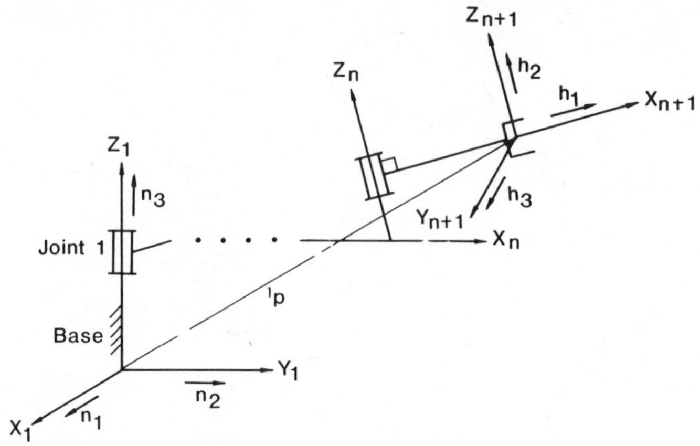

Figure 5 Position and orientation of the end-effector (Link n+1) with respect to the base (Link 1).

$$\begin{bmatrix} C & {}^1p \\ 0 & 1 \end{bmatrix} \underset{(20)}{=} {}^1A^{n+1}\ (q_1,\ldots,q_n) \begin{bmatrix} 1 & 0 & 0 & 0 \\ 0 & 1 & 0 & 0 \\ 0 & 0 & 1 & 0 \\ 0 & 0 & 0 & 1 \end{bmatrix} \tag{22}$$

or

$$ {}^1A^{n+1}\ (q_1,q_2,\ldots,q_n) = \begin{bmatrix} C & {}^1p \\ 0 & 1 \end{bmatrix}. \tag{23}$$

Let

$$H \triangleq \begin{bmatrix} C & {}^1p \\ 0 & 1 \end{bmatrix} \tag{24}$$

where H is a 4×4 matrix characterizing the position† and orientation of the end-effector (hand) with respect to the base. Then (23) can be written as

$$ {}^1A^{n+1}\ (q_1,q_2,\ldots,q_n) = H. \tag{25}$$

Two cases frequently arise in the kinematic analysis of manipulators. In the first case, known as the direct kinematic problem, the joint variables

† The word "position", when used in the context "the position and orientation of the end-effector," always implies "the position of some point of the end-effector (measured) in the base coordinate system and the orientation of the end-effector with respect to the base."

$(q_1,q_2,...,q_n)$ are specified, i.e., ${}^1\boldsymbol{A}^{n+1}$ is known, and the end-effector (hand) position is computed. This is relatively straightforward and can be accomplished by (25) directly or by some algorithmic scheme which avoids all the extra zero elements introduced by the use of homogeneous coordinates. Although the use of (25) is conceptually easy, it is computationally inefficient even after the zeros are eliminated. Position and orientation require only six quantities and the matrices use twelve non-trivial elements. If computation times are important other techniques such as, for example, vector based methods using quaternions [4,27] or dual quaternions [4] are more efficient alternatives. Since efficiency for the direct method is usually not a big problem we will not treat it any further here.

In the second case - known as the inverse kinematic problem - the position of the hand is specified, i.e., $\boldsymbol{H}$ is known, and corresponding joint variables $(q_1,...,q_n)$ to accomplish this are required. This problem and its solution are the subject of Section 3.

2.6 Examples

The concepts discussed in the preceding sections are illustrated with two examples.

Example 2.1 A PLANAR 2R MANIPULATOR (FIGURE 6)

The manipulator is shown in Figure 6a and suitable coordinate systems are shown in Figure 6b. Joint and link parameters are given in Table 1. Using these parameters we obtain the following A matrices:

$$ {}^1\boldsymbol{A}^2 \underset{(17)}{=} \begin{bmatrix} c\theta_1 & -s\theta_1 & 0 & \ell c\theta_1 \\ s\theta_1 & c\theta_1 & 0 & \ell s\theta_1 \\ 0 & 0 & 1 & 0 \\ 0 & 0 & 0 & 1 \end{bmatrix} \tag{26} $$

$$ {}^2\boldsymbol{A}^3 \underset{(17)}{=} \begin{bmatrix} c\theta_2 & -s\theta_2 & 0 & mc\theta_2 \\ s\theta_2 & c\theta_2 & 0 & ms\theta_2 \\ 0 & 0 & 1 & 0 \\ 0 & 0 & 0 & 1 \end{bmatrix} \tag{27} $$

where $c\theta_i = \cos\,\theta_i$ and $s\theta_i = \sin\,\theta_i$; $(i{=}1,2)$.

$$ {}^1\boldsymbol{A}^3 = ({}^1\boldsymbol{A}^2)\ ({}^2\boldsymbol{A}^3) \underset{(19)}{=} \begin{bmatrix} c(\theta_1+\theta_2) & -s(\theta_1+\theta_2) & 0 & mc(\theta_1+\theta_2)+\ell c\theta_1 \\ s(\theta_1+\theta_2) & c(\theta_1+\theta_2) & 0 & ms(\theta_1+\theta_2)+\ell s\theta_1 \\ 0 & 0 & 1 & 0 \\ 0 & 0 & 0 & 1 \end{bmatrix}. \tag{28} $$

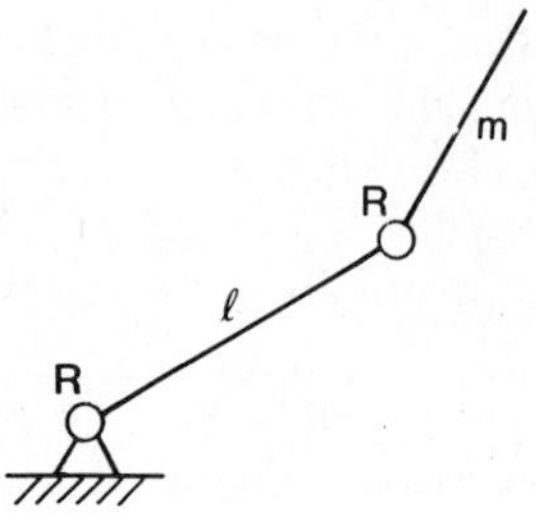

Figure 6a Planar 2R manipulator.

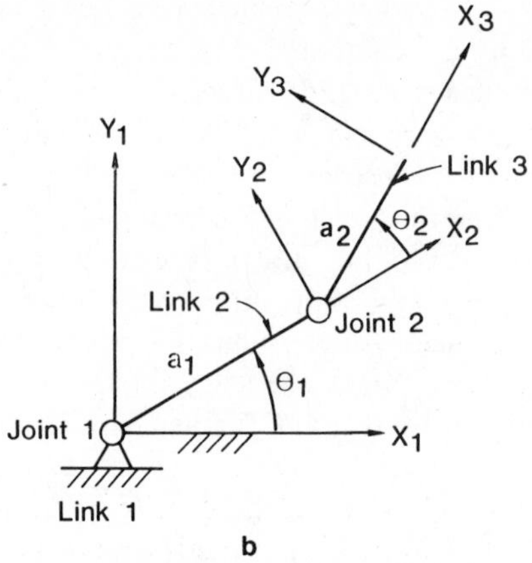

Figure 6b Manipulator of Figure. (6a) showing coordinate systems. Link parameters and joint variables.

Table 1:
Kinematic Parameters for Example 2.1

i	1	2
a_i	ℓ	m
s_i	0	0
α_i	0	0
θ_i	θ_1	θ_2

The position of points of link 3 in link 1 can be expressed as

$$\begin{bmatrix} X_1 \\ Y_1 \\ Z_1 \\ 1 \end{bmatrix} \underset{(20)}{=} {}^1\boldsymbol{A}^3\ (\theta_1,\theta_2) \begin{bmatrix} X_3 \\ Y_3 \\ Z_3 \\ 1 \end{bmatrix}, \tag{29}$$

where ${}^1\boldsymbol{A}^3$ is given by (28).

Example 2.2 A SPATIAL R2P MANIPULATOR (FIGURE 7)

The manipulator is shown in Figure 7a; pertinent coordinate systems and parameters are shown in Figure 7b. The joint and link parameters are given in

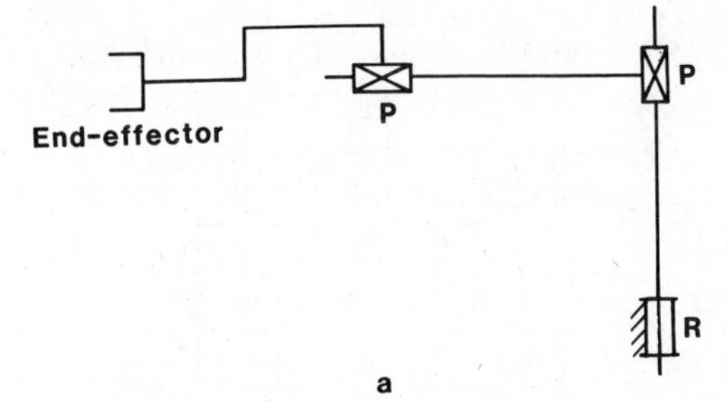

Figure 7a An R2P manipulator.

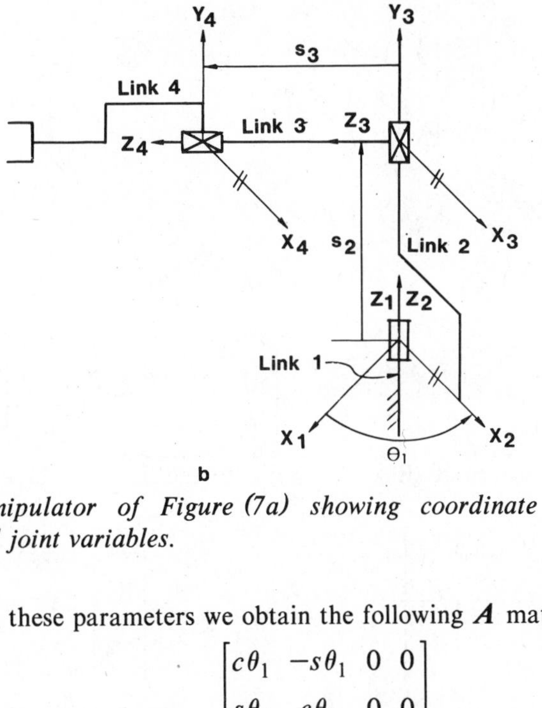

Figure 7b Manipulator of Figure (7a) showing coordinate systems, link parameters and joint variables.

Table 2. Using these parameters we obtain the following $\boldsymbol{A}$ matrices

$$ {}^{1}\boldsymbol{A}^{2} \underset{(17)}{=} \begin{bmatrix} c\theta_1 & -s\theta_1 & 0 & 0 \\ s\theta_1 & c\theta_1 & 0 & 0 \\ 0 & 0 & 1 & 0 \\ 0 & 0 & 0 & 1 \end{bmatrix} \tag{30} $$

where $c\theta_1 = \cos\,\theta_1$, $s\theta_1 = \sin\,\theta_1$

$$ {}^{2}\boldsymbol{A}^{3} \underset{(17)}{=} \begin{bmatrix} 1 & 0 & 0 & 0 \\ 0 & 0 & -1 & 0 \\ 0 & 1 & 0 & s_2 \\ 0 & 0 & 0 & 0 \end{bmatrix}, \tag{31} $$

Table 2:
Kinematic Parameters for Example 2.2

i	1	2	3
a_i	0	0	0
s_i	0	s_2	s_3
α_i	0	90°	0
θ_i	θ_1	0	0

$$ {}^3\boldsymbol{A}^4 \underset{(17)}{=} \begin{bmatrix} 1 & 0 & 0 & 0 \\ 0 & 1 & 0 & 0 \\ 0 & 0 & 1 & s_3 \\ 0 & 0 & 0 & 1 \end{bmatrix}, \tag{32} $$

$$ {}^1\boldsymbol{A}^4 \underset{(18)}{=} ({}^1\boldsymbol{A}^2)\ ({}^2\boldsymbol{A}^3)\ ({}^3\boldsymbol{A}^4), \tag{33} $$

$$ {}^1\boldsymbol{A}^4\ (\theta_1, s_2, s_3) \underset{(30-33)}{=} \begin{bmatrix} c\theta_1 & 0 & s\theta_1 & s_3 s\theta_1 \\ s\theta_1 & 0 & -c\theta_1 & -s_3 c\theta_1 \\ 0 & 1 & 0 & s_2 \\ 0 & 0 & 0 & 1 \end{bmatrix}. \tag{34} $$

The position of the hand (link 4) relative to the base can be expressed in the form,

$$ \begin{bmatrix} X_1 \\ Y_1 \\ Z_1 \\ 1 \end{bmatrix} \underset{(21)}{=} {}^1\boldsymbol{A}^4\ (\theta_1, s_2, s_3) \begin{bmatrix} X_4 \\ Y_4 \\ Z_4 \\ 1 \end{bmatrix}, \tag{35} $$

where ${}^1\boldsymbol{A}^4$ is given by (34).

3. DISPLACEMENT ANALYSIS OF MANIPULATORS

This section discusses the solution of the inverse kinematic problem (see Section 2.5) - a problem of some importance in the kinematic theory of manipulators.

A manipulator capable of placing an end-effector in an arbitrary position and orientation with respect to the base must have (at least) six degrees-of-freedom (d-o-f); the discussion below will focus on six-d-o-f† manipulators. Common examples of six d-o-f industrial manipulators are the Cincinnati

† Degree(s)-of-freedom will be abbreviated as d-o-f.

Common examples of six d-o-f industrial manipulators are the Cincinnati Milacron T^3 and the Puma 600, both of type 6R and the IBM SR-1, of type 3P3R.

We first discuss, in Section 3.1, how a manipulator - an open-loop chain - in a specified position and orientation is equivalent to a mechanism - a closed-loop chain. This enables us to apply methods and results of mechanism analysis to the displacement analysis of manipulators. Section 3.2 looks at different methods of solving the inverse kinematic problem for manipulators and mechanisms. We conclude with examples of the inverse kinematics of simple manipulators (Section 3.3).

3.1 The Manipulator as a Closed-Loop Mechanism

Given an n-d-o-f manipulator $(n \leqslant 6)$ comprising n one-d-o-f joints and n moving links, one can form an equivalent closed-loop single d-o-f mechanism by connecting the free-end (link $n+1$) to the base (link 1) by $(7 - (n+1))$ hypothetical links and $(7-n)$ one d-o-f joints.

For example, a 6R manipulator with specified position and orientation of the hand can be converted into an equivalent 7R mechanism with its "input - crank" (attached to the hand) in a known position with respect to the base (link 1) (see Figure 8).

One result of the "loop-closure" concept is that we immediately know that the number of ways of positioning a manipulator of a given type so that the last link (end-effector) has a specified position and orientation with respect to the base must be equal to the number of ways of assembling the equivalent

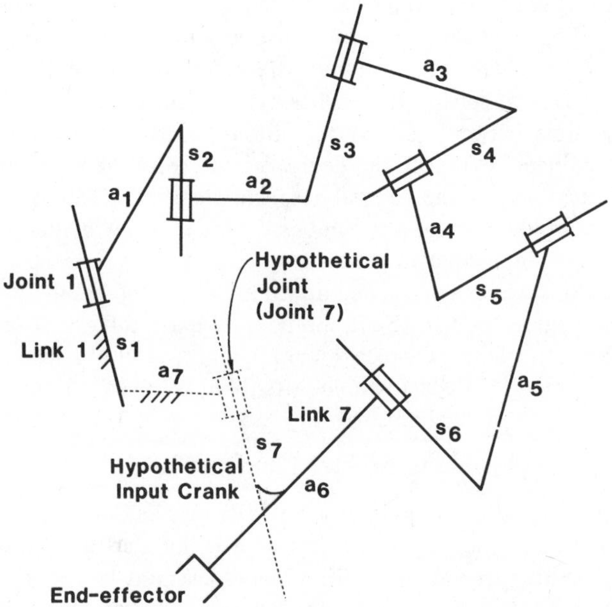

Figure 8 The 6R manipulator as a (closed-loop) mechanism.

linkage at a given input-crank position. For example, the 6R manipulator (Figure 8) yields the following results [20];

1. if $a_1=a_3=a_5=0$† there are at most 4 ways,
2. if $a_3=a_5=0$ there are at most 8 ways,
3. if $a_3=0$ there are at most 16 ways,
4. if none of the parameters are zero, there are at most 16 different ways,

of positioning the manipulator for specified position and orientation of the end-effector with respect to the base.

3.2 Solving the Inverse Kinematic Problem

The inverse kinematic problem may be stated as follows:

> Given the desired end-effector position and orientation with respect to the base, and the various link parameters, find the corresponding joint variables $(q_1,q_2,...,q_n)$.

The inverse kinematic problem therefore involves the solution of equation (2.25),

$$^1\boldsymbol{A}^{n+1}\ (q_1,q_2,...,q_n) = \boldsymbol{H}, \tag{2.25}$$

for the joint variables $(q_1,...,q_n)$, given $\boldsymbol{H}$.

To find the joint variables $q_i=\theta_i$, $i = 1$ to 6, of a general 6R manipulator for a given $\boldsymbol{H}$, $^1\boldsymbol{A}^{n+1}$ can be formed from equation (2.21) and substituted into equation (2.25). By equating like elements on each side of the equations twelve non-trivial scalar equations result from (2.25), nine for orientation and three for position. Since three of the orientation equations should be independent, we should be able to obtain six independent equations in $\theta_1,...,\theta_6$. The direct approach to solving the resultant set of equations is to eliminate one unknown at a time between pairs of equations until a single equation in one unknown is obtained. However the degree of the resulting equations increases with each elimination. In the general case, simple blind elimination, if it could be carried out, would lead to the final polynomial being of degree 524,288 - which is certainly impossible to solve [18]. Clearly other methods have to be sought which eliminate extraneous or unwanted roots and result in polynomials of manageable degree ($\leqslant 32$). Such methods do exist and are discussed below. Because of the equivalence between manipulators and closed loop mechanisms discussed in Section 3.1, considerable attention is given to the methods of displacement analysis of mechanisms.

3.2.1 Methods Applicable to Special Configurations

All cases of the six-d-o-f manipulators, with the restrictions that three adjacent joints be revolute and their axes intersect at a point, can be readily solved for the unknown joint variables [18,19]. The constraint of three intersecting revolute axes kinematically decouples the problem into two parts - one dealing with position and the other with orientation - thereby simplifying its solution.

The most complicated solutions - for these cases - are expressed as a single fourth degree polynomial in one unknown (joint variable), with linear or quadratic equations for the remaining unknown joint variables.

For most existing industrial manipulators, the joints and links are specially configured (for example, $\alpha = 0, 90^\circ$, $a = 0$, $s = 0$) and a "closed-form" solution can be obtained. (A solution is "closed-form" if one can symbolically solve for the unknowns.) Solutions for several commonly used industrial manipulator types are given in [6].

3.2.2 Vector Methods Using Spherical Trigonometry [7]

This method consists of deriving an equivalent spherical mechanism with R joints (or pairs) for a given 1-d-o-f, single-loop, spatial mechanism with R, P, C (cylindric) pairs: each joint of the spatial mechanism maps into an R pair of the equivalent spherical mechanism. To obtain a spherical map - or, spherical indicatrix - one simply uses a set of intersecting lines which are constrained to remain parallel to the corresponding axes in the original spatial mechanism. Using spherical trigonometry (cosine laws for spherical polygons) one obtains the input-output equation for the (derived) spherical mechanism from which the input-output equation for the spatial mechanism can be deduced.

The spatial mechanisms are divided into four groups [7] depending on the mobility (i.e., the number of degrees of freedom) of the equivalent spherical mechanism. For example, the RCCC mechanism is a Group 1 mechanism since it yields an equivalent 4R spherical mechanism which has mobility 1; a general 7R, single loop, spatial mechanism is a Group 4 mechanism since its equivalent spherical mechanism - a 7R - is of mobility 4. The problem of solving the input-output equation of the spatial mechanism is therefore one of solving the input-output equations of the equivalent spherical mechanisms and somehow eliminating unwanted angular displacements. The ease of solving the latter problem clearly depends on the mobility of the equivalent spherical mechanism.

For Group 1 mechanisms, since the equivalent spherical mechanism has mobility 1, the input-output equation of the mechanism's rotation is the same as the input-output equation of the equivalent spherical 4R; the latter is the cosine law for a spherical polygon and can be expressed as a polynomial of degree 2 in the tangent of the half-angle of the output. For a group k spatial mechanism ($k = 2,3,4$), the input and output angular displacements of the equivalent spherical mechanism have $(k-1)$ unwanted angular displacement which must be eliminated to yield the required input-output relation for the spatial mechanism. The degree of the resulting equation is 4 or 8 for Group 2, 16 for Group 3 and 32 for Group 4. The method therefore results in a "closed-form" solution for the input-output equations of the spatial mechanism. For a given input position (joint variable q_7 for a 6R manipulator) there will, in general, be as many possible values of the output (joint variable q_1) as the degree of the input-output equation; once the input-output equation has been obtained and solved, it is relatively simple to compute the remaining joint variables.

The details of implementing the above procedure as well as its application to mechanism and manipulator displacement analysis are well documented in [7].

3.2.3 Method of H. Albala [2]

This relatively new technique [1] develops a special mathematical symbology in order to facilitate the displacement analysis of n-bar single-loop spatial mechanisms. A Cyclic Algebra and an Indicial Trigonometry are constructed; the former arises from having to deal with the cyclic nature of the single-loop chain, and defines cyclic sums and cyclic products, while the latter is an indicial way of handling trigonometric expressions. Displacement Functions - functions of joint variables and linkage parameters - and linkage functions, while are functions of linkage parameters only, are also defined. Using these mathematical tools and starting with a matrix description of displacements, an algebraic form of the displacement equations is obtained. A method is presented to eliminate the intermediate joint variables one at a time, while the remaining variables are kept "sealed inside boxes (symbols)," [2]. It is claimed that the resulting input-output equation is thereby kept at the lowest possible degree. The technique is described in [2] and applied in [3] to a special case of the 7R spatial single-loop mechanism.

3.2.4 Iterative Techniques

The Newton-Raphson iterative technique [28] is widely used for the displacement analysis of general mechanisms (and also manipulators). It is based on linearizing equation (2.26) about some "initial" position and solving the resulting set of equations for the incremental change in the joint variables which are then used to generate a new set of (guesses for) the joint variables. If the displacement from "initial" to desired position is large, as in most manipulator work, it is necessary to break the desired displacement into a series of small steps. The Newton-Raphson technique is then used to find the joint angles to go from one step to the next. This is repeated until the final position and orientation are reached. The iterative velocity method [18], based on expressing changes in position and orientation by a screw displacement, has good convergence properties even for large displacements.

3.2.5 General Comments on the Methods

Most industrial manipulators are designed with dimensions which have the effect of yielding simplified kinematic equations and this facilitates the solution of the inverse kinematic problem. Because of this and the recent work of Albala and Duffy the inverse kinematics solution is now in principle known for all systems. However, in cases where the solution involves solving a polynomial of degree higher than, say, eight there may often be difficulties due to computational instabilities. To avoid these problems, numerical iterations using small step sizes are highly recommended. Although the iterations may not converge for certain singular configurations, small perturbations or other special tricks can almost always be used to get around such difficulties. The one great disadvantage of the iterative techniques is that they do not give all possible solutions. Hence they are not preferred in cases where there is an interest in optimized motions or certain types of obstacle avoidance, or if there are reasons for wanting all the ways to configure a manipulator with the same end-effector position and orientation. In these cases and in most cases where the solution is expressible as a polynomial of degree four or less it is probably best to use non-iterative techniques which provide all the solutions.

It is pointed out in passing that there exist numerical techniques for obtaining all the solutions for manipulator systems; however, they are very time consuming.

Section 3.3: In this section we solve the inverse kinematic problem for the (simple) manipulators of Examples 2.1 and 2.2.

Example 3.1 A PLANAR 2R MANIPULATOR (FIGURES 6,9)

The position $(X_1,Y_1,Z_1,1)$ in link 1 (the base) of point $(X_3,Y_3,Z_3,1)$ fixed in link 3 (end-effector) can be written in the form

$$\begin{bmatrix} X_1 \\ Y_1 \\ Z_1 \\ 1 \end{bmatrix} \underset{(2.18)}{=} {}^1\boldsymbol{A}^3 \begin{bmatrix} X_3 \\ Y_3 \\ Z_3 \\ 1 \end{bmatrix}. \tag{1}$$

If the origin of the $X_3Y_3Z_3$ coordinate system (Figures 6,9) is chosen at the "center" P of the end effector and if P has coordinates $(p_1,p_2,0)$ in the $X_1Y_1Z_1$ coordinate system of link 1, then we can write (1) for the point p as follows:

$$\begin{bmatrix} p_1 \\ p_2 \\ 0 \\ 1 \end{bmatrix} \underset{(2.18)}{=} {}^1\boldsymbol{A}^3 \begin{bmatrix} 0 \\ 0 \\ 0 \\ 1 \end{bmatrix}. \tag{2}$$

Substituting for ${}^1\boldsymbol{A}^3$ from equation (2.28), equation 2 becomes:

$$\begin{bmatrix} p_1 \\ p_2 \\ 0 \\ 1 \end{bmatrix} \underset{(2,2.28)}{=} \begin{bmatrix} c(\theta_1+\theta_2) & -s(\theta_1+\theta_2) & 0 & mc(\theta_1+\theta_2)+\ell c\theta_1 \\ s(\theta_1+\theta_2) & c(\theta_1+\theta_2) & 0 & ms(\theta_1+\theta_2)+\ell s\theta_1 \\ 0 & 0 & 1 & 0 \\ 0 & 0 & 0 & 1 \end{bmatrix} \begin{bmatrix} 0 \\ 0 \\ 0 \\ 1 \end{bmatrix} \tag{3}$$

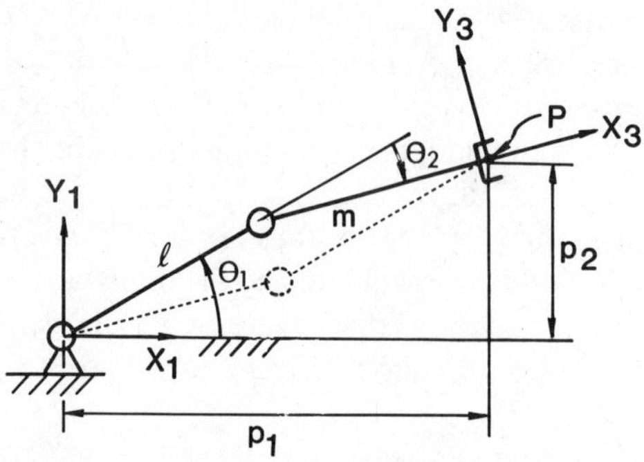

Figure 9 (Example 3.1). Dotted line shows second possible configuration of the manipulator for the position shown.

or

$$p_1 = mc(\theta_1 + \theta_2) + \ell c\theta_1$$

$$p_2 = ms(\theta_1 + \theta_2) + \ell s\theta_1 \tag{4}$$

where $c(\theta_1 + \theta_2)$ denotes $\cos(\theta_1 + \theta_2)$ and $s(\theta_1 + \theta_2)$ denotes $\sin(\theta_1 + \theta_2)$; ℓ, m, θ_1 and θ_2 are shown in Figure 9.

Defining,

$$\phi = \theta_1 + \theta_2$$

$$\psi = \theta_1, \tag{5}$$

(4) can be rewritten as

$$m\, c\phi + \ell\, c\psi \underset{(4,5)}{=} p_1, \tag{6}$$

$$m\, s\phi + \ell\, s\psi \underset{(4,5)}{=} p_2. \tag{7}$$

We first solve for ψ as follows

$$m\, c\phi \underset{(6)}{=} p_1 - \ell c\psi, \tag{8}$$

$$m\, s\phi \underset{(7)}{=} p_2 - \ell s\psi \tag{9}$$

$$m^2 \underset{(8,9)}{=} (p_1 - \ell c\psi)^2 + (p_2 - \ell s\psi)^2$$

$$m^2 = p_1^2 + (\ell c\psi)^2 - 2p_1\ell c\psi + p_2^2 + (\ell s\psi)^2 - 2p_2\ell s\psi. \tag{10}$$

Defining

$$t_1 \triangleq \tan\frac{\psi}{2}, \tag{11}$$

and using the identities

$$\cos\psi = \frac{1 - t_1^2}{1 + t_1^2}, \qquad \sin\psi = \frac{2t_1}{1 + t_1^2}, \tag{12}$$

equation (10) becomes

$$2p_1\ell\,\frac{1 - t_1^2}{1 + t_1^2} + 2p_2\ell\,\frac{2t_1}{1 + t_1^2} \underset{(10,12)}{=} p_1^2 + p_2^2 + \ell^2 - m^2. \tag{13}$$

Defining,

$$\xi \triangleq (p_1^2 + p_2^2 + \ell^2 - m^2)/\ell, \tag{14}$$

we can write (13) in the form

$$(\xi + 2p_1)t_1^2 - 4p_2t_1 + (\xi - 2p_1) = 0. \tag{15}$$

Solving for t_1, we obtain

$$t_1 \underset{(15)}{=} \frac{+2p_2 \pm \sqrt{4p_2^2 - (\xi^2 - 4p_1^2)}}{(\xi + 2p_1)}, \tag{16}$$

or

$$\tan\frac{\psi}{2} = \frac{2p_2 \pm \sqrt{4(p_1^2 + p_2^2) - \xi^2}}{(\xi + 2p_1)}. \tag{17}$$

In a similar manner we obtain the following equation for $\tan\phi/2$:

$$\tan\frac{\phi}{2} = \frac{2p_2 \pm \sqrt{4(p_2^2 + p_1^2) - \eta^2}}{(\eta + 2p_1)} \tag{18}$$

where

$$\eta = (p_1^2 + p_2^2 + m^2 - \ell^2)/m. \tag{19}$$

Equations (17) and (18) can be solved for the joint variables θ_1 and θ_2 as follows:

$$\theta_1 \underset{(5,17)}{=} 2\tan^{-1}\frac{2p_2 \pm \sqrt{4(p_1^2 + p_2^2) - \xi^2}}{(\xi + 2p_1)} \tag{20}$$

$$\theta_2 \underset{(5,18)}{=} 2\tan^{-1}\frac{2p_2 \pm \sqrt{4(p_1^2 + p_2^2) - \eta^2}}{(\eta + 2p_1)} - \theta_1 \tag{21}$$

Note that there will be two possible configurations for a given (p_1, p_2) [see Figure 9].

Example 3.2 A SPATIAL R2P MANIPULATOR (FIGURES 7 AND 10)

The positions $(X_1,Y_1,Z_1,1)$ in link 1 (the base) of points $(X_4,Y_4,Z_4,1)$ fixed in link 4 (the end-effector) can be written in the form

$$\begin{bmatrix} X_1 \\ Y_1 \\ Z_1 \\ 1 \end{bmatrix} \underset{(2.18)}{=} {}^1\boldsymbol{A}^4 \begin{bmatrix} X_4 \\ Y_4 \\ Z_4 \\ 1 \end{bmatrix}. \tag{22}$$

If $(0,0,h)$ are the (X_4,Y_4,Z_4) coordinates of the center P of the end-effector and (p_1,p_2,p_3) its (X_1,Y_1,Z_1) coordinates, then equation (22) can be written for P as follows:

$$\begin{bmatrix} p_1 \\ p_2 \\ p_3 \\ 1 \end{bmatrix} = {}^1\boldsymbol{A}^4 \begin{bmatrix} 0 \\ 0 \\ h \\ 1 \end{bmatrix}. \tag{23}$$

Substituting for ${}^1\boldsymbol{A}^4$ from equation (2.34), equation (23) becomes

$$\begin{bmatrix} p_1 \\ p_2 \\ p_3 \\ 1 \end{bmatrix} \underset{(23,2.34)}{=} \begin{bmatrix} c\theta_1 & 0 & s\theta_1 & s_3 s\theta_1 \\ s\theta_1 & 0 & -c\theta_1 & -s_3 c\theta_1 \\ 0 & 1 & 0 & s_2 \\ 0 & 0 & 0 & 1 \end{bmatrix} \begin{bmatrix} 0 \\ 0 \\ h \\ 1 \end{bmatrix}, \tag{24}$$

or

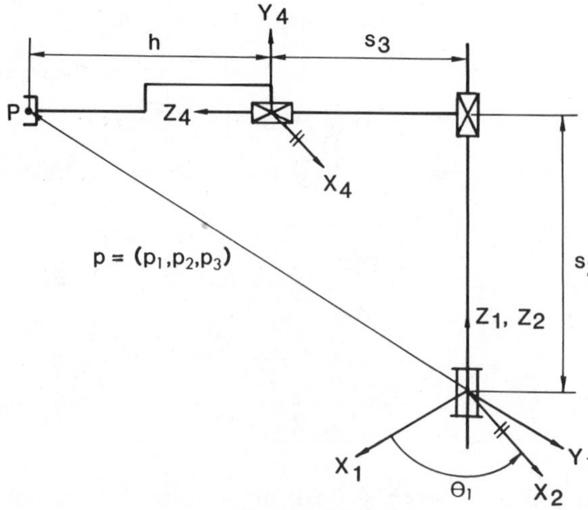

Figure 10 (Example 3.2).

or

$$p_1 = (s_3 + h)\, s\theta_1, \tag{25}$$

$$p_2 = -(s_3 + h)\, c\theta_1, \tag{26}$$

$$p_3 = s_2, \tag{27}$$

which yields the following expressions for the joint variables θ_1, s_2, s_3:

$$s_2 = p_3 \tag{28}$$

$$s_3 = \pm \sqrt{p_1^2 + p_2^2} - h \tag{29}$$

$$\tan \theta_1 = \frac{-p_1}{p_2}. \tag{30}$$

If the manipulator design is such that $s_3 \geqslant 0$, then θ_1, s_2, s_3 are given by

$$s_2 \underset{(28)}{=} p_3 \tag{31}$$

$$s_3 \underset{(29)}{=} \sqrt{p_1^2 + p_2^2} - h \tag{32}$$

$$\theta_1 \underset{(30)}{=} \tan^{-1}\left(\frac{-p_1}{p_2}\right) \tag{33}$$

and

$$0 < \theta_1 < 180^\circ \text{ if } p_1 \text{ is positive}$$

$$180^\circ < \theta_1 < 360^\circ \text{ if } p_1 \text{ is negative.}$$

4. THE WORKSPACE OF A MANIPULATOR

The (total) workspace of the manipulator is the volume of space comprising all points that a reference point attached to the free-end of a manipulator can reach. This reference point is usually the "center" of the end-effector, the tip of the end effector, the end of the manipulator, or some so-called wrist point on the manipulator. Changing the reference point changes the dimensions of the workspace. When not all positions in a workspace can be approached from arbitrary directions, it is useful to single out those portions of the workspace, if any, which the end-effector can approach from any angle. The aggregate of such regions form the "primary" or dextrous workspace; we denote them as W^p. The remainder of the workspace is then referred to as the secondary workspace, W^s, since only limited end-effector orientations are available at points in this region.

In regard to workspace there are two basic questions:

> Given a manipulator with a certain kinematic structure, and ranges of joint variables, what are the corresponding primary and total workspaces (the direct workspace problem)?
>
> Given the primary and total workspaces of the manipulator, what should the structure of the manipulator be (the inverse workspace problem)?

Both these issues are discussed in this section: in Section 4.1 we address the problem of determining the workspaces of the manipulator [16,26]; in Section 4.2 we present theorems related to extreme positions (with respect to the base) of points and lines on the end-effector [23,26]; finally, in Section 4.3, we present material [9] useful for designing manipulators to meet certain workspace requirements. Examples are given to illustrate the basic concepts.

4.1 The Bounding Surfaces of the Manipulator Workspace

4.1.1 Definitions:

Total (or Reachable) workspace, $W(P)$, of a point P of the end-effector of a manipulator is the set of all points (in the fixed frame) which P occupies as the joint variables (θ_i or s_i) are varied through their entire ranges.

Primary (or Dextrous) workspace, $W^p(P)$, of a point P of the end-effector is a subset of $W(P)$ such that at every point (in this subset) the end-effector can assume any orientation with respect to the fixed frame.

Secondary workspace, $W^s(P)$, of a point P is the portion of its total workspace which can be reached with only limited orientations of the end-effector.

Remark 1: The point P is usually chosen as either the center of the end-effector (viz., the center of the "hand"), or the tip of a "finger", or the end of the manipulator itself, or a so-called wrist point.

Remark 2: The reachable workspace may also be defined in terms of a line fixed in the end-effector [23]

4.1.2 Determination of the Bounding Surfaces of a Manipulator Workspace

An algorithm for determining the bounding surface of the manipulator workspace, $W(P)$, is given in [16] and is based on the following rationale:

To determine the extent of the workspace, $W(P)$, of a point P of the end-effector, a force F, is applied to the manipulator - assumed weightless - at P. Under the action of the force the manipulator reaches a (stable) static equilibrium configuration in which it assumes maximum extension (with respect to the base) in the direction of the force; in this configuration, P lies on the bounding surface of $W(P)$. Since the moment of the force about each revolute joint axis must be zero in the static equilibrium configuration, the line of action of the force must intersect all the revolute joint axes. This conclusion forms the basis of the workspace algorithm in [16]. (Another way of stating the zero-moment condition is that the force and the joint axes form a reciprocal screw system [16] in the static equilibrium position.)

For a manipulator with N revolute joints, there will be 2^{N-1} configurations of static equilibrium for a given direction of F, only one of which is stable. For example, the 4 configurations of static equilibrium of a 3R planar manipulator for an applied force F are shown in Figure 11; configuration (a) is the stable one.

4.1.3 Examples

Example 4.1: A PLANAR 3R MANIPULATOR

Using the algorithm it follows there are two extreme positions which are stable. The two extreme positions for the planar 3R manipulator, for any given θ_1, are shown in Figures 12a and 12b. It follows the bounding curves are the circles, C_1 and C_2 of radii $(a_1-(a_2+h))$ and $(a_1+(a_2+h))$ respectively. (Figure 12c); $W(P)$ is the area enclosed by C_1 and C_2. If $a_1 > (a_2+h)$, as in the case of the 3R manipulator shown, then $W(P)$ has a "hole" (enclosed by C_1) as shown.

Example 4.2: A 6R MANIPULATOR WITH ITS LAST THREE AXES INTERSECTING [15]

The 6R manipulator has the following dimensions (see Figure 13):

$$\alpha_1 = -\alpha_3 = \alpha_4 = \alpha_5 = 90^\circ$$

$$a_2 \neq 0, \quad a_3 \neq 0,$$

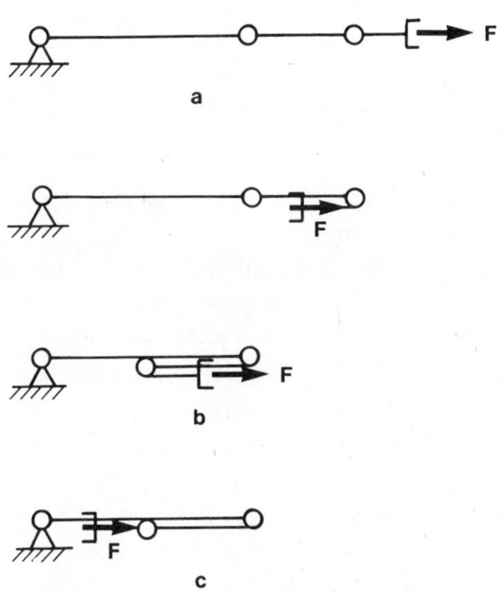

Figure 11

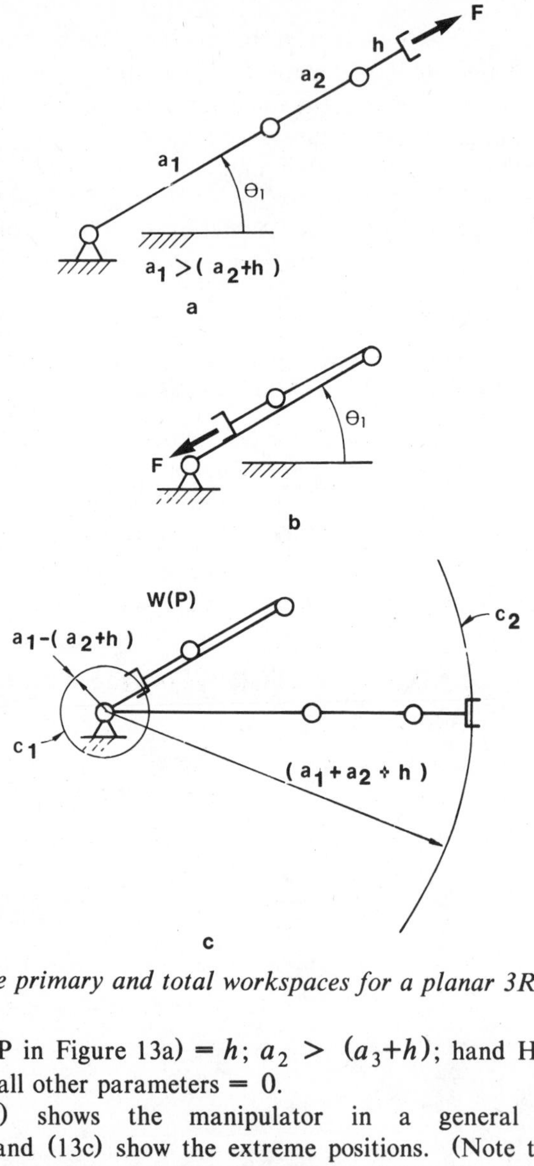

Figure 12 The primary and total workspaces for a planar 3R manipulator.

"hand size" (HP in Figure 13a) = h; $a_2 > (a_3+h)$; hand HP perpendicular to joint axis 6, all other parameters = 0.

Figure (13a) shows the manipulator in a general position while Figures (13b) and (13c) show the extreme positions. (Note that the extreme positions are independent of the first two joint variables θ_1 and θ_2.)

The total workspace, $W(P)$, is the volume enclosed by the spheres V_1, of radius $[a_2+(a_3+h)]$, and V_4, of radius $[a_2-(a_3+h)]$, in Figure (13d).

The dextrous workspace is the volume, $W^p(P)$, enclosed by the spheres V_2, of radius $a_2+(a_3-h)$, and V_3, of radius $a_2-(a_3-h)$, in Figure (13d).

We make the following observations:

1. For the case of zero hand-size, i.e., h=0, the primary workspace is the same as the total workspace.

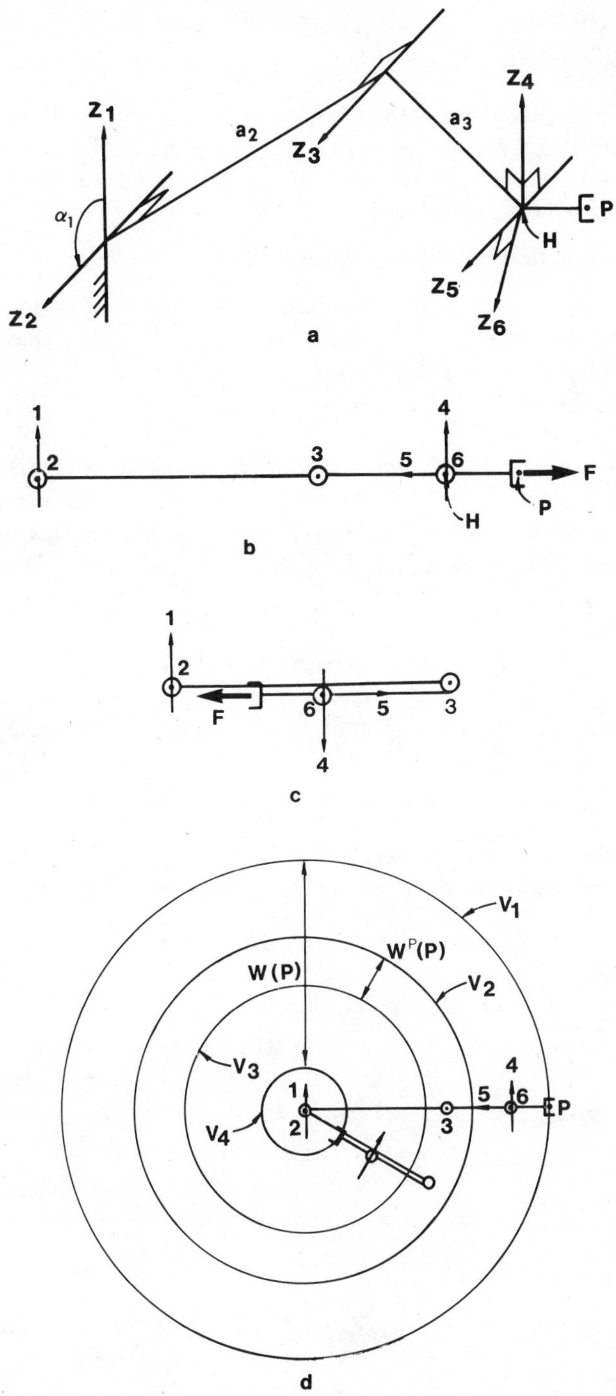

Figure 13 The primary and total workspaces for a 6R manipulator.

2. Finite hand-size, i.e., $h \neq 0$, in general increases the total workspace and reduces the primary workspace.
3. A 6R manipulator, with the first two joint axes intersecting and the last three axes intersecting with neighboring axes orthogonal to each other, would meet the requirements of (i) a spherical workspace for $W^p(P)$ and $W(P)$, and (ii) $W^p(P)$ as large as possible for a given h (as $h \to 0$, $W^p(P) \to W(P)$).

4.4 Extreme Positions of Manipulators

Knowing the extreme positions of the end-effector (with respect to the base) is useful in designing manipulators to meet workspace requirements. Three theorems relating to the extrema of a manipulator† are:

Theorem 1 [23]

The distance (*along the common normal L*), between the axis Z_1, of the first joint and an axis Z_{n+1} fixed in the end-effector, is at an extremum with respect to the joint variables if, and only if, the corresponding normal L to Z_1 and Z_{n+1} intersects all intermediate joint axes Z_i, $(i=2,...,n)$. (Figure 14).

Theorem 2 [26]

The distance between a point P_B fixed in the base and a point P_E fixed in the end effector is an an extremum with respect to the joint variables, if, and only if, the line L joining P_B and P_E intersects all the joint axes Z_i, $(i=1,...,n)$. (Figure 15)

Theorem 3 [26]

The (*normal*) distance between the axis Z_B fixed in the base and a point P_E fixed in the end-effector is at an extremum with respect to the joint

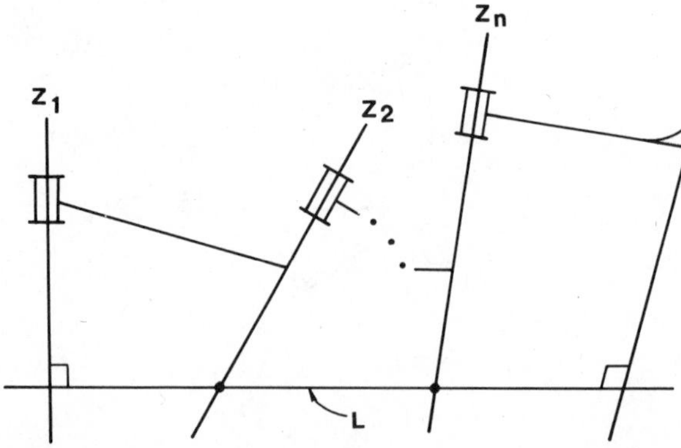

Figure 14 Normal line L between Z_1 and Z_{n+1}, illustrating the condition of Theorem 1.

† Assumed to have only revolute joints.

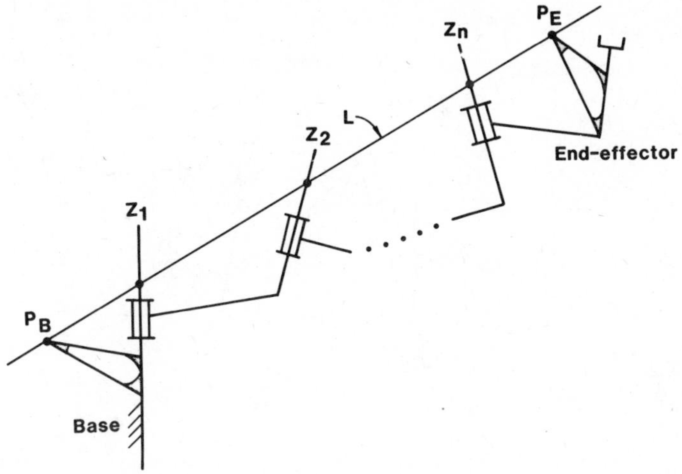

Figure 15 Line L between P_B and P_E, illustrating the condition of Theorem 2.

variables if, and only if, the normal L from P_E to Z_B intersects all the joint axes Z_i, $(i=1,...,n)$. (Figure 16)

A proof of Theorem 1 - using Plücker line-coordinates - is given in [23] while proofs of Theorems 2 and 3 - using vector algebra - are given in [26].

An iterative scheme, using the results of Theorem 1, for determining the positions of the manipulator corresponding to all extrema of the normal distance between Z_1 and Z_{n+1} is given in [23]; for a manipulator with n moving links there will in general be 2^{n-1} such configurations. This same work gives closed-form expressions for determining manipulator parameters so that a manipulator can be designed to have specified values of extreme normal distance between its axes.

An algorithm - based on Theorem 2 - for determining "extreme" distances between P_B and P_E is given in [26]. This algorithm can also be used, with suitable modifications, to calculate extreme distances corresponding to the cases of Theorems 1 and 3.

Remark 1: Mathematical extrema only exist with respect to revolute joint variables θ_i corresponding to revolute joints with theoretically unlimited rotation capability (i.e., $0 \leqslant \theta_i < 360°$). In practice, of course, revolute joints have physical limitations on the allowed rotation and the prismatic joints have limited ranges of travel, and these limits define the actual extrema.

Remark 2: The extrema discussed above include local and global extrema. For example, the iterative schemes and algorithms mentioned in this section will yield all the four configurations (and the corresponding distances), shown in Figure 11, for the 3R manipulator.

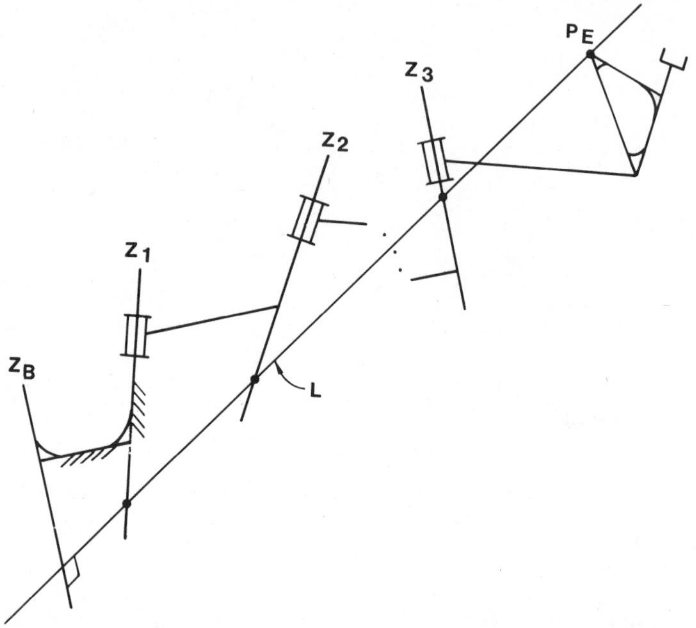

Figure 16 Normal line between Z_B and P_E, illustrating the condition of Theorem 3.

Example 4.3:

The results of Theorems 1, 2, and 3 can readily be visualized for the (global) extreme positions of the 3R manipulator (Figures 12a,b) of Example 4.1 and the 6R spatial manipulator (Figures 13b,c) of Example 4.2.

4.6 Design Considerations for Manipulator Workspace

In this section we address the matter of designing manipulators to avoid holes and voids (defined below) in the workspace [9].

4.6.1 Terminology

The following terminology will be useful in the sequel [9]:

P_m: point P fixed in reference frame m.

$W_k(P_m)$: The reachable workspace, in a reference frame fixed to link k, of a point P fixed in reference frame m.

The workspace $W_n(P_{n+1})$, of a point P† fixed in link $n+1$ (the end-effector) in Figure 17, is a circle, denoted by circle (n). The workspace

† The subscript $n+1$ on P will be dropped hereafter, i.e., P will always denote a point fixed in link $n+1$.

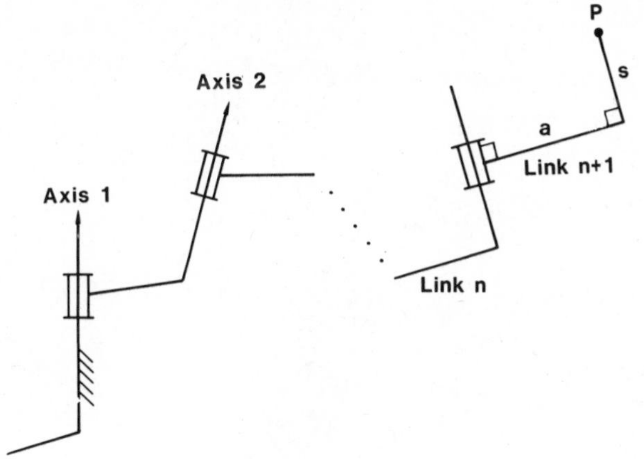

Figure 17

$W_{n-1}(P)$ of P will be the torus, denoted by torus $(n-1)$, generated by rotating $W_n(P)$ (or circle (n)) about the axis $(n-1)$. The workspace $W_{n-2}(P)$ will be the solid of revolution, denoted by $SR(n-2)$, obtained by rotating torus $(n-1)$ about axis $(n-2)$; in general $SR(n-j)$, $(j>2)$, will denote the solid of revolution obtained by rotating $SR(n-j+1)$ about the joint axis $(n-j)$.

4.6.2 Holes and Voids in Workspaces

A basic building block in studying manipulator workspaces is the spatial 2R dyad shown in Figure 18; $W_n(P)$ is a circle and $W_{n-1}(P)$ is the surface of a torus described by the following equation (see Figure 18):

$$[x_{n-1}^2 + y_{n-1}^2 + (z_{n-1} - s_{n-1})^2 - (a_{n-1}^2 + a^2 + (s + s_n)^2)]^2$$

$$= 4a_{n-1}^2 \left[a^2 - \left(\frac{(z_{n-1} - s_{n-1}) - (s + s_n) \cos \alpha_{n-1}}{\sin \alpha_{n-1}} \right)^2 \right]. \qquad (1)$$

The parameters a/a_{n-1}, α_{n-1} and $(s + s_n)$ determine the shape of the workspace - four forms of the torus [8], with the corresponding values of the parameters, are shown (in section) in Figure 19. The garden variety (donut-shaped) torus has $a/a_{n-1} < 1$, $s+s_n=0$ and $\alpha_{n-1} = 90$.

Definition: $W_k(P)$ is said to have a *hole* if there is a neighborhood intersecting the axis k which is not reachable by P.

Definition: $W_k(P)$ is said to have a *void* if there is a closed region R such that all points "inside" the bounding surface of R are not reachable by the manipulator. (A void is simply an internal hole in $W_k(P)$.)

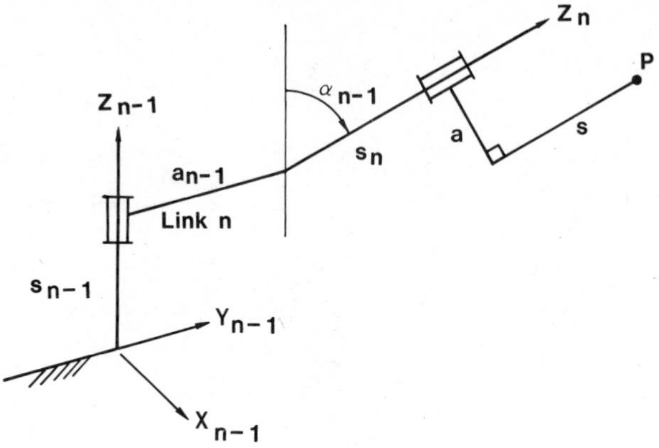

Figure 18 2R dyad.

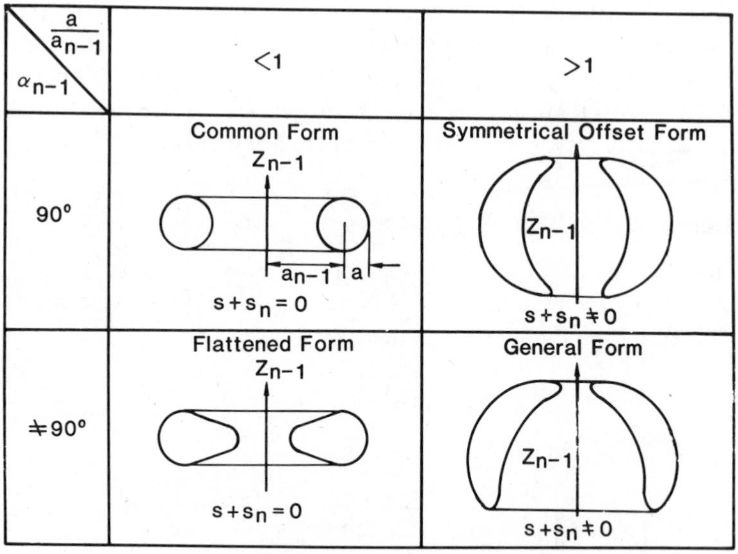
a / a n-1
α n-1
<1
>1
90°
∓90°
Common Form
Z n-1
a n-1
a
s + s n = 0
Symmetrical Offset Form
Z n-1
s + s n ∓ 0
Flattened Form
Z n-1
s + s n = 0
General Form
Z n-1
s + s n ∓ 0

Figure 19 Different forms of the torus.

We now enumerate some useful properties of $W_{n-1}(P)$:

1. In general $W_{n-1}(P)$, being the surface of a torus, will always have a void (see Figure 20).
2. If $a/a_{n-1} << 1$, $W_{n-1}(P)$ will always have a hole.
3. As a/a_{n-1} increases, the torus may intersect itself; the condition for this is [9],

$$\left(\frac{a}{a_{n-1}}\right)^2 \geqslant 1 + \left[\frac{s+s_n}{a_{n-1}}\right]^2 \tan^2 \alpha_{n-1} \tag{2}$$

4. A necessary condition, from (2), for the absence of a hole is $(a/a_{n-1}) > 1$.
5. If $a/a_{n-1} > 1$ and $\alpha_{n-1} = 0$, or if $a/a_{n-1} > 1$ and $(s+s_n) = 0$, then (from equation (2)) there will be no hole in the workspace $W_{n-1}(P)$.

Next consider W_{n-2} or the solid of revolution generated by revolving $W_{n-1}(P)$ (=torus $(n-1)$) about axis $n-2$ (Figure 21). Some useful properties of $W_{n-2}(P)$ or $SR(n-2)$ are the following [9]:

1. $SR(n-2)$ will not have a hole if axis $n-2$ intersects torus $(n-1)$.
2. $SR(n-2)$ will *not* have a void if
   a. torus $(n-1)$ has a hole and α_{n-2} is small,
   b. axis $(n-2)$ does not pass through the void in torus $(n-1)$, and
   c. axes $(n-1)$ and $(n-2)$ are distinct.

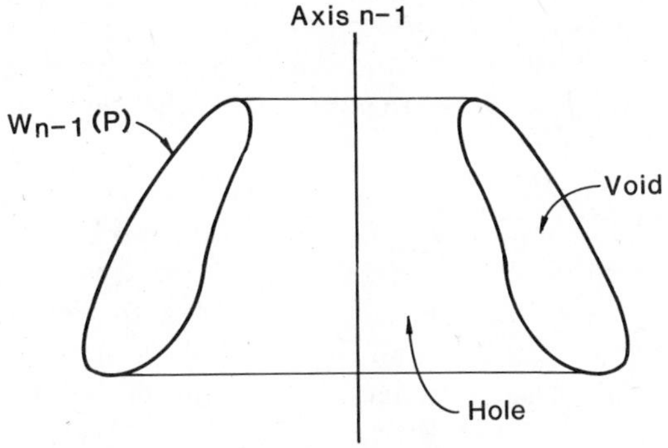

Figure 20 Holes and Voids for $W_{n-1}(P)$. (The figure shows an axial cross section of the workspace.)

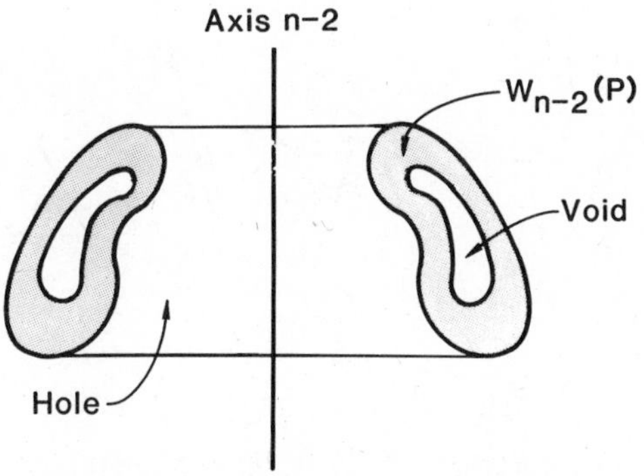

Figure 21 Holes and Voids for $W_{n-2}(P)$. (The figure shows an axial cross section of the workspace.)

Generalizing the above properties to $SR(k)$ and the $SR(k-1)$ obtained from it, by rotation about axis k-1, we have the following:

1. $SR(k-1)$ does not have hole if axis $(k-1)$ passes through $SR(k)$.
2. $SR(k-1)$ does not have a void if
   a. $SR(k)$ has a hole and α_{k-1} is small
   b. axis $(k-1)$ does not pass through the void in $SR(k)$, and
   c. axes $(k-1)$ and (k) are distinct.

4.6.3 Design Considerations

Let $V_1(P)$ denote the volume of the total workspace $W_1(P)$ and $V_1^p(P)$ the volume of the primary workspace $W^p(P)$. General design objectives for manipulators are to maximize $V_1(P)$ and $V_1^p(P)/V_1(P)$. In many cases workspace $W_1(P)$ is obtained with three (suitably positioned) axes. Use is also made of the fact that any 6R manipulator with the three outermost revolute axes intersecting, at the point P, such that neighboring axes are orthogonal, will have the property $W_1^p(P) = W_1(P)$ (Example 4.2) provided the joint angles are not mechanically limited. The case when P is not at the point of intersection is discussed below.

Let the point of intersection of the last three axes be H, and let P be at a distance h from H; h can be thought of as the "hand-size" and we now examine the effect of hand-size on workspace.

For the point H, $W^p(H) = W(H)$†, while the point, P, $W^p(P) < W(P)$; also $W(H) < W(P)$ and $W^p(P) < W^p(H)$. In words, a finite "hand"-size increases the total workspace and decreases the primary workspace (see Example 4.2). Regarding the effect of hand-size on holes and voids in the manipulator workspace we have the following result [9]:

For the case of the last three axes intersecting, if $W(H)$ has a void, then for $h \neq 0$ (finite hand-size), $W^p(P)$ has a void even though the total

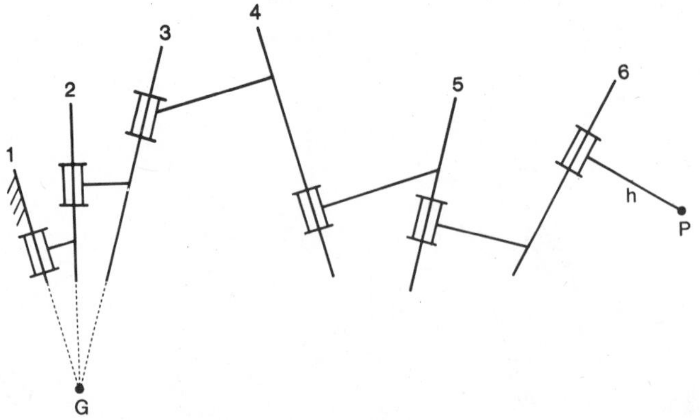

Figure 22 6R manipulator with first three axes intersecting at a point.

† The subscript 1 will be dropped hereafter, i.e, $W(P) \equiv W_1(P)$. For the last three axes, we require the neighboring axes to intersect at right angles.

workspace $W(P)$ may or may not have a void; therefore a necessary condition for the absence of voids in $W^p(P)$ is that $W(H)$ should not have any voids.

For the case of a 6R manipulator where the first three axes intersect at a point G, and the neighboring axes are orthogonal (i.e., $\alpha_1 = \pm 90°$, $\alpha_2 = \pm 90°$), we have the following design conclusion (see Figure 22):

1. If P is a point on the end-effector (or hand) and h the hand-size (in Figure 17, if $s = 0$ then $h = a$), then for $W^p(P)$ to exist, point G, must be inside the workspace $W(P)$, i.e., G must be a reachable point.

The case when three intermediate axes intersect is discussed in [9].

We conclude with an example [9] to illustrate the concepts presented in this section.

Example 4.4: 6R manipulator with the last three axes forming two orthogonal pairs and intersecting at H; the three intersecting joints are replaced by a spherical joint at H. Let $s_1 = s_2 = s = 0$, $\alpha_1 = 90°$, α_2 arbitrary, $a_1 = c$, $a_2 = b$, $a \leqslant b \leqslant c$ (see Figure 23a). The workspace $W_3(H)$ is circle (3), $W_2(H)$ is

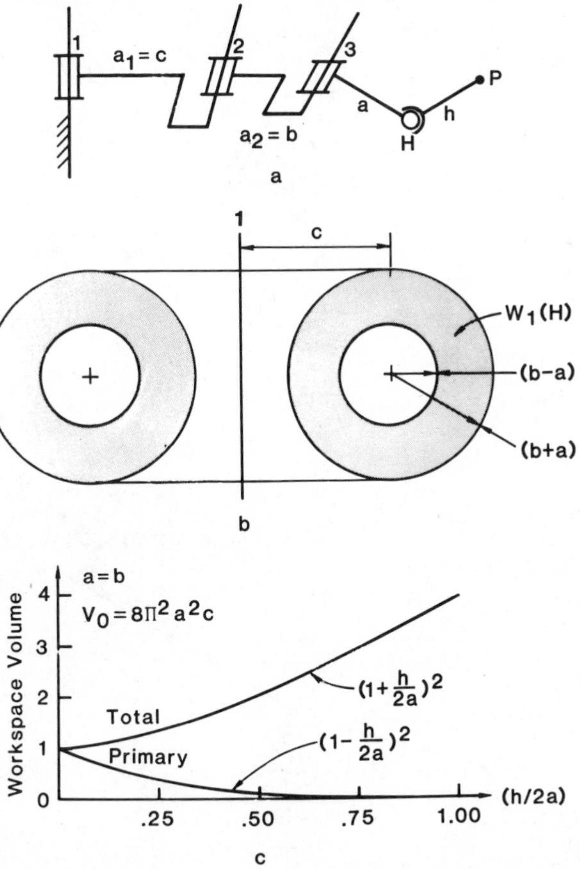

Figure 23 (Example 4.4). V_0 is the volume when $h = 0$.

torus (2) and $W_1(H)$ is $SR(1)$. An axial section of workspace $W_1(H)$ is shown in Figure (23b). It has an internal void if $a \neq b$. The volume of $W_1(H)$ is $8\pi^2 abc$. The shape and volume of $W_1(H)$ are not influenced by the value of the angle α_2. The volumes of workspaces $W_1^p(P)$ and $W_1(P)$ are plotted with respect to hand size in Figure (23c) for the case $a=b$ (no void in $W_1(H)$) in the range $0 \leqslant h \leqslant 2a$.

5. SMALL MOTIONS

In studying the kinematics of small (or fine) motions of the manipulator, the "Manipulator Jacobian" is a useful concept. We present a simple derivation of the Manipulator Jacobian and discuss some of its uses.

Given an n-d-o-f manipulator, with n 1-d-o-f joints (revolute or prismatic) - as shown in Figure 24 - the velocity ${}^N\boldsymbol{V}^P$, with respect to the base

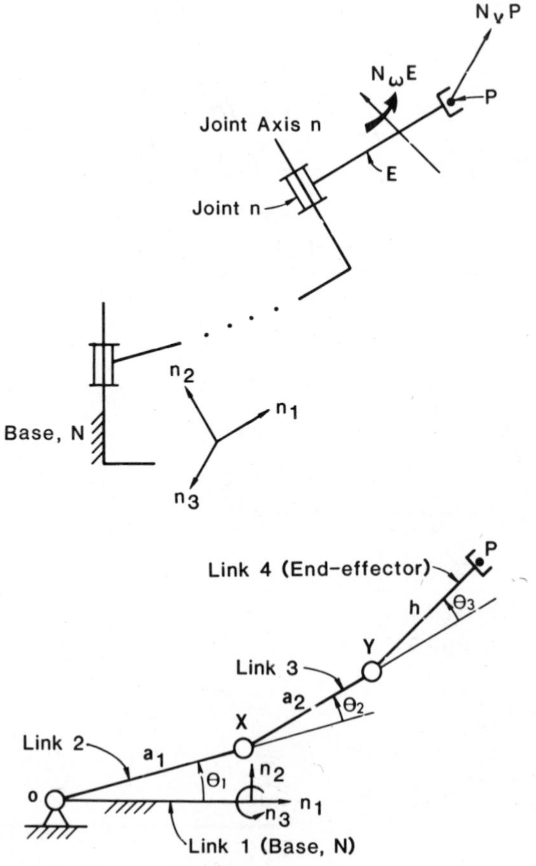

Figure 24

(reference frame N), of a point, P of the end-effector (i.e., link $n+1$) can be written in the form [12],

$$^N\boldsymbol{V}^P = \sum_{r=1}^{n} {}^N\boldsymbol{V}_r^P \dot{q}_r + {}^N\boldsymbol{V}_t^P \tag{1}$$

The angular velocity, $^N\boldsymbol{\omega}^E$, of link $n+1$ denoted as E (for end-effector), with respect to the base can be written in the form [12],

$$^N\boldsymbol{\omega}^E = \sum_{r=1}^{n} {}^N\boldsymbol{\omega}_r^E \dot{q}_r + {}^N\boldsymbol{\omega}_t^E, \tag{2}$$

Where $^N\boldsymbol{V}_r^P$ is called the partial velocity of P in N and $^N\boldsymbol{\omega}_r^E$ is called the partial angular velocity of E in N [14]; partial velocities and partial angular velocities are simply the coefficients in equations (1) and (2), respectively, of the derivatives of the joint variables q_r†. In Example 5.1, below, we show how to determine these partials. (When the reference frames and points are clear from the context, as in this section, we will drop the corresponding superscripts.)

Let $\boldsymbol{n}_i$, $i=1,2,3$, be a dextral orthogonal set of unit vectors fixed in the base N and define the $\boldsymbol{n}_1$, $\boldsymbol{n}_2$, $\boldsymbol{n}_3$ measure numbers of $\boldsymbol{V}$ and $\boldsymbol{\omega}$ as follows:

$$v_i = \boldsymbol{n}_i \cdot \boldsymbol{V}\ ;\ (i=1,2,3) \tag{3}$$

$$\omega_i = \boldsymbol{n}_i \cdot \boldsymbol{\omega}\ ;\ (i=1,2,3) \tag{4}$$

Equations (1) and (3) and equations (2) and (4) can be combined to yield (after setting $\boldsymbol{\omega}_t^P = \boldsymbol{V}_t^P = 0$).

$$v_i \underset{(1,3)}{=} \sum_{r=1}^{n} (\boldsymbol{n}_i \cdot \boldsymbol{V}_r)\, \dot{q}_r; \qquad (i=1,2,3) \tag{5}$$

$$\omega_i \underset{(2,4)}{=} \sum_{r=1}^{n} (\boldsymbol{n}_i \cdot \boldsymbol{\omega}_r)\, \dot{q}_r; \qquad (i=1,2,3) \tag{6}$$

Defining quantities J_{ir} as:

$$J_{ir} = \boldsymbol{n}_i \cdot \boldsymbol{V}_r; \qquad (i=1,2,3); \qquad (r=1,...,n) \tag{7a}$$

$$J_{ir} = \boldsymbol{n}_{(i-3)} \cdot \boldsymbol{\omega}_r\ ; \qquad (i=4,5,6); \qquad (r=1,...,n) \tag{7b}$$

† In most manipulator work, $^N\boldsymbol{V}_t^P = {}^N\boldsymbol{\omega}_t^E = 0$; these are the terms in the velocity that depend only on time and not on the rates of change of the joint variables.

we can write (5) and (6) as

$$v_i \underset{(5)}{=} \sum_{r=1}^{n} J_{ir}\, \dot{q}_r; \qquad (i=1,2,3) \tag{8a}$$

$$\omega_{i-3} \underset{(6)}{=} \sum_{r=1}^{n} J_{ir}\, \dot{q}_r; \qquad (i=4,5,6) \tag{8b}$$

Equation (8) can be written in the matrix form

$$[v_1\ v_2\ v_3\ \omega_1\ \omega_2\ \omega_3]^{\mathrm{T}} \underset{(8)}{=} \boldsymbol{J}\, [\dot{q}_1\ \dot{q}_2\ \cdots\ \dot{q}_n]^{\mathrm{T}}, \tag{9}$$

where $\boldsymbol{J}$ is a $(6Xn)$ matrix with elements J_{ij} defined by equation (7). $\boldsymbol{J}$ is often called the Manipulator Jacobian; for a manipulator with six 1-d-o-f joints, $\boldsymbol{J}$ is a 6×6 square matrix.

For small motions about a given position, we can make the following approximations

$$\delta q_r = \dot{q}_r\, \Delta t; \qquad (r=1,\dots,n), \tag{10}$$

$$\delta x_i = v_i\, \Delta t; \qquad (i=1,2,3), \tag{11}$$

$$\delta \phi_i = \omega_i\, \Delta t; \qquad (i=1,2,3), \tag{12}$$

where δq_r is a small motion (rotation or translation) at joint i, δx_i is a small translation of P in the direction $\boldsymbol{n}_i$ and $\delta\phi_i$ is a small rotation of the hand about the $\boldsymbol{n}_i$ axis. Using equations (10,11,12), we can rewrite equation (9) as

$$[\delta X|\delta\phi]^{\mathrm{T}} \underset{(9-12)}{=} \boldsymbol{J}\, [\delta q]^{\mathrm{T}}, \tag{13}$$

where

$$[\delta X] = [\delta x_i\ \delta x_2\ \delta x_3], \tag{14}$$

$$[\delta\phi] = [\delta\phi_1\ \delta\phi_2\ \delta\phi_3], \tag{15}$$

$$[\delta q] = [\delta q_1\ \dots\ \delta q_n]. \tag{16}$$

If J^{-1} exists then (13) can be rewritten as

$$[\delta q]^{\mathrm{T}} = \boldsymbol{J}^{-1}\, [\delta X|\delta\phi]^{\mathrm{T}}. \tag{17}$$

Equations (13) and (17) relate small joint motions, $\delta q_1,\dots,\delta q_n$, to the corresponding small translations of P and small rotations of the end-effector. Clearly, any set of joint motions $[\delta q]$ will yield an end-effector motion $[\delta X|\delta\phi]$. However, it is not always possible to determine a set of joint motions, $[\delta q]$, corresponding to an arbitrary end-effector motion $[\delta X|\delta\phi]$. This problem arises when $\boldsymbol{J}^{-1}$ does not exist. For a 6 d-o-f manipulator, the

positions for which $|\boldsymbol{J}| = 0$ are called singular positions. Singular positions are those positions for which the manipulator transitorily loses one or more d-o-f. This occurs at the extreme positions and when two or more joints move into dependent configurations. For systems with less than six degrees of freedom it is customary to use the pseudo inverses instead of $\boldsymbol{J}^{-1}$. For such systems we use the following equation in place of equation (17):

$$[\delta q]^{\mathrm{T}} = (\boldsymbol{J}^{\mathrm{T}}\boldsymbol{J})^{-1}\boldsymbol{J}^{\mathrm{T}}[\delta X|\delta\phi]^{\mathrm{T}}. \tag{18}$$

Example 5.1 A 3R PLANAR MANIPULATOR

The manipulator and pertinent dimensions and variables are shown in Figure 25.

Let $\boldsymbol{a}_1$ denote the position vector of X relative to 0, $\boldsymbol{a}_2$ the position vector of Y relative to X and $\boldsymbol{h}$ the position vector of P relative to Y (see Figure 25). Let $\boldsymbol{\omega}^2$, $\boldsymbol{\omega}^3$, and $\boldsymbol{\omega}^4$ denote the angular velocities in N (base) of links 2,

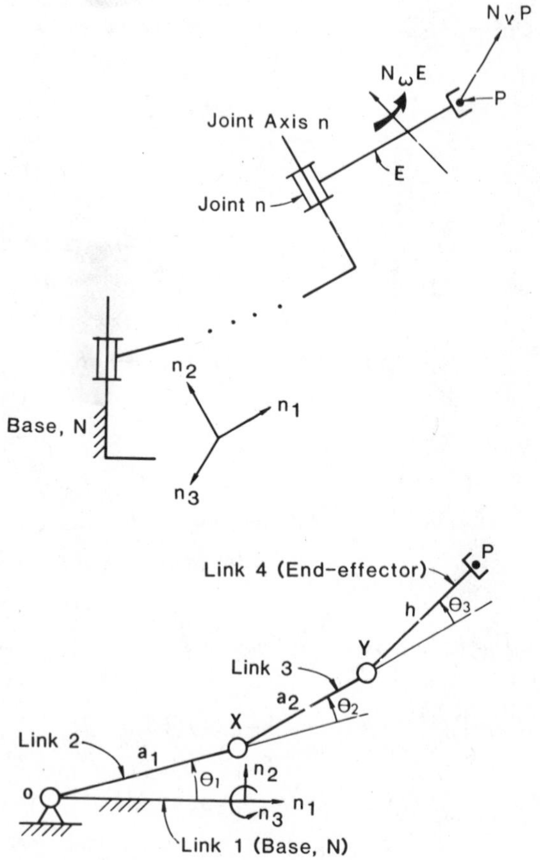

Figure 25 (Example 5.1).

3, and 4, respectively, and let $(\boldsymbol{n}_1, \boldsymbol{n}_2, \boldsymbol{n}_3)$ be unit vectors fixed in N. We then have the following relationships:

$$\omega^2 = \dot{\theta}_1 \, \boldsymbol{n}_3, \tag{a}$$

$$\omega^3 = (\dot{\theta}_1 + \dot{\theta}_2) \, \boldsymbol{n}_3, \tag{b}$$

$${}^N\boldsymbol{\omega}^E = \omega^4 = (\dot{\theta}_1 + \dot{\theta}_2 + \dot{\theta}_3) \, \boldsymbol{n}_3, \tag{c}$$

$$\boldsymbol{a}_1 = a_1(c\theta_1 \, \boldsymbol{n}_1 + s\theta_1 \, \boldsymbol{n}_2), \tag{d}$$

$$\boldsymbol{a}_2 = a_2\Big[(c\,(\theta_1+\theta_2) \, \boldsymbol{n}_1 + s\,(\theta_1+\theta_2) \, \boldsymbol{n}_2)\Big], \tag{e}$$

$$\boldsymbol{h} = h\Big[(c\,(\theta_1+\theta_2+\theta_3) \, \boldsymbol{n}_1 + s\,(\theta_1+\theta_2+\theta_3) \, \boldsymbol{n}_2)\Big], \tag{f}$$

where $s\,(\theta_i+\theta_j)$ is an abbreviation for sin $(\theta_i+\theta_j)$, etc., and a_1, a_2, and h are the magnitudes of $\boldsymbol{a}_1$, $\boldsymbol{a}_2$, and $\boldsymbol{h}$ respectively.

The velocities of points, X,Y,P (Figure 25) can be written as

$$\begin{aligned} \boldsymbol{V}^X &= \boldsymbol{\omega}^2 \times \boldsymbol{a}_1 \\ &\underset{(a,d)}{=} \dot{\theta}_1 \, a_1 \, (c\theta_1\boldsymbol{n}_2 - s\theta_1\boldsymbol{n}_1) \end{aligned} \tag{g}$$

$$\begin{aligned} \boldsymbol{V}^Y &= \boldsymbol{V}^X + \boldsymbol{\omega}^3 \times \boldsymbol{a}_2 \\ &\underset{(g,b,e)}{=} \dot{\theta}_1 \, [\boldsymbol{n}_1 \, \{-a_1 s\theta_1 - a_2 s\,(\theta_1+\theta_2)\} + \boldsymbol{n}_2 \, \{a_1 c\theta_1 \\ &\quad + a_2 c\,(\theta_1+\theta_2)\}] + \dot{\theta}_2 \, [-a_2 s\,(\theta_1+\theta_2) \, \boldsymbol{n}_1 \\ &\quad + a_2 c\,(\theta_1+\theta_2) \, \boldsymbol{n}_2] \end{aligned} \tag{h}$$

$$\boldsymbol{V}^P = \boldsymbol{V}^Y + {}^B\boldsymbol{\omega}^E \times \boldsymbol{h}$$

$$\begin{aligned} \boldsymbol{V}^P &\underset{(h,c,f)}{=} \dot{\theta}_1[\boldsymbol{n}_1\{-a_1 s\theta_1 - a_2 s\,(\theta_1+\theta_2) - hs\,(\theta_1+\theta_2+\theta_3)\} \\ &\quad + \boldsymbol{n}_2\{a_1 c\theta_1 + a_2 c\,(\theta_1+\theta_2) + hc\,(\theta_1+\theta_2+\theta_3)\}] \\ &\quad + \dot{\theta}_2[\boldsymbol{n}_1\{-a_2 s\,(\theta_1+\theta_2) - hs\,(\theta_1+\theta_2+\theta_3)\} \\ &\quad + \boldsymbol{n}_2\{a_2 c\,(\theta_1+\theta_2) + hc\,(\theta_1+\theta_2+\theta_3)\}] \\ &\quad + \dot{\theta}_3[\boldsymbol{n}_1\{-hs\,(\theta_1+\theta_2+\theta_3)\} + \boldsymbol{n}_2\{hc\,(\theta_1+\theta_2+\theta_3)\}]. \end{aligned} \tag{i}$$

In this example, the joint variable $q_i = \theta_i$, $i=1,2,3$. Hence,

$$\dot{q}_i = \dot{\theta}_i \quad i=1,2,3, \tag{k}$$

The partial velocities and partial angular velocities in equations (1) and (2) can be determined by inspection from equations (i) and (c). For example,

$$\begin{aligned} V_3^P &= \text{coeff. of } \dot{\theta}_3 \text{ in equation (i)} \\ &= \boldsymbol{n}_1\{-hs\,(\theta_1+\theta_2+\theta_3)\}+\boldsymbol{n}_2\{hc\,(\theta_1+\theta_2+\theta_3)\} \,, \end{aligned} \tag{l}$$

and, from equation (c),

$$\boldsymbol{\omega}_1^E = \boldsymbol{\omega}_2^E = \boldsymbol{\omega}_3^E = \boldsymbol{n}_3. \tag{m}$$

For planar manipulators, equations (5) to (8) can be replaced with

$$\boldsymbol{n}_i\cdot\boldsymbol{V} \underset{(5)}{=} \sum_{r=1}^{n} (\boldsymbol{n}_i\cdot\boldsymbol{V}_r)\,\dot{q}_r; \qquad (i=1,2), \tag{n}$$

$$\boldsymbol{n}_3\cdot\boldsymbol{\omega}^E = \omega_3 \underset{(6)}{=} \sum_{r=1}^{n} (\boldsymbol{n}_3\cdot\boldsymbol{\omega}_r)\,\dot{q}_r, \tag{o}$$

$$J_{ir} \underset{(7a)}{=} \boldsymbol{n}_i\cdot\boldsymbol{V}_r; \qquad (i=1,2); \qquad (r=1,...,n) \tag{p}$$

$$J_{3r} \underset{(7b)}{=} \boldsymbol{n}_3\cdot\boldsymbol{\omega}_r^E; \qquad (r=1,...,n), \tag{q}$$

$$v_i \underset{(8a)}{=} \sum_{r=1}^{n} J_{ir}\,\dot{q}_r; \qquad (i=1,2), \tag{r}$$

$$\omega \underset{(8b)}{=} \sum_{r=1}^{n} J_{3r}\,\dot{q}_r. \tag{s}$$

The elements J_{ij} are as follows:

$$\begin{aligned} J_{11} &\underset{(p,i)}{=} -[a_1 s\theta_1+a_2 s\,(\theta_1+\theta_2)+hs\,(\theta_1+\theta_2+\theta_3)] \\ J_{12} &= -[a_2 s\,(\theta_1+\theta_2)+hs\,(\theta_1+\theta_2+\theta_3)] \\ J_{13} &= -hs\,(\theta_1+\theta_2+\theta_3) \\ J_{21} &\underset{(p,i)}{=} a_1 c\theta_1+a_2 c\,(\theta_1+\theta_2)+hc\,(\theta_1+\theta_2+\theta_3) \\ J_{22} &= a_2 c\,(\theta_1+\theta_2)+hc\,(\theta_1+\theta_2+\theta_3) \end{aligned} \tag{t}$$

$$J_{23} = hc\,(\theta_1+\theta_2+\theta_3)$$

$$J_{31} \underset{(q,m)}{=} 1$$

$$J_{32} = 1$$

$$J_{33} = 1$$

We can therefore write

$$\begin{bmatrix} v_1 \\ v_2 \\ \omega \end{bmatrix} \underset{(9)}{=} \boldsymbol{J} \begin{bmatrix} \dot{\theta}_1 \\ \dot{\theta}_2 \\ \dot{\theta}_3 \end{bmatrix}, \tag{u}$$

or

$$\begin{bmatrix} \delta x_1 \\ \delta x_2 \\ \delta \phi \end{bmatrix} \underset{(13)}{=} \boldsymbol{J} \begin{bmatrix} \delta\theta_1 \\ \delta\theta_2 \\ \delta\theta_3 \end{bmatrix}, \tag{v}$$

where $\delta\theta_i = \dot{\theta}_i \Delta t$, $\delta x_i = v_i \Delta t$, $(i{=}1,2)$, and $\delta\phi = \omega\Delta t$, and the elements of $\boldsymbol{J}$ are given be equation (t). For the position $\theta_2 = \theta_3 = 0$, (v) becomes

$$\begin{bmatrix} \delta x_1 \\ \delta x_2 \\ \delta \phi \end{bmatrix} = -s\theta_1 \begin{bmatrix} (a_1+a_2+h) & (a_2+h) & h \\ (a_1+a_2+h) & (a_2+h) & h \\ 1 & 1 & 1 \end{bmatrix} \begin{bmatrix} \delta\theta_1 \\ \delta\theta_2 \\ \delta\theta_3 \end{bmatrix}. \tag{w}$$

Note that the Jacobian, $\boldsymbol{J}$, is singular for this extreme position (see Section 4, Example 4.2).

6. STATICS

In this section we deal with the following issues:

1. Given a load (force and torque) on the end-effector, what are the joint torques and the joint forces needed to maintain a state of static equilibrium.
2. Given a set of torque and force measurements at the driven joints, what is the load (force and torque) at the end-effector for a state of static equilibrium.

The above issues are important in the design and force-control of manipulators [23]. When manipulators move slowly, dynamic effects can be neglected and control can be based on purely static considerations. This type of situation almost always pertains at the beginning and end of a motion as an object is being moved from one position to another.

Nominally, the manipulator is assumed to be in a state of static equilibrium under the action of the gravitational forces exerted on the links by the earth. An external load is then applied to the manipulator and additional (over the zero external load case) joint forces and torques necessary to maintain a state of static equilibrium are required.

The condition for static equilibrium for an n d-o-f system is

$$F_r = 0; \qquad (r = 1,..,n), \tag{1}$$

where F_r is the generalized active force [12].

The generalized active force is defined as follows [12]:

Consider a system S of ν particles, and let $\boldsymbol{R}_i$ be the resultant of all contact and body forces acting on a generic particle P_i of S; then the rth generalized active force F_r for S is defined as

$$F_r = \sum_{i=1}^{\nu} \boldsymbol{V}_r^{P_i} \cdot \boldsymbol{R}_i; \qquad (r=1,..,n),$$

where $\boldsymbol{V}_r^{P_i}$ is the rth partial velocity of P_i (see Section 5.1) in a Newtonian Reference Frame N.

If R is a rigid body belonging to S, then the following fact is useful in constructing expressions for F_r [14]:

If a set of contact and/or body forces acting on R is equivalent to a couple of torque $\boldsymbol{T}$ together with a force $\boldsymbol{R}$ applied at a point Q of R, then $(F_r)_R$, the contribution of this set of forces to F_r, is given by

$$(F_r)_R = \boldsymbol{\omega}_r \cdot \boldsymbol{T} + \boldsymbol{V}_r^Q \cdot \boldsymbol{R}, \tag{2}$$

where $\boldsymbol{\omega}_r$ is the rth partial angular velocity of R in N and $\boldsymbol{V}_r^Q$ is the $r-th$ partial velocity of point Q in N.

It is possible to replace the external load on the end-effector, E, by a couple of torque $\boldsymbol{T}^E$ together with a force $\boldsymbol{F}^E$ applied to E at a point P designated as the center of the end-effector. The contribution of the external load to the generalized active force F_r, denoted as $(F_r)_E$, is given by

$$(F_r)_E \underset{(2)}{=} \boldsymbol{\omega}_r^E \cdot \boldsymbol{T}^E + \boldsymbol{V}_r^P \cdot \boldsymbol{F}^E, \tag{3}$$

where $\boldsymbol{V}_r^P$ is the rth partial velocity of P in the base N, and $\boldsymbol{\omega}_r^E$ is the rth partial angular velocity of E in N. (Note that the base N is assumed to be a Newtonian reference frame.)

If $\boldsymbol{n}_1,\boldsymbol{n}_2,\boldsymbol{n}_3$ is a dextral orthogonal set of unit vectors fixed in the base, and

$$F_i \triangleq \boldsymbol{F}^E \cdot \boldsymbol{n}_i; \qquad (1=1,2,3)$$

$$T_{i-3} \triangleq \boldsymbol{T}^E \cdot \boldsymbol{n}_{i-3}; \qquad (i=4,5,6)$$

$$J_{ri}{}' \triangleq \boldsymbol{V}_r^P \cdot \boldsymbol{n}_i; \qquad (i=1,2,3);\ (r=1,...,n)$$

$$J_{ri}{}' \triangleq \boldsymbol{\omega}_r^E \cdot \boldsymbol{n}_{i-1}; \qquad (i=4,5,6);\ (r=1,...,n) \tag{3'}$$

then

$$(F_r)_E \underset{(3,3')}{=} \sum_{i=1}^{3} (F_i\, J_{ri}{}') + \sum_{i=4}^{6} T_{i-3}\, J_{ri}{}'; \qquad (r=1,...,n). \tag{4}$$

The set of contact forces exerted by link $i+1$ on link i, $i=1,...,n$, can be replaced by a couple of torque $\boldsymbol{T}^{i+1/i}$ together with a force $\boldsymbol{F}^{i+1/i}$ applied to link i at a point $Q_i{}'$ (fixed in link i) situated on joint axis i. Similarly, the set of contact forces exerted by link i on link $i+1$ can be replaced by a couple of torque $\boldsymbol{T}^{i/i+1}$ $(= -\boldsymbol{T}^{i+1/i})$ together with a force $\boldsymbol{F}^{i/i+1}$ $(= -\boldsymbol{F}^{i+1/i})$ applied to link $i+1$ at point Q'_i (fixed in link $i+1$) coincident with Q_i. The contribution, denoted by $(F_r)_i$, of $(\boldsymbol{F}^{i+1/i}, \boldsymbol{T}^{i+1/i}, \boldsymbol{F}^{i/i+1}, \boldsymbol{T}^{i/i+1})$ to the generalized active force F_r is given by

$$(F_r)_i \underset{(2)}{=} \boldsymbol{\omega}_r^i \cdot \boldsymbol{T}^{i+1/i} + \boldsymbol{V}_r^{Q_i} \cdot \boldsymbol{F}^{i+1/i}$$

$$+ \boldsymbol{\omega}_r^{i+1} \cdot \boldsymbol{T}^{i/i+1} + \boldsymbol{V}_r^{Q_i'} \cdot \boldsymbol{F}^{i/i+1}; \ (i=1,...,6);\ (r=1,...,n)$$

or

$$(F_r)_i = (\boldsymbol{\omega}_r^i - \boldsymbol{\omega}_r^{i+1}) \cdot \boldsymbol{T}^{i+1/i}$$

$$+ (\boldsymbol{V}_r^{Q_i} - \boldsymbol{V}_r^{Q_i'}) \cdot \boldsymbol{F}^{i+1/i}; \ (i=1,...,6);\ (r=1,...,n). \tag{5}$$

For a revolute joint,

$$\boldsymbol{V}^{Q_i'} = \boldsymbol{V}^{Q_i} \Rightarrow \boldsymbol{V}_r^{Q_i'} = \boldsymbol{V}_r^{Q_i}\ ;\ (r=1,..,n), \tag{6}$$

$$\boldsymbol{\omega}^{i+1} = \boldsymbol{\omega}^i + \dot{q}_i \hat{\boldsymbol{s}}_1 \Rightarrow \boldsymbol{\omega}_r^{i+1} = \boldsymbol{\omega}_r^i + \hat{\boldsymbol{s}}_i\, \delta_{ir};\ (r=1,...,n), \tag{7}$$

where $\hat{\boldsymbol{s}}_i$ is a unit vector along the joint axis i and $\dot{q}_i = \dot{\theta}_i$ (θ_i being the joint variable "at" axis i); $\delta_{ir} = 1$, for $i = r$ and $\delta_{ir} = 0$, for $i \neq r$. When these relations are used, equation (5) assumes the following form (for a revolute joint):

$$(F_r)_i \underset{(5,6,7)}{=} \boldsymbol{T}^{i+1/i} \cdot (-\hat{\boldsymbol{s}}_i\ \delta_{ir}). \tag{8}$$

If axis k is a prismatic joint axis, then

$$\boldsymbol{V}^{Q_i'} = \boldsymbol{V}^{Q_i} + \dot{q}_i \, \hat{\boldsymbol{s}}_i \Rightarrow \boldsymbol{V}_r^{Q_i'} = \boldsymbol{V}_r^{Q_i} + \hat{\boldsymbol{s}}_i \, \delta_{ir}; \; (r=1,...,n), \tag{9}$$

$$\boldsymbol{\omega}^{i+1} = \boldsymbol{\omega}^i \Rightarrow \boldsymbol{\omega}_r^{i+1} = \boldsymbol{\omega}_r^i; \; (r=1,...,n), \tag{10}$$

where $\hat{\boldsymbol{s}}_i$ is a unit vector along the joint axis i, and $\dot{q}_i = \dot{s}_i$ (s_i being the joint variable associated with axis i) and δ_{ir} is as previously defined. Using these relations, equation (5) assumes the following form (for a prismatic joint):

$$(F_r)_i \underset{(5,9,10)}{=} \boldsymbol{F}^{i+1/i} \cdot (-\hat{\boldsymbol{s}}_i \, \delta_{ir}). \tag{11}$$

If we define τ_i as

$$\tau_i = \boldsymbol{T}^{i+1/i} \cdot \hat{\boldsymbol{s}}_i, \tag{12}$$

for a revolute joint, and as

$$\tau_i = \boldsymbol{F}^{i+1/i} \cdot \hat{\boldsymbol{s}}_i, \tag{13}$$

for a prismatic joint, then (8) and (11) can be written in the form

$$(F_r)_i \underset{(8,12);(11,13)}{=} -\tau_i \, \delta_{ir}; \; (i=1,...,6); \; (r=1,...,n). \tag{14}$$

[Note that τ_i is the measure number, along joint axis i, of either a torque $\boldsymbol{T}^{i+1/i}$ (in the case of a revolute joint) or a force $\boldsymbol{F}^{i+1/i}$ (in the case of a prismatic joint)].

If $(F_r)_M$ is the contribution to the generalized active force, F_r, of the contact forces that the links exert on each other, we obtain from (14):

$$(F_r)_M = \sum_{i=1}^{n} (F_r)_i \; ; \; (r=1,...,n)$$

$$(F_r)_M = \sum_{i=1}^{N} -\tau_i \, \delta_{ir} \; ; \; (r=1,...,n) \tag{15}$$

$$(F_r)_M = -\tau_r \; ; \; (r=1,...,n) \tag{16}$$

The generalized active force, F_r, can now be written as

$$F_r = (F_r)_M + (F_r)_E \tag{17}$$

or,

$$F_r \underset{(17,16,4)}{=} -\tau_r + \sum_{i=1}^{3} (J_{ri}' \, F_i) + \sum_{i=4}^{6} (J_{ri}' \, T_{(i-3)}) \tag{18}$$

Static equilibrium, i.e., equation (1), requires that

$$-\tau_r + \sum_{i=1}^{3} (J_{ri}' F_i) + \sum_{i=4}^{6} (J_{ri}' T_{(i-3)}) \underset{(1)}{=} 0 \tag{19}$$

Equation (19) can be written in the matrix form

$$\begin{bmatrix} \tau_1 \\ \tau_2 \\ \cdot \\ \cdot \\ \cdot \\ \tau_6 \end{bmatrix} \underset{(19)}{=} \boldsymbol{J}^{\mathrm{T}} \begin{bmatrix} F_1 \\ F_2 \\ F_3 \\ T_1 \\ T_2 \\ T_3 \end{bmatrix} \tag{20}$$

where the matrix $\boldsymbol{J}^{\mathrm{T}}$ (whose (r,i) element is J_{ri}' defined by equation (3') is the transpose of the Jacobain matrix $\boldsymbol{J}$ (whose (i,r) element, J_{ir}, is defined by equation (5.7)).

Given the $\boldsymbol{n}_1,\boldsymbol{n}_2,\boldsymbol{n}_3$ measure numbers of the equivalent force and torque applied at point P of E, viz. $(F_1,F_2,F_3,T_1,T_2,T_3)$, equation (20) can be used to compute the "joint-torques" and "joint-forces" τ_i defined by equations (12) and (13).

Alternatively, given a set of forces and torques applied at the joints, the corresponding "loads" $(F_1,F_2,F_3,T_1,T_2,T_3)$, experienced by the hand, for a state of static equilibrium can be obtained from (20) as

$$\begin{bmatrix} F_1 \\ F_2 \\ F_3 \\ T_1 \\ T_2 \\ T_3 \end{bmatrix} = (\boldsymbol{J}^{\mathrm{T}})^{-1} \begin{bmatrix} \tau_1 \\ \tau_2 \\ \tau_3 \\ \tau_4 \\ \tau_5 \\ \tau_6 \end{bmatrix} \tag{21}$$

Equation (20) is used in determining the motor or actuator torques needed to exert a given load at the end effector, whereas, equation (21) is useful in predicting the effects of the various joint torques on the applied load.

7. DYNAMICS

A dynamic analysis of a manipulator is useful for the following purposes:

1. It determines the joint forces and torques required to produce specified end-effector motions (the direct dynamic problem).
2. It produces a mathematical model which simulates the motion of the manipulator under various loading conditions (the inverse dynamic problem) and/or control schemes.

3. It provides a dynamic model for use in the control of the actual manipulator.

The classical methods of Lagrange and Newton can both be applied to manipulators [11,25,10, etc.]; however it is often necessary to solve the dynamic equations in real time (or "on-line"), and in these cases computational speed and efficiency are prime concerns in choosing a suitable dynamic formulation. With this in mind the computational efficiencies of these methods has been compared in [10] and [24]. Both the Lagrangian and Newton-Euler approach are very laborious and hence not well suited for generating the dynamic equations in explicit form. One is therefore forced to generate implicit forms of the equations numerically. Extreme care must be taken with the bookkeeping of terms developed numerically or else all the methods will involve large amounts of unnecessary computations [14]. In addition to the Lagrangian and Newton-Euler methods there are several others which are not as widely known. Several of these are discussed in a recent book [29]. Here we will describe yet another method, one that is based on the use of Kane's Dynamical Equations [13]. This method facilitates the generation of dynamic equations in an *explicit* and computationally efficient form; recently Kane's approach has been applied to the dynamic analysis of manipulators [14].

We outline a procedure for generating Kane's dynamical equations and then apply the procedure to a simple example (a 2-d-o-f manipulator). Since several of the concepts may be unfamiliar, the reader may find it helpful to study the procedure and the example concurrently. The application of the method to a more complex problem (viz., the 2RP3R "Stanford Arm") is the subject of [14].

Kane's dynamical equations [12,14] for an n-d-o-f holonomic dynamic system are

$$F_r + F_r^* = 0; \qquad (r=1,...,n) \tag{1}$$

where F_r is the generalized active force and F_r^* is the generalized inertia force. A procedure for generating these equations for an n-d-o-f manipulator is as follows:

1. Define n *generalized speeds* for the manipulator. Generalized speeds are quantities associated with the motion of a system and can be introduced as follows [14]:

$$u_r \triangleq \sum_{s=1}^{n} A_{rs}\, \dot{q}_s + B_r; \qquad (r=1,...,n) \tag{2}$$

   where A_{rs} and B_r are functions of the joint variables $(q_1,..,q_n)$ and time t; useful candidates for u_r are angular velocity measure numbers, velocity measure numbers or simply $\dot{q}_r$.

2. Obtain expressions for the angular velocities ${}^N\boldsymbol{\omega}^k$†, $k=2,...,n+1$, of each

† As before ${}^N\boldsymbol{\omega}^k$ denotes the velocity of link k in N; henceforward the superscript N will be dropped.

of the n moving links with respect to the base, N (link 1), which is assumed to be a Newtonian reference frame.

3. Determine the *partial angular velocities* $\boldsymbol{\omega}_r^k$, $(r=1,...,n)$, of the kth link $(k=2,...,n+1)$ in N.

   The rth partial velocity $\boldsymbol{\omega}_r^k$ of link k in reference frame N, is defined by the following equation:

$$\boldsymbol{\omega}^k = \sum_{r=1}^{n} \boldsymbol{\omega}_r^k \, u_r + \boldsymbol{\omega}_t^k. \tag{3}$$

   $\boldsymbol{\omega}_r^k$ and $\boldsymbol{\omega}_t^k$ are functions of $q_1,...q_n$, and t; the $\boldsymbol{\omega}_r^k$ are simply the coefficients of the generalized speeds u_r in the expression for $\boldsymbol{\omega}^k$.

4. Obtain expressions for (i) the *velocities* $\boldsymbol{V}^k$ $(k=2,...,n+1)$, in N, of the mass-centers of each of the moving links and (ii) the velocity $\boldsymbol{V}^P$ in N of the "center" of the end-effector P.

5. Determine the *partial velocities* $\boldsymbol{V}_r^k$, $(r=1,...,n)$, in N, of the mass-centers of the kth link $(k=2,...,n+1)$ and the partial velocities $\boldsymbol{V}_r^P$ of P in N.

   The rth partial velocity $\boldsymbol{V}_r^Q$ of a point Q in reference frame N, is defined by the following equation:

$$\boldsymbol{V}^Q = \sum_{r=1}^{n} \boldsymbol{V}_r^Q \, u_r + \boldsymbol{V}_t^Q, \tag{4}$$

   Where $\boldsymbol{V}_r^Q$ and $\boldsymbol{V}_t^Q$ are functions of $q_1,...,q_n$, and t; the $\boldsymbol{V}_r^Q$ are simply the coefficients of the generalized speeds u_r in the expression for $\boldsymbol{V}^Q$.

   Note: The definitions of the partial velocities and partial angular velocities given above coincide with those given in Section 5 (equations (5.1) and (5.2)) if we set u_r (in equations (3) and (4)) equal to $\dot{q}_r$.

6. Obtain expressions for the *angular accelerations* $\boldsymbol{\alpha}^k$ $(k=2,...,n+1)$, in N, of the n moving links.

7. Obtain expressions for (i) the *accelerations* $\boldsymbol{a}^k$ $(k=2,...,n+1)$, in N, of the mass centers of the moving links and (ii) the acceleration $\boldsymbol{a}^P$, in N, of the center of the end-effector.

8. Determine the generalized inertia forces F_r^* $(r=1,...,n)$.

   For a system S comprising λ particles, the generalized inertia force F_r^* for S is defined as [14]

$$F_r = \sum_{i=1}^{\lambda} (-m_i \boldsymbol{a}^{P_i}) \cdot \boldsymbol{V}_r^{P_i}; \qquad (r=1,...,n) \tag{5}$$

   where m_i, $\boldsymbol{a}^{P_i}$ and $\boldsymbol{V}_r^{P_i}$ are the mass, acceleration and rth partial velocity of the generic particle P_i of S.

If R is a rigid body belonging to S, then the contribution to F_r made by R, denoted by $(F_r^*)_R$, is given by [12]

$$(F_r^*)_R = \boldsymbol{\omega}_r \cdot \boldsymbol{T}^* + \boldsymbol{V}_r \cdot \boldsymbol{R}^*; \quad (r=1,...,n) \tag{6}$$

where $\boldsymbol{\omega}_r$ and $\boldsymbol{V}_r$ are, respectively, the rth partial angular velocity of R in N and the rth partial velocity of the mass center of R in N. $\boldsymbol{T}^*$ and $\boldsymbol{R}^*$ are, respectively, the inertia torque for R and the inertia force for R. $\boldsymbol{T}^*$ and $\boldsymbol{R}^*$ are given by

$$\boldsymbol{T}^* = -\boldsymbol{\alpha} \cdot \boldsymbol{I} - \boldsymbol{\omega} \times \boldsymbol{I} \cdot \boldsymbol{\omega} \tag{7}$$

and

$$\boldsymbol{R}^* = -M\boldsymbol{a}^* \tag{8}$$

where $\boldsymbol{\omega}$ and $\boldsymbol{\alpha}$ are the angular velocity and angular acceleration, respectively, of R in N; $\boldsymbol{a}^*$ is the acceleration of the mass center of R in N; M is the mass of R and $\boldsymbol{I}$ is the central inertia dyadic of R; $\boldsymbol{I}$ is given by

$$\boldsymbol{I} = I_1\boldsymbol{n}_1\boldsymbol{n}_1 + I_2\boldsymbol{n}_2\boldsymbol{n}_2 + I_3\boldsymbol{n}_3\boldsymbol{n}_3 \tag{9}$$

where $\boldsymbol{n}_1$, $\boldsymbol{n}_2$, and $\boldsymbol{n}_3$ are a dextral orthogonal set of unit vectors fixed in R and parallel to the central principal axes of inertia of R, and I_1, I_2, I_3 are the central principal moments of inertia of R.

Equations (6) to (9) suggest the following procedure for forming the generalized active forces:

a. determine the inertia force $(\boldsymbol{R}^*)_k$ for the kth link, $(k=2,...,n+1)$, as follows:

$$(\boldsymbol{R}^*)_k \underset{(8)}{=} -M_k\boldsymbol{a}_k^*; \qquad k=2,...,n+1 \tag{10}$$

where M_k is the mass of the kth link and $\boldsymbol{a}_k^*$ is the acceleration of the mass center of the kth link.

b. Form the central inertia dyadic $\boldsymbol{I}_k$ for the kth link from the expression

$$\boldsymbol{I}_k \underset{(9)}{=} I_1^k\boldsymbol{n}_1^k\boldsymbol{n}_1^k + I_2^k\boldsymbol{n}_2^k\boldsymbol{n}_2^k + I_3^k\boldsymbol{n}_3^k\boldsymbol{n}_3^k; \quad (k=1,...,n) \tag{11}$$

where I_i^k, $(i=1,2,3)$, are the central principal moments of inertia of link k and $\boldsymbol{n}_i^k$, $(i=1,2,3)$, are unit vector fixed in k and parallel to the central principal axes of inertia for k.

c. Form the inertia torque $(T^*)_k$ for the kth link from

$$(\boldsymbol{T}^*)_k \underset{(7,11)}{=} -\boldsymbol{\alpha}^k \cdot \boldsymbol{I}_k - \boldsymbol{\omega}^k \times \boldsymbol{I}_k \cdot \boldsymbol{\omega}^k \tag{12'}$$

where $\boldsymbol{\omega}^k$ and $\boldsymbol{\alpha}^k$ have been determined in steps 2 and 6 respectively.

If $\boldsymbol{\omega}^k$ and $\boldsymbol{\alpha}^k$ are expressed as

$$\boldsymbol{\omega}^k = \omega_1^k \boldsymbol{n}_1^k + \omega_2^k \boldsymbol{n}_2^k + \omega_3^k \boldsymbol{n}_3^k$$

$$\boldsymbol{\alpha}^k = \alpha_1^k \boldsymbol{n}_1^k + \alpha_2^k \boldsymbol{n}_2^k + \alpha_3^k \boldsymbol{n}_3^k$$

then (12′) becomes

$$\begin{aligned}(\boldsymbol{T}_R^*)_k &= [\omega_2^k \omega_3^k \, (I_2^k - I_3^k) - \alpha_1^k I_1^k] \, \boldsymbol{n}_1^k \\ &+ [\omega_3^k \omega_1^k \, (I_3^k - I_1^k) - \alpha_2^k I_2^k] \, \boldsymbol{n}_2^k \\ &+ [\omega_1^k \omega_2^k \, (I_1^k - I_2^k) - \alpha_3^k I_3^k] \, \boldsymbol{n}_3^k \end{aligned} \tag{12}$$

d. The contribution to the generalized inertia force F_r^* made by link k, denoted by $(F_r^*)_k$, is given by

$$(F_r^*)_k \underset{(6,10,12')}{=} \boldsymbol{\omega}_r^k \cdot (\boldsymbol{T}^*)_k + \boldsymbol{V}_r^k \cdot (\boldsymbol{R}^*)_k \tag{13}$$

where the $\boldsymbol{\omega}_r^k$ and $\boldsymbol{V}_r^k$ have been determined in Steps 3 and 5 respectively.

e. The contribution to the generalized inertia force F_r made by the load, denoted by $(F_r^*)_L$, is given by

$$(F_r^*)_L = -M_L \, \boldsymbol{a}^P \cdot \boldsymbol{V}_r^P \tag{14}$$

where we have assumed that the load can be approximated as a mass M_L concentrated at the "center" P of the end-effector; $\boldsymbol{a}^P$ and $\boldsymbol{V}_r^P$ are obtained in Steps 7 and 4 respectively.†

f. The generalized inertia forces F_r^*, $(r=1,...,n)$, are given by

$$F_r^* \underset{(13,14)}{=} \sum_{k=2}^{n+1} (F_r^*)_k + (F_r^*)_L \; ; \qquad (r=1,...,n) \tag{15}$$

9. Determine the generalized active forces F_r.

For the system S of ν particles, let R_i be the resultant of all contact and body forces acting on a generic particle P_i of S; the rth *generalized active force* for S, denoted by F_r, is then given by [12,14]

$$F_r \triangleq \sum_{i=1}^{\nu} \boldsymbol{V}^{P_i} \cdot \boldsymbol{R}_i \; ; \qquad (r=1,...n). \tag{16}$$

If R is a rigid body belonging to S, then the total contribution to F_r of all gravitational forces exerted by particles of R on each other is equal to zero. If a set of contact and/or body forces acting on R is equivalent to a couple of torque $\boldsymbol{T}$ together with a force $\boldsymbol{R}$ applied at a point Q of

† If the load has a non-negligible inertia dyadic we would add a term as in (13).

R, the contribution, denoted by $(F_r)_R$, of this set of forces to F_r, is given by [12,14]

$$(F_r)_R = \boldsymbol{\omega}_r \cdot \boldsymbol{T} + \boldsymbol{V}_r^Q \cdot \boldsymbol{R} \tag{17}$$

where $\boldsymbol{\omega}_r$ and $\boldsymbol{V}_r^Q$ are, respectively, the rth partial angular velocity of R in N and the rth partial velocity of Q in N.

For manipulators there are generally two types of forces which contribute to the generalized active force: contact forces at the joints and gravitational forces exerted on the links by the earth.

It is convenient in the dynamic analysis of manipulators to replace the set of contact forces exerted by link $k+1$ on link k, $k=1,...,n$, by a couple of torque $\boldsymbol{T}^{k+1/k}$ together with a force $\boldsymbol{F}^{k+1/k}$ applied to link k at a point Q_k (of k) situated on joint axis k.† Similarly the set of contact forces exerted by link k on link $k+1$ can be replaced by a couple of torque $\boldsymbol{T}^{k/k+1}$ $(= -\boldsymbol{T}^{k+1/k})$ together with a force $\boldsymbol{F}^{k/k+1}$ $(= -\boldsymbol{F}^{k+1/k})$ applied to link $k+1$ at a point $Q_k{}'$ (of $k+1$) coincident with Q_k. The contribution of $\boldsymbol{T}^{k+1/k}$, $\boldsymbol{F}^{k+1/k}$, $\boldsymbol{T}^{k/k+1}$, and $\boldsymbol{F}^{k/k+1}$ to the generalized active force F_r will be denoted by $(F_r)_k$ and is given by

$$\begin{aligned}(F_r)_k &\underset{(17)}{=} \boldsymbol{F}^{k+1/k}\cdot\boldsymbol{V}_r^{Q_k}+\boldsymbol{T}^{k+1/k}\cdot\boldsymbol{\omega}_r^{k}+\boldsymbol{F}^{k/k+1}\cdot\boldsymbol{V}_r^{Q_k''}+\boldsymbol{T}^{k/k+1}\cdot\boldsymbol{\omega}_r^{k+1} \\ &= \boldsymbol{F}^{k/k+1}\cdot(\boldsymbol{V}_r^{Q_k'}-\boldsymbol{V}_r^{Q_k})+\boldsymbol{T}^{k/k+1}\cdot(\boldsymbol{\omega}_r^{k+1}-\boldsymbol{\omega}_r^{k}); \quad (k=1,...,n).\end{aligned} \tag{18}$$

If axis k is the axis of a revolute joint, then

$$\boldsymbol{V}^{Q_k'} = \boldsymbol{V}^{Q_k} \Rightarrow \boldsymbol{V}_r^{Q_k''} = \boldsymbol{V}_r^{Q_k}$$

and (18) becomes

$$(F_r)_k = \boldsymbol{T}^{k/k+1} \cdot (\boldsymbol{\omega}_r^{k+1} - \boldsymbol{\omega}_r^{k}). \tag{19}$$

If axis k is the axis of a prismatic joint then,

$$\boldsymbol{\omega}^{k} = \boldsymbol{\omega}^{k+1} \Rightarrow \boldsymbol{\omega}_r^{k} = \boldsymbol{\omega}_r^{k+1}$$

and (18) becomes

$$(F_r)_k = \boldsymbol{F}^{k/k+1} \cdot (\boldsymbol{V}_r^{Q_k''} - \boldsymbol{V}_r^{Q_k}). \tag{20}$$

Furthermore, for a revolute joint,

$$\boldsymbol{\omega}^{k+1} = \boldsymbol{\omega}^{k} + \dot{q}_k\, \hat{\boldsymbol{s}}_k \tag{21}$$

while for a prismatic joint,

† In [14] Q is chosen to be the mass-center of the link; such a choice, in general, does not yield the simplest expressions for F_r.

$$V^{Q'} = V^{Q} + \dot{q}_k \, \hat{s}_k \tag{22}$$

where the joint variable $q_r = \theta_r$ in (21) and $q_r = s_r$ in (22), and $\hat{s}_k$ is a unit vector parallel to joint axis k. If the generalized speeds u_r are chosen such that $u_r = \dot{q}_r$, then

$$\omega_r^{k+1} \underset{(21)}{=} \omega_r^{k} + \hat{s}_k \, \delta_{kr} \; ; \qquad (r=1,...,n) \tag{23}$$

for a revolute joint, and

$$V_r^{Q_k''} \underset{(22)}{=} V_r^{Q_k} + \hat{s}_k \, \delta_{kr} \; ; \qquad (r=1,...,n) \tag{24}$$

for a prismatic joint, where

$$\delta_{kr} = \begin{cases} 1 \text{ if } k = r \\ 0 \text{ if } k \neq r \end{cases} \tag{25}$$

Defining

$$\tau_k \triangleq T^{k/k+1} \cdot \hat{s}_k \tag{26}$$

for a revolute joint, and

$$\tau_k \triangleq F^{k/k+1} \cdot \hat{s}_k \tag{27}$$

for a prismatic joint†, and using equations (23) and (24) equations (19) and (20) become

$$(F_r)_k \underset{(19,20,23,24,26,27)}{=} \tau_k \, \delta_{kr}; \qquad (r=1,...,n) \tag{28}$$

If $G_k = (-gM_k \hat{h})$ is the gravitational force exerted on link k $(k=2,..,n+1)$ by the earth, $\hat{h}$ being a unit vector directed vertically upward, then the contribution to (F_r) made by G_k, $(k=2,..,n+1)$, denoted by $(F_r)_G$, is given by

$$(F_r)_G = \sum_{k=2}^{n+1} (-gM_k \hat{h}) \cdot V_r^k; \qquad (r=1,...,n) \tag{29}$$

where V_r^k is the rth partial velocity of the mass center of link k (determined in Step 5). Similarly the contribution to F_r made by the gravitational force G_L exerted on the load by the earth is given by

$$(F_r)_L = (-gM_L \hat{h}) \cdot V_r^P; \qquad (r=1,...,n) \tag{30}$$

The generalized active force can now be obtained as

† Note that τ_k as defined in this section is the negative of τ_k as defined by equations (12) and (13) of Section 6.

$$(F_r) = \left[\sum_{k=1}^{n} (F_r)_k\right] + (F_r)_G + (F_r)_L. \quad (31)$$

For a revolute joint $(F_r)_k$ is given by (19), whereas for a prismatic joint it is given by (20). $(F_r)_G$ is given by (29) and $(F_r)_L$ is given by (30).

For the case where $u_r = \dot{q}_r$, (31) takes the form

$$(F_r) \underset{(31,28,29,30)}{=} \sum_{k=1}^{n} \tau_k\, \delta_{kr} + \sum_{k=2}^{n+1} (-gM_k\hat{\boldsymbol{h}}) \cdot \boldsymbol{V}_r^k - gM_L\hat{\boldsymbol{h}}\cdot\boldsymbol{V}_r^P$$

$$= \tau_r + (\sum_{k=2}^{n+1} - gM_k\hat{\boldsymbol{h}}\cdot\boldsymbol{V}_r^k) - gM_L\hat{\boldsymbol{h}}\cdot\boldsymbol{V}_r^P. \quad (32)$$

10. Obtain Kane's Dynamical equations:

$$F_r^* + F_r = 0; \qquad r=1,...,n \quad (33)$$

where F_r^* is determined in Step 8 and F_r in Step 9.

The dynamical behavior of a manipulator is thus defined by the $2n$ first-order differential equations [eqns. (2) and (33)] in the variables u_r, q_r, $(r=1,...,n)$.

EXAMPLE 7.1: To illustrate the use of the above procedure, we derive the dynamical equations of a 2-degree-of-freedom (2R) manipulator. We assume there is no applied load on the end-effector (Figure 26).

To help the reader, the example is solved in steps (1 to 10) corresponding to the steps (1 to 10) in the above procedure. The reader may find it convenient, at each stage of the example, to refer back to the relevant step in the above procedure.

For simplicity we adopt the following notation:

Links 2 and 3 will be denoted by A and B respectively (Figure 26); $(\boldsymbol{a}_1,\boldsymbol{a}_2,\boldsymbol{a}_3)$ and $(\boldsymbol{b}_1,\boldsymbol{b}_3,\boldsymbol{b}_3)$ are, respectively, dextral orthogonal sets

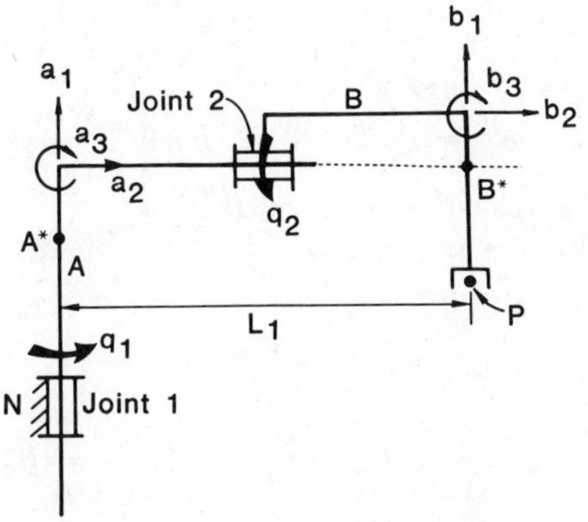

Figure 26 (Example 7.1).

fixed in A and B. (Link 1 will, as usual, be denoted by N). This notation avoids unnecessary superscripts/subscripts and also conforms with the notation used in [14]

Let joint variables q_1 and q_2 be the generalized coordinates describing the configuration of the system (Figure 26).

The sets $(\boldsymbol{a}_1, \boldsymbol{a}_2, \boldsymbol{a}_3)$ and $(\boldsymbol{b}_1, \boldsymbol{b}_2, \boldsymbol{b}_3)$ are related as follows

$$[\boldsymbol{b}_1 \boldsymbol{b}_2 \boldsymbol{b}_3] = [\boldsymbol{a}_1 \boldsymbol{a}_2 \boldsymbol{a}_3] \begin{bmatrix} c_2 & 0 & s_2 \\ 0 & 1 & 0 \\ -s_2 & 0 & c_2 \end{bmatrix} \tag{34}$$

where $c_2 \triangleq \cos q_2$ and $s_2 \triangleq \sin q_2$

1. Generalized speeds

We define generalized speeds u_1, u_2 as

$$u_1 \triangleq \dot{q}_1 \tag{35}$$

$$u_2 \triangleq \dot{q}_2. \tag{36}$$

2. Angular velocities of A and B in N

$$\boldsymbol{\omega}^A = \dot{q}_1 \, \boldsymbol{a}_1 \underset{(35)}{=} u_1 \, \boldsymbol{a}_1 \tag{37}$$

$$\boldsymbol{\omega}^B = \dot{q}_1 \, \boldsymbol{a}_1 + \dot{q}_2 \, \boldsymbol{a}_2 \tag{38}$$

$$\boldsymbol{\omega}^B \underset{(38,34,35,36)}{=} u_1(c_2 \boldsymbol{b}_1 + s_2 \boldsymbol{b}_3) + u_2 \boldsymbol{b}_2. \tag{39}$$

Defining

$$Z_1 \triangleq u_1 \, c_2, \; Z_2 \triangleq u_2, \; Z_3 \triangleq u_1 \, s_2 \tag{40}$$

we have

$$\boldsymbol{\omega}^B = Z_1 \, \boldsymbol{b}_1 + Z_2 \, \boldsymbol{b}_2 + Z_3 \, \boldsymbol{b}_3. \tag{41}$$

3. Partial angular velocities of A and B in N

The coefficients of u_1 and u_2 in (37) and (39) are the required partial angular velocities

$$\boldsymbol{\omega}_1^A \underset{(37)}{=} \boldsymbol{a}_1; \quad \boldsymbol{\omega}_2^A \underset{(37)}{=} \boldsymbol{0} \tag{42}$$

$$\boldsymbol{\omega}_1^B \underset{(39)}{=} c_2 \boldsymbol{b}_1 + s_2 \boldsymbol{b}_3; \quad \boldsymbol{\omega}_2^B \underset{(39)}{=} \boldsymbol{b}_2. \tag{43}$$

4. Velocities of A^* (mass center of A) and B^* (mass center of B) in N

$$\boldsymbol{V}^{A^*} = \boldsymbol{0} \tag{44}$$

$$\boldsymbol{V}^{B^*} = \boldsymbol{\omega}^A \times L_1\, \boldsymbol{a}_2$$

$$\boldsymbol{V}^{B^*} \underset{(37)}{=} u_1 L_1 \boldsymbol{a}_3 \tag{45}$$

$$\boldsymbol{V}^{B^*} = Z_4 \boldsymbol{a}_3 \tag{46}$$

where

$$Z_4 \triangleq u_1 L_1. \tag{47}$$

5. Partial velocities of A^* and B^* in N

The coefficients of u_1 and u_2 in (44) and (45) are the required partial velocities

$$\boldsymbol{V}_1^{A^*} \underset{(44)}{=} 0; \qquad \boldsymbol{V}_2^{A^*} \underset{(44)}{=} 0 \tag{48}$$

$$\boldsymbol{V}_1^{B^*} \underset{(45)}{=} L_1 \boldsymbol{a}_3; \quad \boldsymbol{V}_2^{B^*} \underset{(45)}{=} 0. \tag{49}$$

6. Angular accelerations of A and B in N

$$\boldsymbol{\alpha}^A = \frac{{}^N d}{dt}\, \boldsymbol{\omega}^A \underset{(37)}{=} \dot{u}_1 \boldsymbol{a}_1 \tag{50}$$

$$\boldsymbol{\alpha}^B = \frac{{}^N d}{dt}\, \boldsymbol{\omega}^B = \frac{{}^B d}{dt}\, \boldsymbol{\omega}^B \tag{51}$$

$$\underset{(39)}{=} (\dot{u}_1 c_2 - u_1 u_2 s_2)\, \boldsymbol{b}_1 + \dot{u}_2 \boldsymbol{b}_2 + (\dot{u}_1 s_2 + u_1 u_2 c_2)\, \boldsymbol{b}_3$$

or

$$\boldsymbol{\alpha}^B = (\dot{u}_1 c_2 + Z_5)\, \boldsymbol{b}_1 + \dot{u}_2 \boldsymbol{b}_2 + (\dot{u}_1 s_2 + Z_6)\, \boldsymbol{b}_3 \tag{52}$$

where

$$Z_5 \triangleq -u_1 u_2 s_2 \underset{(40)}{=} -Z_3 u_2 \tag{53}$$

$$Z_6 \triangleq u_1 u_2 c_2 \underset{(40)}{=} Z_1 u_2. \tag{54}$$

7. Accelerations of A^* and B^* in N

$$\boldsymbol{a}^{A^*} = \boldsymbol{0} \tag{55}$$

$$\boldsymbol{a}^{B^*} = \frac{d}{dt}(\boldsymbol{V}^{B^*}) \underset{(45)}{=} -L_1 u_1^2 \boldsymbol{a}_2 + L_1 \dot{u}_1 \boldsymbol{a}_3 \tag{56}$$

$$\boldsymbol{a}^{B^*} = Z_7 \boldsymbol{a}_2 + L_1 \dot{u}_1 \boldsymbol{a}_3 \tag{57}$$

where

$$Z_7 \triangleq -Lu_1^2 \underset{(47)}{=} -Z_4 u_1 \tag{58}$$

8. Generalized inertia forces, F_r^*
   a. Inertia forces $(\boldsymbol{R}^*)_A$ for A and $(\boldsymbol{R}^*)_B$ for B

$$(\boldsymbol{R}^*)_A = -m_A \boldsymbol{a}^{A*} \underset{(55)}{=} 0 \tag{59}$$

$$(\boldsymbol{R}^*)_B = -m_B \boldsymbol{a}^{B*} \underset{(57)}{=} -m_B(Z_7 \boldsymbol{a}_2 + L_1 \dot{u}_1 \boldsymbol{a}_3) \tag{60}$$

m_A and m_B are, respectively, the masses of A and B.

   b. Central inertia dyadics $\boldsymbol{I}_A$ (for A) and $\boldsymbol{I}_B$ (for B).

Assuming that $(\boldsymbol{a}_1\boldsymbol{a}_2\boldsymbol{a}_3)$ and $(\boldsymbol{b}_1\boldsymbol{b}_2\boldsymbol{b}_3)$ are, respectively, parallel to the central principal axes of inertia for A and B we have

$$\boldsymbol{I}_A = A_1 \boldsymbol{a}_1 \boldsymbol{a}_1 + A_2 \boldsymbol{a}_2 \boldsymbol{a}_2 + A_3 \boldsymbol{a}_3 \boldsymbol{a}_3 \tag{61}$$

$$\boldsymbol{I}_B = B_1 \boldsymbol{b}_1 \boldsymbol{b}_1 + B_2 \boldsymbol{b}_2 \boldsymbol{b}_2 + B_3 \boldsymbol{b}_3 \boldsymbol{b}_3 \tag{62}$$

where (A_1, A_2, A_3) and (B_1, B_2, B_3) are, respectively, the central principal moments of inertia of A and B.

   c. Inertia Torques $(\boldsymbol{T}^*)_A$ for A and $(\boldsymbol{T}^*)_B$ for B.

Using equation (12) we can write the following expressions for $(\boldsymbol{T}^*)_A$ and $(\boldsymbol{T}^*)_B$:

$$\boldsymbol{T}_A^* \underset{(12,37,50,61)}{=} -\dot{u}_1 A_1 \boldsymbol{a}_1 \tag{63}$$

$$T_B^* \underset{(12,39,52,62)}{=} (Z_8+\dot{u}_1 Z_9)\boldsymbol{b}_1+(Z_{10}-\dot{u}_2 B_2)\boldsymbol{b}_2+(Z_{11}+\dot{u}_1 Z_{12})\boldsymbol{b}_3 \tag{64}$$

where

$$Z_8 \triangleq Z_2 Z_3(B_2-B_3) - Z_5 B_1 \tag{65}$$

$$Z_9 \triangleq -c_2 B_1 \tag{66}$$

$$Z_{10} \triangleq Z_3 Z_1 (B_3-B_1) \tag{67}$$

$$Z_{11} \triangleq Z_1 Z_2 (B_1-B_2) - Z_6 B_3 \tag{68}$$

$$Z_{12} \triangleq -s_2 B_3. \tag{69}$$

d. Contributions of A and B to the generalized inertia forces.

The contribution, $(F_r^*)_A$, of A to the generalized inertia force F_r^* is given by

$$(F_1^*)_A \underset{(13)}{=} V_1^A \cdot R_A^* + \omega_1^A \cdot T_A^*$$

$$\underset{(48,49,42,63)}{=} (0)\ (0) + (a_1)\cdot(-\dot{u}_1 A_1 a_1) = -\dot{u}_1 A_1 \tag{70}$$

$$(F_2^*)_A = V_2^A \cdot R_A^* + \omega_2^A \cdot T_A^* \underset{(48,59,42,63)}{=} 0 \tag{71}$$

Similarly for B we obtain

$$(F_1^*)_B = (-m_B L_1^2 + Z_9 c_2 + Z_{12} s_2)\dot{u}_1 + Z_8 c_2 + Z_{11} s_2 \tag{72}$$

$$(F_2^*)_B = -B_2 \dot{u}_2 + Z_{10} \tag{73}$$

e. Contribution $(F_r^*)_L$ of the load to the generalized inertia force. Since

$$M_L = 0 \ (\textit{no load}) \tag{74}$$

$$(F_r^*)_L \underset{(14)}{=} 0; \quad (r = 1,2) \tag{75}$$

f. Generalized inertia forces F_r, $r = 1,2$.

The generalized inertia forces are

$$(F_1^*) \underset{(15)}{=} (F_1^*)_A + (F_1^*)_B + (F_1^*)_L$$

$$\underset{(70,72,75)}{=} -A_1 \dot{u}_i + (-m_B L_1^2 + Z_9 c_2 + Z_{12} s_2)\dot{u}_1 + Z_8 c_2 + Z_{11} s_2 \tag{76}$$

$$(F_2^*) \underset{(15)}{=} (F_2^*)_A + (F_2^*)_B + (F_2^*)_L$$

$$\underset{(71,73,75)}{=} -B_2 \dot{u}_2 + Z_{10}. \tag{77}$$

9. Generalized active force, F_r.

The contact forces exerted by N on A can be replaced by a couple of torque $\boldsymbol{T}^{N/A}$ and a force $\boldsymbol{F}^{N/A}$ at the center Q_1 of joint 1. Similarly the contact forces exerted by B on A can be replaced by $(\boldsymbol{T}^{B/A}, \boldsymbol{F}^{B/A})$ acting at the center Q_2 of joint 2, and the contact forces exerted by A on B by $(\boldsymbol{T}^{A/B}, \boldsymbol{F}^{A/B})$, also acting at Q_2. The gravitational forces acting on A and B are

$$\boldsymbol{G}_A = -m_A g \boldsymbol{a}_1 \tag{78}$$

$$\boldsymbol{G}_B = -m_B g \boldsymbol{a}_1 \tag{79}$$

where m_A and m_B are, respectively, the masses of A and B.

Defining†

$$\tau_1 \underset{(26)}{\triangleq} \boldsymbol{T}^{N/A} \cdot \boldsymbol{a}_1 \tag{80}$$

$$\tau_2 \underset{(26)}{\triangleq} \boldsymbol{T}^{A/B} \cdot \boldsymbol{b}_2, \tag{81}$$

the generalized active forces F_r, $r = 1,2$, are given by

$$F_1 \underset{(32)}{=} \tau_1 + (-g m_A \boldsymbol{a}_1 \cdot \boldsymbol{V}_1^{A^*} - g m_B \boldsymbol{a}_1 \cdot \boldsymbol{V}_1^{B^*}) - g M_L \boldsymbol{a}_1 \cdot \boldsymbol{V}_1^P \tag{82}$$

which for this example yields

$$F_1 \underset{(48,49,74)}{=} \tau_1. \tag{83}$$

Similarly

$$F_2 = \tau_2 + (-g m_A \boldsymbol{a}_1 \cdot \boldsymbol{V}_2^{A^*} - g m_B \boldsymbol{a}_1 \cdot \boldsymbol{V}_2^{B^*}) - g M_L \boldsymbol{a}_1 \cdot \boldsymbol{V}_2^P \tag{84}$$

which yields

$$F_2 \underset{(48,49,74)}{=} \tau_2. \tag{85}$$

(P in equations (82,84) denotes the point of the end-effector which coincides with the mass-center of the load.)

Note that the contributions of G_A and G_B to F_r are zero because the mass centers A^* (of A) and B^* (of B) are, respectively, located on joint axes 1 and 2.

10. *Kane's dynamical equations.*

Using (33) we obtain the following dynamical equations

$$(F_1^* + F_1 = 0):\ (-A_1 - m_B L_1^2 + Z_9 c_2 + Z_{12} s_2)\dot{u}_1 + Z_8 c_2 + Z_{11} s_2 + \tau_1 \underset{(33,76,83)}{=} 0 \tag{86}$$

$$(F_2^* + F_2 = 0):\ -B_2 \dot{u}_2 + Z_{10} + \tau_2 \underset{(33,77,85)}{=} 0. \tag{87}$$

† Note that τ_1 and τ_2 as defined here are usually referred to as the joint torques.

Defining

$$Z_{13} \triangleq A_1 + m_B L_1^2 - Z_9 c_2 - Z_{12} s_2 \tag{88}$$

$$Z_{14} \triangleq Z_8 c_2 + Z_{11} s_2 \tag{89}$$

we have the following *equations of motion:*

$$\dot{q}_1 \underset{(35)}{=} u_1 \tag{90}$$

$$Z_{13}\dot{u}_1 \underset{(86,88,89)}{=} \tau_1 + Z_{14} \tag{91}$$

$$\dot{q}_2 \underset{(36)}{=} u_2 \tag{92}$$

$$B_2\dot{u}_2 \underset{(87)}{=} \tau_2 + Z_{10}. \tag{93}$$

Equations (90-93) are a set of four first-order differential equations in the variables q_1, u_1, q_2, u_2.

Remark: In more complex problems, the quantities Z_1, Z_2,... are extremely helpful (i) in minimizing the labor involved in writing equations and (ii) in the generation of computer codes [14].

The dynamical equations (33) are, in general, a set of highly coupled non-linear differential equations. To see this, for the present example, substitute for all the Z's in (91,93) to obtain

$$(A_1+m_B L_1^2+B_1 \cos^2 q_2+B_3 \sin^2 q_2)\dot{u}_1 \underset{(91)}{=} \tau_1+[(B_1-B_3) \sin 2 q_2]u_1 u_2 \tag{94}$$

$$B_2\dot{u}_2 \underset{(93)}{=} \tau_2 - [(B_1 - B_3) \sin 2q_2]u_1^2/2. \tag{95}$$

Acknowledgments

The authors thank Ms. Donalda Speight for drawing the figures and Professor Thomas R. Kane for critically reviewing certain sections of this chapter. The financial support of the National Science Foundation and the Systems Development Foundation is acknowledged.

REFERENCES

[1] H. Albala, *Displacement Analysis of the N-Bar, Single Loop, Spatial Linkage, Application to the 7R, Single-Degree-of-Freedom, Spatial Mechanism,* D.Sc. Thesis, Faculty of Mechanical Engineering, Technion, Haifa, Israel, June, 1976.

[2] H. Albala, "Displacement Analysis of the General N-Bar, Single-Loop Spatial Linkage," Parts I, II *Trans. ASME, J. Mech. Design,* Vol. 104, April, 1982, pp. 504-525.

[3] H. Albala and D. Pessen, "Displacement Analysis of a Special Case of the 7R, Single-Loop, Spatial Mechanism," *Trans. ASME; J. Mechanisms Transmissions and Automation in Design,* Vol. 105, No. 1, March, 1983, pp. 78-87.

[4] O. Bottema, and B. Roth, *Theoretical Kinematics,* North-Holland, Amsterdam, 1979.

[5] M. Brady, et al (ed.), *Robot Motion,* MIT Press, Cambridge, Mass., 1982.

[6] S. J. Derby, "General Robot Arm Simulation Program (GRASP)," Parts I, II, submitted to the *Computer Eng. Division, ASME,* 1982.

[7] J. Duffy, *Analysis of Mechanisms and Robot Manipulators,* John Wiley and Sons, New York, 1980.

[8] E. F. Fichter and K. H. Hunt, "The Fecund Torus, its Bitangent-Circles and Derived Linkages," *Mechanism and Machine Theory,* Vol. 10, No. 4, 1975, pp. 167-176.

[9] K. C. Gupta and B. Roth, "Design Considerations for Manipulator Workspace," *Trans. ASME, J. Mech. Design,* Vol. 104, October, 1982, pp. 704-711.

[10] J. M. Hollerbach, "A Recursive Lagrangian Formulation of Manipulator Dynamics and a Comparative Study of Dynamics Formulation Complexity," *IEEE Trans. on Systems, Man and Cybernetics,* Vol. SMC-10, No. 11, 1980, pp. 730-736.

[11] M. E. Kahn, *The Near-Minimum-Time Control of Open-Loop Articulated Kinematic Chains,* Stanford AI Laboratory Memo AIM-106, December, 1969.

[12] T. R. Kane, *Dynamics,* Holt, Rinehart and Winston, 1968.

[13] T. R. Kane, P. W. Likins, and D. A. Levinson, *Spacecraft Dynamics,* McGraw-Hill, New York, 1982.

[14] T. R. Kane and D. A. Levinson, "The Use of Kane's Dynamical Equations in Robotics," *The International Journal of Robotics Research,* Vol. 2, No. 3, Fall 1983, pp. 3-21.

[15] A. Kumar and K. J. Waldron, "The Dextrous Workspace," *ASME Paper No. 80-DET-108,* 1980.

[16] A. Kumar and K. J. Waldron, "The Workspace of a Mechanical Manipulator," *Trans. ASME, J. Mech. Design,* Vol. 103, No. 3, July, 1981, pp. 665-672.

[17] R. P. Paul, *Robot Manipulators,* MIT Press, Cambridge, Mass., 1981.

[18] D. L. Pieper, *The Kinematics of Manipulators Under Computer Control,* Stanford AI Laboratory Memo AIM-72, 1968.

[19] D. L. Pieper and B. Roth, "The Kinematics of Manipulators Under Computer Control," *Proceedings of the Second International Congress on the Theory of Mechanisms and Machines,* Zakopane, Poland, Vol. 2, 1969, pp. 159-169.

[20] B. Roth, "Performance Evaluation of Manipulators From a Kinematic Viewpoint," *Performance Evaluation of Programmable Robots and Manipulators,* NBS Special Publication 459, 1976, pp. 39-62.

[21] B. Roth, "Robots," *Applied Mechanics Reviews,* Vol. 31, No. 11, November, 1978, pp. 1511-1519.

[22] B. Roth, "Robots - State of the Art in Regard to Mechanisms Theory," *Trans. ASME, J. Mechanisms, Transmissions and Automation in Design,* Vol. 105, No. 1, March, 1983, pp. 11-12.

[23] B. E. Shimano, *Kinematic Design and Force Control of Computer Controlled Manipulators,* Stanford AI Laboratory Memo AIM-313, 1978.

[24] W. M. Silver, "On the Equivalence of Lagrangian and Newton-Euler Dynamics for Manipulators," *The International Journal of Robotics Research,* Vol. 1, No. 2, 1982, pp 60-70.

[25] Y. Stepanenko and M. Vukobratovic, "Dynamics of Articulated Open-Chain Active Mechanisms," *Math. Biosc.,* Vol. 28, 1976, pp. 137-170.

[26] K. Sugimoto and J. Duffy, "Determination of Extreme Distances of a Robot Hand - Part 1: A General Theory," *Trans. ASME, J. Mech. Design,* Vol. 103, No. 4, July, 1981, pp. 631-636.

[27] R. H. Taylor, "Planning and Execution of Straight-Line Manipulator Trajectories," *IBM J. Res. Develop.,* Vol. 23, No. 4, July, 1979, pp. 424-436.

[28] J. J. Uicker, Jr., J. Denavit, and R. S. Hartenberg, "An Iterative Method for the Displacement Analysis of Spatial Mechanisms," *Trans. ASME, J. App. Mech.,* Vol. 86, Series E, 1964, pp. 309-314.

[29] M. Vukobratovic and V. Potkonjak, *Scientific Fundamentals of Robotics, 1: Dynamics of Manipulation Robots,* Springer-Verlag, Berlin, Heidelberg, 1982.

4

Kinematic and Force Analysis of Articulated Hands

J. KENNETH SALISBURY, JR.

1. INTRODUCTION

As the application of mechanical manipulators has grown in the fields of industrial automation, prosthetics and remote manipulation, the need for more versatile and adaptable end effectors has become increasingly apparent. Current manipulation practice is severly limited by our inability to adapt to a variety of parts and the lack of fidelity in force control. It has become clear that articulated end effectors, or mechanical hands, can be used to extend manipulation capability in terms of cost effectiveness and in terms of the overall complexity of tasks that may be performed. Coupled with more intelligent control systems, articulated hands promise to be of major importance in the future of robotics.

This chapter deals with issues central to extending our use and understanding of articulated hands in manipulation. First we establish a rational basis for analyzing the kinematics or geometry of articulated hand designs and grasps upon objects. We define acceptable designs as those which may completely restrain a grasped object as well as impart arbitrary forces and small motions to it. A group of 600 potential hand designs was analyzed and here we identify the subset of acceptable designs. Secondly we introduce a systematic method for formulating force and position control matrices for articulated hands. This includes controlling internal as well as external forces on the grasped object.

As an application of the ideas developed herein, a three-finger articulated hand, suitable for a computer control, was designed and constructed. The description of this device, known as the Stanford/JPL hand, includes details on its active force sensing capability as well as its unique tendon actuation system.

The robots we find today usually consist of six-jointed "arms" with simple hands or "end effectors" for grasping objects. In a typical application they are fixed firmly to the floor and commanded by computer to perform some physical operation on the environment. These may range from simple pick and place operations to such motions as moving a spray painter about. Other applications may range from moving cameras and inspection equipment to performing delicate assembly. While these devices are a far cry from the wonderful machines of science fiction, they are doing useful work. The evolution of robots has not been driven by their ability to emulate human behavior but by their capacity to do useful work. This usually includes some physical interaction with objects in the environment such as exerting forces or causing motions. The understanding of how bodies interact and the mechanics of force and motion is essential to improving robots' capacity to do useful work.

This chapter deals with one aspect of the mechanics of manipulation, the interface between the mechanical arm and the objects it interacts with. We use the term *hand* loosely to describe this interface. It includes the traditional concepts of hands, sometimes called end effectors, grippers, pincers and tongs, as well as more complex devices. Specifically this chapter is concerned with *articulated hands*, devices with two or more powered joints that can grip and manipulate objects. Experience with manipulators has pointed to a need for hands that can adapt to a variety of grasps and augment the arm's manipulative capacity with fine position and force control. The ability of an articulated hand to reconfigure itself into a variety of grasps reduces the need for specialized grippers. The proximity of low mass, powered joints to the objects being manipulated reduces modeling errors and dynamic complexity which facilitates achieving high bandwidth and fine control of motions.

1.1 Overview of Existing Hands

Probably the first occurrence of mechanical hands was in prosthetic devices to replace lost limbs. The literature abounds with descriptions of many clever hand designs ranging from a gripping device designed in 1509 for a knight who had lost his hand in battle [1] to the sophisticated myoelectrically controlled devices developed at the University of Utah [2]. Almost without exception prosthetic hands have been designed to simply grip objects. This shifts the burden of motion control to the remaining upper arm joints of the amputee. The dexterity with which an amputee may employ a prosthetic device comes from his ability to use direct visual information to guide him and not from any inherent dexterity in the artificial hand itself.

The second related field in which artificial hands have found use is in the remote manipulation. The need to work with hazardous materials or environments has spawned the development of "master-slave" or "teleoperator" systems. These devices permit a user to perform simple manipulations from a safe remote location by employing mechanical arms directly controlled by the operator by electric and mechanical means. Applications in space, nuclear and undersea environments are typical uses of teleoperators.

Various studies have been made to develop improved hands for prosthetic and teleoperator use. Crossley [3] enumerated a number of useful hand functions ranging from simple grasping to complex manipulations. He designed an anthropomorphic hand with three fingers and a thumb intended for teleoperator use. Various multi-finger hands, intended to emulate human functions, have been designed by Skinner [4], Rovetta [5], Mori [6], and Okada [7], among others.

An amazing variety of grippers has been designed for industrial use. These special purpose devices use not only fingers but suction cups, magnets, tentacles and adhesives to achieve their gripping function [8]. Although a few attempts have been made to design grippers that can actively move the grasped object, there has been a notable lack of kinematic analysis to guide us in their design.

1.2 Preview

This chapter is intended to explain how articulated hands may be used to securely grasp objects and apply arbitrary forces and small motions to these objects. Among the principal developments of this work are 1) an approach to the analysis of articulated hand kinematics using screw theory and, 2) a formal procedure for controlling forces and small motions imparted by an articulated hand upon grasped objects. We have used screws to represent forces and motions in our kinematic analysis because of their generality. In formulating the control equations we have used more traditional vector representations for force and velocity because it leads to more compact representation of the relationships.

In performing the kinematic analysis our approach has not been to survey desirable human functions and try to emulate them. Rather we have taken a more abstract kinematic viewpoint. We have asked what are the requirements for securing and manipulating objects in the environment. Our path leads from an analysis of the effect of contacts between bodies through an enumeration and evaluation of various hand designs to a control framework for commanding hand motions. Our analysis deals with the statics and instantaneous kinematics of grasping and manipulating of rigid bodies. By instantaneous kinematics we refer to infinitesimal displacements or velocities of a body at a particular instant of concern and not to finite displacements or trajectories. Our work does, however, apply directly to small motions and lays some of the groundwork for the investigation of large motions.

In Section 2 we are concerned with the details of how the motion of a body is constrained by various types of contacts with other bodies, including the effect of friction. The constraints imposed by each type of contact are discussed in terms of the degrees-of-freedom allowed and in terms of screw systems of motion and force associated with each contact type. The basics of screw systems, wrench systems and twist systems are discussed. The formalization developed will be used in Sections 4 and 5 to evaluate the potentially acceptable designs identified in Section 3.

In Section 3 we look at a large class of mechanisms which are potentially suitable hand designs. Here we introduce the concept that a suitable hand design must be able to immobilize a grasped object and impart arbitrary motions to it. At the first step in this process we use a modified form of Grübler's criterion to identify a group of 39 mechanisms which may be suitable articulated hands.

In Section 4 we use the elements of screw theory developed in Section 2 to perform the next step in refining our list of suitable hands. We look in greater detail at the effects of multiple contacts between fingers and grasped object. We have identified groupings of contacts that are suitable for restraining a body and the geometric conditions for independence of these contact constraints.

In Section 5 we explore the role of contacts which may exert forces in one direction only, known as *unisense* constraints. The concept of complete restraint is extended to include all the types of contact that are commonly encountered. Methods for algebraically and geometrically identifying internal grasp forces and an explanation of their use in grasp stabilization are given. The material in this section provides the final analytical tools necessary for selecting the basic kinematics of a real hand (as described in Section 7) and puts us in a position to describe the control of forces and velocities Section 6.

In Section 6 we derive the *grip transformation,* G, which, like the Jacobian transformation in classical manipulator control, is used to map forces (and velocities) between fingertip and grasped object coordinates. The introduction of this powerful concept provides a basis for applying forces and controlling small motions with an articulated hand.

Finally, in Section 7 we describe the novel *Stanford/JPL Hand,* a force-controlled, 3-finger hand based upon the author's research.

2. CONTACT - FREEDOM AND CONSTRAINT

An unconstrained rigid body, B, has six degrees freedom (D.O.F.), i.e, six independent parameters are required to completely specify B. For infinitesimal displacements these six parameters may be considered to be the three components of the body's angular velocity and the three components of the translational velocity of a point fixed in B. (For a finite displacement these six parameters could be the three Euler angles describing the orientation change, and the components of the displacement of a particular point on the body.) If B is brought into contact with another rigid body that is fixed in reference frame R, the motion of B will be restricted. The degree of which the motion of B is restricted depends on the nature of the contact. If B is brought into contact with additional bodies fixed in R, the motion available to B is further reduced until ultimately B may be fully constrained and unable to move in any manner relative to R. The degree to which each contact restricts motion depends not only on the nature of each contact but also on the position and orientation of the contacts relative to each other.

In this section we seek to make precise the meaning of contact, identify the effects of common types of contact between bodies and develop a mathematical basis for determining quantitatively the motion available to a body constrained by one or more contacts. We will be considering instantaneous motion properties of rigid bodies. This means we are concerned with the velocities (angular and translational) of a body and with the forces that can be exerted upon it. When a body is subject to a set of constraints, the set of possible motions allowed can be found by an analysis of the geometry of the constraints. When the body undergoes a finite displacement, it is possible that the nature of the constraints will change and a new analysis will have to be made in order to determine the new set of possible motions. In some cases the new constraints will be exactly the same and the resulting motion allowed will be the same. For example, a cylinder rotating in a hole will be subject to the same constraints throughout its motion. In other cases the constraint situation will change once the body moves even a small amount. This can occur when the contact points between bodies change position as a result of motion as in Figure 2-1 or when the actual number of contact points changes during the motion as in Figure 2-2.

Though our analysis of constraints on a body imposed by contact with the environment will be strictly valid only for the instant under consideration, in practice it is often the case that the instantaneous mobility allowed by a set of constraints will not change drastically for small (finite) motions of the body. We will find this useful when considering small motions of bodies constrained by a set of fingers.

2.1 Contact

When two rigid bodies are brought together until they touch each other at one or more points, there will be a one-to-one correspondence between the points on

Figure 2.1 Constraint change due to displacement. The contact normals change orientation during motion.

Figure 2.2 Constraint change due to additional contact. The shape of the contact area changes with motion.

body B that touch points on body R. We will use the word *contact* to denote a collection of adjacent points where touching occurs over a contiguous area. Two disjoint areas of contact will be considered to be two distinct *contacts.* In the limit an isolated point of touching is considered to be a contact area. In this section we look at the possibilities for motion that arise when one contact occurs between two bodies. In Sections 3 and 5 we will study in detail the effect and consequences of multiple contacts between bodies.

The extent to which relative motion between two bodies is limited by one or more contacts depends upon four things:

1. The constraint imposed by the shape of each contact area.
2. The relative locations of the contact areas.
3. The relative orientation of each contact area.
4. The effect of friction at each contact area.

When contact occurs between two bodies, the subsequent relative motions between them will become limited to directions that do not attempt to merge the material of the two bodies. The bodies may move in a manner that maintains contact and possibly moves the contact points on the bodies to a new set of coincident points, or they may move such that contact is broken. If there is a sufficient friction between the two contact areas, then motion in directions resisted by the friction forces (in the contact tangent plane) will also be precluded.

The geometric properties of a surface in the neighborhood of a particular point are characterized to the second order (position, orientation, curvature) by giving the coordinates of the point, the direction of the surface normal, the maximum and minimum radii of curvature and the orientations of the two planes which contain osculating circles of these radii at the contact point. For a convex surface we define both radii of curvature to be positive. For a plane surface both radii are infinite. A cylinder has one positive and one infinite radius of curvature. A infinitesimal sphere or point can be thought of as having two zero radii of curvature (i.e., corner of a cube or tip of a cone). The surface normal at a point of interest is the outward-pointing normal to the tangent plane at that point. Since the tangent plane will not be defined when either radius of curvature is zero, we cannot uniquely define the surface normal at such points. The outward pointing surface normals of two bodies at a point of contact will have opposite directions. We will select one of these directions to be the *contact normal.* For our purposes it is immaterial which sense we choose for the contact normal. If one of the contacting surface normals is not defined, then the contact normal is that of the other surface. If both surface normals are undefined, then the contact normal is undefined. The direction of the contact normal will be needed to quantify the nature of motion allowed by a particular contact.

Some kinematic analyses [9,10,11,12] treat each contact as frictionless and rely only on structural restraint between bodies. Since many common manipulation situations rely on frictional restraint, it will be included in our analysis. In cases where friction is active (i.e., acts as constraint) we make the assumption that the contact forces and coefficient of friction between bodies are sufficiently large to create frictional forces larger than the forces they must resist. Although this would have to be checked in each case to make sure the assumed constraints are actually active, it is a useful assumption for the purpose of analysis. In Section 5 we will see how extra joint freedoms in a grasping situation can be used to maintain positive contact forces that will ensure that frictional restraints will be active.

2.2 Types of Contact between Bodies

We now consider the possible types of contact or pairings of two surfaces. For each surface we will consider three possible types of contact: A surface has *point contact* if at the point under consideration both of its principal radii of curvature are very small or zero. A surface has *line contact* if at the point under consideration one of its principal radii of curvature is very small or zero and the other radius of curvature is infinite. A surface has *plane contact* if both its principal radii of curvature are infinite. With these three types of features we can analyze the instantaneous mobility of the most common pairings between bodies. It should be emphasized that we are looking only at single contacts for the moment; later we will look at the effect of multiple contacts.

Of these pairings, three are only transient conditions - a point on a point, a point on a line and a line on a point. The constraints imposed by such matings are not stable and will be ignored in the analysis. Next we notice that, in terms of relative freedom of motion between the bodies, it does not matter which feature is on which body. A point of body B touching a plane in R imposes the same constraint as a plane in B touching a point in R. Similarly a line in B touching a plane in R imposes the same constraint as a plane in B

TABLE 2-1

SURFACE FEATURE PAIRINGS

		Body B Feature		
		Point	Line	Plane
Body R Feature	Point			
	Line			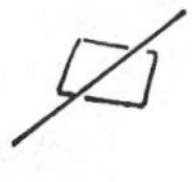
	Plane		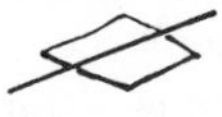	

touching a line in R. Finally we note that a line touching a line imposes the same constraint as a point on a plane or a plane on a point (unless the lines are colinear, in which case the constraint is transient and will be ignored). From the above we can conclude that there are essentially three types of stable pairings between the bodies B and R. We call these

1. Point contact (point on plane, plane on point or line on non-parallel line).
2. Line contact (line on plane or plane on line).
3. Planar contact (plane on plane).

2.3 Effect of Single Contacts Between Bodies

Without looking at the details of each contact type, we can classify them into one of five categories according to the relative freedom of motion allowed between the bodies. A 1-degree-of-freedom contact implies that only one parameter is needed to specify the relative motions of the bodies if contact is maintained. A 2-degree-of-freedom contact requires two parameters to specify subsequent relative motions, and so on up to five parameters for a 5-degree-of-freedom contact. Within each of these five classifications there may be a variety of possible physical configurations that give rise to the particular number of freedoms. For example, a ball-and-socket connection and a planar connection both have three degrees of freedom while the actual types of motions allowed are quite different. In the next section we sill use this number classification of contacts to help us identify a group of mechanism designs that satisfy certain useful mobility conditions. First we will look at the effect of various contact types on the relative mobility between contacting objects.

The relative freedom of motion allowed by each type of contact will depend on whether we consider frictional forces to be significant. Table 2-2 shows the relative number of freedoms possible for the basic contact geometries described in the previous section.

There is another contact type frequently encountered which we will call a *soft finger*. It behaves to the first order as a point contact with friction, except that its contact area is large enough that it is able to resist moments about the contact normal. It is a 2-degree-of-freedom contact.

TABLE 2-2.

FREEDOM OF MOTION FOR BASIC CONTACT GEOMETRIES.

	Without Friction	With Friction
Point Contact	5	3
Line Contact	4	1
Planar Contact	3	0

It is important to emphasize that the above freedoms obtain as long as the particular contact situation is maintained. Of course the contact may be broken by moving the bodies apart, but we assume for the moment that the forces act to maintain contact throughout subsequent motions. Reuleaux called this *force closure* [13]. Subsequent motion may move the bodies into a different constraint situation without breaking contact. For example if the edge of a cube is resting on a plane, it is subject to the constraints of line contact. If the cube rolls about this edge until one of its faces meets the plane, it will then be subject to the constraints of planar contact.

2.4 Screws, Twists, and Wrenches

We now develop the mathematical tools necessary for quantitatively describing the instantaneous, spatial motions of bodies and the exertion of forces upon them. While many of the analyses in this chapter could be performed using conventional force and velocity vectors we have chosen to use the screw system approach for several reasons. Screw representation (developed below) of forces and velocities emphasizes the underlying dependence of their representation upon line geometry. A collection of forces and moments acting upon a body can always be reduced to a force along a unique line of action and a moment about the line. Similarly, the velocity of a body at a particular instant may be represented by an angular velocity about a unique line and a translational velocity along the line. With each particular type of contact there is associated an ensemble of lines about which generalized forces or *wrenches* may be exerted. For each contact there is another ensemble of lines about which generalized displacements or *twists* may occur. By understanding the geometries of these line ensembles we will be able to evaluate more easily the effects of multiple contacts on a body than would be possible with a strict vector approach. The wrench representation treats forces and moments merely as extremes of the more general entity, the wrench. Similarly, the twist representation treats angular and translational velocities as extremes of the more general entity, the twist. The wrench and twist representations have the advantage of allowing us to treat their elements homogeneously. In Section 4 we will see examples using these ensembles or *screw systems* to determine the conditions for constraint of an object. In Section 5 we will use them further to identify the internal forces encountered in certain grasping situations. The power of screw systems is demonstrated in Eqn. (5.8), where we show the equivalence of two methods of finding internal forces in grasping. In the remainder of this section we will explore the wrench and twist system geometries of single contacts for use in later sections. Although we develop relatively simple screw systems for use here, they provide a powerful and elegant means for understanding the geometry of constraint and force exertion with articulated hands. A much more thorough treatment of screw systems may be found in [9].

The following material is based upon more rigorous developments by Hunt, Waldron and Ohwovoriole [10,11,12]. A screw is defined by a straight line in space known as its *axis* and an associated *pitch*, p. A screw may be described with a six-vector of *screw coordinates*, $\boldsymbol{s} = (S_1, S_2, S_3, S_4, S_5, S_6)$, with the following interpretations. The Plücker line coordinates of the axis are

$$
\begin{aligned}
L &= S_1 \\
M &= S_2 \\
N &= S_3 \\
P &= S_4 - pS_1 \\
Q &= S_5 - pS_2 \\
R &= S_6 - pS_3.
\end{aligned}
\tag{2.1}
$$

L, M, and N are proportional to the direction cosines of the line forming the axis while P, Q, and R are proportional to the moment of the line about the origin of the reference frame. If each of the components of the Plücker coordinates is divided by $(L^2 + M^2 + N^2)^{1/2}$ then the first three will be the direction cosines of the line and the last three components will then be the moment of the line. The moment of the line is the cross product of a vector from the origin of the reference frame to any point on the line with the unit vector in the direction of the line. The pitch of the screw is

$$
p = \frac{S_1 S_4 + S_2 S_5 + S_3 S_6}{S_1^2 + S_2^2 + S_2^3}, \tag{2.2}
$$

and the magnitude of the screw is $(S_1^2 + S_2^2 + S_3^2)^{1/2}$ unless the pitch is infinite in which case the magnitude is $(S_4^2 + S_5^2 + S_6^2)^{1/2}$. Scalar multiplication and vector addition are valid for infinitesimal screws so that two screws, $\boldsymbol{s}_1$ and $\boldsymbol{s}_2$, are considered linearly dependent if we can find non-zero scalars, c_1 and c_2 such that $c_1 \boldsymbol{s}_1 + c_2 \boldsymbol{s}_2 = 0$.

With any particular infinitesimal motion or velocity of a body in space there is associated a unique line, the twist axis, about which the body rotates and along which it translates. We call this motion a *twist* and identify it with a six-vector of twist coordinates, $\boldsymbol{t} = (T_1, T_2, T_3, T_4, T_5, T_6)$. T_1, T_2 and T_3 are the components of the angular velocity, $\boldsymbol{\omega}$, of the body and T_4, T_5 and T_6 are the components of the velocity $\boldsymbol{v}$, of a point fixed in the body and lying at the origin of the coordinate system. The Plücker coordinates of the *twist axis* are

$$
\begin{aligned}
L &= T_1 \\
M &= T_2 \\
N &= T_3
\end{aligned}
\tag{2.3}
$$

$$P = T_4 - pT_1$$

$$Q = T_5 - pT_2$$

$$R = T_6 - pT_3.$$

The pitch of the twist is

$$p = \frac{T_1T_4 + T_2T_5 + T_3T_6}{T_1^2 + T_2^2 + T_3^2} = \frac{\boldsymbol{\omega} \cdot \boldsymbol{v}}{\boldsymbol{\omega} \cdot \boldsymbol{\omega}}. \tag{2.4}$$

The pitch of the twist is the ratio of the magnitude of the velocity of a point on the twist axis to the magnitude of the angular velocity about the twist axis. A *zero pitch twist is a pure rotation and an infinite pitch twist is a pure translation.* The magnitude of the twist is $(T_1^2 + T_2^2 + T_2^3)^{1/2} = ||\boldsymbol{\omega}||$ unless the pitch is infinite in which case the magnitude is $(T_4^2 + T_5^2 + T_6^2)^{1/2} = ||\boldsymbol{v}||$.

With any particular set of forces and moments applied to a rigid body there is associated a unique line known as the *wrench axis* a pitch, p and a magnitude. The set of forces and moments acting on the body is equivalent to a single force acting along this wrench axis and a moment exerted about the axis. We call this equivalent force and moment a *wrench* and identify it with a six-vector of wrench coordinates, $\boldsymbol{w} = (W_1, W_2, W_3, W_4, W_5, W_6)$. W_1, W_2, and W_3 are the components of the net force, $\boldsymbol{f}$, exerted on the body and W_4, W_5, and W_6 are the components of the net moment, $\boldsymbol{m}$, resolved at the origin of the reference frame. The Plücker coordinates of the *wrench axis* are

$$L = W_1$$

$$M = W_2$$

$$N = W_3 \tag{2.5}$$

$$P = W_4 - pW_1$$

$$Q = W_5 - pW_2$$

$$R = W_6 - pW_3.$$

The pitch of the wrench is

$$p = \frac{W_1W_4 + W_2W_5 + W_3W_6}{W_1^2 + W_2^2 + W_3^2} = \frac{\boldsymbol{f} \cdot \boldsymbol{m}}{\boldsymbol{f} \cdot \boldsymbol{f}}. \tag{2.6}$$

The pitch of the wrench is the ratio of the magnitude of the moment applied about a point on the axis to the magnitude of the force applied along the wrench axis. A *zero pitch wrench is a pure force and an infinite pitch wrench is a pure moment.* The magnitude of the wrench is $(W_1^2 + W_2^2 + W_3^2)^{1/2} = ||\boldsymbol{f}||$ unless pitch is infinite in which case the magnitude is $(W_4^2 + W_5^2 + W_6^2)^{1/2} = ||\boldsymbol{m}||$.

When the magnitude of a screw, twist or wrench is 1, it is known as a *unit* screw, twist, or wrench. With this one condition imposed upon the six screw coordinates, we see that five independent parameters remain, hence there are ∞^5 different unit screws, twists or wrenches. Often with twist and wrenchs we wish to associate a magnitude other than 1 so that there are ∞^6 different twists and wrenches if we consider their magnitudes as well. With this interpretation we can think of a unit screw as defining only an axis and a pitch. Then we can think of a given magnitude as defining a twist (if its units are rotation/time) or a wrench (if its units are force) acting along the screw. In both cases the pitch is expressed in length units. These six-spaces of (infinitesimal) twists or wrenches can be considered to be vector spaces in that they are closed under vector addition and scalar multiplication.

Two screws, $\boldsymbol{s}_1 = (\alpha_1,\alpha_2,\alpha_3,\alpha_4,\alpha_5,\alpha_6)$ and $\boldsymbol{s}_2 = (\beta_1,\beta_2,\beta_3,\beta_4,\beta_5,\beta_6)$, are said to be *reciprocal* if a number known as their *virtual coefficient*, $\alpha_1\beta_4 + \alpha_2\beta_5 + \alpha_3\beta_6 + \alpha_4\beta_1 + \alpha_5\beta_2 + \alpha_6\beta_3$, is zero. One physical interpretation of the virtual coefficient is that it is the rate of work done by a wrench, $\boldsymbol{w} = (\boldsymbol{f},\boldsymbol{m})$, on a body moving with twist, $\boldsymbol{t} = (\boldsymbol{\omega},\boldsymbol{v})$, i.e., $\boldsymbol{f} \cdot \boldsymbol{v} + \boldsymbol{m} \cdot \boldsymbol{\omega}$. No work is done when the twist and wrench are reciprocal. Note that the condition for reciprocity is independent of the twist and wrench magnitudes.

An ensemble of screws is known as a *screw system*. For example Ball [9] and others have shown that reciprocal to a given unit screw there are ∞^4 unit screws comprising the reciprocal *5-system*. This 5-system can be completely described by a set of 5 *basis screws* (mutually reciprocal unit screws). By taking appropriate linear combinations of these basis screws (and after normalization) we may construct any unit screw in the ensemble. Two basis screws define a 2-system. Reciprocal of this 2-system is a 4-system of screws which may be completely described by giving a set of four basis screws. Every screw in the 4-system is reciprocal to every screw in the reciprocal 2-system. Finally, to a 3-system defined by three basis screws there is another reciprocal 3-system, every screw of which is reciprocal to every screw in the first 3-system. The order of a screw system, n, is equal to the number of basis screws required to define it. The order of a screw system reciprocal to an n-system is $6 - n$ [9].

The foregoing ideas apply as well to ensembles of wrenches or twists acting along the screws. The constraints imposed by a particular contact can be described in two equivalent ways. We may give either the *twist system* or motions available to the body or the *wrench system* of forces and moments (sometimes called forces collectively) that can be resisted by the contact. These two systems are reciprocal to each other in that any wrench that can be applied to the body through a contact can do no work about twists that are allowed by the contact.

2.5 Geometry of Contact Twist and Wrench Systems

Table 2-3 lists suitable sets of unit basis wrenches for each of the commonly encountered types of contact. The freedom allowed by each contact is equal to the order of the associated twist system.

The following list describes the geometries of the contact twist and wrench systems in Table 2-3. Though essential to understanding fully the geometric methods presented in later sections, it may be skipped by the reader until later. It is based upon the more general descriptions of screw system geometries given by Hunt [10]. Refer to Table 2-3 for axes definitions.

No Contact The twist 6-system consists of twists of all pitches about all lines in space. The wrench system is empty.

Frictionless Point Contact The wrench 1-system is a single zero pitch wrench along the line forming the contact normal. The associated twist 5-system contains all twists reciprocal to this wrench.

Frictionless Line Contact The twist 4-system consists of all infinite pitch twists parallel to the $x-y$ plane, all zero pitch twists normal to the $x-y$ plane, all zero pitch twists in the $x-z$ plane and twists of all pitches along the line of contact (y-axis). The reciprocal wrench 2-system for this contact consists of all zero pitch wrenches through the line of contact and normal to the plane, plus an infinite pitch wrench that is mutually perpendicular to the line of contact (y-axis) and the contact normal (z-axis).

Friction Point Contact The twist 3-system contains all zero pitch twists through the point of contact. The reciprocal wrench 3-system is composed of all zero pitch wrenches through the point of contact.

Frictionless Plane Contact The twist 3-system allowed is composed of all zero pitch twists normal to the $x-y$ plane (plane of contact) plus all infinite pitch twists parallel to the plane of contact. There are no finite pitch twists allowed. The reciprocal wrench 3-system is composed of all infinite pitch wrenches parallel to the plane of contact ($x-y$ plane) and all zero pitch wrenches normal to the plane of contact. There are no finite pitch wrenches allowed.

Soft Finger The twist 2-system contains all zero pitch twists through the point of contact in the contact tangent plane ($x-y$ plane). The wrench 4-system that can be exerted by the soft finger contact consists of all zero pitch wrenches acting about all lines in the tangent plane ($x-y$ plane), wrenches of infinite pitch acting about any line normal to he tangent plane and wrenches of any pitch acting about the contact normal (z-axis). In addition, for each finite pitch p there are ∞^2 lines about which a wrench of that pitch may act. For each finite value of p there are ∞^1 concentric hyperboloids of revolution, the

TABLE 2-3

BASIS SCREWS OF COMMON CONTACT TWIST AND WRENCH SYSTEMS

Contact Type	D.O.F.	Twist System	Wrench System
No Contact	6	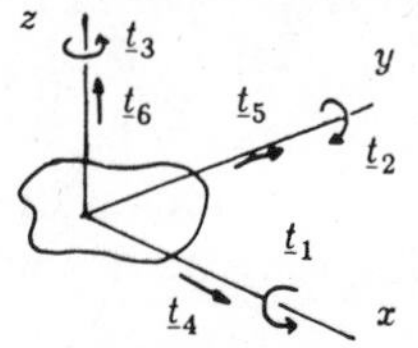 $\underline{t}_1 = (1,0,0,0,0,0)$ $\underline{t}_2 = (0,1,0,0,0,0)$ $\underline{t}_3 = (0,0,1,0,0,0)$ $\underline{t}_4 = (0,0,0,1,0,0)$ $\underline{t}_5 = (0,0,0,0,1,0)$ $\underline{t}_6 = (0,0,0,0,0,1)$	 (null wrench system)
Point Contact Without Friction	5	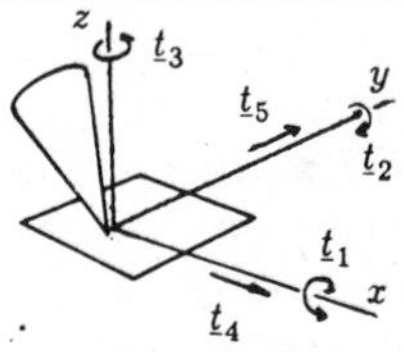 $\underline{t}_1 = (1,0,0,0,0,0)$ $\underline{t}_2 = (0,1,0,0,0,0)$ $\underline{t}_3 = (0,0,1,0,0,0)$ $\underline{t}_4 = (0,0,0,1,0,0)$ $\underline{t}_5 = (0,0,0,0,1,0)$	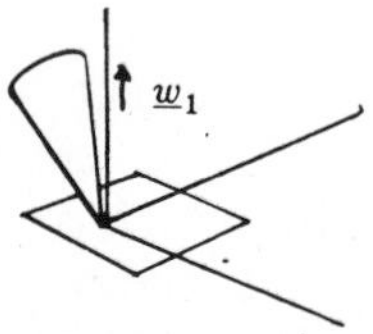 $\underline{w}_1 = (0,0,1,0,0,0)$
Line Contact Without Friction	4	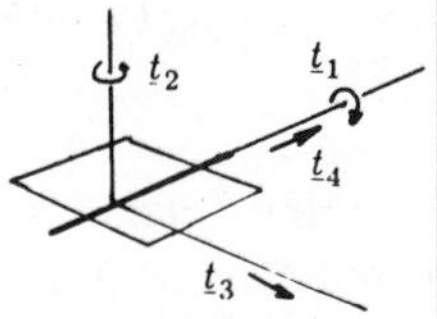 $\underline{t}_1 = (0,1,0,0,0,0)$ $\underline{t}_2 = (0,0,1,0,0,0)$ $\underline{t}_3 = (0,0,0,1,0,0)$ $\underline{t}_4 = (0,0,0,0,1,0)$	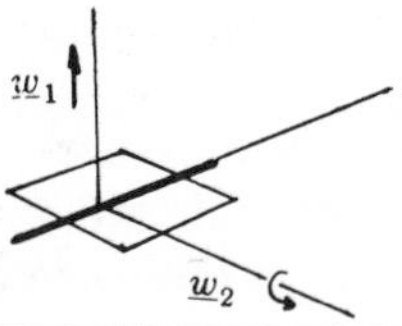 $\underline{w}_1 = (0,0,1,0,0,0)$ $\underline{w}_2 = (0,0,0,1,0,0)$
Point Contact With Friction	3	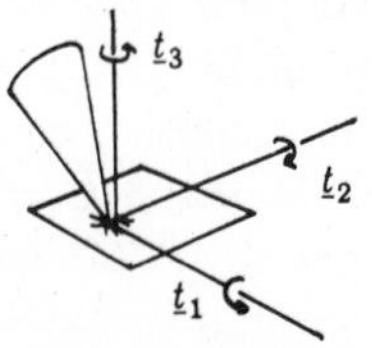 $\underline{t}_1 = (1,0,0,0,0,0)$ $\underline{t}_2 = (0,1,0,0,0,0)$ $\underline{t}_3 = (0,0,1,0,0,0)$	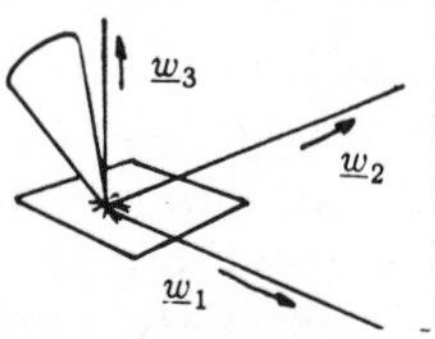 $\underline{w}_1 = (1,0,0,0,0,0)$ $\underline{w}_2 = (0,1,0,0,0,0)$ $\underline{w}_3 = (0,0,1,0,0,0)$

Planar Contact Without Friction	3	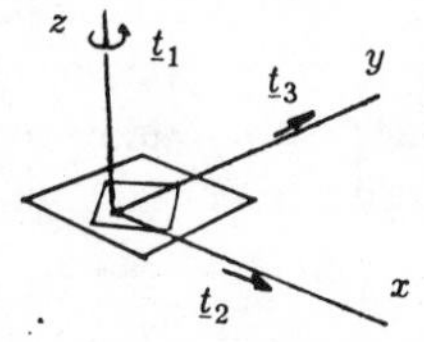 $\underline{t}_1 = (0,0,1,0,0,0)$ $\underline{t}_2 = (0,0,0,1,0,0)$ $\underline{t}_3 = (0,0,0,0,1,0)$	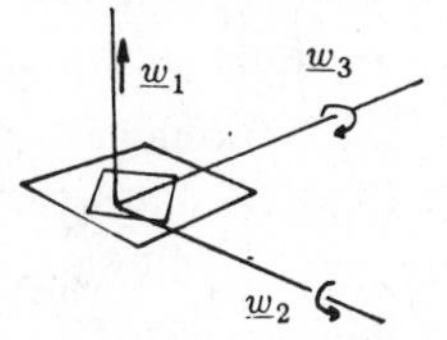 $\underline{w}_1 = (0,0,1,0,0,0)$ $\underline{w}_2 = (0,0,0,1,0,0)$ $\underline{w}_3 = (0,0,0,0,1,0)$
Soft Finger	2	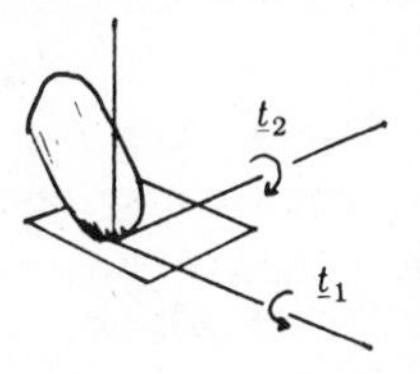 $\underline{t}_1 = (1,0,0,0,0,0)$ $\underline{t}_2 = (0,1,0,0,0,0)$	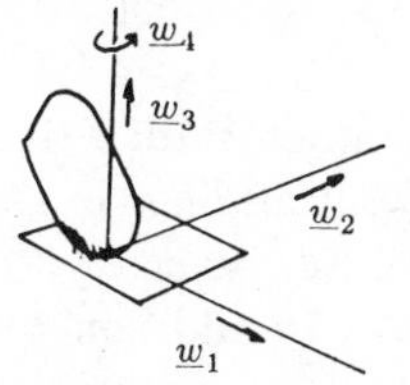 $\underline{w}_1 = (1,0,0,0,0,0)$ $\underline{w}_2 = (0,1,0,0,0,0)$ $\underline{w}_3 = (0,0,1,0,0,0)$ $\underline{w}_4 = (0,0,0,0,0,1)$
Line Contact With Friction	1	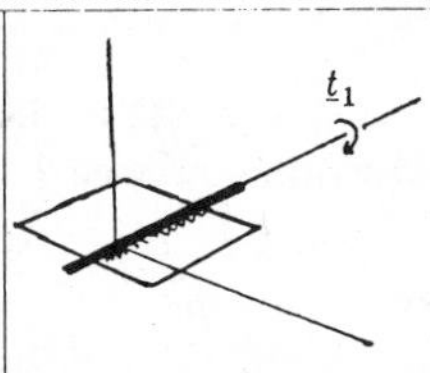 $\underline{t}_1 = (0,1,0,0,0,0)$	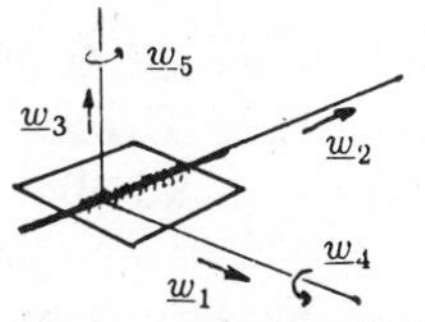 $\underline{w}_1 = (1,0,0,0,0,0)$ $\underline{w}_2 = (0,1,0,0,0,0)$ $\underline{w}_3 = (0,0,1,0,0,0)$ $\underline{w}_4 = (0,0,0,1,0,0)$ $\underline{w}_5 = (0,0,0,0,0,1)$
Planar Contact With Friction	0	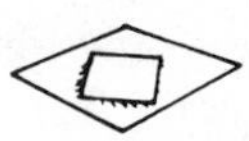 (null twist system)	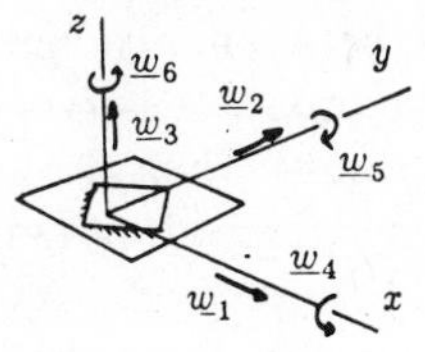 $\underline{w}_1 = (1,0,0,0,0,0)$ $\underline{w}_2 = (0,1,0,0,0,0)$ $\underline{w}_3 = (0,0,1,0,0,0)$ $\underline{w}_4 = (0,0,0,1,0,0)$ $\underline{w}_5 = (0,0,0,0,1,0)$ $\underline{w}_6 = (0,0,0,0,0,1)$

∞^1 reguli of which serve as axes for wrenches of pitch p. The equations of these hyperboloids are given by

$$-p(x^2 + y^2) + +gz^2 + p^2g = 0 \tag{2.7}$$

where g is a free variable that may take on ∞^1 different values (see [10], Eqn. (12.33)). Wrenches of zero pitch may act about all lines of the $x-y$ plane as well as all lines through the contact point. Wrenches of infinite pitch may act about all lines parallel to the z-axis. Finally wrenches of any pitch may act about the z-axis.

Friction Line Contact The twist 1-system consists of a single zero pitch twist along the line of contact. The wrench 5-system is composed of all wrenches reciprocal to this twist.

Planar Contact With Friction The wrench 6-system contains wrenches of all pitches about all lines in space. The twist system is empty.

Of course other types of contacts may be synthesized with different geometries. In addition special, mechanisms such as rollers and slides may be added to the contact to alter the nature of the constraint. The above, however, are the most commonly encountered contact types.

3. NUMBER SYNTHESIS OF HANDS

We now look at the effect of multiple contacts between bodies. By applying a modified version of Grübler's formula we can determine the mobilities and connectivities of hand-object systems. A systemically generated group of mechanisms has been investigated in terms of these properties. Based upon this evaluation we identify a set of mechanism designs having motion properties desirable for mechanical "hands."

A hand mechanism is composed of a collection of rigid bodies called links. One link, sometimes referred to as the palm, is fixed in a reference frame. We also count the object being manipulated as one of the links. The links in the mechanism are connected by joints and contacts. A joint is a permanent connection between two links that is designed to control the relative motion between the links. It is typically a lower pair of the revolute or prismatic type. A contact results when the surface of any link touches the surface of the object that is being grasped or manipulated. The contact may be located anywhere along a link's surface. As discussed in Section 2 the nature of a contact depends upon the shape of the mating surfaces and their friction properties.

The design of mechanisms may be approached in at least three ways. *Number synthesis* deals with the number of degrees of freedom (or simply freedoms) in a mechanism by looking only at the freedoms in the joints and contacts connecting the links. *Type synthesis* looks at the type of relative motion allowed by each connection within a mechanism and the net effect of all such connections on the motion of each link within the mechanism. At this level the screw system description of motion, as seen in Section 2.4, is useful in identifying the types of motion possible. *Dimensional synthesis* deals with the specification of the major dimensions of a mechanism, such as link lengths, and their effect on motion of links. Of these three design approaches number synthesis is the simplest and can provide us with useful information on conditions necessary for a hand mechanism to manipulate grasped objects in desirable ways.

3.1 Mobility and Connectivity

To apply number synthesis to hand mechanisms we first define *mobility* and *connectivity*. The mobility, M, of a kinematic system is defined to be the number of independent parameters that are necessary to specify completely (or to within a finite set) the position of *every* body in the system at the instant of concern. To compute the mobility we use the following version of Grübler's formula [10].

$$M \geqslant \sum f_i + \sum g_j - 6L \tag{3.1}$$

$$M' \geqslant \sum g_j - 6L \tag{3.2}$$

where: M = mobility of system with finger joints free to move

M' = mobility of system with finger joints locked

f_i = degrees of freedom in ith joint (considered to be 1 here)

g_j = degrees of freedom of motion at jth contact point (1-5)

L = number of independent loops in system

The inequality in these relations results from the fact that constraints on the motion of a body in the system may not be independent. In this case, strict equality would indicate fewer degrees of freedom of motion than are actually possible.

In manipulating objects with a hand mechanisms we are concerned with the relative motion (or lack of) between a grasped object and the palm of the hand. The connectivity, C, between two particular bodies in a kinematic system is defined to be the number of independent parameters necessary to specify completely the *relative* positions of the two bodies at the instant of concern. We will use C to denote the connectivity between the palm and the grasped object.

To illustrate the difference between mobility and connectively consider the mechanism in Figure 3.1. Two ball-and-socket joints (3 degrees of freedom each) are connected by a rigid link. M is 6 because 6 parameters will locate completely all the parts of the mechanism. The connectivity between link 0 and link 2, however, is 5 because the parameter specifying the rotation of link 1 about its x axis is not needed to locate link 2 relative to link 0.

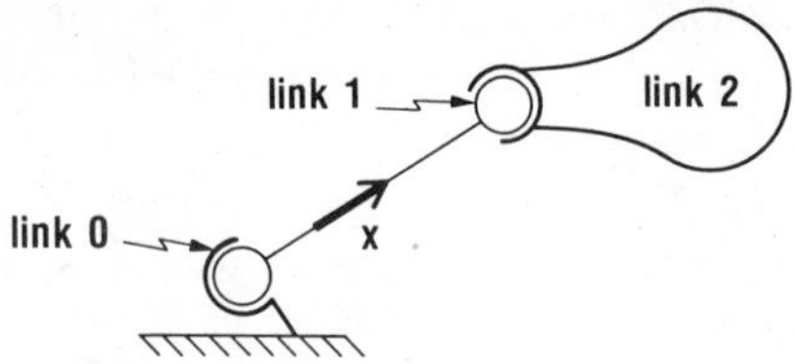

Figure 3.1 Ball and socket mechanism with $M = 6$ and $C = 5$. The connectivity is less than the mobility because link 1 may rotate without affecting link 2.

The connectivity between two bodies in a mechanism may be derived from the overall mobility of the mechanism. First we consider the two bodies in question to be fixed and determine the mobility of each sub-chain connecting them. In the case of a hand, this will mean fixing the object relative to the palm and determining the mobility of each finger sub-chain. Sub-chain mobilities greater than 0 are then subtracted from overall mechanism to yield the connectivity. This eliminates from consideration motions in the mechanism that can be made without affecting the motion of the grasped object.

3.2 Enumeration of Hand Mechanisms

It could be instructive at this point to make a systematic listing of all possible hand designs and then classify them according to their connectivities, both with joints active and joints locked. This, however, would be an unending task if we were to consider the dimensions of each mechanism in detail. If we approach this listing from a number-synthesis point of view the problem becomes more tractable. By limiting ourselves to hand mechanisms with a maximum of three fingers and a maximum of three joints per finger we can generate a manageable set of designs to analyze. The choice of a maximum of three fingers per hand and three joints per finger covers a wide variety of designs, some of which are potentially quite useful.

Many different hand mechanism designs are possible, and for each design a rigid object may be grasped in many different ways. From the number-synthesis point of view it does not matter where on the link a particular contact occurs, it only matters that it does occur. Similarly we may ignore the link dimensions. What must be included is the number of freedoms allowed by each connection between links. Thus a single link of a finger may contact an object with 0 through 6 freedoms, as shown in Table 2-3, in a total of seven ways. A finger with three links, for example, can touch an object in 343 (7^3) unique ways. To determine the number of different grasps possible for a hand composed of k fingers, each of which can contact an object in n ways, we use the formula for the number of combinations, with repetitions, of n things taken k at a time:

$$\binom{n+k-1}{k} = \frac{(n)(n+1)\cdots(n+k-1)}{k!}$$

[13]. For a three finger-hand with three links on each finger this number is 6,784,540. This enumeration includes all 1-, 2- or 3-finger designs with 1, 2 or 3 joints on each finger. Although all these mechanisms could be examined for acceptable designs, we feel that the large number of special cases revealed would not yield significant insight on the problem. To simplify the problem we assume that all the contacts in a given design allow the same freedom of motion (1-5 d.o.f.). With three links per finger each finger can touch the object in one of the eight configurations shown in Figure 3.2.

We may systemically combine these configurations for multiple fingers to define many different grasping situations. For example, a two-finger design with two joints on each finger touching an object only on the last link would have a 2-2-0 configuration. Other examples are shown in Figure 3.3. With three fingers the number of unique grasps ignoring contact type is

$$\binom{8+3-1}{3} = 120.$$

For 5 different contact types (ignoring 0 and 6 d.o.f. contacts) this yields 600 different designs to be investigated.

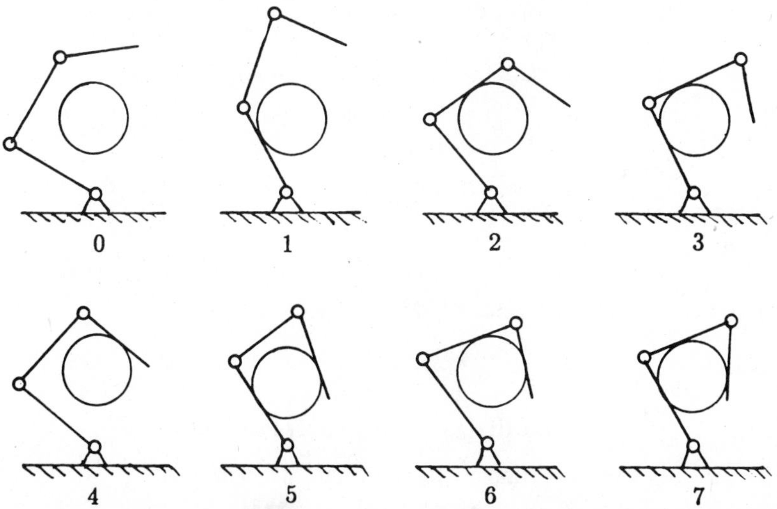

Figure 3.2 Contact configurations. The eight ways in which a 3.link finger may touch an object are shown.

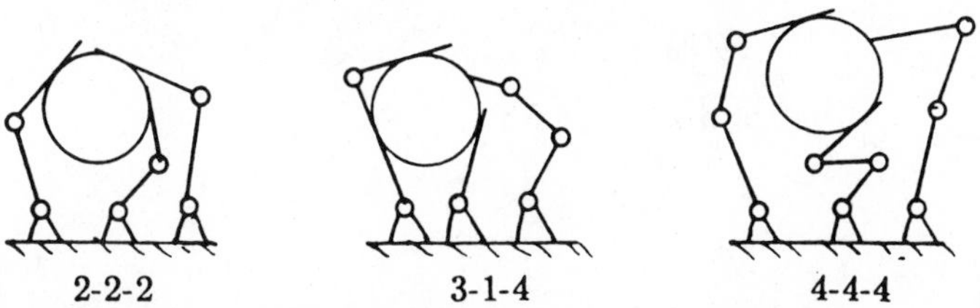

Figure 3.3 Examples of hand configurations. Several of the configurations used in Table 3.1 are illustrated.

For a manipulator or hand mechanism to do useful work it must be able to predictably cause motion of, or apply force to, grasped objects. Unexpected and unobservable motions due to slippage or incomplete constraint of a grasped object place serious limitations on a mechanism's utility. Therefore, an important subset of hand mechanisms are those that are able to: 1) exert arbitrary forces or impress arbitrary small motions on the grasped object when the joints are allowed to move, and 2) constrain a grasped object by fixing (locking) all the joints. We can identify acceptable designs in our listing by looking at the connectivities of each mechanism. Requirement 1 means that it is necessary for the connectivity, C, between the grasped object and the palm to be 6 with the joints active. If the hand were required to move objects in an environment where the mobility of grasped objects is limited (e.g., planar tasks or spherical tasks), this number would be less than 6. Lacking any foreknowledge of the task environment we must seek designs with $C = 6$. Requirement 2 dictates that with the finger joints locked the new connectivity, C', must be less than or equal to 0. This is a necessary condition for a grasped object to be completely constrainable at will.

Of the 600 designs considered, 39 were found to be acceptable with $C = 6$ and $C' \leqslant 0$. Thirty-three of these were based on 5 freedoms per contact with the grasped object, 4 based on 4 freedoms per contact and 2 based on 3 freedoms per contact. These 39 designs are given in Table 3.1, below. It is important to understand that these are only necessary conditions. Each design

TABLE 3-1

MOBILITIES AND CONNECTIVES FOR ACCEPTABLE HAND MECHANISMS

0,1,2 FREEDOMS PER CONTACT:

...0...

3 FREEDOMS PER CONTACT:

Configuration	Joints	M, M'	C, C'
4-4-0	6	6, 0	6, 0
4-4-4	9	6, -3	6, -3

4 FREEDOMS PER CONTACT:

Configuration	Joints	M, M'	C, C'
2-2-2	6	6, 0	6, 0
4-2-2	7	7, 0	6, 0
4-4-2	8	8, 0	6, 0
4-4-4	9	9, 0	6, 0

5 FREEDOMS PER CONTACT:

Configuration	Joints	M, M'	C, C'
7-7-0	6	6, 0	6, 0
7-3-1	6	6, 0	6, 0
7-5-1	7	7, 0	6, 0
7-6-1	7	7, 0	6, 0
7-7-1	7	6, -1	6, -1
7-3-2	7	7, 0	6, 0
7-5-2	8	8, 0	6, 0
7-6-2	8	8, 0	6, 0
7-7-2	8	7, -1	6, -1
3-3-3	6	6, 0	6, 0
5-3-3	7	7, 0	6, 0
6-3-3	7	7, 0	6, 0
7-3-3	7	6, -1	6, -1
7-4-3	8	8, 0	6, 0
5-5-3	8	8, 0	6, 0
6-5-3	8	8, 0	6, 0
7-5-3	8	7, -1	6, -1
6-6-3	8	8, 0	6, 0
7-6-3	8	7, -1	6, -1
7-7-3	8	6, -2	6, -2
7-5-4	9	9, 0	6, 0
7-6-4	9	9, 0	6, 0
7-7-4	9	8, -1	6, -1
5-5-5	9	9, 0	6, 0
6-5-5	9	9, 0	6, 0
7-5-5	9	8, -1	6, -1
6-6-5	9	9, 0	6, 0
7-6-5	9	8, -1	6, -1
7-7-5	9	7, -2	6, -2
6-6-6	9	9, 0	6, 0
7-6-6	9	8, -1	6, -1
7-7-6	9	7, -2	6, -2
7-7-7	9	6, -3	6, -3

found to satisfy these conditions will be analyzed further to determine the appropriate joint and contact geometries. In Section 4 we will assess the effect of these geometries on the 39 potentially acceptable designs by further application of the screw system elements developed in Section 2.

The negative connectivities imply a degree of excess constraint and allow control of internal forces necessary to keep frictional constraints active as will be shown in Section 5. In order for a grasped object to have full 6-degree-of-freedom motion, each subchain (finger) connecting it to the reference frame (palm) must have a sum of freedoms (joint plus contact freedoms) equal to or greater than 6. In addition, the total number of active freedoms (joints) in the hand must be equal to or (as we will see in Section 5) greater than 6.

4. CONTACTS IN GROUPS

To determine if a particular grasp on a body imposes sufficient constraint to immobilize it completely, we must go beyond the simple freedom-counting of Grübler's equation. Each contact between a grasped object and a finger reduces the body's freedom of motion and allows forces to be applied as described in Section 2. In this section we will describe two methods of assessing the effect of groups of contacts on a grasped object. These methods then will be applied to several important cases.

The effect of several contacts on a body can be viewed from a motion or a force point of view. In terms of motion each contact limits the body to executing a particular system of twists. When several contacts are simultaneously imposed on a body, the net motion twist system available to the body will be the intersection of the twist systems allowed by each contact alone. Geometrically, the resulting twist system is that sub-set composed of those twists which have the same axis and pitch in the systems determined by each contact.

From a force point of view, each contact with a body can exert a system of wrenches on the body. When several contacts act simultaneously on a body, the net wrench system which can be applied is the union of the wrench systems of all the contacts. Algebraically the net wrench is given by $\boldsymbol{w}$, as defined below when $\boldsymbol{c}$ takes on all possible values,

$$\boldsymbol{w} = \left[\boldsymbol{w}_1, \boldsymbol{w}_2, \ldots, \boldsymbol{w}_n\right] \boldsymbol{c} \tag{4.1}$$

where the n $\boldsymbol{w}_i$s are the screw coordinates of the principal wrenches of each contact wrench system and $\boldsymbol{c}$ is an arbitrary n-vector of coefficients corresponding to each wrench intensity. For the moment we will assume that all wrenches may be applied in a positive or a negative sense. In Section 5 we develop a method of analyzing grasps that include contact wrenches which may be applied only in the sense that does not break contact.

The motion and force approaches are equivalent ways of looking at the same information. If s_i is the system of screws about which the body may execute twists, then the reciprocal screw system, designated by s_i'', is the system of screws about which wrenches may be applied to the body. Later these will be referred to simply as wrench systems (w) and twist systems (t). Finding the reciprocal of a screw system is analogous to taking its complement in the homogeneous 6-space $(\mathscr{R}^6)$ of all non-normalized screws; the following two properties are true:

$$s_i \bigcup s_i'' = \mathscr{R}^6$$

$$s_i \bigcap s_i'' = \phi.$$

To immobilize a body completely it is necessary and sufficient that the intersection of all the contact twist systems be the null set:

$$s_1 \bigcap s_2 \bigcap \cdots \bigcap s_n = \phi. \qquad (4.2)$$

Application of DeMorgan's theorem to this equation yields:

$$s_1'' \bigcup s_2'' \bigcup \cdots \bigcup s_n'' = \mathscr{R}^6. \qquad (4.3)$$

Eqn. (4.3) states that an arbitrary wrench may be applied to a body (thus immobilizing it) if the union of all the contact wrench systems comprises the full 6-space of all wrenches. Thus Eqn. (4.2) is satisfied if and only if Eqn. (4.3) is satisfied. To determine in practice if an object is immobilized by a set of contacts we use either Eqns. (4.2) or (4.3) depending on which is easier to apply. In simple cases we will be able to geometrically visualize the unions and intersections of the screw systems; in more complex or general cases the algebraic method of Eqn. (4.1) is more useful. If the rank of the matrix formed from the screw coordinates of the principal contact wrenches is 6, the body will be immobilized.

4.1 Constraint by Groups of Contacts

We now proceed to analyze the conditions for constraint of a body touched by contacts which allow 3, 4 or 5 degrees-of-freedom. For conciseness we will refer to these as 3-, 4- and 5-freedom contacts.

4.1.1 3-Freedom Contact

Theoretically, two 3-freedom contacts could constrain an object. It would be necessary for the twist systems of each contact to be reciprocal to each other. This is not true for any combination of the common 3-freedom contacts shown in Table 2-3. A point contact with friction allows the body to move within a 3-system of twists consisting of all the zero-pitch twists through the point of contact. The twist systems of two such contacts will always have one zero-pitch screw in common along the line connecting the two points and therefore will not fully constrain the object.

A planar contact without friction allows the body to move within a 3-system of twists consisting of all infinite-pitch screws parallel to the plane of contact and all zero-pitch screws normal to the plane of contact. The intersection of two planar contact twist systems will always have one infinite pitch twist in common in the direction of the line of intersection of the two planes. A body contacted by a point with friction and a planar contact without friction will always be able to move about one zero pitch twist through the point and normal to the plane. Thus the two common 3-freedom contacts cannot immobilize the contacted body, and we must look at the various combinations of three 3-freedom contacts to determine if any combinations can immobilize it.

Three point contacts with friction will fully constrain the object if the three points are not co-linear. Any two of the contacts' systems intersect in a single zero pitch screw, and if the third contact does not lie on the line connecting the other two contact points there is no mutual intersection of the three twist systems. An object touched by two point contacts with friction and a plane without friction will also be constrained fully, unless the line connecting the points is normal to the plane. In this case there would be a single zero pitch screw in common. An object contacted by two non-parallel planes and a point with friction will be immobilized. The two planes will have in common all infinite-pitch twists in the direction of their line of intersection. Since the point with friction will allow only zero pitch twists there will be no mutual intersection of these three contact twist systems regardless of the location of the point. Three planar contacts without friction, no two of which are parallel, will immobilize an object if no two of the three lines of intersection have the same direction. Thus, all the possible combinations of common three 3-freedom contacts will immobilize the object if the aforementioned geometric conditions are met.

4.1.2 4-Freedom Contact

The only commonly encountered 4-freedom contact is a frictionless line contact as defined in Table 2-3. It consists of a line on one body touching a planar surface, or plane, on another body. A single such contact can apply a 2-system of wrenches with $\boldsymbol{w}_1$ and $\boldsymbol{w}_2$ as principal wrenches as shown in Figure 4-1. $\boldsymbol{w}_1$ is an infinite-pitch wrench (moment) with wrench coordinates (0,0,0,0,1,0) and $\boldsymbol{w}_2$ is a zero-pitch wrench (force) with wrench coordinates (0,0,1,0,0,0).

Since through a line contact a 2-system of wrenches can be applied to a body, theoretically three line contacts on a body are sufficient to immobilize it.

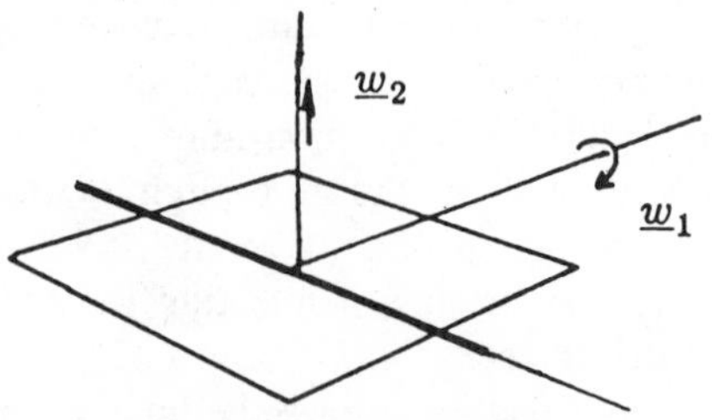

Figure 4.1 Frictionless line contact. The principal wrenches are shown.

For the body to be constrained the matrix formed from the wrench coordinates of the principal wrenches which can be applied by each line must have rank 6. This condition is

$$\operatorname{rank}\begin{bmatrix} 0 & 0 & 0 & 0 & \gamma_1 & \delta_1 \\ 0 & 0 & 0 & 0 & \gamma_2 & \delta_2 \\ 0 & 0 & 0 & 1 & \gamma_3 & \delta_3 \\ 0 & \alpha_4 & \beta_4 & 0 & \gamma_4 & \delta_4 \\ 1 & \alpha_5 & \beta_5 & 0 & \gamma_5 & \delta_5 \\ 0 & \alpha_6 & \beta_6 & 0 & \gamma_6 & \delta_6 \end{bmatrix} = 6. \tag{4.4}$$

For the matrix in Eqn. (4.4) to have full rank it is sufficient that none of the rows be linearly dependent on each other. In particular this means that, if no two of the line contact wrench systems intersect, the constraints imposed will be independent and the body will be immobilized. It is easiest to determine the conditions for independent line contacts by looking at the geometry of wrench systems that each constraint can apply. The wrench system associated with a single line contact wrench system consists of all zero-pitch wrenches (forces) that pass through the line of contact normal to the plane plus all infinite-pitch wrenches (moments) lying in the plane and perpendicular to the line of contact.

By examining wrench systems associated with two line contacts we determine that for independence the following three conditions must be true:

1. The lines of contact must not be parallel.
2. The planes that the lines lie on must not be parallel.
3. The common normal of the lines of contact must not be parallel to the line of intersection of the two planes.

Thus for a body to be constrained by three such line contacts, each pair formed from the three contacts must obey the above conditions.

4.1.3 5-Freedom Contact

The only common 5-freedom contact is a point contact without friction. Groups of such point contacts have been well studied in the literature [9] and

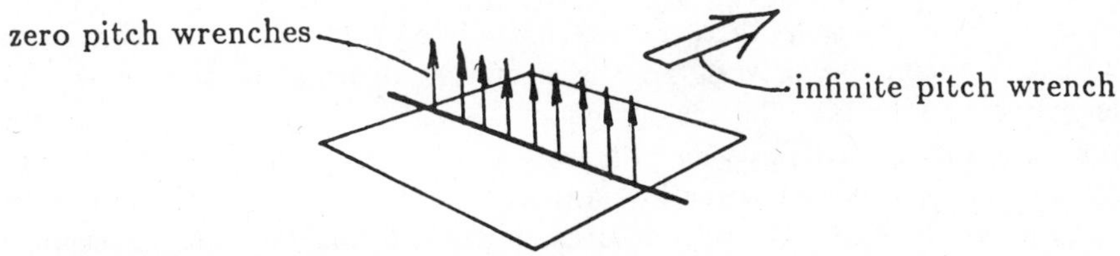

Figure 4.2 Frictionless line contact wrench system. Examples of wrenches contained in the system are shown.

need not be examined in detail here. Since each such contact can apply only a 1-system of wrenches, at least six are needed to constrain a body. If the matrix of principal wrenches coordinates for each contact,

$$\begin{bmatrix} 0 & \alpha_1 & \beta_1 & \gamma_1 & \delta_1 & \epsilon_1 \\ 0 & \alpha_2 & \beta_2 & \gamma_2 & \delta_2 & \epsilon_2 \\ 1 & \alpha_3 & \beta_3 & \gamma_3 & \delta_3 & \epsilon_3 \\ 0 & \alpha_4 & \beta_4 & \gamma_4 & \delta_4 & \epsilon_4 \\ 0 & \alpha_5 & \beta_5 & \gamma_5 & \delta_5 & \epsilon_5 \\ 0 & \alpha_6 & \beta_6 & \gamma_6 & \delta_6 & \epsilon_6 \end{bmatrix},$$

has rank 6 then the body will be constrained. The lines of action of five independent contact wrenches define a linear complex, and sixth contact wrench will be independent if its line of action does not belong to this linear complex [10]. The shapes of certain objects impose conditions as to which forces may be applied with point contact. In particular a sphere requires that all lines of action of applied forces pass through the center of the sphere, and a cylinder requires that all applied forces either intersect or are parallel to its axis. Thus it is impossible to constrain these two shapes with point contacts without friction.

It should be emphasized that the requirement of rank 6 on the matrix of principal wrenches is only a boundary condition. A (6×6) matrix may indeed have full-rank and still represent a set of wrenches which are badly conditioned for immobilizing an object. The question of the relative quality of various grasps is still a topic open to further research.

5. COMPLETE RESTRAINT AND INTERNAL FORCES

Up to this point we have been tacitly assuming that the wrenches that may be applied by the fingers to an object to restrain it can act in either a positive or negative sense. In reality this is not the case with many types of contacts between bodies. Two bodies contacting at a point can only push on each other; they cannot pull on each other. Forces that rely on friction between contacting bodies can be present only if there is a sufficient normal force to make the friction active. In fact the magnitude of a friction force depends upon the product of the normal force and the coefficient of friction, but it is at least necessary that the normal force by positive. If such *unisense* force limitations are imposed on some of the wrenches acting to constrain an object in a particular grasp, even if the matrix of screw coordinates of these wrenches has rank 6, it can happen that only a subset of all possible disturbance forces

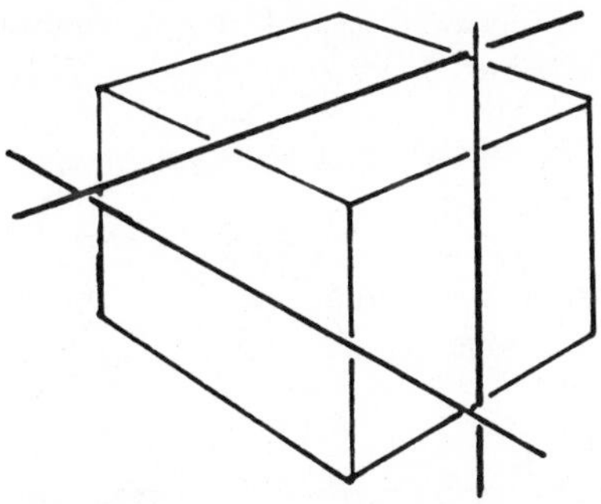

Figure 5.1 Cube constrained by three frictionless line contacts. The cube will be held securely only against forces which do not cause contact with the lines to be broken.

acting on the body can be resisted by the grasp. The grasp can be maintained even in this case if the disturbance forces on the object (its weight, assembly forces, etc.), act to maintain contact between the fingers and the object. This is known as *force closure*, a term coined by Reuleaux [14]. If, however, the collection of wrenches acting on the object (uni- and bisense) can resist arbitrary disturbance wrenches, we have *form closure* or *complete restrait* [15]. As an example consider Figure 5.1.

By the conditions given in Section 4.1.2 the cube should be immobilized by the three frictionless line contacts shown. It is, however, easy to visualize that the cube will be held securely against disturbance forces which tend to push the cube into the lines. Any other forces will cause contact to be broken, allowing the cube to move.

Reuleaux showed that for planar motion at least four point contacts (without friction) are needed to completely restrain an object against all disturbance forces in the plane. Somov [16] and later Lakshminarayana [15] showed that a minimum of seven point contacts (without friction) are needed to completely restrain an object in a 3-dimensional space. Both of these results were based upon the fact that wrenches exerted on a body by point contacts have only one sense. As will be shown below, when an object is completely restrained by a grasp there will be certain sets of contact and frictional forces which can be applied to the object without disturbing its equilibrium. We will show algebraic and geometric approaches for identifying these *internal forces*. The algebraic approach will allow us to evaluate whether or not a particular grasp can completely restrain an object. The geometric approach gives a simple way to visualize the physical meaning of these internal forces.

5.1 Algebraic Analysis

Assume n wrenches act on an objects, p of which have only one sense such that they must be kept positive, while $n - p$ are bidirectional and can act in either sense. As in Section 4, we form the $(6 \times n)$ matrix, $\mathcal{W}$, this time with the first p row vectors representing the screw coordinates of the wrenches which have only a single sense:

$$\mathcal{W} = \left[\boldsymbol{w}_1 \boldsymbol{w}_2 \cdots \boldsymbol{w}_p \boldsymbol{w}_{p+1} \cdots \boldsymbol{w}_n \right]. \tag{5.1}$$

For an arbitrary net wrench $\boldsymbol{w}$ to be applied to the object (i.e., to resist externally applied disturbance wrenches) we must be able to find $\boldsymbol{c}$, a vector of contact wrench intensities, which satisfies

$$\mathcal{W}\boldsymbol{c} = \boldsymbol{w}. \tag{5.2}$$

We must make the first p elements of $\boldsymbol{c}$ positive because they are the coefficients of the wrenches with only positive senses. If $n = 6$ and $\mathcal{W}$ has rank 6, then $\boldsymbol{c} = \mathcal{W}^{-1}\boldsymbol{w}$. However, in general the first p elements of $\boldsymbol{c}$ will not all be positive. Hence an arbitrarily applied disturbance wrench will cause contacts to be broken or cause slip to occur at the point of contact of some or all the unisense wrenches. A grasp can resist some wrenches on an object so long as it is within the column space of $\mathcal{W}$ in Eqn. (5.2), subject to the limitation that the first p elements of $\boldsymbol{c}$ are positive. It follows that at least seven wrenches must be applied to a body in order to restrain it completely with some or all unisense wrenches. For in the case $n = 7$, solutions to equation 5.2 will have the form

$$\boldsymbol{c} = \boldsymbol{c}_p + \lambda \boldsymbol{c}_h \tag{5.3}$$

where $\boldsymbol{c}_p$ is a particular solution to Eqn. (5.2) and $\boldsymbol{c}_h$ is a homogeneous solution to it. The scalar, λ, is an arbitrary free variable, which determines the magnitude of the "internal force". $\boldsymbol{c}_h$ lies in the null space of $\mathcal{W}$. As noticed by Lakshminarayana, if all the elements of $\lambda\boldsymbol{c}_h$ are positive then, for any value $\boldsymbol{c}_p$, an arbitrarily large λ can be found that will make all the elements of $\boldsymbol{c}$ positive while still satisfying Eqn. (5.2). In our case we require that the first p elements of $\lambda\boldsymbol{c}_h$ be positive. The homogeneous solution can be thought of as a set of bias forces which, without disturbing the object's equilibrium, can be used to increase some or all the contact forces until they are positive.

Designating q as the dimension of the null space of $\mathcal{W}$, we have $q \leqslant n - 6$. Thus it is possible that the dimension of the null space of $\mathcal{W}$ will be greater than 1 when more than seven wrenches, $n > 7$, act upon the body. The homogeneous solution will then have up to q free variables which can be chosen so as to make the first p elements of $\boldsymbol{c}$ positive. Thus there will be more than one set of independent internal forces.

Whether these q-dimensional homogeneous solutions will permit us to make the first p elements of $\boldsymbol{c}$ positive must be investigated in each case. The q-dimensional homogeneous solution can be represented as

$$\boldsymbol{c}_h = \left(\boldsymbol{c}_{1,h}\lambda_1 + \boldsymbol{c}_{2,h}\lambda_2 + \cdots + \boldsymbol{c}_{q-1,h}\lambda_{q-1} + \boldsymbol{c}_{q,h} \right)\lambda_q \tag{5.4}$$

where the q $\boldsymbol{c}_{i,h}$s are a set of n-element basis vectors which span the q-dimensional null space. There are two cases where we are assured that the first

p elements of $\boldsymbol{c}$ in Eqn. (5.4) can be made positive. If the first p elements of any of the $\boldsymbol{c}_{i,h}s$ are all the same sign then the appropriate choice of the corresponding λ_i will ensure the positive sense of the first p elements of $\boldsymbol{c}$. The second case occurs if there are $q \geqslant p$ free variables and the first p elements of the q $\boldsymbol{c}_{i,h}$ components of the homogeneous solution span a p-dimensional space. This, in effect, allows us to arbitrarily select the value of each of the first p elements of $\boldsymbol{c}_h$ with appropriate choices of the $\lambda_i s$. In all other cases we must test for complete restraint by solving a set of simultaneous linear inequalities to determine if values of the $\lambda_i s$ exist which will make $\boldsymbol{c}'s$ first p elements positive. This process is illustrated in the following examples.

Example 5.1

Consider a two dimensional homogeneous solution in which we want to make the first 4 elements positive. We need look only at the first 4 elements since the other $n - 4$ elements may have arbitrary sign. We designate the subvector consisting of the first p elements of $\boldsymbol{c}_h$ as $\boldsymbol{c}_h^*$. With $p = 4$ we have

$$\boldsymbol{c}_h^* = \left(\begin{bmatrix} a \\ b \\ c \\ d \end{bmatrix} \lambda_1 + \begin{bmatrix} e \\ f \\ g \\ h \end{bmatrix} \right) \lambda_2 \tag{5.5}$$

If all the elements of the quantity in the parentheses above have the same sign, the elements of $\boldsymbol{c}_h^*$ can be made arbitrarily positive. Thus, if we can find a λ_1 which satisfies either of the following groups of inequalities, complete restraint will be possible. Either

$$\begin{aligned} a\lambda_1 + e &> 0 \\ b\lambda_1 + f &> 0 \\ c\lambda_1 + g &> 0 \\ d\lambda_1 + h &> 0 \end{aligned} \tag{5.6a}$$

or

$$\begin{aligned} a\lambda_1 + e &< 0 \\ b\lambda_1 + f &< 0 \\ c\lambda_1 + g &< 0 \\ d\lambda_1 + h &< 0. \end{aligned} \tag{5.6b}$$

For example, with $a = c = d = f = g = h = 1$ and $b = e = -1$, there is

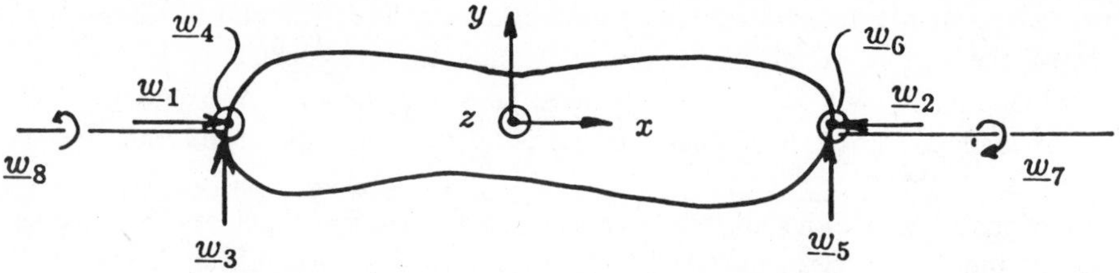

Figure 5.2 Body contacted by two soft fingers.

no value of λ_1 that solves either Eqn. (5.6a) or (5.6b) and complete restraint would be impossible. On the other hand if $a = b = c = d = e = f = g = h = 1$ then $\lambda_1 = 1$ satisfies Eqn. (5.6.a), and we have complete restraint with the given grip.

Example 5.2

Consider a body contacted by two soft fingers with corresponding wrenches as shown in Figure 5-2. $\boldsymbol{w}_1$ and $\boldsymbol{w}_2$ are contact forces which must be maintained positive to assure contact and allow friction to be active. Thus, $p = 2$. $\boldsymbol{w}_3 \cdots \boldsymbol{w}_6$ are forces while $\boldsymbol{w}_7$ and $\boldsymbol{w}_8$ are moments.

The other wrenches are the same as in example 5.1, with the addition of another moment applied as wrench $\boldsymbol{w}_8$. Then

$$\mathcal{W} = \begin{bmatrix} 1 & -1 & 0 & 0 & 0 & 0 & 0 & 0 \\ 0 & 0 & 1 & 0 & 1 & 0 & 0 & 0 \\ 0 & 0 & 0 & 1 & 0 & 1 & 0 & 0 \\ 0 & 0 & 0 & 0 & 0 & 0 & 1 & -1 \\ 0 & 0 & 0 & 1 & 0 & -1 & 0 & 0 \\ 0 & 0 & -1 & 0 & 1 & 0 & 0 & 0 \end{bmatrix}$$

and

$$\boldsymbol{c}_h = ([1\ 1\ 0\ 0\ 0\ 0\ 0\ 0]^T\lambda_1 + [0\ 0\ 0\ 0\ 0\ 0\ 1\ 1]^T)\lambda_2.$$

Thus the first $p = 2$ elements of $\boldsymbol{c}_h$ can be made arbitrarily positive and the grasp shown in Figure 5-2 can completely restrain the object. The grasp has 2 independent internal forces, one parameterization of which is given in the homogeneous solution above.

5.2 Geometric Analysis

Each contact with an object allows a system of wrenches to be applied to it. When two contacts act on an object, the corresponding *wrench systems*, say w_a and w_b, may or may not have some wrenches in common (non-empty intersection). If the two systems do intersect, the effect of a wrench from w_a, belonging to the intersection, can be completely counteracted by a wrench from w_b, also in the intersection and having the same pitch and axis, applied with equal intensity in the opposite sense. This of course assumes that these two wrenches may be applied without breaking contact. This pair of equal and opposite wrenches corresponds to one of the homogeneous solutions of Eqn. (5.4). The fact that the two magnitudes may be increased or decreased by the same amount without disturbing the object's net equilibrium corresponds to the freedom we have in choosing the value of the corresponding λ_i in Eqn. (5.4). The application of such equal and opposite wrenches to an object by separate contacts does affect the state of internal stress in the object, but since we consider the object rigid, no net motion results. The self-canceling effect of these two wrenches holds even if other wrenches or constraints act upon the object. As in Section 5.1, by properly choosing the magnitude of the internal forces, we can ensure that contact will be maintained and frictional restraints kept active even in the presence of external forces on the object.

It is possible that the wrench system that comprises the intersection of w_a and w_b will be of order greater than 1. If

$$w_h = w_a \cap w_b \tag{5.7}$$

is of order q (a q-system) then there will be q linearly independent wrenches in w_h that can be balanced exactly by q linearly independent wrenches in w_b. The wrenches in w_h are the same as those in the homogeneous solution, Eqn. (5.4). There will be q free variables or internal forces which can be specified independently. The actual choice of these q pairs of equal and opposite wrenches is not unique in the same way that the choice of the q basis vectors

$(c_{i,h}s)$ is not unique in Equation (5.4).

From Table 2-3 we know that the *twist system,* t_a, which an object may execute as a consequence of a particular contact is equal to the reciprocal of the system of wrenches, w_a, that may be exerted on the object by the same contact (or set of contacts). Denoting the reciprocal system with the prime (′) notation as before, $w_a = t_a''$ and $t_a = w_a''$. Thus, by application of DeMorgan's theorem, w_h may be written in terms of the contact twist systems:

$$w_h = (w_a \cap w_b) = (t_a'' \cap t_b'') = (t_a \cup t_b)'. \tag{5.8}$$

This rather unexpected result provides an alternative (though equivalent) method of identifying internal forces.

The wrenches comprising w_a need not necessarily come from a single contact wrench system. Several contact wrench systems may be amalgamated to form w_a. The same holds for the composition of w_b as long as it results from the union of contact wrench systems different from those used to form w_a. In searching for internal forces that may be applied to an object, we look for intersections between different groupings of wrench systems acting on the body. If n is the number of wrenches acting on a body (n is the sum of the orders of all contact wrench systems acting on the object), then the number of independent internal forces (pairs of wrenches of equal and opposite intensities that may be applied simultaneously) will be $q \leqslant n - 6$ as in Section 5.1.

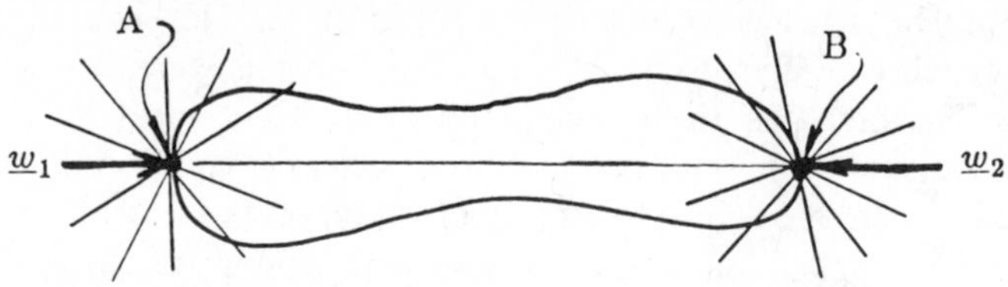

Figure 5.3 Body contacted by two friction point contacts.

Example 5.3

Suppose two friction point contact wrench systems act on an object as in Figure 5.3. Each contact wrench system is a 3-system of all zero-pitch wrenches passing through the point of contact.

Two such systems will always intersect along the line joining the two contact points. Thus the action of wrench (force) $\boldsymbol{w}_1$ at A can be counteracted by an equal and opposite wrench (force) $\boldsymbol{w}_2$ acting as B. Since the intensities of $\boldsymbol{w}_1$ and $\boldsymbol{w}_2$ are equal and opposite at equilibrium, the contact forces at A and B can be arbitrarily large (subject to structural limitations). Note that since these two 3-systems intersect (in a 1-system) their union will be only a 5-system and full immobilization of the object will be impossible according to Eqn. (4.4).

Example 5.4

If instead two soft fingers act on the body as in Figure 5-4 there will be two internal forces as follows. The system of wrenches which may be exerted by each soft finger is a 4-system as shown in Table 2-3. The intersection of two of these systems with co-axial surface normals as shown in Figure 5-2 consists of wrenches with any pitch along the line $A-B$ connecting the contact points. A

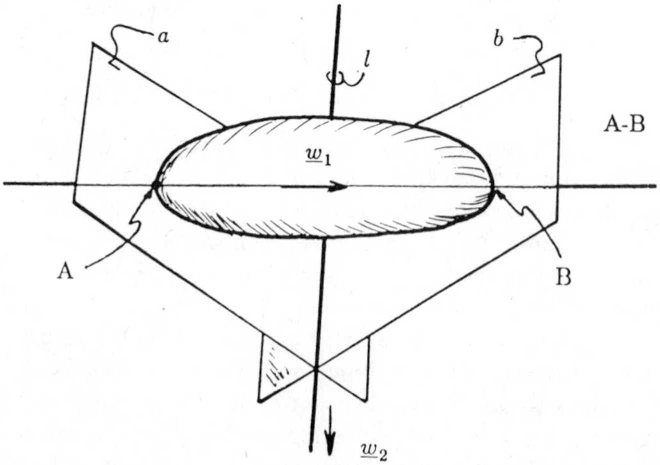

Figure 5.4 Body held by two soft fingers with non-parallel surface normals.

suitable choice of the 2 internal forces is a pair of equal and opposite forces directed along $A-B$ and a pair of equal and opposite moments exerted about $A-B$. These would cause the object to be under compression and torsion respectively.

If the contact normals were not co-axial we would have the situation depicted in Figure 5.3. The contact tangent planes, a and b, and their line of intersection, l, are shown. The intersection of the two wrench systems is again a 2-system, this time containing a zero-pitch wrench along $A-B$ and another zero-pitch wrench along line l.

If the order of the intersection of these two 4th-order wrench systems was greater than 2 we would have to find other independent screws in the intersection to completely describe the internal forces. However, from Eqn. (5.8) we can show that the intersection of the 2 soft-finger wrench systems is a 2-system. The systems of twists allowed by contact A or B alone are 2-systems (degenerate cylindroid) consisting of all zero-pitch wrenches in the tangent plane and passing through the point of contact. In the current example these 2-systems are disjoint. Their union is a 4-system and its reciprocal system (w_h) is a 2-system.

6. FORCE APPLICATION AND VELOCITY ANALYSIS

In practice an object grasped by an articulated hand will encounter external forces as well as internal forces. These external forces will result from contact with bodies in the environment other than the fingers as well as from the weight of the grasped object and dynamic forces on it. To do useful work we will want to measure forces acting on the body, such as its weight or contact with obstacles. In other instances we will want to control the forces exerted by the grasped object on bodies in the environment, as in fitting parts together during assembly tasks. Alternatively we also want to monitor and control displacements of the grasped object. Monitoring displacements will allow us to assess the effect of external disturbances on the object's position. We can also determine the constraints imposed upon the object by contact with bodies in the environment by observing the motion that results when the hand applies forces to the object. Finally by controlling the object's displacement we could reorient or reposition it according to task requirements.

The following development begins with a static force analysis to determine the net external and internal forces that result when finger contacts act to exert a set of wrenches on the object. We then derive the three of the important relationships between forces and velocities in a hand-object system. The relationships derived are all embodied in the *grip transform*, G, or functions of it. G is a square matrix having dimension and rank equal to the number, n, of finger contact wrenches acting on the object, and is constant for a particular grasp on an object. G need be recomputed only when the grasp on the object changes (i.e., location of contact points in object frame of reference changes significantly).

We treat the fingers as "black boxes" which allow us to sense and control the wrenches exerted by the finger link or links on the object as well as sense and control the motion of links at their points of contact.

In this section we will use a slightly different nomenclature for forces and velocities. $\mathcal{F}$ will be used to represent generalized forces acting upon the body being manipulated. In the case of an open chain manipulator it is simply the net wrench exerted on the object. In the case of an object held by an articulated hand it is composed of the components of the net wrench on the object and the intensities of the internal forces. $\mathcal{V}$ will be used to represent generalized velocities of the object being manipulated. In the case of an open chain manipulator it is composed of the Cartesian components of the net twist of the body with the first and second three elements interchanged. (This interchange leads to more compact formulations.) In the case of an articulated hand it is composed of the net twist of the body (with the first and second three elements interchanged) and the intensities of the internal (virtual) velocities.

6.1 The Grip Transform G

The role played by the grip transformation, G, in controlling manipulation with a multi-fingered (or multi-armed for that matter) manipulation system is similar to the role played by the Jacobian matrix, J, in controlling manipulation with an open chain arm. Before introducing G we will review the use of J. The relationship

$$\boldsymbol{\tau} = J^{\mathrm{T}}\mathcal{F} \tag{6.1}$$

is often used to determine what torques (or forces), $\boldsymbol{\tau}$, must be applied at the (controlled) joints of an open chain manipulator to cause it to exert a net generalized force (or wrench), $\mathcal{F}$, at the end effector. The Jacobian matrix, J, relates the vector of joint angular and linear velocities, $\boldsymbol{\Omega}$, to the end effector's velocity (angular and linear), $\mathcal{V}$, such that

$$\mathcal{V} = J\boldsymbol{\Omega} \tag{6.2}$$

where $J_{ij} = \partial \mathcal{V}_i / \partial \Omega_j$. Equation (6.1) can be derived from Eqn. (6.2) by assuming no energy storage or dissipation occurs in the system and equating input power with output power. Thus

$$\boldsymbol{\Omega}^{\mathrm{T}}\boldsymbol{\tau} = \mathcal{V}^{\mathrm{T}}\mathcal{F} = (\boldsymbol{\Omega}^{\mathrm{T}}J^{\mathrm{T}})\mathcal{F}. \tag{6.3}$$

Assuming $\boldsymbol{\Omega} \neq 0$, this yields Eqn. (6.1). A geometric method for finding J is given by Shimano [17]. The procedure entails finding the columns of J by projecting the effect of each isolated joint motion onto a coordinate system based in the hand. The columns of J^T can also be found directly by projecting wrenches exerted on the end effector onto each joint axis. Eqn. (6.2) could then be derived from (6.1) again using energy conservation principles.

In a similar way the grip matrix, G, introduced here, allows us to determine what finger contact wrenches must be exerted in order to apply a desired net force on and in a grasped object as well as to determine net object velocities from sensed contact velocities. Because forces are exerted on the object by links connected in parallel, rather then serially, the sense in which we may make projections to geometrically find the elements of G is reversed. G^{-T} and G^{-1} may be found geometrically; G^{T} and G must be found by matrix inversion. (We use $G^{-\mathrm{T}}$ to mean the *inverse* of the *transpose* of G.)

6.2 Control and Sensing of External and Internal Forces

To construct $G^{-\mathrm{T}}$ we first form the $\mathcal{W}$ matrix as in Eqn. (5.1) and check for complete restraint as described in Section 5.1. This requires that the rank of $\mathcal{W}$ be six and that the first p elements, corresponding to the p unisense contact wrenches, of the homogeneous solution of Eqn. (5.1) can be made arbitrarily positive. $G^{-\mathrm{T}}$ is formed by augmenting the $(6 \times n)\, \mathcal{W}$ matrix with the $(n - 6)$, n-element basis vectors of the homogeneous solution (found either geometrically or algebraically as shown in Section 5):

$$G^{-\mathrm{T}} = \begin{bmatrix} \mathcal{W} \\ \hline \boldsymbol{c}_{1,h}^{\mathrm{T}} \\ \vdots \\ \boldsymbol{c}_{n-6,h}^{T} \end{bmatrix}. \tag{6.4}$$

This matrix will be square, and assuming that the $(n - 6)\boldsymbol{c}_{i,h}s$ are linearly independent and span the $(n - 6)$ dimensional null space of $\mathcal{W}$, $G^{-\mathrm{T}}$ will be invertible. We define the generalized force,

$$\mathcal{F} = \left[f_x f_y f_z m_x m_y m_z \lambda_1 \cdots \lambda_{n-6}\right]^{\mathrm{T}}, \tag{6.5}$$

to be an n-vector with the first six elements being the net wrench (external force) applied to the object expressed as three force and three moments components and the last $n - 6$ elements being the magnitudes of the internal forces. If $\boldsymbol{F}$ is defined to be a n-vector of contact wrench intensities for the n contact wrenches (it is equivalent to $\boldsymbol{c}$ in Eqn. [5.2]), then

$$\boldsymbol{F} = [c_1 \cdots c_n]^{\mathrm{T}}, \tag{6.6}$$

$$\mathcal{F} = G^{-\mathrm{T}}\boldsymbol{F} \tag{6.7}$$

and

$$\boldsymbol{F} = G^{\mathrm{T}}\mathcal{F}. \tag{6.8}$$

(In the case of a grip using three friction point contacts to secure the object, $\boldsymbol{F}$ would be composed of the nine components of the finger tip contact forces.) Equations (6.7) and (6.8) are useful in sensing and control as follows. By using Eqn. (6.7) we can determine what external forces are acting on the object and what internal forces are supported by the fingers from a set of measurements (i.e., from sensors) of forces and moments acting on the fingers. On the other hand, if we wanted to push on a grasped object with a particular net external wrench, say for example during an assembly task, and create a controlled set of internal forces, λ_i, Eqn. (6.8) would be used. The values of the $\lambda_i s$ would be used to keep the unisense wrenches positive and maintain a secure grasp.

6.3 Control and Sensing of Velocities

Before using energy conservation principles to derive the grip velocity relationships, we must clarify the meaning of input and output power in grip systems. We define $\boldsymbol{V}$ to be a vector of twist intensities per unit time occurring along the *wrench* axes at each contact.

$$\boldsymbol{V} = [d_1 \cdots d_n]^{\mathrm{T}}. \tag{6.9}$$

(Again, in the case of a grip using three friction point contacts this vector would be comprised of the nine components of the fingertip velocities.) Since the elements of $\boldsymbol{V}$ are the intensities of motion along the axes of the corresponding contact wrenches

$$\boldsymbol{F}^{\mathrm{T}}\boldsymbol{V} = \sum_{i=1}^{n} c_i d_i \tag{6.10}$$

is the power being input to the object. It is the sum of the power being input at the contact points as a result of motion along each axis of force application.

We define $\mathscr{V}$ to be an n-vector, the first six components of which represent the body's linear and angular velocity and the remaining $n - 6$, γ_is are its (virtual) velocities resulting from deformation of the body. Thus

$$\mathscr{V} = [v_x v_y v_z \omega_x \omega_y \omega_z \gamma_1 \cdots \gamma_{n-6}]^{\mathrm{T}} \tag{6.11}$$

and

$$\mathscr{F}^{\mathrm{T}}\mathscr{V} = f_x v_x + f_y v_y + f_z v_z + m_x \omega_x + m_y \omega_y + m_z \omega_z + \lambda_1 \gamma_1 + \cdots + \lambda_{n-6}\gamma_{n-6} \tag{6.12}$$

is the sum of power that is used to do work on the environment and to store energy as the object deforms. However, since we consider objects rigid and massless no energy can be stored. The internal forces, λ_i, may be non-zero but they can do no work without violating our assumptions. Thus we expect the $\gamma_i s$ to be zero and if we do encounter non-zero values for these internal displacements it will mean that our grip has slipped and/or our assumptions about contact constraints have been violated.

Equating power input with power output and rate of storage [Eqn. (6.10) and (6.12)] and using Eqn. (6.7), we have

$$\boldsymbol{F}^{\mathrm{T}}\boldsymbol{V} = \mathscr{F}^{\mathrm{T}}\mathscr{V} = (\boldsymbol{F}^{\mathrm{T}}G^{-1})\mathscr{V}, \tag{6.13}$$

from which it follows that if $\boldsymbol{F} \neq 0$

$$\boldsymbol{V} = G^{-1}\boldsymbol{\mathcal{V}} \tag{6.14}$$

and finally, by inversion, that

$$\boldsymbol{\mathcal{V}} = G\boldsymbol{V}. \tag{6.15}$$

Equation (6.14) allows us to take a desired set of body velocities (with the (γ_is = 0) and determine what set of finger velocities (contact twists) are required. Eqn. (6.15) permits us to determine the body's velocity (as well as check for slipping grip) from a set of finger velocity measurements. An example of the grip matrix and its use in hand control system may be found in [19].

Thus we have determined the four basic relationships necessary for sensing and control in manipulation of an object grasped by an articulated hand. It is seen that the linear transformation, G, is of fundamental importance. The duality of forces and velocities, which is a principle in mechanics, is evident in the way in which force and velocity transformations may be derived from each other.

7. CONCLUSIONS

This chapter has been concerned with two aspects of articulated hands: (a) their kinematic structure and (b) the control of forces and small motions. We began with an analysis of the constraints imposed by contacts between bodies including the effects of friction. We determined the nature of the constraints imposed by different types of contact, both in terms of the number of freedoms of motion allowed by each contact and more quantitatively in terms of screw systems. An enumeration of a large number of potential hand designs and grasps was then subjected to mobility and connectivity analysis and a group of acceptable designs was identified. Acceptable designs were those that, in terms of connectivity, could immobilize a grasped object with the finger joints locked while also having the ability to impart arbitrary velocities (and implicitly small displacements) to the grasped object.

The ability to move the body actively with arbitrary velocity also implies that arbitrary forces can be exerted. Because the connectivity analysis only establishes a set of necessary design requirements, further analysis, in terms of screw systems, was made on each of the potentially acceptable designs. By looking at the constraints imposed by groups of contacts, we were able to reduce the set of acceptable designs to the two most suitable designs - a three-finger hand with three joints in each finger and a three-finger hand with two joints in each finger. The first touches the object with 3-freedom contacts at each fingertip and the second with 4-freedom contacts at each fingertip. An analysis of the effects of unisense contacts was then used to show the need for more than six active joints in acceptable articulated hand designs. Our analysis thus points quite directly at a particular hand design - one with three 3-jointed fingers with friction point contacts at each fingertip. Thus of the 600 hand designs originally considered we have shown that only one is truly satisfactory for our purposes - a rather remarkable conclusion.

In broadening our understanding of the mechanics of grasping we have developed the basic relationships necessary for performing force and velocity control with an articulated hand. All the relationships apply equally well to hands with more than three fingers.

7.1 Stanford/JPL Hand

As an application of the ideas developed here a three-finger hand, known as the *Stanford/JPL Hand,* was designed. It is intended for use as a research tool in the control and design of articulated hands. Ultimately it is intended for industrial use in conjunction with existing mechanical arms.

In order to minimize the weight and volume of the hand the motors were located on the forearm of the manipulator and teflon coated cables in flexible conduits were used to transmit forces to the finger joints. By reasoning similar to that shown in Section 5, it was determined that $n + 1$ unisense actuators (tension cables) were needed to control n bidirectional degrees of freedom (finger joints). Morecki [18] used similar reasoning to propose a six-degree-of-freedom manipulator actuated by seven tension cables. Here it indicates that to power the nine finger joints a minimum of ten actuators is needed. This, however, requires that each finger joint be coupled to the other eight joints. To reduce this coupling and to make the finger systems modular, it was decided to use four cables for each three-degree-of-freedom finger making each finger identical and independently controllable.

This has the benefit of reducing the number of different parts, making the design more economical, and permits additional fingers to be added if desired. The arrangement of cables is shown in Figure 7-1. By applying appropriate combinations of cable tensions, $T_1, \ldots, T_4$, arbitrary torques may be exerted about axes z_1, z_2 and z_3. The homogeneous solution in this case corresponds to a set of tensions that result in no net torque about any of the three joints. This null vector of cable tension affects only the bearing reaction forces on axis z_1. An appropriate control algorithm would seek to minimize this null vector subject to the constraint that all the cable tensions remain non-negative. Thus by using the (3×3) Jacobian matrix of the finger, a desired set of forces to be exerted at the fingertip at point P can be mapped into joint torques which can in turn be mapped into tendon coordinates. Once we are able to exert controllable forces at each fingertip, the methods of force and position control outlined in Section 6 can be applied. A more detailed description of the control algorithms for the Stanford/JPL hand is given in [19].

Our conviction that accurate exertion of forces is central of dextrous manipulation led us to place a tendon tension sensing mechanism, shown in Figure 7-2, on each cable where it enters a finger. Tension in the cable will

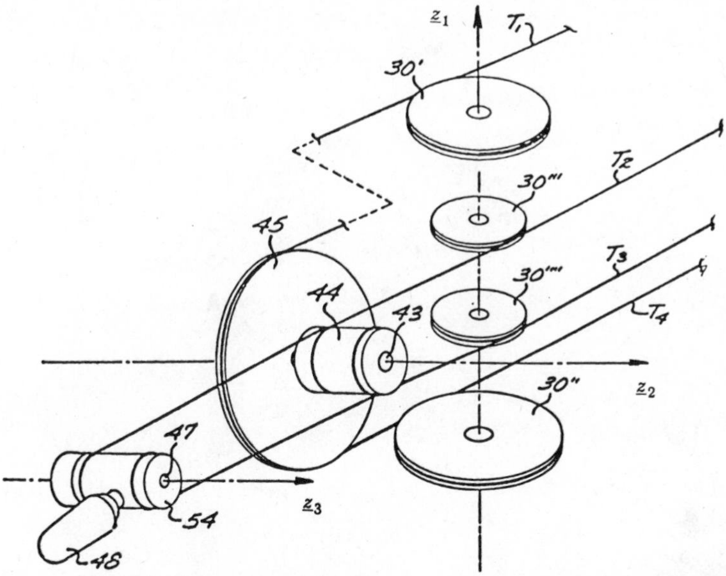

Figure 7.1 Stanford/JPL Finger Cabling. The appropriate combinations of cable tensions, T_1 through T_4, can be found to achieve desired moments about the three finger joints. The sum of the cable tensions is a free variable not affecting the joint moments.

induce strain at the base of the idler pulley support which can me measured with a strain gauge.

This provides two advantages: we can accurately sense forces applied to the finger tip by monitoring the cable tensions. (Because the sensors are close to the site of force application, the readings will be more accurate than if the sensors were placed at the motors.) And by placing the tension sensors after the major sources of friction in the system (brush, gear and cable conduit friction), we may close a control loop around these non-linearities and minimize their effect on system performance. We should thus be able to sense and apply forces accurately with the fingertip.

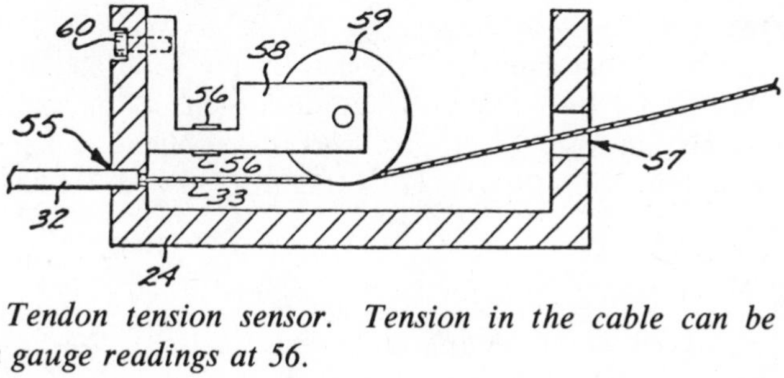

Figure 7.2 Tendon tension sensor. Tension in the cable can be determined from strain gauge readings at 56.

Figure 7.3 Stanford/JPL Hand with 4 tendons installed.

Position and velocity information is obtained by using incremental optical encoders placed directly on the motor shafts. As described in [19] actual joint and fingertip positions can be derived from these encoder readings and corrected for cable stretch by using the tension readings.

The relative placement of fingers on the palm was based on heuristic information obtained from a physical mockup of the hand and from a partial optimization performed by Craig [19]. A discussion of the choice of finger dimensions and placements to maximize workspace volume and force exertion accuracy may be found in [19] and [20]. Figures 7-3 and 7-4 show the hand as fabricated.

7.2 The Future

This chapter has been based upon instantaneous analysis. We have not really dealt with the problems of large motions of grasped objects. It may, for

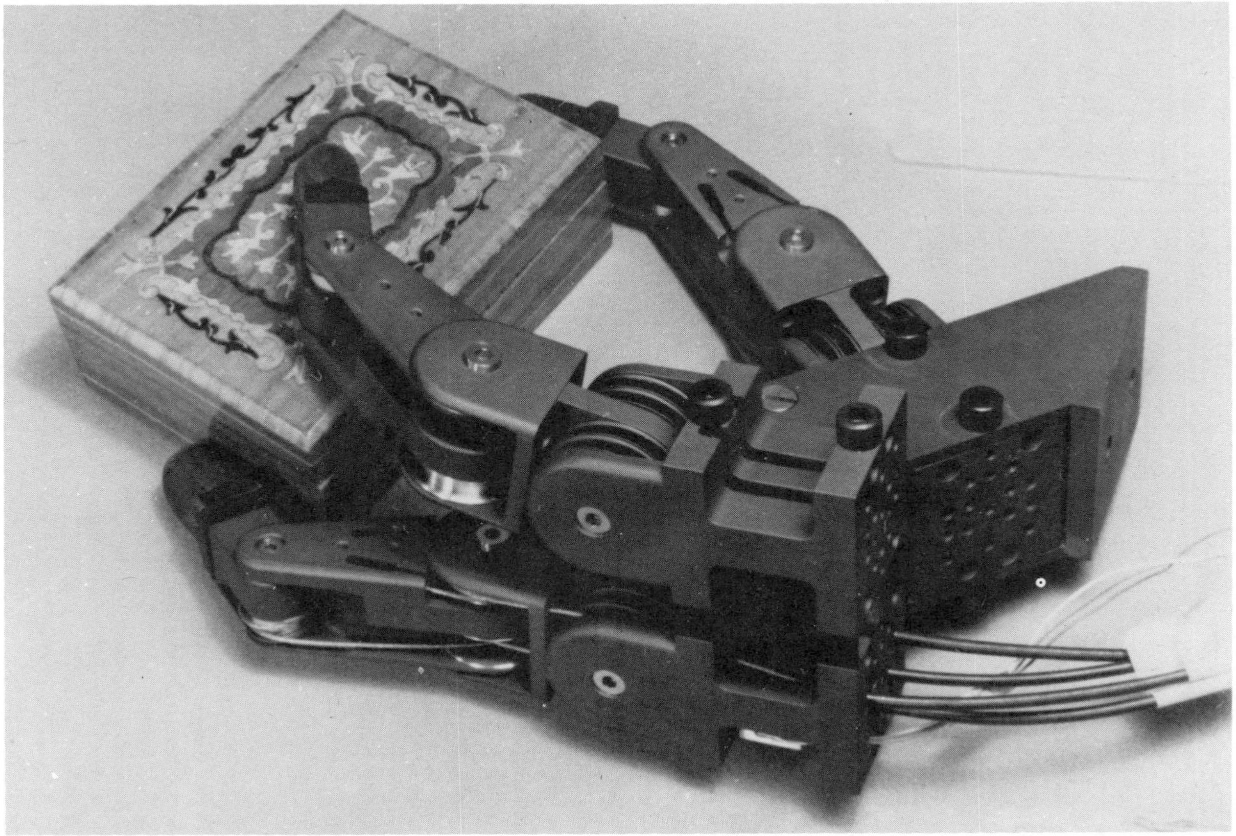

Figure 7.4 Stanford/JPL Hand holding a cube in tip prehension.

example, be possible to characterize the workspace of several fingers (or manipulators for that matter) working together by looking at intersections of the reachable volumes, for single-chain mechanisms described by Roth and Gupta [21,22]. Although we feel that a hand is most appropriate for the control of small motions, it is obvious that human hands can do much more. Such operations as performing large or even continuous reorientation of objects are trivial for humans. This requires repositioning some of the fingers into new grasps, and the transition from one grasp to another necessitates having enough fingers to stably secure the object during the transitions. Thus four or more fingers may be required to extend the utility of a mechanical hand beyond our current design. The value of more fingers is qualitatively obvious if one attempts to manipulate objects with one's own hands using two, three, four and five fingers. With two fingers very little more than pick-and-place operations can be performed. With three fingers more stable grasping and limited reorientations can be performed. With four and five fingers we can make the transitions between different grasping postures without losing complete restraint of the object, and we can even use some of the fingers to secure the object to the palm while the remaining fingers can be used to perform limited operations on the grasped object.

Our criterion for identifying acceptable hands was based upon the assumption that a hand must be able to restrain completely a grasped object as well as impart arbitrary forces and small motions to it. It may be, however, that a hand can usefully augment a manipulator by providing accurate force and small motion control only in particular directions. The arm joints could then be used to orient the sensitive axes of the hand into directions suitable for each particular task. Finally it may be that the boundary between the hand and the arm should be less distinct than we have implied. Hybrid designs combining limited-degree-of-freedom hands with limited-degree-of-freedom arms could provide full mobility and force application capability to grasped objects. Such designs could combine the adaptive grasping capability of an articulated hand with the greater range of motion of more traditional arm designs while reducing the number of joints necessary.

REFERENCES

[1] D. S. Childress, "Artificial Hand Mechanisms," Mechanisms Conference and Internal Symposium on Gearing and Transmissions, San Francisco, CA, October, 1972.

[2] S. C. Jacobsen, et al. "Development of the Utah Artificial Arm," *IEEE Transactions on Biomedical Engineering,* Vol. BME-29, No. 4, April, 1982.

[3] F. R. E. Crossley and F. G. Umholtz, "Design for a Three Fingered Hand," *Mechanism and Machine Theory,* Vol. 12, 1977.

[4] F. Skinner, "Designing a Multiple Prehension Manipulator," *Mechanical Engineering,* September, 1975.

[5] A. Rovetta, "On Specific Problems of Design of Multipurpose Mechanical Hands in Industrial Robots," *Proc. 7th ISIR,* Tokyo, 1977.

[6] M. Mori and Yamashita, "Mechanical Fingers as Control Organ and its Fundamental Analysis," JSME Conf. Proc. (conference unknown), Session 3, Paper 4.

[7] T. Okada, "Computer Control of Multi-Jointed Finger System," 6th International Joint Conference on Artificial Intelligence, Tokyo, Japan, 1979.
[8] B. Lündstrom, *Industrial Robot - Gripper Review,* Published by International Fuidics Services, Ltd., Bedford, England, 1977.
[9] R. S. Ball, *A Treatise on the Theory of Screws,* Cambridge University Press, 1900.
[10] K. H. Hunt, *Kinematic Geometry of Mechanisms,* Oxford University Press, 1978.
[11] K. J. Waldron, *The Mobility of Linkages,* Ph.D. Dissertation, Stanford University, 1982.
[12] M. S. Ohwovoriole, *An Extension of Screw Theory and its Application to the Automation of Industrial Assemblies,* Stanford Artificial Intelligence Laboratory Memo AIM-338, April, 1980.
[13] D. E. Knuth, *Fundamental Algorithms,* Volume 1, Addison-Wesley Publishing Co., 1973, pg. 488.
[14] F. Reuleaux, *The Kinematics of Machinery,* Dover Publications, 1963.
[15] K. Lakshminarayana, "Mechanics of Form Closure," *ASME* Paper No. 78-DET-32, 1978.
[16] P. Somov, "Über Schraubengeschwindigkeiten eines festen Körpers bei verschiedener Zahl von Stützflächen," *Zeitschrift für* Mathematik und Physik, V. 42, 1897, pp. 133-153, 161-182.
[17] B. Shimano, *The Kinematic Design and Force Control of Computer Controlled Manipulators,* Stanford Artificial Intelligence Laboratory Memo AIM-313, March, 1978.
[18] A. Morecki, "Synthesis and Control of the Anthropomorphic Two-Handed Manipulator," *Proc 10th International Symp. on Industrial Robots,* Milan, Italy, 1980.
[19] J. K. Salisbury and J. J. Craig, "Articulated Hands: Force Control and Kinematic Issues," *International Journal of Robotics Research,* Vol. 1, No. 1 Spring 1982, MIT Press.
[20] J. K. Salisbury, *Kinematic and Force Analysis of Articulated Hands,* Ph.D. dissertation, Stanford University, 1982. Also Stanford Computer Science Department Report No. STAN-CS-82-921.
[21] B. Roth, "Performance Evaluation of Manipulators from a Kinematic Viewpoint," *NBS Special Publication, Performance Evaluation of Programmable Robots and Manipulators,* pp. 39-61, 1975.
[22] K. C. Gupta and B. Roth, "Design Considerations for Manipulator Workspace," *Trans ASME Journal of Mechanical Design,* 1982.

NOMENCLATURE

L,M,N,P,Q,R	Plücker line coordinates
p	screw, wrench or twist pitch
$(S_1,S_2,S_3,S_4,S_5,S_6),\boldsymbol{s},\boldsymbol{s}_i$	screw
$\boldsymbol{w},(w_x,w_y,w_z)$	angular velocity body
$\boldsymbol{v},(v_x,v_y,v_z)$	velocity of a point
$(T_1,T_2,T_3,T_4,T_5,T_6),\boldsymbol{t},\boldsymbol{t}_i$	twist
$\boldsymbol{f}$	force
$\boldsymbol{m}$	moment
$(W_1,W_2,W_3,W_4,W_5,W_6),\boldsymbol{w},\boldsymbol{w}_i$	wrench
s_i,w_i,t_i	screw, wrench or twist system

M	mobility
$\cup$	union
$\cap$	intersection
$\mathcal{R}^n$	n-dimensional space
ϕ	empty set
$\mathcal{W}$	matrix of contact wrench coordinates
λ, λ_i	free variable, "internal force"
$\boldsymbol{\tau}$	vector of joint forces and torques
$\boldsymbol{\Omega}$	vector of joint translational and angular velocities
G	grip transform
$\boldsymbol{\mathcal{F}}$	generalized force on grasped object
$\boldsymbol{\mathcal{V}}$	generalized velocity of grasped object
$\boldsymbol{x}$	generalized displacement of grasped object
$\boldsymbol{F}$	vector of contact wrench intensities
$\boldsymbol{V}$	vector of twist intensities along contact wrench axes
J	Jacobian matrix open chain mechanism

5

A Vector Analysis Of Robot Manipulators

HARVEY LIPKIN
AND
JOSEPH DUFFY

1. INTRODUCTION

Robot manipulators are complex mechanical devices. The computation of joint displacements is determined by highly non-linear trigonometrical equations. Here, a novel method is developed for generating these trigonometrical equations using elementary vector operations. *A major goal is accomplished in that the geometry of a robot is related directly to the trigonometrical formulation.* This provides a basis for a deeper understanding of robot geometry and equally important it facilitates the identification of singular configurations which cause problems in the control of robots. A comprehensive understanding of the geometry of singular configurations is fundamental to the process of robot design.

In Section 2 an algorithm based on simple vector operations is developed to model a 6 degree-of-freedom robot by an equivalent one degree-of-freedom closed-loop spatial mechanism. The concept of closing the loop is not new. Pieper and Roth [1] were the first to point out the implicit correspondence between manipulators and spatial mechanisms using homogeneous transfer matrices. A set of explicit parameters for closing the loop are derived in Ref. [2] and an alternative derivation is presented in Section 2.

The geometry of the equivalent closed-loop spatial mechanism (Section 2) provides the framework for the development of the recursive notation which is developed in Section 3 using elementary vector operations and 3×3 rotation matrices. Briefly, the recursive notation expresses the direction cosines of unit vectors which specify the directions of robot links and offsets and is extended to include expressions for the mutual moments of these unit vectors. This novel derivation of mutual moments is an alternative to introducing dual numbers [3,4,5] into trigonometrical equations for spherical polygons [6,7] in order to obtain equations for corresponding spatial polygons. The recursive notation developed here has not only the advantage that lengthy trigonometrical equations can be written in a concise form but also that their geometric meaning is apparent. This is because the trigonometrical equations represent alternative expressions for direction cosines and mutual moments.

In Section 4 a number of solutions of trigonometrical equations are presented, and the advantages and disadvantages when applied to the computation of the joint displacements of robot manipulators is discussed. As far as the authors are aware no such comparison has appeared in the literature, and the results presented here indicate that one method of solution may well be preferred to another. For instance, trigonometrical equations can be transformed into algebraic equations using tan-half angle substitutions. Although this method facilitates the elimination of unwanted joint angles [2] it is shown here that indeterminacies are inherent in the formulation.

The recursive notation developed in Section 3 is a powerful analytical tool and by way of example it is used to analyze representative types of industrial robots including the PUMA 600 in Section 5. The analysis of such robot manipulators with special geometry enables a comparison to be made with other methods[8,9]. The analysis presented in this section yields a geometric identification of indeterminate manipulator configurations.

Finally in Section 6 a linearized singularity analysis using the theory of screws is presented for the industrial robots previously analyzed in Section 5. This section clearly illustrates the importance of singular configurations and their relevance in the control of robot manipulators.

2. MOBILITY AND COMPLETION OF THE SPATIAL LOOP

A kinematic chain may be described as a collection of rigid bodies that are connected together by joints which restrict the relative freedom of movement between the bodies. These chains are frequently formed using binary joints which connect two adjacent bodies. The most fundamental binary pair is the helical joint of which the revolute and prismatic joints are special cases where the helix pitches are zero and infinite respectively. Combinations of these two joints permit simple modeling of cylindrical, Hooke, spherical, and planar pair joints for both finite and instantaneous motion.

Kinematic chains are generally classified as open, closed or combinations of both. Broadly speaking, closed chains exhibit a loop structure where the beginning and end of the chain may be considered coincident. Open chains may be described as articulated in nature. Usually, serial manipulators can be modeled as open kinematic chains employing revolute and prismatic joints exclusively.

The object of this section is to detail the algorithm which completes or closes the open chain modeling a six degree-of-freedom serial manipulator into a single loop spatial mechanism of mobility one. This completion permits the investigation of serial manipulators via the existing body of literature that has been developed for single loop spatial mechanisms, Ref. [2].

The mobility of a general kinematic chain may be defined as the total degrees-of-freedom of the constituent rigid bodies relative to one body designated as ground and diminished by the total number of constraints imposed by the connecting joints. This may be formulated by the equation,

$$M = m(n-1) - \sum_{i=1}^{j} (m-f_i), \tag{2.1}$$

where,

M = mobility
m = degrees of freedom of single unconstrained rigid body
n = number of rigid bodies
j = number of joints
f_i = relative freedom afforded by joint i.

For spatial chains m=6 whereas for the special cases of spherical and planar motion m=3. For one degree-of-freedom joints (revolute, prismatic, helical) f_i=1.

The terminal body of a spatial open kinematic chain can be connected back to ground using a one degree-of-freedom joint, and therefore imposes an additional 6−1=5 constraints. Thus by Eqn. (2.1) a serial manipulator of mobility six is converted into a single loop spatial mechanism of mobility one. For the purpose of analysis, the additional joint is considered hypothetical, not physical.

The spatial loop can be completed when the position and orientation of the manipulator end-effector is specified. As the end-effector changes its location in space the hypothetical dimensions corresponding to the loop completion change in value. Thus the loop completion is instantaneous in nature and must be repeatedly updated when applied to a manipulator traversing space.

The six independent input parameters specifying the end-effector location are transformed into five constraint parameters together with a single input parameter corresponding to the relative displacement of the hypothetical joint. In Ref. [2] and in this study the hypothetical joint pair is conceptualized as a revolute however, the completion algorithm is identical if a prismatic or helical joint is used instead.

Figure 2.1 illustrates a skeletal model of a serially connected six revolute manipulator. The directions of the joint axes are labeled sequentially with the unit vectors $\boldsymbol{s}_i$ (i=1,2...7, where $\boldsymbol{s}_7$ denotes the hypothetical joint). The directions of the common normal between two successive joint axes $\boldsymbol{s}_i$ and $\boldsymbol{s}_j$ are labeled with the unit vectors $\boldsymbol{a}_{ij}$ (ij=12,23...67) and the lengths of the common normals (link lengths) are labeled a_{ij}. The mutual perpendicular distances between pairs of successive links $\boldsymbol{a}_{ij}$ and $\boldsymbol{a}_{jk}$ are labeled S_{jj} (jj=11,22...66) and are called (joint) offsets.

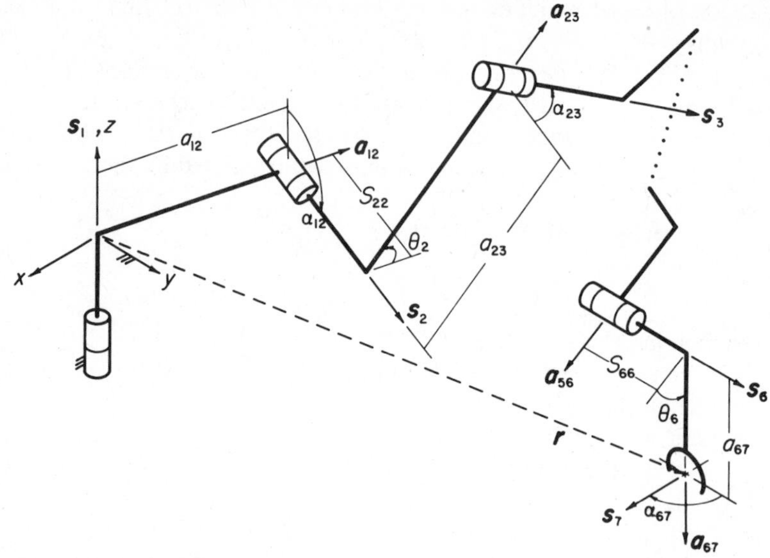

Figure 2.1 A general six revolute manipulator. Directions of the joint axes and their common normals are labeled with unit vectors $\mathbf{s}_i$ and $\mathbf{a}_{ij}$. Physical dimensions in these directions are S_{ii} and a_{ij}. Joint angles θ_i and twist angles α_{ij} are measured respectively about $\mathbf{s}_i$ and $\mathbf{a}_{ij}$. The end-effector is located by position vector $\mathbf{r}$ and $\mathbf{a}_{67}$, $\mathbf{s}_7$.

Twist angles α_{ij} are measured by a right-handed rotation of $\mathbf{s}_i$ into $\mathbf{s}_j$ about $\mathbf{a}_{ij}$ and

$$c_{ij} = \mathbf{s}_i \cdot \mathbf{s}_j \tag{2.2}$$

$$s_{ij} = \mathbf{s}_i \times \mathbf{s}_j \cdot \mathbf{a}_{ij} \tag{2.3}$$

where the abbreviations $c_{ij} \equiv \cos\alpha_{ij}$ and $s_{ij} \equiv \sin\alpha_{ij}$ are introduced. The relative angular joint displacements θ_j are measured by a right-handed rotation of $\mathbf{a}_{ij}$ into $\mathbf{a}_{jk}$ about $\mathbf{s}_j$ and,

$$c_j = \mathbf{a}_{ij} \cdot \mathbf{a}_{jk} \tag{2.4}$$

$$s_j = \mathbf{a}_{ij} \times \mathbf{a}_{jk} \cdot \mathbf{s}_j \tag{2.5}$$

where the abbreviations $c_j \equiv \cos\theta_j$ and $s_j \equiv \sin\theta_j$ are introduced.

The perpendicular distance from any target point on the end-effector to the joint axis $\mathbf{s}_6$, is labeled a_{67}. It is assumed that a_{67} is known together with the offset S_{66} and the twist angle α_{67} (see Figure 2.1).

A fixed coordinate system is located with the z axis coaxial with $\mathbf{s}_1$ and the x and y axes are in the plane of rotation of $\mathbf{a}_{12}$. The location of the end-effector can be specified in this fixed frame by the position vector to the target point $\boldsymbol{r}$, and the pair of orthogonal unit vectors $\boldsymbol{a}_{67}$ and $\boldsymbol{s}_7$. Although $\boldsymbol{r}$, $\boldsymbol{a}_{67}$ and $\boldsymbol{s}_7$ have a total of nine components, the latter two are related by the three conditions,

$$\begin{aligned} \boldsymbol{a}_{67} \cdot \boldsymbol{a}_{67} &= 1, \\ \boldsymbol{a}_{67} \cdot \boldsymbol{s}_7 &= 0, \\ \boldsymbol{s}_7 \cdot \boldsymbol{s}_7 &= 1 \end{aligned} \tag{2.6}$$

so that the three vectors $(\boldsymbol{a}_{67}, \boldsymbol{s}_7, \boldsymbol{r})$ represent the 9−3=6 independent parameters necessary to locate a rigid body in space.

Figure 2.2 shows the completion of the spatial loop where $\boldsymbol{s}_7$ designates the axis of the hypothetical joint. It is now necessary to determine the five constraint parameters S_{77}, a_{71}, S_{11}, α_{71} and $(\theta_1 - \phi_1)$ that complete the loop together with the input angle of the spatial mechanism, θ_7. Angle ϕ_1 is the first manipulator *actuator angle* and is measured from the x axis to $\boldsymbol{a}_{12}$ about

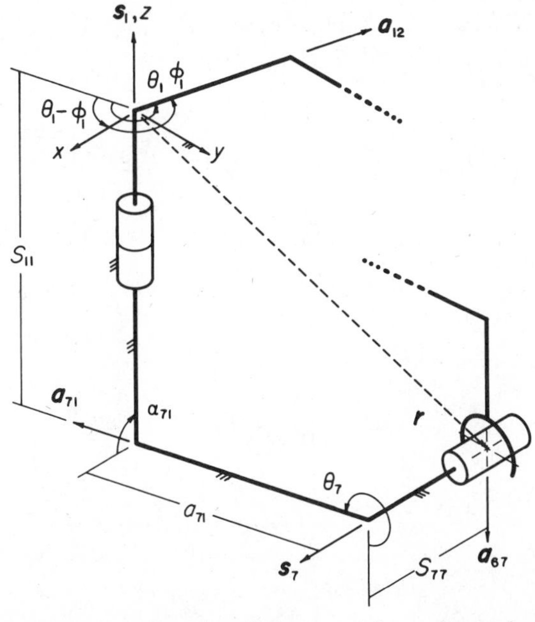

Figure 2.2 Completion of the spatial loop. A hypothetical seventh joint connects the end-effector to ground and transforms a 6 DOF manipulator into a single loop mechanism of mobility one (1 DOF). The location of the end-effector $(\boldsymbol{r}, \boldsymbol{a}_{67}, \boldsymbol{s}_7)$ *is transformed into 5 constraint parameters* $S_{77}, a_{71}, S_{11}, \alpha_{71}, (\theta_1 - \phi_1)$ *and the input angle of the mechanism* θ_7. *Angle* ϕ_1 *is the first manipulator actuator angle and is measured relative to ground (x axis).*

$\boldsymbol{s}_1$. It can be computed from $(\theta_1-\phi_1)$ and θ_1 once the latter is determined from a displacement analysis (Section 5) of the equivalent spatial mechanism.

The initial calculation in the completion algorithm is

$$c_{71} = \boldsymbol{s}_7 \cdot \boldsymbol{s}_1. \tag{2.7}$$

A value of $c_{71} = \pm 1$ immediately flags the necessary and sufficient condition for singularities which will be discussed subsequently. The unit vector $\boldsymbol{a}_{71}$ is now defined by,

$$\boldsymbol{a}_{71} = \frac{\boldsymbol{s}_7 \times \boldsymbol{s}_1}{\|\boldsymbol{s}_7 \times \boldsymbol{s}_1\|} \tag{2.8}$$

and is used in the computation,

$$s_{71} = |\boldsymbol{a}_{71}\ \boldsymbol{s}_7\ \boldsymbol{s}_1| \tag{2.9}$$

where determinant notation is used for the scalar triple product (see Eqn. (2.3)). Solving the vector loop equation (see Figure 2.2),

$$\boldsymbol{r} + S_{77}\,\boldsymbol{s}_7 + a_{71}\,\boldsymbol{a}_{71} + S_{11}\,\boldsymbol{s}_1 = \boldsymbol{0} \tag{2.10}$$

and using Eqn. (2.9) results in explicit expressions for the hypothetical link a_{71} and hypothetical offsets S_{77} and S_{11},

$$S_{77} = |\boldsymbol{r}\ \ \boldsymbol{a}_{71}\ \boldsymbol{s}_1|/s_{71} \tag{2.11}$$

$$a_{71} = |\boldsymbol{s}_7\,\boldsymbol{r}\ \ \boldsymbol{s}_1|/s_{71} \tag{2.12}$$

$$S_{11} = |\boldsymbol{s}_7\,\boldsymbol{a}_{71}\,\boldsymbol{r}|/s_{71}. \tag{2.13}$$

Using Eqns. (2.2-5), the remainder of the completion algorithm is given by the following expressions for θ_7 and $(\theta_1-\phi_1)$,

$$c_7 = \boldsymbol{a}_{67} \cdot \boldsymbol{a}_{71} \tag{2.14}$$

$$s_7 = |\boldsymbol{s}_7\,\boldsymbol{a}_{67}\,\boldsymbol{a}_{71}| \tag{2.15}$$

and,

$$\cos(\theta_1-\phi_1) = \boldsymbol{a}_{71} \cdot \boldsymbol{i} \tag{2.16}$$

$$\sin(\theta_1-\phi_1) = |\boldsymbol{s}_1\,\boldsymbol{a}_{71}\,\boldsymbol{i}| \tag{2.17}$$

where $\boldsymbol{i}$ is the unit vector in the direction of the x axis.

It should be noted that the definition of the unit vector $\boldsymbol{a}_{71}$ in Eqn. (2.8) is somewhat arbitrary since $\boldsymbol{a}_{71}$ will point in one direction when $0 < \alpha_{71} < \pi$ and $\boldsymbol{a}_{71}$ will point in the opposite direction when $-\pi < \alpha_{71} < 0$. Alternatively, if $\boldsymbol{a}_{71}$ is defined by

$$\boldsymbol{a}_{71} \equiv \frac{\boldsymbol{s}_1 \times \boldsymbol{s}_7}{\|\boldsymbol{s}_1 \times \boldsymbol{s}_7\|} \tag{2.18}$$

the preceding completion algorithm leads to a second parameterization. The two parameterizations are,

$$\boldsymbol{a}_{71} \equiv \frac{\boldsymbol{s}_7 \times \boldsymbol{s}_1}{\|\boldsymbol{s}_7 \times \boldsymbol{s}_1\|}: S_{77},\ a_{71},\ S_{11},\ \alpha_{71},\ (\theta_1-\phi_1),\ \theta_7$$

and

$$\boldsymbol{a}_{71} \equiv \frac{\boldsymbol{s}_1 \times \boldsymbol{s}_7}{\|\boldsymbol{s}_1 \times \boldsymbol{s}_7\|}: S_{77},\ -a_{71},\ S_{11},\ -\alpha_{71},\ (\theta_1-\phi_1)\pm\pi,\ \theta_7\pm\pi$$

and although they are different the corresponding completions are identical in structure.

It was mentioned earlier that a singular condition is flagged when $c_{71} = \pm 1$ (and therefore $s_{71} = 0$). This is because the vectors $\boldsymbol{s}_7$ and $\boldsymbol{s}_1$ are now either parallel or antiparallel and there are a single infinity of possible links a_{71} mutually perpendicular to $\boldsymbol{s}_7$ and $\boldsymbol{s}_1$. However, a convenient arbitrary constraint $S_{77} = 0$ can be imposed to obtain a unique result (see Figure 2.3). Forming the inner product of Eqn. (2.10) with $\boldsymbol{s}_1$ yields,

$$S_{11} = -\boldsymbol{r} \cdot \boldsymbol{s}_1. \tag{2.19}$$

Further, from Eqn. (2.10),

$$a_{71} = \|\boldsymbol{r} + S_{11}\, \boldsymbol{s}_1\| \tag{2.20}$$

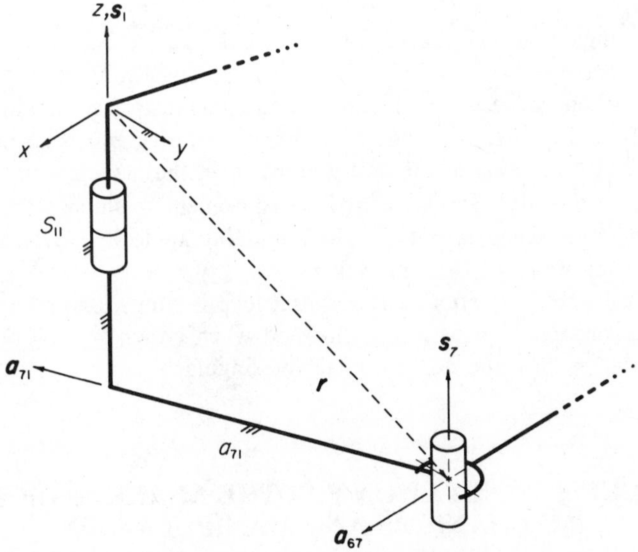

Figure 2.3 First singularity of the completion algorithm. When $\boldsymbol{s}_7$ and $\boldsymbol{s}_1$ are parallel there is a single infinity of common normals $\boldsymbol{a}_{71}$. Imposing the constraint S_{77}=0 determines a unique parameterization.

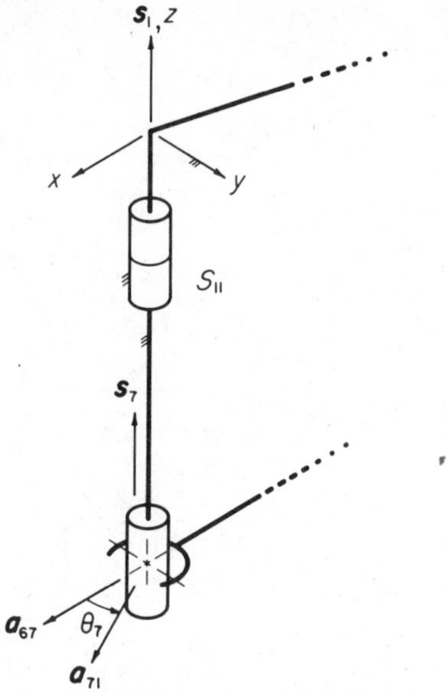

Figure 2.4 Second singularity of the completion algorithm. When $\boldsymbol{s}_7$ and $\boldsymbol{s}_1$ are collinear the direction $\boldsymbol{a}_{71}$ is indeterminate. Imposing the constraint θ_7=0 yields the unique result $\boldsymbol{a}_{71} = \boldsymbol{a}_{67}$.

and provided $a_{71} \neq 0$,

$$\boldsymbol{a}_{71} = -(\boldsymbol{r} + S_{11}\,\boldsymbol{s}_1)/a_{71}. \tag{2.21}$$

The remaining angles θ_7 and $(\theta_1 - \phi_1)$ can again be calculated using Eqns. (2.14-17).

Finally, when the axes of $\boldsymbol{s}_7$ and $\boldsymbol{s}_1$ are collinear, the condition a_{71}=0 (resulting from Eqn. (2.20)) flags a further singularity which is illustrated in Figure 2.4. The direction of the unit vector $\boldsymbol{a}_{71}$ in the plane normal to the axis of $\boldsymbol{s}_1$ is now arbitrary. In this case it is convenient to impose the additional constraint θ_7=0 making $\boldsymbol{a}_{71}$=$\boldsymbol{a}_{67}$. The remaining angle $(\theta_1 - \phi_1)$ can again be calculated using Eqns. (2.16) and (2.17).

In general, the completion of the spatial loop is simple and straightforward. The completion algorithm is easily modified when either or both of the scalar conditions $c_{71} = \pm 1$ and $a_{71} = 0$ flag the singular cases.

3. A RECURSIVE NOTATION FOR THE ANALYSIS OF SPATIAL MECHANISMS AND MANIPULATORS

A recursive notation is developed which provides compact expressions for the unit vector components used in the analysis of single loop spatial mechanisms and manipulators. The development of the recursive notation is based on

3 × 3 rotation matrices and is an extension of the recursive notation based on spherical polygons in Refs. [2,6,7] and the formulation developed in Ref. [10]. It provides an efficient method for writing highly non-linear expressions and facilitates their geometrical interpretation. This is particularly important for the recognition of uncertainty configurations where the expressions for the joint angles become indeterminate.

In developing the subsequent recursion formulas for a general single loop mechanism with seven single degree-of-freedom joints, it will be useful to employ a cyclically increasing sequence of seven positive integers,

$$(h\ i\ j\ k\ l\ m\ n). \tag{3.1}$$

It is implicit in the term "cyclically increasing" that n is followed by h in the sequence.

As described in the preceding section, unit vectors along the joint axes are labeled sequentially $\boldsymbol{s}_1, \boldsymbol{s}_2 \cdots \boldsymbol{s}_7$. Unit vectors in the direction of the common normal between adjacent joint axes are labeled $\boldsymbol{a}_{12}, \boldsymbol{a}_{23} \cdots \boldsymbol{a}_{71}$. Joint angles $\theta_1, \theta_2 \cdots \theta_7$ are measured using a right-handed rotation about the axes and are defined by,

$$c_i = \boldsymbol{a}_{hi} \cdot \boldsymbol{a}_{ij} \tag{3.2}$$

$$s_i = \boldsymbol{a}_{hi} \times \boldsymbol{a}_{ij} \cdot \boldsymbol{s}_i. \tag{3.3}$$

Similarly, twist angles $\alpha_{12}, \alpha_{23} \cdots \alpha_{71}$ are measured about the common normals and are defined by

$$c_{hi} = \boldsymbol{s}_h \cdot \boldsymbol{s}_i \tag{3.4}$$

$$s_{hi} = \boldsymbol{s}_h \times \boldsymbol{s}_i \cdot \boldsymbol{a}_{hi}. \tag{3.5}$$

It is convenient to use mutually orthogonal triples of unit vectors to specify sets of dextral bases which can be written in the matrix form,

$$[B_h] \equiv [\boldsymbol{a}_{hi}\ \boldsymbol{b}_{hi}\ \boldsymbol{s}_h], \tag{3.6}$$

where,

$$\boldsymbol{b}_{hi} \equiv \boldsymbol{s}_h \times \boldsymbol{a}_{hi}. \tag{3.7}$$

Using base $[B_h]$ as a reference frame, the next successive base $[B_i]$ (Figure 3.1a) is expressed by,

$$[B_h]^{\mathrm{T}}[B_i] = [\boldsymbol{a}_{hi}\ \boldsymbol{b}_{hi}\ \boldsymbol{s}_h]^{\mathrm{T}}[\boldsymbol{a}_{ij}\ \boldsymbol{b}_{ij}\ \boldsymbol{s}_i] \tag{3.8}$$

and further,

$$[M_i] \equiv [B_h]^{\mathrm{T}}[B_i] = \begin{bmatrix} \boldsymbol{a}_{hi} \cdot \boldsymbol{a}_{ij} & \boldsymbol{a}_{hi} \cdot \boldsymbol{b}_{ij} & \boldsymbol{a}_{hi} \cdot \boldsymbol{s}_i \\ \boldsymbol{b}_{hi} \cdot \boldsymbol{a}_{ij} & \boldsymbol{b}_{hi} \cdot \boldsymbol{b}_{ij} & \boldsymbol{b}_{hi} \cdot \boldsymbol{s}_i \\ \boldsymbol{s}_h \cdot \boldsymbol{a}_{ij} & \boldsymbol{s}_h \cdot \boldsymbol{b}_{ij} & \boldsymbol{s}_h \cdot \boldsymbol{s}_i \end{bmatrix}. \tag{3.9}$$

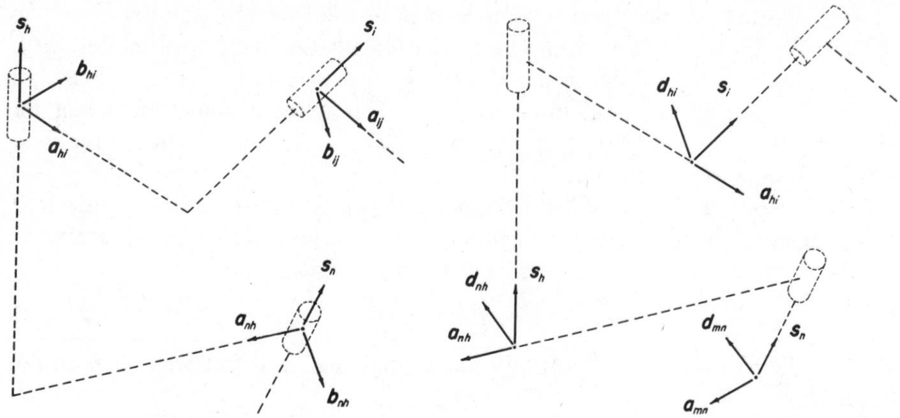

Figure 3.1 Triples of orthogonal unit vectors form bases which denote directions associated with mechanism dimensions. a) For dextral bases the subscript of the **s** *vector is the same as the first subscript of the* **a** *and* **b** *vectors. b) For sinistral bases the subscript of the* **s** *vector is the same as the second subscript of the* **a** *and* **d** *vectors.*

This latter expression shows explicitly that $[M_i]$ represents the direction cosines of base $[B_i]$ in the directions of $[B_h]$. The rotation matrix $[M_i]$ can be expressed as the product of a pair of elementary rotation matrices,

$$[M_i] = ([\alpha_{hi}][\theta_i]) \tag{3.10}$$

where,

$$[\alpha_{hi}] \equiv \begin{bmatrix} 1 & 0 & 0 \\ 0 & c_{hi} & -s_{hi} \\ 0 & s_{hi} & c_{hi} \end{bmatrix}, \quad [\theta_i] \equiv \begin{bmatrix} c_i & -s_i & 0 \\ s_i & c_i & 0 \\ 0 & 0 & 1 \end{bmatrix}. \tag{3.11}$$

Proceeding inductively, Eqns. (3.9) and (3.10) can be used to express all seven bases in terms of $[B_h]$,

$$[I] \equiv [B_h]^{\mathrm{T}}[B_h] \tag{3.12}$$

$$[M_i] \equiv [B_h]^{\mathrm{T}}[B_i] = ([\alpha_{hi}][\theta_i]) \tag{3.13}$$

$$[M_{ij}] \equiv [B_h]^{\mathrm{T}}[B_j] = ([\alpha_{hi}][\theta_i])([\alpha_{ij}][\theta_j]) = [M_i][M_j] \tag{3.14}$$

$$\vdots$$

$$[M_{ij..mn}] \equiv [B_h]^{\mathrm{T}}[B_n] = ([\alpha_{hi}][\theta_i]) \cdots ([\alpha_{mn}][\theta_n]) \tag{3.15}$$

$$= [M_i] \cdots [M_n]$$

where $[I]$ is the identity matrix. The following two fundamental recursive expressions can be deduced from these relationships and are written in the general forms,

$$[M_{ij..mn}] = [M_i][M_{j..mn}] = ([\alpha_{hi}][\theta_i])[M_{j..mn}] \tag{3.16}$$

and

$$[M_{ij..mn}] = [M_{ij..m}][M_n] = [M_{ij..m}]([\alpha_{mn}][\theta_n]). \tag{3.17}$$

These will be referred to as the backward and forward recursive definitions for the direction cosines.

Recursive definitions for the elements of the $[M]$ matrices are now formulated. Firstly, substituting Eqn. (3.11) into (3.13) yields,

$$[M_i] = \begin{bmatrix} c_i & -s_i & 0 \\ c_{hi}s_i & c_{hi}c_i & -s_{hi} \\ s_{hi}s_i & s_{hi}c_i & c_{hi} \end{bmatrix} = \begin{bmatrix} P_i & U_i & 0 \\ Q_i & V_i & -s_{hi} \\ R_i & W_i & c_{hi} \end{bmatrix}. \tag{3.18}$$

The subscript i on the right side of Eqn. (3.18) denotes that the elements in the first two columns are functions of the joint variable θ_i. Secondly, by making the substitution $(h\ i) \leftarrow (i\ j)$ in Eqn. (3.18) the elements of the matrix $[M_j]$ are expressed as,

$$[M_j] = \begin{bmatrix} P_j & U_j & 0 \\ Q_j & V_j & -s_{ij} \\ R_j & W_j & c_{ij} \end{bmatrix} = ([\alpha_{ij}][\theta_j]). \tag{3.19}$$

Then substituting Eqns. (3.11) and (3.19) into (3.14) and using the backward definition (see Eqn. (3.16)) yields

$$[M_{ij}] = \begin{bmatrix} (c_iP_j-s_iQ_j) & (c_iU_j-s_iV_j) & s_{ij}s_i \\ c_{hi}(s_iP_j+c_iQ_j)-s_{hi}R_j & c_{hi}(s_iU_j+c_iV_j)-s_{hi}W_j & -(s_{hi}c_{ij}+c_{hi}s_{ij}c_i) \\ s_{hi}(s_iP_j+c_iQ_j)+c_{hi}R_j & s_{hi}(s_iU_j+c_iV_j)+c_{hi}W_j & (c_{hi}c_{ij}-s_{hi}s_{ij}c_i) \end{bmatrix}. \tag{3.20}$$

The elements of the right side of Eqn. (3.20) are now defined by,

$$[M_{ij}] = \begin{bmatrix} P_{ij} & U_{ij} & X_i \\ Q_{ij} & V_{ij} & Y_i \\ R_{ij} & W_{ij} & Z_i \end{bmatrix}. \tag{3.21}$$

The above subscripts indicate that the first two columns are functions of θ_i and θ_j whereas the last column is a function of θ_i only. Proceeding inductively, Eqn. (3.15) can be expressed in the form,

$$[M_{ij..mn}] = \begin{bmatrix} P_{ij..mn} & U_{ij..mn} & X_{ij..m} \\ Q_{ij..mn} & V_{ij..mn} & Y_{ij..m} \\ R_{ij..mn} & W_{ij..mn} & Z_{ij..m} \end{bmatrix}. \tag{3.22}$$

Table 3.1a illustrates the projection of base $[B_m]$ on the base $[B_h]$ in terms of the matrix $[M_{i..m}]$, the elements of which are expressed in the recursive notation. *The subscripts of the vectors have a simple and elegant relationship with the subscripts of the elements; this relationship greatly facilitates the geometric interpretation of the recursive notation.* The definitions of the single subscripted elements are listed in Table 3.1b. Equations (3.16) and (3.17) are used to formulate the backward and forward definitions for multi-subscripted elements which are listed in Tables 3.1c and 3.1d.

Alternative expressions for the preceding direction cosines are now derived from the transpose of the next term in the sequence Eqn. (3.12)-(3.15),

$$[I] = [M_{ij..nh}]^{\mathrm{T}} = [B_h]^T[B_h]. \tag{3.23}$$

Using the basic property of rotation matrices $[M]^{-1} = [M]^T$ it follows that,

$$[M_i] = [M_{j..nh}]^{\mathrm{T}} \tag{3.24}$$

$$[M_{ij}] = [M_{k..nh}]^{\mathrm{T}} \tag{3.25}$$

$$\vdots$$

$$[M_{ij..mn}] = [M_h]^{\mathrm{T}} \tag{3.26}$$

The left sides of Eqns. (3.23) - (3.26) are the original expressions for the direction cosines and the right sides are the alternative expressions for the direction cosines. The definitions for the elements of the alternative expressions are the same as those already given in Table 3.1.

By making the substitution $(h\ i...n) \leftarrow (1\ 2...7)$ in Eqns. (3.23) – (3.26) the original and alternative expressions for the direction cosines are determined in the reference frame $[B_1] = [\boldsymbol{a}_{12}\ \boldsymbol{b}_{12}\ \boldsymbol{s}_1]$. The respective elements are listed in Table 3.3, Sets 1a and 1b. (Table 3.2 is described subsequently.) A further six sets of direction cosines can easily be obtained by substitution of successive cyclic permutations of subscripts, $(h\ i...n) \leftarrow (2\ 3...1),\ (3\ 4...2),...(7\ 1...6)$. These are listed in Table 3.3 Sets 2-7. All subscripts contained in Sets 1-7 are in ascending order.

Expressions for sets of direction cosines with subscripts in descending order are now derived. This is accomplished by first using sets of dextral bases which are expressed in the matrix form,

$$[D_h{}'] \equiv [\boldsymbol{a}_{nh}\ -\boldsymbol{d}_{nh}\ \boldsymbol{s}_h] \tag{3.27}$$

where,

$$-\boldsymbol{d}_{nh} \equiv \boldsymbol{s}_h \times \boldsymbol{a}_{nh}. \tag{3.28}$$

Using $[D_h{}']$ as the reference frame, the direction cosines of the preceding base $[D_n{}']$ are expressed by,

$$[\overline{M}_n'] \equiv [D_h']^T[D_n'] = [\boldsymbol{a}_{nh} \; -\boldsymbol{d}_{nh} \; \boldsymbol{s}_h]^T[\boldsymbol{a}_{mn} \; -\boldsymbol{d}_{mn} \; \boldsymbol{s}_n]. \tag{3.29}$$

The rotation matrix $[\overline{M}_n']$ can be formulated as the product of a pair of elementary rotation matrices with negative angles,

$$[\overline{M}_n'] = ([-\alpha_{nh}][-\theta_n]). \tag{3.30}$$

Equating the right-hand sides of Eqns. (3.29) and (3.30) and then expanding, yields

$$\begin{bmatrix} \boldsymbol{a}_{nh} \cdot \boldsymbol{a}_{mn} & -\boldsymbol{a}_{nh} \cdot \boldsymbol{d}_{mn} & \boldsymbol{a}_{nh} \cdot \boldsymbol{s}_n \\ -\boldsymbol{d}_{nh} \cdot \boldsymbol{a}_{mn} & \boldsymbol{d}_{nh} \cdot \boldsymbol{d}_{mn} & -\boldsymbol{d}_{nh} \cdot \boldsymbol{s}_n \\ \boldsymbol{s}_h \cdot \boldsymbol{a}_{mn} & -\boldsymbol{s}_h \cdot \boldsymbol{d}_{mn} & \boldsymbol{s}_h \cdot \boldsymbol{s}_n \end{bmatrix} = \begin{bmatrix} c_n & s_n & 0 \\ -c_{nh}s_n & c_{nh}c_n & s_{nh} \\ s_{nh}s_n & -s_{nh}c_n & c_{nh} \end{bmatrix}. \tag{3.31}$$

Negating the 12, 21, 23, and 32 elements on both sides of Eqn. (3.31) results in the expressions,

$$[\overline{M}_n] \equiv [D_h]^T[D_n] = [\boldsymbol{a}_{nh} \; \boldsymbol{d}_{nh} \; \boldsymbol{s}_h]^T[\boldsymbol{a}_{mn} \; \boldsymbol{d}_{mn} \; \boldsymbol{s}_n] \tag{3.32}$$

and,

$$[\overline{M}_n] = ([\alpha_{nh}][\theta_n]). \tag{3.33}$$

This negation has simply transformed the dextral bases $[D']$ into sinistral bases $[D]$ and therefore (see Figure 3.1b and compare with Figure 3.1a),

$$[D_h] \equiv [\boldsymbol{a}_{nh} \; \boldsymbol{d}_{nh} \; \boldsymbol{s}_h] \tag{3.34}$$

where

$$\boldsymbol{d}_{nh} = -\boldsymbol{s}_h \times \boldsymbol{a}_{nh}. \tag{3.35}$$

The sinistral bases $[D]$ are used from here on since $[\overline{M}_n]$ is similar in form to $[M_i]$ (compare Eqn. (3.10) with (3.33)). This yields a recursive notation which is virtually identical to the preceding recursive notation developed for the dextral bases $[B]$. The subscripts of these newly defined recursive elements will now be in descending order. However, the ascending order of the subscripts labeling $\boldsymbol{a}$, $\boldsymbol{d}$ and α will be preserved. Therefore, the geometrical definitions of $\boldsymbol{a}$, $\boldsymbol{s}$, α and θ will remain unchanged. (It is noted that if the definitions $\boldsymbol{a}_{hi} \equiv \boldsymbol{a}_{ih}$ and $\alpha_{hi} \equiv \alpha_{ih}$ are made, then the recursive notation becomes identical for the elements with subscripts in either ascending or descending order. This convention, however, will not be used here.)

The backward and forward recursion formulas can now be expressed by,

$$[M_{nm..ji}] = [\overline{M}_n][M_{m..ji}] = ([\alpha_{nh}][\theta_n])[M_{m..ji}] \tag{3.36}$$

$$[M_{nm..ji}] = [M_{nm..j}][\overline{M}_i] = [M_{nm...j}]([\alpha_{ij}][\theta_i]). \tag{3.37}$$

A bar is introduced in the single subscripted $[M]$ terms to denote that they belong to sets of descending subscripts. The same convention is used for single subscripted matrix elements.

TABLE 3.1

RECURSIVE DEFINITIONS FOR CYCLICALLY ASCENDING SUBSCRIPTS $(h\ i\ j\ k\ l\ m\ n)$

a. Scalar Products in the Dextral Bases $[B_h]^T[B_m]=[M_{i..m}]$

$\cdot$	$\boldsymbol{a}_{mn}$	$\boldsymbol{b}_{mn}$	$\boldsymbol{s}_m$
$\boldsymbol{a}_{hi}$	$P_{i..m}$	$U_{i..m}$	$X_{i..l}$
$\boldsymbol{b}_{hi}$	$Q_{i..m}$	$V_{i..m}$	$Y_{i..l}$
$\boldsymbol{s}_h$	$R_{i..m}$	$W_{i..m}$	$Z_{i..l}$

b. Single Subscripted Elements

$$
\begin{array}{lll}
P_i = c_i & U_i = -s_i & X_i = s_{ij}\,s_i \\
Q_i = c_{hi}s_i & V_i = c_{hi}c_i & Y_i = -(s_{hi}c_{ij} + c_{hi}s_{ij}c_i) \\
R_i = s_{hi}s_i & W_i = s_{hi}c_i & Z_i = (c_{hi}c_{ij} - s_{hi}s_{ij}c_i)
\end{array}
$$

Note that these nine elements do not all belong to the same base.

c. Backward Recursions (2 or more subscripts)

$$E_{ij..n} = (c_i E_{j..n} - s_i F_{j..n})$$

$$F_{ij..n} = c_{hi}(s_i E_{j..n} + c_i F_{j..n}) - s_{hi} G_{j..n}$$

$$G_{ij..n} = s_{hi}(s_i E_{j..n} + c_i F_{j..n}) + c_{hi} G_{j..n}$$

where $(E\ F\ G) \equiv (P\ Q\ R)$, $(U\ V\ W)$ or $(X\ Y\ Z)$. Note that these nine elements do not all belong to the same base.

TABLE 3.1 (cont'd)

d. Forward Recursions (3 or more subscripts) $[M_{i..mn}] = [M_{i..m}]([\alpha_{mn}][\theta_n])$

$$
\begin{aligned}
E_{i..mn} &= E_{i..m}c_n + (F_{i..m}c_{mn} + G_{i..l}s_{mn})s_n \\
F_{i..mn} &= -E_{i..m}s_n + (F_{i..m}c_{mn} + G_{i..l}s_{mn})c_n \\
G_{i..m} &= (-F_{i..m}s_{mn} + G_{i..l}c_{mn})
\end{aligned}
$$

where $(E\ F\ G) \equiv (P\ U\ X)$, $(Q\ V\ Y)$ or $(R\ W\ Z)$

e. Scalar Products in the Associated Sinistral Bases $[B_h']^T[B_m'] = [M'_{i..m}]$

$\cdot$	$\mathbf{a}_{mn}$	$-\mathbf{b}_{mn}$	$\mathbf{s}_m$
$\mathbf{a}_{hi}$	$P_{i..m}$	$-U_{i..m}$	$X_{i..l}$
$-\mathbf{b}_{hi}$	$-Q_{i..m}$	$V_{i..m}$	$-Y_{i..l}$
$\mathbf{s}_i$	$R_{i..m}$	$-W_{i..m}$	$Z_{i..l}$

f. Trigonometric Laws $[M_{hij}] = [M_{klmn}]^T$

$$
\begin{bmatrix} P_{hij} & U_{hij} & X_{hi} \\ Q_{hij} & V_{hij} & Y_{hi} \\ R_{hij} & W_{hij} & Z_{hi} \end{bmatrix} = \begin{bmatrix} P_{klmn} & Q_{klmn} & R_{klmn} \\ U_{klmn} & V_{klmn} & W_{klmn} \\ X_{klm} & Y_{klm} & Z_{klm} \end{bmatrix}
$$

TABLE 3.2

RECURSIVE DEFINITIONS FOR CYCLICALLY DESCENDING SUBSCRIPTS $(n\ m\ l\ k\ j\ i\ h)$

a. Scalar Products in the Sinistral Bases $[D_n]^T[D_i]=[M_{m..i}]$

$\cdot$	$\mathbf{a}_{hi}$	$\mathbf{d}_{hi}$	$\mathbf{s}_i$
$\mathbf{a}_{mn}$	$P_{m..i}$	$U_{m..i}$	$X_{m..j}$
$\mathbf{d}_{mn}$	$Q_{m..i}$	$V_{m..i}$	$Y_{m..j}$
$\mathbf{s}_n$	$R_{m..i}$	$W_{m..i}$	$Z_{m..j}$

b. Single Subscripted Elements

$$\begin{aligned} \bar{P}_m &= c_m & \bar{U}_m &= -s_m & \bar{X}_m &= s_{lm}s_m \\ \bar{Q}_m &= c_{mn}s_m & \bar{V}_m &= c_{mn}c_m & \bar{Y}_m &= -(s_{mn}c_{lm}+c_{mn}s_{lm}c_m) \\ \bar{R}_m &= s_{mn}s_m & \bar{W}_m &= s_{mn}c_m & \bar{Z}_m &= (c_{mn}c_{lm}-s_{mn}s_{lm}c_m) \end{aligned}$$

Note that these nine elements do not all belong to the same base.

c. Backward Recursions (2 or more subscripts)

$$E_{ml..h} = (c_m E_{l..h} - s_m F_{l..h})$$

$$F_{ml..h} = c_{mn}(s_m E_{l..h} + c_m F_{l..h}) - s_{mn} G_{l..h}$$

$$G_{ml..h} = s_{mn}(s_m E_{l..h} + c_m F_{l..h}) + c_{mn} G_{l..h}$$

where $(E\ F\ G) \equiv (P\ Q\ R)$, $(U\ V\ W)$ or $(X\ Y\ Z)$. Note that these nine elements do not all belong to the same base.

d. Forward Recursions (3 or more subscripts) $[M_{m..ih}] = [M_{m..i}]$ $([\alpha_{hi}][\theta_h])$

$$\begin{aligned}
E_{m..ih} &= E_{m..i}c_h + (F_{m..i}c_{hi} + G_{m..j}s_{hi})s_h \\
F_{m..ih} &= -E_{m..i}s_h + (F_{m..i}c_{hi} + G_{m..j}s_{hi})c_h \\
G_{m..i} &= (-F_{m..i}s_{hi} + G_{m..j}c_{hi})
\end{aligned}$$

where $(E\ F\ G) \equiv (P\ U\ X)$, $(Q\ V\ Y)$ or $(R\ W\ Z)$.

e. Scalar Products in the Associated Dextral Bases $[D_n']^{\mathrm{T}}[D_i'] = [M'_{m..i}]$

$\cdot$	$\boldsymbol{a}_{hi}$	$-\boldsymbol{d}_{hi}$	$\boldsymbol{s}_i$
$\boldsymbol{a}_{mn}$	$P_{m..i}$	$-U_{m..i}$	$X_{m..j}$
$-\boldsymbol{d}_{mn}$	$-Q_{m..i}$	$V_{m..i}$	$-Y_{m..j}$
$\boldsymbol{s}_n$	$R_{m..i}$	$-W_{m..i}$	$Z_{m..j}$

f. Trigonometric Laws $[M_{nml}] = [M_{kjih}]^{\mathrm{T}}$

$$\begin{bmatrix} E_{nml} & U_{nml} & X_{nm} \\ F_{nml} & V_{nml} & Y_{nm} \\ G_{nml} & W_{nml} & Z_{nm} \end{bmatrix} = \begin{bmatrix} E_{kjih} & F_{kjih} & G_{kjih} \\ U_{kjih} & V_{kjih} & W_{kjih} \\ X_{kji} & Y_{kji} & Z_{kji} \end{bmatrix}.$$

TABLE 3.3

DIRECTION COSINES AND THE ALTERNATIVE FORMS

	SET 1a			*SET* 1b		
	$\boldsymbol{a}_{12}$	$\boldsymbol{b}_{12}$	$\boldsymbol{s}_1$	$\boldsymbol{a}_{12}$	$\boldsymbol{b}_{12}$	$\boldsymbol{s}_1$
$\boldsymbol{a}_{12}$	(1 ,	0 ,	0)	($P_{2345671}$,	$U_{2345671}$,	X_{234567})
$\boldsymbol{a}_{23}$	(P_2 ,	Q_2 ,	R_2)	(P_{345671} ,	U_{345671} ,	X_{34567})
$\boldsymbol{a}_{34}$	(P_{23} ,	Q_{23} ,	R_{23})	(P_{45671} ,	U_{45671} ,	X_{4567})
$\boldsymbol{a}_{45}$	(P_{234} ,	Q_{234} ,	R_{234})	(P_{5671} ,	U_{5671} ,	X_{567})
$\boldsymbol{a}_{56}$	(P_{2345} ,	Q_{2345} ,	R_{2345})	(P_{671} ,	U_{671} ,	X_{67})
$\boldsymbol{a}_{67}$	(P_{23456} ,	Q_{23456} ,	R_{23456})	(P_{71} ,	U_{71} ,	X_7)
$\boldsymbol{a}_{71}$	(P_{234567} ,	Q_{234567} ,	R_{234567})	(P_1 ,	U_1 ,	0)
$\boldsymbol{b}_{12}$	(0 ,	1 ,	0)	($Q_{2345671}$,	$V_{2345671}$,	Y_{234567})
$\boldsymbol{b}_{23}$	(U_2 ,	V_2 ,	W_2)	(Q_{345671} ,	V_{345671} ,	Y_{34567})
$\boldsymbol{b}_{34}$	(U_{23} ,	V_{23} ,	W_{23})	(Q_{45671} ,	V_{45671} ,	Y_{4567})
$\boldsymbol{b}_{45}$	(U_{234} ,	V_{234} ,	W_{234})	(Q_{5671} ,	V_{5671} ,	Y_{567})
$\boldsymbol{b}_{56}$	(U_{2345} ,	V_{2345} ,	W_{2345})	(Q_{671} ,	V_{671} ,	Y_{67})
$\boldsymbol{b}_{67}$	(U_{23456} ,	V_{23456} ,	W_{23456})	(Q_{71} ,	V_{71} ,	Y_7)
$\boldsymbol{b}_{71}$	(U_{234567} ,	V_{234567} ,	W_{234567})	(Q_1 ,	V_1 ,	$-s_{71}$)
$\boldsymbol{s}_1$	(0 ,	0 ,	1)	($R_{2345671}$,	$W_{2345671}$,	Z_{234567})
$\boldsymbol{s}_2$	(0 ,	$-s_{12}$,	c_{12})	(R_{345671} ,	W_{345671} ,	Z_{34567})
$\boldsymbol{s}_3$	(X_2 ,	Y_2 ,	Z_2)	(R_{45671} ,	W_{45671} ,	Z_{4567})
$\boldsymbol{s}_4$	(X_{23} ,	Y_{23} ,	Z_{23})	(R_{5671} ,	W_{5671} ,	Z_{567})
$\boldsymbol{s}_5$	(X_{234} ,	Y_{234} ,	Z_{234})	(R_{671} ,	W_{671} ,	Z_{67})
$\boldsymbol{s}_6$	(X_{2345} ,	Y_{2345} ,	Z_{2345})	(R_{71} ,	W_{71} ,	Z_7)
$\boldsymbol{s}_7$	(X_{23456} ,	Y_{23456} ,	Z_{23456})	(R_1 ,	W_1 ,	c_{71})

TABLE 3.3, Continued

	SET 2a			*SET* 2b		
	$\mathbf{a}_{23}$	$\mathbf{b}_{23}$	$\mathbf{s}_{2}$	$\mathbf{a}_{23}$	$\mathbf{b}_{23}$	$\mathbf{s}_{2}$
$\mathbf{a}_{23}$	1	0	0	$P_{3456712}$	$U_{3456712}$	X_{345671}
$\mathbf{a}_{34}$	P_{3}	Q_{3}	R_{3}	P_{456712}	U_{456712}	X_{45671}
$\mathbf{a}_{45}$	P_{34}	Q_{34}	R_{34}	P_{56712}	U_{56712}	X_{5671}
$\mathbf{a}_{56}$	P_{345}	Q_{345}	R_{345}	P_{6712}	U_{6712}	X_{671}
$\mathbf{a}_{67}$	P_{3456}	Q_{3456}	R_{3456}	P_{712}	U_{712}	X_{71}
$\mathbf{a}_{71}$	P_{34567}	Q_{34567}	R_{34567}	P_{12}	U_{12}	X_{1}
$\mathbf{a}_{12}$	P_{345671}	Q_{345671}	R_{345671}	P_{2}	U_{2}	0
$\mathbf{b}_{23}$	0	1	0	$Q_{3456712}$	$V_{3456712}$	Y_{345671}
$\mathbf{b}_{34}$	U_{3}	V_{3}	W_{3}	Q_{456712}	V_{456712}	Y_{45671}
$\mathbf{b}_{45}$	U_{34}	V_{34}	W_{34}	Q_{56712}	V_{56712}	Y_{5671}
$\mathbf{b}_{56}$	U_{345}	V_{345}	W_{345}	Q_{6712}	V_{6712}	Y_{671}
$\mathbf{b}_{67}$	U_{3456}	V_{3456}	W_{3456}	Q_{712}	V_{712}	Y_{71}
$\mathbf{b}_{71}$	U_{34567}	V_{34567}	W_{34567}	Q_{12}	V_{12}	Y_{1}
$\mathbf{b}_{12}$	U_{345671}	V_{345671}	W_{345671}	Q_{2}	V_{2}	$-s_{12}$
$\mathbf{s}_{2}$	0	0	1	$R_{3456712}$	$W_{3456712}$	Z_{345671}
$\mathbf{s}_{3}$	0	$-s_{23}$	c_{23}	R_{456712}	W_{456712}	Z_{45671}
$\mathbf{s}_{4}$	X_{3}	Y_{3}	Z_{3}	R_{56712}	W_{56712}	Z_{5671}
$\mathbf{s}_{5}$	X_{34}	Y_{34}	Z_{34}	R_{6712}	W_{6712}	Z_{671}
$\mathbf{s}_{6}$	X_{345}	Y_{345}	Z_{345}	R_{712}	W_{712}	Z_{71}
$\mathbf{s}_{7}$	X_{3456}	Y_{3456}	Z_{3456}	R_{12}	W_{12}	Z_{1}
$\mathbf{s}_{1}$	X_{34567}	Y_{34567}	Z_{34567}	R_{2}	W_{2}	c_{12}

TABLE 3.3, Continued

	SET 3a			*SET* 3b		
	$\boldsymbol{a}_{34}$	$\boldsymbol{b}_{34}$	$\boldsymbol{s}_3$	$\boldsymbol{a}_{34}$	$\boldsymbol{b}_{34}$	$\boldsymbol{s}_3$
$\boldsymbol{a}_{34}$	(1 ,	0 ,	0)	($P_{4567123}$,	$U_{4567123}$,	X_{456712})
$\boldsymbol{a}_{45}$	(P_4 ,	Q_4 ,	R_4)	(P_{567123} ,	U_{567123} ,	X_{56712})
$\boldsymbol{a}_{56}$	(P_{45} ,	Q_{45} ,	R_{45})	(P_{67123} ,	U_{67123} ,	X_{6712})
$\boldsymbol{a}_{67}$	(P_{456} ,	Q_{456} ,	R_{456})	(P_{7123} ,	U_{7123} ,	X_{712})
$\boldsymbol{a}_{71}$	(P_{4567} ,	Q_{4567} ,	R_{4567})	(P_{123} ,	U_{123} ,	X_{12})
$\boldsymbol{a}_{12}$	(P_{45671} ,	Q_{45671} ,	R_{45671})	(P_{23} ,	U_{23} ,	X_2)
$\boldsymbol{a}_{23}$	(P_{456712} ,	Q_{456712} ,	R_{456712})	(P_3 ,	U_3 ,	0)
$\boldsymbol{b}_{34}$	(0 ,	1 ,	0)	($Q_{4567123}$,	$V_{4567123}$,	Y_{456712})
$\boldsymbol{b}_{45}$	(U_4 ,	V_4 ,	W_4)	(Q_{567123} ,	V_{567123} ,	Y_{56712})
$\boldsymbol{b}_{56}$	(U_{45} ,	V_{45} ,	W_{45})	(Q_{67123} ,	V_{67123} ,	Y_{6712})
$\boldsymbol{b}_{67}$	(U_{456} ,	V_{456} ,	W_{456})	(Q_{7123} ,	V_{7123} ,	Y_{712})
$\boldsymbol{b}_{71}$	(U_{4567} ,	V_{4567} ,	W_{4567})	(Q_{123} ,	V_{123} ,	Y_{12})
$\boldsymbol{b}_{12}$	(U_{45671} ,	V_{45671} ,	W_{45671})	(Q_{23} ,	V_{23} ,	Y_2)
$\boldsymbol{b}_{23}$	(U_{456712} ,	V_{456712} ,	W_{456712})	(Q_3 ,	V_3 ,	$-s_{23}$)
$\boldsymbol{s}_3$	(0 ,	0 ,	1)	($R_{4567123}$,	$W_{4567123}$,	Z_{456712})
$\boldsymbol{s}_4$	(0 ,	$-s_{34}$,	c_{34})	(R_{567123} ,	W_{567123} ,	Z_{56712})
$\boldsymbol{s}_5$	(X_4 ,	Y_4 ,	Z_4)	(R_{67123} ,	W_{67123} ,	Z_{6712})
$\boldsymbol{s}_6$	(X_{45} ,	Y_{45} ,	Z_{45})	(R_{7123} ,	W_{7123} ,	Z_{712})
$\boldsymbol{s}_7$	(X_{456} ,	Y_{456} ,	Z_{456})	(R_{123} ,	W_{123} ,	Z_{12})
$\boldsymbol{s}_1$	(X_{4567} ,	Y_{4567} ,	Z_{4567})	(R_{23} ,	W_{23} ,	Z_2)
$\boldsymbol{s}_2$	(X_{45671} ,	Y_{45671} ,	Z_{45671})	(R_3 ,	W_3 ,	c_{23})

TABLE 3.3, Continued

	SET 4a			*SET* 4b		
	$\boldsymbol{a}_{45}$	$\boldsymbol{b}_{45}$	$\boldsymbol{s}_4$	$\boldsymbol{a}_{45}$	$\boldsymbol{b}_{45}$	$\boldsymbol{s}_4$
$\boldsymbol{a}_{45}$	(1 ,	0 ,	0)	($P_{5671234}$,	$U_{5671234}$,	X_{567123})
$\boldsymbol{a}_{56}$	(P_5 ,	Q_5 ,	R_5)	(P_{671234} ,	U_{671234} ,	X_{67123})
$\boldsymbol{a}_{67}$	(P_{56} ,	Q_{56} ,	R_{56})	(P_{71234} ,	U_{71234} ,	X_{7123})
$\boldsymbol{a}_{71}$	(P_{567} ,	Q_{567} ,	R_{567})	(P_{1234} ,	U_{1234} ,	X_{123})
$\boldsymbol{a}_{12}$	(P_{5671} ,	Q_{5671} ,	R_{5671})	(P_{234} ,	U_{234} ,	X_{23})
$\boldsymbol{a}_{23}$	(P_{56712} ,	Q_{56712} ,	R_{56712})	(P_{34} ,	U_{34} ,	X_3)
$\boldsymbol{a}_{34}$	(P_{567123} ,	Q_{567123} ,	R_{567123})	(P_4 ,	U_4 ,	0)
$\boldsymbol{b}_{45}$	(0 ,	1 ,	0)	($Q_{5671234}$,	$V_{5671234}$,	Y_{567123})
$\boldsymbol{b}_{56}$	(U_5 ,	V_5 ,	W_5)	(Q_{671234} ,	V_{671234} ,	Y_{67123})
$\boldsymbol{b}_{67}$	(U_{56} ,	V_{56} ,	W_{56})	(Q_{71234} ,	V_{71234} ,	Y_{7123})
$\boldsymbol{b}_{71}$	(U_{567} ,	V_{567} ,	W_{567})	(Q_{1234} ,	V_{1234} ,	Y_{123})
$\boldsymbol{b}_{12}$	(U_{5671} ,	V_{5671} ,	W_{5671})	(Q_{234} ,	V_{234} ,	Y_{23})
$\boldsymbol{b}_{23}$	(U_{56712} ,	V_{56712} ,	W_{56712})	(Q_{34} ,	V_{34} ,	Y_3)
$\boldsymbol{b}_{34}$	(U_{567123} ,	V_{567123} ,	W_{567123})	(Q_4 ,	V_4 ,	$-s_{34}$)
$\boldsymbol{s}_4$	(0 ,	0 ,	1)	($R_{5671234}$,	$W_{5671234}$,	Z_{567123})
$\boldsymbol{s}_5$	(0 ,	$-s_{45}$,	c_{45})	(R_{671234} ,	W_{671234} ,	Z_{67123})
$\boldsymbol{s}_6$	(X_5 ,	Y_5 ,	Z_5)	(R_{71234} ,	W_{71234} ,	Z_{7123})
$\boldsymbol{s}_7$	(X_{56} ,	Y_{56} ,	Z_{56})	(R_{1234} ,	W_{1234} ,	Z_{123})
$\boldsymbol{s}_1$	(X_{567} ,	Y_{567} ,	Z_{567})	(R_{234} ,	W_{234} ,	Z_{23})
$\boldsymbol{s}_2$	(X_{5671} ,	Y_{5671} ,	Z_{5671})	(R_{34} ,	W_{34} ,	Z_3)
$\boldsymbol{s}_3$	(X_{56712} ,	Y_{56712} ,	Z_{56712})	(R_4 ,	W_4 ,	c_{34})

TABLE 3.3, Continued

	SET 5a			*SET* 5b		
	$\boldsymbol{a}_{56}$	$\boldsymbol{b}_{56}$	$\boldsymbol{s}_5$	$\boldsymbol{a}_{56}$	$\boldsymbol{b}_{56}$	$\boldsymbol{s}_5$
$\boldsymbol{a}_{56}$	(1 ,	0 ,	0)	($P_{6712345}$,	$U_{6712345}$,	X_{671234})
$\boldsymbol{a}_{67}$	(P_6 ,	Q_6 ,	R_6)	(P_{712345} ,	U_{712345} ,	X_{71234})
$\boldsymbol{a}_{71}$	(P_{67} ,	Q_{67} ,	R_{67})	(P_{12345} ,	U_{12345} ,	X_{1234})
$\boldsymbol{a}_{12}$	(P_{671} ,	Q_{671} ,	R_{671})	(P_{2345} ,	U_{2345} ,	X_{234})
$\boldsymbol{a}_{23}$	(P_{6712} ,	Q_{6712} ,	R_{6712})	(P_{345} ,	U_{345} ,	X_{34})
$\boldsymbol{a}_{34}$	(P_{67123} ,	Q_{67123} ,	R_{67123})	(P_{45} ,	U_{45} ,	X_4)
$\boldsymbol{a}_{45}$	(P_{671234} ,	Q_{671234} ,	R_{671234})	(P_5 ,	U_5 ,	0)
$\boldsymbol{b}_{56}$	(0 ,	1 ,	0)	($Q_{6712345}$,	$V_{6712345}$,	Y_{671234})
$\boldsymbol{b}_{67}$	(U_6 ,	V_6 ,	W_6)	(Q_{712345} ,	V_{712345} ,	Y_{71234})
$\boldsymbol{b}_{71}$	(U_{67} ,	V_{67} ,	W_{67})	(Q_{12345} ,	V_{12345} ,	Y_{1234})
$\boldsymbol{b}_{12}$	(U_{671} ,	V_{671} ,	W_{671})	(Q_{2345} ,	V_{2345} ,	Y_{234})
$\boldsymbol{b}_{23}$	(U_{6712} ,	V_{6712} ,	W_{6712})	(Q_{345} ,	V_{345} ,	Y_{34})
$\boldsymbol{b}_{34}$	(U_{67123} ,	V_{67123} ,	W_{67123})	(Q_{45} ,	V_{45} ,	Y_4)
$\boldsymbol{b}_{45}$	(U_{671234} ,	V_{671234} ,	W_{671234})	(Q_5 ,	V_5 ,	$-s_{45}$)
$\boldsymbol{s}_5$	(0 ,	0 ,	1)	($R_{6712345}$,	$W_{6712345}$,	Z_{671234})
$\boldsymbol{s}_6$	(0 ,	$-s_{56}$,	c_{56})	(R_{712345} ,	W_{712345} ,	Z_{71234})
$\boldsymbol{s}_7$	(X_6 ,	Y_6 ,	Z_6)	(R_{12345} ,	W_{12345} ,	Z_{1234})
$\boldsymbol{s}_1$	(X_{67} ,	Y_{67} ,	Z_{67})	(R_{2345} ,	W_{2345} ,	Z_{234})
$\boldsymbol{s}_2$	(X_{671} ,	Y_{671} ,	Z_{671})	(R_{345} ,	W_{345} ,	Z_{34})
$\boldsymbol{s}_3$	(X_{6712} ,	Y_{6712} ,	Z_{6712})	(R_{45} ,	W_{45} ,	Z_4)
$\boldsymbol{s}_4$	(X_{67123} ,	Y_{67123} ,	Z_{67123})	(R_5 ,	W_5 ,	c_{45})

TABLE 3.3, Continued

	SET 6a			*SET* 6b		
	$\boldsymbol{a}_{67}$	$\boldsymbol{b}_{67}$	$\boldsymbol{s}_6$	$\boldsymbol{a}_{67}$	$\boldsymbol{b}_{67}$	$\boldsymbol{s}_6$
$\boldsymbol{a}_{67}$	(1	, 0	, 0)	($P_{7123456}$	, $U_{7123456}$	, X_{712345})
$\boldsymbol{a}_{71}$	(P_7	, Q_7	, R_7)	(P_{123456}	, U_{123456}	, X_{12345})
$\boldsymbol{a}_{12}$	(P_{71}	, Q_{71}	, R_{71})	(P_{23456}	, U_{23456}	, X_{2345})
$\boldsymbol{a}_{23}$	(P_{712}	, Q_{712}	, R_{712})	(P_{3456}	, U_{3456}	, X_{345})
$\boldsymbol{a}_{34}$	(P_{7123}	, Q_{7123}	, R_{7123})	(P_{456}	, U_{456}	, X_{45})
$\boldsymbol{a}_{45}$	(P_{71234}	, Q_{71234}	, R_{71234})	(P_{56}	, U_{56}	, X_5)
$\boldsymbol{a}_{56}$	(P_{712345}	, Q_{712345}	, R_{712345})	(P_6	, U_6	, 0)
$\boldsymbol{b}_{67}$	(0	, 1	, 0)	($Q_{7123456}$	, $V_{7123456}$	, Y_{712345})
$\boldsymbol{b}_{71}$	(U_7	, V_7	, W_7)	(Q_{123456}	, V_{123456}	, Y_{12345})
$\boldsymbol{b}_{12}$	(U_{71}	, V_{71}	, W_{71})	(Q_{23456}	, V_{23456}	, Y_{2345})
$\boldsymbol{b}_{23}$	(U_{712}	, V_{712}	, W_{712})	(Q_{3456}	, V_{3456}	, Y_{345})
$\boldsymbol{b}_{34}$	(U_{7123}	, V_{7123}	, W_{7123})	(Q_{456}	, V_{456}	, Y_{45})
$\boldsymbol{b}_{45}$	(U_{71234}	, V_{71234}	, W_{71234})	(Q_{56}	, V_{56}	, Y_5)
$\boldsymbol{b}_{56}$	(U_{712345}	, V_{712345}	, W_{712345})	(Q_6	, V_6	, $-s_{56}$)
$\boldsymbol{s}_6$	(0	, 0	, 1)	($R_{7123456}$	, $W_{7123456}$	, Z_{712345})
$\boldsymbol{s}_7$	(0	, $-s_{67}$	, c_{67})	(R_{123456}	, W_{123456}	, Z_{12345})
$\boldsymbol{s}_1$	(X_7	, Y_7	, Z_7)	(R_{23456}	, W_{23456}	, Z_{2345})
$\boldsymbol{s}_2$	(X_{71}	, Y_{71}	, Z_{71})	(R_{3456}	, W_{3456}	, Z_{345})
$\boldsymbol{s}_3$	(X_{712}	, Y_{712}	, Z_{712})	(R_{456}	, W_{456}	, Z_{45})
$\boldsymbol{s}_4$	(X_{7123}	, Y_{7123}	, Z_{7123})	(R_{56}	, W_{56}	, Z_5)
$\boldsymbol{s}_5$	(X_{71234}	, Y_{71234}	, Z_{71234})	(R_6	, W_6	, c_{56})

TABLE 3.3, Continued

	SET 7a			*SET* 7b		
	$\boldsymbol{a}_{71}$	$\boldsymbol{b}_{71}$	$\boldsymbol{s}_{7}$	$\boldsymbol{a}_{71}$	$\boldsymbol{b}_{71}$	$\boldsymbol{s}_{7}$
$\boldsymbol{a}_{71}$	(1 ,	0 ,	0)	($P_{1234567}$,	$U_{1234567}$,	X_{123456})
$\boldsymbol{a}_{12}$	(P_1 ,	Q_1 ,	R_1)	(P_{234567} ,	U_{234567} ,	X_{23456})
$\boldsymbol{a}_{23}$	(P_{12} ,	Q_{12} ,	R_{12})	(P_{34567} ,	U_{34567} ,	X_{3456})
$\boldsymbol{a}_{34}$	(P_{123} ,	Q_{123} ,	R_{123})	(P_{4567} ,	U_{4567} ,	X_{456})
$\boldsymbol{a}_{45}$	(P_{1234} ,	Q_{1234} ,	R_{1234})	(P_{567} ,	U_{567} ,	X_{56})
$\boldsymbol{a}_{56}$	(P_{12345} ,	Q_{12345} ,	R_{12345})	(P_{67} ,	U_{67} ,	X_6)
$\boldsymbol{a}_{67}$	(P_{123456} ,	Q_{123456} ,	R_{123456})	(P_7 ,	U_7 ,	0)
$\boldsymbol{b}_{71}$	(0 ,	1 ,	0)	($Q_{1234567}$,	$V_{1234567}$,	Y_{123456})
$\boldsymbol{b}_{12}$	(U_1 ,	V_1 ,	W_1)	(Q_{234567} ,	V_{234567} ,	Y_{23456})
$\boldsymbol{b}_{23}$	(U_{12} ,	V_{12} ,	W_{12})	(Q_{34567} ,	V_{34567} ,	Y_{3456})
$\boldsymbol{b}_{34}$	(U_{123} ,	V_{123} ,	W_{123})	(Q_{4567} ,	V_{4567} ,	Y_{456})
$\boldsymbol{b}_{45}$	(U_{1234} ,	V_{1234} ,	W_{1234})	(Q_{567} ,	V_{567} ,	Y_{56})
$\boldsymbol{b}_{56}$	(U_{12345} ,	V_{12345} ,	W_{12345})	(Q_{67} ,	V_{67} ,	Y_6)
$\boldsymbol{b}_{67}$	(U_{123456} ,	V_{123456} ,	W_{123456})	(Q_7 ,	V_7 ,	$-s_{67}$)
$\boldsymbol{s}_{7}$	(0 ,	0 ,	1)	($R_{1234567}$,	$W_{1234567}$,	Z_{123456})
$\boldsymbol{s}_{1}$	(0 ,	$-s_{71}$,	c_{71})	(R_{234567} ,	W_{234567} ,	Z_{23456})
$\boldsymbol{s}_{2}$	(X_1 ,	Y_1 ,	Z_1)	(R_{34567} ,	W_{34567} ,	Z_{3456})
$\boldsymbol{s}_{3}$	(X_{12} ,	Y_{12} ,	Z_{12})	(R_{4567} ,	W_{4567} ,	Z_{456})
$\boldsymbol{s}_{4}$	(X_{123} ,	Y_{123} ,	Z_{123})	(R_{567} ,	W_{567} ,	Z_{56})
$\boldsymbol{s}_{5}$	(X_{1234} ,	Y_{1234} ,	Z_{1234})	(R_{67} ,	W_{67} ,	Z_6)
$\boldsymbol{s}_{6}$	(X_{12345} ,	Y_{12345} ,	Z_{12345})	(R_7 ,	W_7 ,	c_{67})

TABLE 3.3, Continued

	SET 8a			*SET* 8b		
	$\boldsymbol{a}_{71}$	$\boldsymbol{d}_{71}$	$\boldsymbol{s}_1$	$\boldsymbol{a}_{71}$	$\boldsymbol{d}_{71}$	$\boldsymbol{s}_1$
$\boldsymbol{a}_{71}$	(1	, 0	, 0)	($P_{7654321}$	, $U_{7654321}$	, X_{765432})
$\boldsymbol{a}_{67}$	($\bar{P}_7$	, $\bar{Q}_7$	, $\bar{R}_7$)	(P_{654321}	, U_{654321}	, X_{65432})
$\boldsymbol{a}_{56}$	(P_{76}	, Q_{76}	, R_{76})	(P_{54321}	, U_{54321}	, X_{5432})
$\boldsymbol{a}_{45}$	(P_{765}	, Q_{765}	, R_{765})	(P_{4321}	, U_{4321}	, X_{432})
$\boldsymbol{a}_{34}$	(P_{7654}	, Q_{7654}	, R_{7654})	(P_{321}	, U_{321}	, X_{32})
$\boldsymbol{a}_{23}$	(P_{76543}	, Q_{76543}	, R_{76543})	(P_{21}	, U_{21}	, X_2)
$\boldsymbol{a}_{12}$	(P_{765432}	, Q_{765432}	, R_{765432})	($\bar{P}_1$	, $\bar{U}_1$	, 0)
$\boldsymbol{d}_{71}$	(0	, 1	, 0)	($Q_{7654321}$	, $V_{7654321}$	, Y_{765432})
$\boldsymbol{d}_{67}$	($\bar{U}_7$	, $\bar{V}_7$	, $\bar{W}_7$)	(Q_{654321}	, V_{654321}	, Y_{65432})
$\boldsymbol{d}_{56}$	(U_{76}	, V_{76}	, W_{76})	(Q_{54321}	, V_{54321}	, Y_{5432})
$\boldsymbol{d}_{45}$	(U_{765}	, V_{765}	, W_{765})	(Q_{4321}	, V_{4321}	, Y_{432})
$\boldsymbol{d}_{34}$	(U_{7654}	, V_{7654}	, W_{7654})	(Q_{321}	, V_{321}	, Y_{32})
$\boldsymbol{d}_{23}$	(U_{76543}	, V_{76543}	, W_{76543})	(Q_{21}	, V_{21}	, $\bar{Y}_2$)
$\boldsymbol{d}_{12}$	(U_{765432}	, V_{765432}	, W_{765432})	($\bar{Q}_1$	, $\bar{V}_1$	, $-s_{12}$)
$\boldsymbol{s}_1$	(0	, 0	, 1)	($R_{7654321}$	, $W_{7654321}$	, Z_{765432})
$\boldsymbol{s}_7$	(0	, $-s_{71}$	, c_{71})	(R_{654321}	, W_{654321}	, Z_{65432})
$\boldsymbol{s}_6$	($\bar{X}_7$	, $\bar{Y}_7$	, $\bar{Z}_7$)	(R_{54321}	, W_{54321}	, Z_{5432})
$\boldsymbol{s}_5$	(X_{76}	, Y_{76}	, Z_{76})	(R_{4321}	, W_{4321}	, Z_{432})
$\boldsymbol{s}_4$	(X_{765}	, Y_{765}	, Z_{765})	(R_{321}	, W_{321}	, Z_{32})
$\boldsymbol{s}_3$	(X_{7654}	, Y_{7654}	, Z_{7654})	(R_{21}	, W_{21}	, $\bar{Z}_2$)
$\boldsymbol{s}_2$	(X_{76543}	, Y_{76543}	, Z_{76543})	($\bar{R}_1$	, $\bar{W}_1$	, c_{12})

TABLE 3.3, Continued

	SET 9a			SET 9b		
	$\boldsymbol{a}_{67}$	$\boldsymbol{d}_{67}$	$\boldsymbol{s}_7$	$\boldsymbol{a}_{67}$	$\boldsymbol{d}_{67}$	$\boldsymbol{s}_7$
$\boldsymbol{a}_{67}$	1	0	0	$P_{6543217}$	$U_{6543217}$	X_{654321}
$\boldsymbol{a}_{56}$	$\bar{P}_6$	$\bar{Q}_6$	$\bar{R}_6$	P_{543217}	U_{543217}	X_{54321}
$\boldsymbol{a}_{45}$	P_{65}	Q_{65}	R_{65}	P_{43217}	U_{43217}	X_{4321}
$\boldsymbol{a}_{34}$	P_{654}	Q_{654}	R_{654}	P_{3217}	U_{3217}	X_{321}
$\boldsymbol{a}_{23}$	P_{6543}	Q_{6543}	R_{6543}	P_{217}	U_{217}	X_{21}
$\boldsymbol{a}_{12}$	P_{65432}	Q_{65432}	R_{65432}	P_{17}	U_{17}	$\bar{X}_1$
$\boldsymbol{a}_{71}$	P_{654321}	Q_{654321}	R_{654321}	$\bar{P}_7$	$\bar{U}_7$	0
$\boldsymbol{d}_{67}$	0	1	0	$Q_{6543217}$	$V_{6543217}$	Y_{654321}
$\boldsymbol{d}_{56}$	$\bar{U}_6$	$\bar{V}_6$	$\bar{W}_6$	Q_{543217}	V_{543217}	Y_{54321}
$\boldsymbol{d}_{45}$	U_{65}	V_{65}	W_{65}	Q_{43217}	V_{43217}	Y_{4321}
$\boldsymbol{d}_{34}$	U_{654}	V_{654}	W_{654}	Q_{3217}	V_{3217}	Y_{321}
$\boldsymbol{d}_{23}$	U_{6543}	V_{6543}	W_{6543}	Q_{217}	V_{217}	Y_{21}
$\boldsymbol{d}_{12}$	U_{65432}	V_{65432}	W_{65432}	Q_{17}	V_{17}	$\bar{Y}_1$
$\boldsymbol{d}_{71}$	U_{654321}	V_{654321}	W_{654321}	$\bar{Q}_7$	$\bar{V}_7$	$-s_{71}$
$\boldsymbol{s}_7$	0	0	1	$R_{6543217}$	$W_{6543217}$	Z_{654321}
$\boldsymbol{s}_6$	0	$-s_{67}$	c_{67}	R_{543217}	W_{543217}	Z_{54321}
$\boldsymbol{s}_5$	$\bar{X}_6$	$\bar{Y}_6$	$\bar{Z}_6$	R_{43217}	W_{43217}	Z_{4321}
$\boldsymbol{s}_4$	X_{65}	Y_{65}	Z_{65}	R_{3217}	W_{3217}	Z_{321}
$\boldsymbol{s}_3$	X_{654}	Y_{654}	Z_{654}	R_{217}	W_{217}	Z_{21}
$\boldsymbol{s}_2$	X_{6543}	Y_{6543}	Z_{6543}	R_{17}	W_{17}	$\bar{Z}_1$
$\boldsymbol{s}_1$	X_{65432}	Y_{65432}	Z_{65432}	$\bar{R}_7$	$\bar{W}_7$	c_{71}

TABLE 3.3, Continued

	SET 10a			*SET* 10b		
	$\boldsymbol{a}_{56}$	$\boldsymbol{d}_{56}$	$\boldsymbol{s}_6$	$\boldsymbol{a}_{56}$	$\boldsymbol{d}_{56}$	$\boldsymbol{s}_6$
$\boldsymbol{a}_{56}$	(1 ,	0 ,	0)	($P_{5432176}$,	$U_{5432176}$,	X_{543217})
$\boldsymbol{a}_{45}$	($\bar{P}_5$,	$\bar{Q}_5$,	$\bar{R}_5$)	(P_{432176} ,	U_{432176} ,	X_{43217})
$\boldsymbol{a}_{34}$	(P_{54} ,	Q_{54} ,	R_{54})	(P_{32176} ,	U_{32176} ,	X_{3217})
$\boldsymbol{a}_{23}$	(P_{543} ,	Q_{543} ,	R_{543})	(P_{2176} ,	U_{2176} ,	X_{217})
$\boldsymbol{a}_{12}$	(P_{5432} ,	Q_{5432} ,	R_{5432})	(P_{176} ,	U_{176} ,	X_{17})
$\boldsymbol{a}_{71}$	(P_{54321} ,	Q_{54321} ,	R_{54321})	(P_{76} ,	U_{76} ,	$\bar{X}_7$)
$\boldsymbol{a}_{67}$	(P_{543217} ,	Q_{543217} ,	R_{543217})	($\bar{P}_6$,	$\bar{U}_6$,	0)
$\boldsymbol{d}_{56}$	(0 ,	1 ,	0)	($Q_{5432176}$,	$V_{5432176}$,	Y_{543217})
$\boldsymbol{d}_{45}$	($\bar{U}_5$,	$\bar{V}_5$,	$\bar{W}_5$)	(Q_{432176} ,	V_{432176} ,	Y_{43217})
$\boldsymbol{d}_{34}$	(U_{54} ,	V_{54} ,	W_{54})	(Q_{32176} ,	V_{32176} ,	Y_{3217})
$\boldsymbol{d}_{23}$	(U_{543} ,	V_{543} ,	W_{543})	(Q_{2176} ,	V_{2176} ,	Y_{217})
$\boldsymbol{d}_{12}$	(U_{5432} ,	V_{5432} ,	W_{5432})	(Q_{176} ,	V_{176} ,	Y_{17})
$\boldsymbol{d}_{71}$	(U_{54321} ,	V_{54321} ,	W_{54321})	(Q_{76} ,	V_{76} ,	$\bar{Y}_7$)
$\boldsymbol{d}_{67}$	(U_{543217} ,	V_{543217} ,	W_{543217})	($\bar{Q}_6$,	$\bar{V}_6$,	$-s_{67}$)
$\boldsymbol{s}_6$	(0 ,	0 ,	1)	($R_{5432176}$,	$W_{5432176}$,	Z_{543217})
$\boldsymbol{s}_5$	(0 ,	$-s_{56}$,	c_{56})	(R_{432176} ,	W_{432176} ,	Z_{43217})
$\boldsymbol{s}_4$	($\bar{X}_5$,	$\bar{Y}_5$,	$\bar{Z}_5$)	(R_{32176} ,	W_{32176} ,	Z_{3217})
$\boldsymbol{s}_3$	(X_{54} ,	Y_{54} ,	Z_{54})	(R_{2176} ,	W_{2176} ,	Z_{217})
$\boldsymbol{s}_2$	(X_{543} ,	Y_{543} ,	Z_{543})	(R_{176} ,	W_{176} ,	Z_{17})
$\boldsymbol{s}_1$	(X_{5432} ,	Y_{5432} ,	Z_{5432})	(R_{76} ,	W_{76} ,	$\bar{Z}_7$)
$\boldsymbol{s}_7$	(X_{54321} ,	Y_{54321} ,	Z_{54321})	($\bar{R}_6$,	$\bar{W}_6$,	c_{67})

TABLE 3.3, Continued

	SET 11a			*SET* 11b		
	$\boldsymbol{a}_{45}$	$\boldsymbol{d}_{45}$	$\boldsymbol{s}_{5}$	$\boldsymbol{a}_{45}$	$\boldsymbol{d}_{45}$	$\boldsymbol{s}_{5}$
$\boldsymbol{a}_{45}$	(1 ,	0 ,	0)	($P_{4321765}$,	$U_{4321765}$,	X_{432176})
$\boldsymbol{a}_{34}$	($\overline{P}_{4}$,	$\overline{Q}_{4}$,	$\overline{R}_{4}$)	(P_{321765} ,	U_{321765} ,	X_{32176})
$\boldsymbol{a}_{23}$	(P_{43} ,	Q_{43} ,	R_{43})	(P_{21765} ,	U_{21765} ,	X_{2176})
$\boldsymbol{a}_{12}$	(P_{432} ,	Q_{432} ,	R_{432})	(P_{1765} ,	U_{1765} ,	X_{176})
$\boldsymbol{a}_{71}$	(P_{4321} ,	Q_{4321} ,	R_{4321})	(P_{765} ,	U_{765} ,	X_{76})
$\boldsymbol{a}_{67}$	(P_{43217} ,	Q_{43217} ,	R_{43217})	(P_{65} ,	U_{65} ,	$\overline{X}_{6}$)
$\boldsymbol{a}_{56}$	(P_{432176} ,	Q_{432176} ,	R_{432176})	($\overline{P}_{5}$,	$\overline{U}_{5}$,	0)
$\boldsymbol{d}_{45}$	(0 ,	1 ,	0)	($Q_{4321765}$,	$V_{4321765}$,	Y_{432176})
$\boldsymbol{d}_{34}$	($\overline{U}_{4}$,	$\overline{V}_{4}$,	$\overline{W}_{4}$)	(Q_{321765} ,	V_{321765} ,	Y_{32176})
$\boldsymbol{d}_{23}$	(U_{43} ,	V_{43} ,	W_{43})	(Q_{21765} ,	V_{21765} ,	Y_{2176})
$\boldsymbol{d}_{12}$	(U_{432} ,	V_{432} ,	W_{432})	(Q_{1765} ,	V_{1765} ,	Y_{176})
$\boldsymbol{d}_{71}$	(U_{4321} ,	V_{4321} ,	W_{4321})	(Q_{765} ,	V_{765} ,	Y_{76})
$\boldsymbol{d}_{67}$	(U_{43217} ,	V_{43217} ,	W_{43217})	(Q_{65} ,	V_{65} ,	$\overline{Y}_{6}$)
$\boldsymbol{d}_{56}$	(U_{432176} ,	V_{432176} ,	W_{432176})	($\overline{Q}_{5}$,	$\overline{V}_{5}$,	$-s_{56}$)
$\boldsymbol{s}_{5}$	(0 ,	0 ,	1)	($R_{4321765}$,	$W_{4321765}$,	Z_{432176})
$\boldsymbol{s}_{4}$	(0 ,	$-s_{45}$,	c_{45})	(R_{321765} ,	W_{321765} ,	Z_{32176})
$\boldsymbol{s}_{3}$	($\overline{X}_{4}$,	$\overline{Y}_{4}$,	$\overline{Z}_{4}$)	(R_{21765} ,	W_{21765} ,	Z_{2176})
$\boldsymbol{s}_{2}$	(X_{43} ,	Y_{43} ,	Z_{43})	(R_{1765} ,	W_{1765} ,	Z_{176})
$\boldsymbol{s}_{1}$	(X_{432} ,	Y_{432} ,	Z_{432})	(R_{765} ,	W_{765} ,	Z_{76})
$\boldsymbol{s}_{7}$	(X_{4321} ,	Y_{4321} ,	Z_{4321})	(R_{65} ,	W_{65} ,	$\overline{Z}_{6}$)
$\boldsymbol{s}_{6}$	(X_{43217} ,	Y_{43217} ,	Z_{43217})	($\overline{R}_{5}$,	$\overline{W}_{5}$,	c_{56})

TABLE 3.3, Continued

	SET 12a			*SET* 12b		
	$\boldsymbol{a}_{34}$	$\boldsymbol{d}_{34}$	$\boldsymbol{s}_4$	$\boldsymbol{a}_{34}$	$\boldsymbol{d}_{34}$	$\boldsymbol{s}_4$
$\boldsymbol{a}_{34}$	1	0	0	$P_{3217654}$	$U_{3217654}$	X_{321765}
$\boldsymbol{a}_{23}$	$\bar{P}_3$	$\bar{Q}_3$	$\bar{R}_3$	P_{217654}	U_{217654}	X_{21765}
$\boldsymbol{a}_{12}$	P_{32}	Q_{32}	R_{32}	P_{17654}	U_{17654}	X_{1765}
$\boldsymbol{a}_{71}$	P_{321}	Q_{321}	R_{321}	P_{7654}	U_{7654}	X_{765}
$\boldsymbol{a}_{67}$	P_{3217}	Q_{3217}	R_{3217}	P_{654}	U_{654}	X_{65}
$\boldsymbol{a}_{56}$	P_{32176}	Q_{32176}	R_{32176}	P_{54}	U_{54}	$\bar{X}_5$
$\boldsymbol{a}_{45}$	P_{321765}	Q_{321765}	R_{321765}	$\bar{P}_4$	$\bar{U}_4$	0
$\boldsymbol{d}_{34}$	0	1	0	$Q_{3217654}$	$V_{3217654}$	Y_{321765}
$\boldsymbol{d}_{23}$	$\bar{U}_3$	$\bar{V}_3$	$\bar{W}_3$	Q_{217654}	V_{217654}	Y_{21765}
$\boldsymbol{d}_{12}$	U_{32}	V_{32}	W_{32}	Q_{17654}	V_{17654}	Y_{1765}
$\boldsymbol{d}_{71}$	U_{321}	V_{321}	W_{321}	Q_{7654}	V_{7654}	Y_{765}
$\boldsymbol{d}_{67}$	U_{3217}	V_{3217}	W_{3217}	Q_{654}	V_{654}	Y_{65}
$\boldsymbol{d}_{56}$	U_{32176}	V_{32176}	W_{32176}	Q_{54}	V_{54}	$\bar{Y}_5$
$\boldsymbol{d}_{45}$	U_{321765}	V_{321765}	W_{321765}	$\bar{Q}_4$	$\bar{V}_4$	$-s_{45}$
$\boldsymbol{s}_4$	0	0	1	$R_{3217654}$	$W_{3217654}$	Z_{321765}
$\boldsymbol{s}_3$	0	$-s_{34}$	c_{34}	R_{217654}	W_{217654}	Z_{21765}
$\boldsymbol{s}_2$	$\bar{X}_3$	$\bar{Y}_3$	$\bar{Z}_3$	R_{17654}	W_{17654}	Z_{1765}
$\boldsymbol{s}_1$	X_{32}	Y_{32}	Z_{32}	R_{7654}	W_{7654}	Z_{765}
$\boldsymbol{s}_7$	X_{321}	Y_{321}	Z_{321}	R_{654}	W_{654}	Z_{65}
$\boldsymbol{s}_6$	X_{3217}	Y_{3217}	Z_{3217}	R_{54}	W_{54}	$\bar{Z}_5$
$\boldsymbol{s}_5$	X_{32176}	Y_{32176}	Z_{32176}	$\bar{R}_4$	$\bar{W}_4$	c_{45}

TABLE 3.3, Continued

	SET 13a			*SET* 13b		
	$\boldsymbol{a}_{23}$	$\boldsymbol{d}_{23}$	$\boldsymbol{s}_3$	$\boldsymbol{a}_{23}$	$\boldsymbol{d}_{23}$	$\boldsymbol{s}_3$
$\boldsymbol{a}_{23}$	(1 ,	0 ,	0)	($P_{2176543}$,	$U_{2176543}$,	X_{217654})
$\boldsymbol{a}_{12}$	($\bar{P}_2$,	$\bar{Q}_2$,	$\bar{R}_2$)	(P_{176543} ,	U_{176543} ,	X_{17654})
$\boldsymbol{a}_{71}$	(P_{21} ,	Q_{21} ,	R_{21})	(P_{76543} ,	U_{76543} ,	X_{7654})
$\boldsymbol{a}_{67}$	(P_{217} ,	Q_{217} ,	R_{217})	(P_{6543} ,	U_{6543} ,	X_{654})
$\boldsymbol{a}_{56}$	(P_{2176} ,	Q_{2176} ,	R_{2176})	(P_{543} ,	U_{543} ,	X_{54})
$\boldsymbol{a}_{45}$	(P_{21765} ,	Q_{21765} ,	R_{21765})	(P_{43} ,	U_{43} ,	$\bar{X}_4$)
$\boldsymbol{a}_{34}$	(P_{217654} ,	Q_{217654} ,	R_{217654})	($\bar{P}_3$,	$\bar{U}_3$,	0)
$\boldsymbol{d}_{23}$	(0 ,	1 ,	0)	($Q_{2176543}$,	$V_{2176543}$,	Y_{217654})
$\boldsymbol{d}_{12}$	($\bar{U}_2$,	$\bar{V}_2$,	$\bar{W}_2$)	(Q_{176543} ,	V_{176543} ,	Y_{17654})
$\boldsymbol{d}_{71}$	(U_{21} ,	V_{21} ,	W_{21})	(Q_{76543} ,	V_{76543} ,	Y_{7654})
$\boldsymbol{d}_{67}$	(U_{217} ,	V_{217} ,	W_{217})	(Q_{6543} ,	V_{6543} ,	Y_{654})
$\boldsymbol{d}_{56}$	(U_{2176} ,	V_{2176} ,	W_{2176})	(Q_{543} ,	V_{543} ,	Y_{54})
$\boldsymbol{d}_{45}$	(U_{21765} ,	V_{21765} ,	W_{21765})	(Q_{43} ,	V_{43} ,	$\bar{Y}_4$)
$\boldsymbol{d}_{34}$	(U_{217654} ,	V_{217654} ,	W_{217654})	($\bar{Q}_3$,	$\bar{V}_3$,	$-s_{34}$)
$\boldsymbol{s}_3$	(0 ,	0 ,	1)	($R_{2176543}$,	$W_{2176543}$,	Z_{217654})
$\boldsymbol{s}_2$	(0 ,	$-s_{23}$,	c_{23})	(R_{176543} ,	W_{176543} ,	Z_{17654})
$\boldsymbol{s}_1$	($\bar{X}_2$,	$\bar{Y}_2$,	$\bar{Z}_2$)	(R_{76543} ,	W_{76543} ,	Z_{7654})
$\boldsymbol{s}_7$	(X_{21} ,	Y_{21} ,	Z_{21})	(R_{6543} ,	W_{6543} ,	Z_{654})
$\boldsymbol{s}_6$	(X_{217} ,	Y_{217} ,	Z_{217})	(R_{543} ,	W_{543} ,	Z_{54})
$\boldsymbol{s}_5$	(X_{2176} ,	Y_{2176} ,	Z_{2176})	(R_{43} ,	W_{43} ,	$\bar{Z}_4$)
$\boldsymbol{s}_4$	(X_{21765} ,	Y_{21765} ,	Z_{21765})	($\bar{R}_3$,	$\bar{W}_3$,	c_{34})

TABLE 3.3, Continued

	SET 14a				*SET* 14b		
	$\boldsymbol{a}_{12}$	$\boldsymbol{d}_{12}$	$\boldsymbol{s}_2$		$\boldsymbol{a}_{12}$	$\boldsymbol{d}_{12}$	$\boldsymbol{s}_2$
$\boldsymbol{a}_{12}$	(1	, 0	, 0	)	($P_{1765432}$	, $U_{1765432}$	, X_{176543})
$\boldsymbol{a}_{71}$	($\bar{P}_1$	, $\bar{Q}_1$	, $\bar{R}_1$	)	(P_{765432}	, U_{765432}	, X_{76543})
$\boldsymbol{a}_{67}$	(P_{17}	, Q_{17}	, R_{17}	)	(P_{65432}	, U_{65432}	, X_{6543})
$\boldsymbol{a}_{56}$	(P_{176}	, Q_{176}	, R_{176}	)	(P_{5432}	, U_{5432}	, X_{543})
$\boldsymbol{a}_{45}$	(P_{1765}	, Q_{1765}	, R_{1765}	)	(P_{432}	, U_{432}	, X_{43})
$\boldsymbol{a}_{34}$	(P_{17654}	, Q_{17654}	, R_{17654}	)	(P_{32}	, U_{32}	, $\bar{X}_3$)
$\boldsymbol{a}_{23}$	(P_{176543}	, Q_{176543}	, R_{176543}	)	($\bar{P}_2$	, $\bar{U}_2$	, 0)
$\boldsymbol{d}_{12}$	(0	, 1	, 0	)	($Q_{1765432}$	, $V_{1765432}$	, Y_{176543})
$\boldsymbol{d}_{71}$	($\bar{U}_1$	, $\bar{V}_1$	, $\bar{W}_1$	)	(Q_{765432}	, V_{765432}	, Y_{76543})
$\boldsymbol{d}_{67}$	(U_{17}	, V_{17}	, W_{17}	)	(Q_{65432}	, V_{65432}	, Y_{6543})
$\boldsymbol{d}_{56}$	(U_{176}	, V_{176}	, W_{176}	)	(Q_{5432}	, V_{5432}	, Y_{543})
$\boldsymbol{d}_{45}$	(U_{1765}	, V_{1765}	, W_{1765}	)	(Q_{432}	, V_{432}	, Y_{43})
$\boldsymbol{d}_{34}$	(U_{17654}	, V_{17654}	, W_{17654}	)	(Q_{32}	, V_{32}	, $\bar{Y}_3$)
$\boldsymbol{d}_{23}$	(U_{176543}	, V_{176543}	, W_{176543}	)	($\bar{Q}_2$	, $\bar{V}_2$	, $-s_{23}$)
$\boldsymbol{s}_2$	(0	, 0	, 1	)	($R_{1765432}$	, $W_{1765432}$	, Z_{176543})
$\boldsymbol{s}_1$	(0	, $-s_{12}$	, c_{12}	)	(R_{765432}	, W_{765432}	, Z_{76543})
$\boldsymbol{s}_7$	($\bar{X}_1$	, $\bar{Y}_1$	, $\bar{Z}_1$	)	(R_{65432}	, W_{65432}	, Z_{6543})
$\boldsymbol{s}_6$	(X_{17}	, Y_{17}	, Z_{17}	)	(R_{5432}	, W_{5432}	, Z_{543})
$\boldsymbol{s}_5$	(X_{176}	, Y_{176}	, Z_{176}	)	(R_{432}	, W_{432}	, Z_{43})
$\boldsymbol{s}_4$	(X_{1765}	, Y_{1765}	, Z_{1765}	)	(R_{32}	, W_{32}	, $\bar{Z}_3$)
$\boldsymbol{s}_3$	(X_{17654}	, Y_{17654}	, Z_{17654}	)	($\bar{R}_2$	, $\bar{W}_2$	, c_{23})

Analogous with Tables 3.1a-d, the Tables 3.2a-d list the various definitions for the recursive notation with descending subscripts. Additionally, Table 3.2e gives the recursive elements in the associated dextral bases $[D']$ and can be used when a sinistral base is inconvenient. (For completeness the recursive elements in the associated sinistral bases $[B']$ are listed in Table 3.1e.) Expressions for the direction cosines are listed in Table 3.3 (Sets 8a-14a) together with the alternative expressions (Sets 8b-14b).

In general, Table 3.3 can be used to establish trigonometric laws for application in displacement analysis by equating a direction cosine element to its alternative expression in the same set. In analogy with the laws of spherical polygons, equating the three elements of an $\boldsymbol{a}$ vector yields the polar sine, sine-cosine and cosine laws respectively. Similarly, equating the elements of $\boldsymbol{b}$ (or $\boldsymbol{d}$) and $\boldsymbol{s}$ vectors yield the complementary-polar and spherical sine, sine-cosine and cosine laws. These laws are listed in Tables 3.1f and 3.2f. It should be noted that four laws contain seven variables, four laws contain six variables and a single law contains only five variables. When expanding (defining) all three elements of a vector, it is more efficient to use the backward recursion formulas. Similarly, the alternative expressions for the three elements of a vector are most easily expanded using the forward recursion formulas.

Since both the dextral bases $[B]$ and the sinistral bases $[D]$ have been established using the same $\boldsymbol{a}$ and $\boldsymbol{s}$ vectors it is possible to deduce three elementary symmetrical identities using subscripts in both ascending and descending order, the general form of which are,

$$P_{i..m} = P_{m..i} \tag{3.38}$$

$$R_{i..m} = X_{m..i} \quad (X_{i..m} = R_{m..i}) \tag{3.39}$$

$$Z_{i..m} = Z_{m..i} \tag{3.40}$$

These expressions are easily verified by inspection of Table 3.3.

The labeling of the direction cosine matrices ($[M]$, $[\overline{M}]$, $[M']$ and $[\overline{M}']$) with the ordered sequence of subscripts yields a simple multiplicative property which can be expressed generally as,

$$[M_{i..kl..n}] = [M_{i..k}][M_{l..n}]. \tag{3.41}$$

Such expressions can be used to establish useful identities between the elements. For example, the 11 element of

$$[M_{hijk}] = [M_{hi}][M_{jk}] \tag{3.42}$$

yields the identity,

$$P_{hijk} = P_{hi}P_{jk} + U_{hi}Q_{jk} + X_h R_{jk}, \tag{3.43}$$

which could be used to reduce highly complex non-linear expressions.

The recursive notation has been developed for a mechanism with seven joints. It is readily adapted to the analysis of structures and mechanisms with less than seven joints, and it is easily extended by induction to the analysis of mechanisms with eight or more joints. Tables 3.3 sets 1-14 are directly applicable to the analysis of a general seven-link spatial mechanism. Equations relating various joint angles are simply obtained by equating each direction cosine element to its corresponding alternative expression. However, direction cosine equations are in general not sufficient to perform a displacement analysis and further equations are required which contain the link lengths a_{ij} and the offsets S_{jj}. These are the mutual moment equations, special cases of which are the scalar components of the vector loop equation. The vector loop equation can be expressed in the form

$$\boldsymbol{R} \equiv S_{11}\boldsymbol{s}_1 + a_{12}\boldsymbol{a}_{12} + S_{22}\boldsymbol{s}_2 + \cdots + S_{77}\boldsymbol{s}_7 + a_{71}\boldsymbol{a}_{71} = \boldsymbol{0}. \tag{3.44}$$

Any of the sets of direction cosines listed in Table 3.3 can be used to express the three scalar components of Eqn. (3.44) in terms of the various bases, and one base is inevitably preferred to another when solving for the joint angles of a particular mechanism. This is mainly because all current industrial robots have special dimensions (twist angles α_{ij} are usually zero or ninety degrees and, various link lengths a_{ij} and offsets S_{jj} are zero). *The introduction of special dimensions leads to considerable simplification of the vector loop equation and of its three scalar components which when expressed in a particular base contain the least number of joint variables.* The selection of the preferred base is more often than not easily deduced by consideration of the special geometry since the projection of the vector loop equation in one or more directions will preclude unwanted joint variables. For instance, when three joint axes $\boldsymbol{s}_i$, $\boldsymbol{s}_j$ and $\boldsymbol{s}_k$ are parallel a projection of the vector loop equation in any of these directions will yield a scalar equation which does not contain the joint variables θ_i, θ_j and θ_k.

The self-scalar product, $\boldsymbol{R} \cdot \boldsymbol{R} = \boldsymbol{0}$, of the vector loop equation is frequently used in the analysis of spatial mechanisms. The sets of direction cosines listed in Table 3.3 can be used to advantage to obtain the simplest expressions for various scalar products. For instance, scalar products such as $\boldsymbol{s}_1 \cdot (\boldsymbol{s}_2, \boldsymbol{a}_{23}, \ldots \boldsymbol{s}_7)$ are most easily determined using the bases in Sets 1 or 8 where $\boldsymbol{s}_1 = (0,0,1)$.

Figure 3.2 illustrates the dextral base $[B_h] = [\boldsymbol{a}_{hi}\ \boldsymbol{b}_{hi}\ \boldsymbol{s}_h]$ located at the origin 0_h which will be used as a reference frame together with a second base $[B_m] = [\boldsymbol{a}_{mn}\ \boldsymbol{b}_{mn}\ \boldsymbol{s}_m]$ located at an origin 0_m. The vector $\boldsymbol{r}_{hm}$ is measured from 0_h to 0_m and,

$$\boldsymbol{r}_{hm} = a_{hi}\boldsymbol{a}_{hi} + S_{ii}\boldsymbol{s}_i + \cdots a_{lm}\boldsymbol{a}_{lm} + S_{mm}\boldsymbol{s}_m. \tag{3.45}$$

For the three vectors in the base $[B_m]$, their moments about 0_h are given by

$$[B_{om}] \equiv [\boldsymbol{r}_{hm}\times\boldsymbol{a}_{mn}\ \ \boldsymbol{r}_{hm}\times\boldsymbol{b}_{mn}\ \ \boldsymbol{r}_{hm}\times\boldsymbol{s}_m]. \tag{3.46}$$

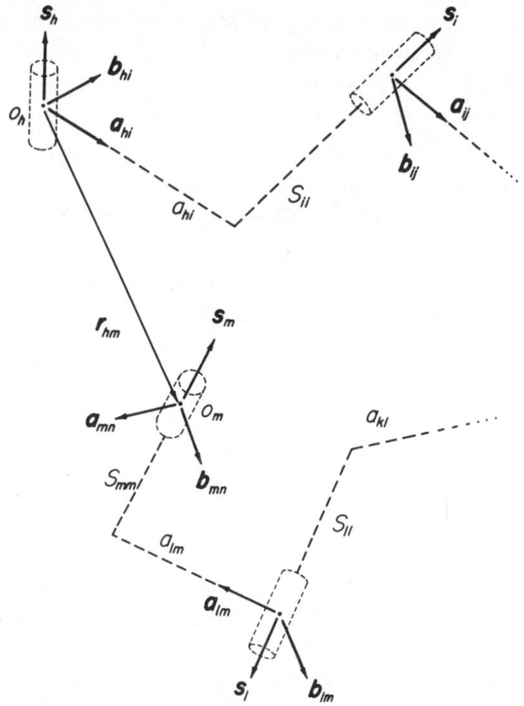

Figure 3.2 Mutual moments for dextral bases include the mechanism dimensions a and S whereas direction cosine expressions do not. A unit force along a line develops a torque about a second line which is equivalent to their mutual moment. Sinistral bases (not shown) are developed analogously (see Figure 3.1).

The projection of $[B_{om}]$ in the directions $[B_h]$ can be expressed in the following form analogous to that of the direction cosines (compare Eqns. (3.9) and (3.47)),

$$[M_{oij..m}] \equiv [B_h]^{\mathrm{T}}[B_{om}] = \begin{bmatrix} \boldsymbol{a}_{hi}\cdot\boldsymbol{r}_{hm}\times\boldsymbol{a}_{mn} & \boldsymbol{a}_{hi}\cdot\boldsymbol{r}_{hm}\times\boldsymbol{b}_{mn} & \boldsymbol{a}_{hi}\cdot\boldsymbol{r}_{hm}\times\boldsymbol{s}_{m} \\ \boldsymbol{b}_{hi}\cdot\boldsymbol{r}_{hm}\times\boldsymbol{a}_{mn} & \boldsymbol{b}_{hi}\cdot\boldsymbol{r}_{hm}\times\boldsymbol{b}_{mn} & \boldsymbol{b}_{hi}\cdot\boldsymbol{r}_{hm}\times\boldsymbol{s}_{m} \\ \boldsymbol{s}_{h}\cdot\boldsymbol{r}_{hm}\times\boldsymbol{a}_{mn} & \boldsymbol{s}_{h}\cdot\boldsymbol{r}_{hm}\times\boldsymbol{b}_{mn} & \boldsymbol{s}_{h}\cdot\boldsymbol{r}_{hm}\times\boldsymbol{s}_{m} \end{bmatrix}. \tag{3.47}$$

Each of the elements on the right side of Eqn. (3.47) is the mutual moment of a vector in $[B_m]$ and a vector in $[B_h]$. The reader who is not familiar with the term mutual moment should consult a text on vector mechanics, for instance Refs. [5,11]. Briefly, a meaningful interpretation of mutual moment is that a unit force on $\boldsymbol{a}_{mn}$ produces a torque $\boldsymbol{a}_{hi}\cdot\boldsymbol{r}_{hm}\times\boldsymbol{a}_{mn}$ about the vector $\boldsymbol{a}_{hi}$. Conversely a unit force on $\boldsymbol{a}_{hi}$ produces the same torque $\boldsymbol{a}_{mn}\cdot\boldsymbol{r}_{mh}\times\boldsymbol{a}_{hi} \equiv \boldsymbol{a}_{hi}\cdot\boldsymbol{r}_{hm}\times\boldsymbol{a}_{mn}$ about the vector $\boldsymbol{a}_{mn}$, where $\boldsymbol{r}_{mh} = -\boldsymbol{r}_{hm}$.

Further, analogous to the elements of the direction cosine matrix $[M_{ij..m}]$, the elements of the mutual moment matrix $[M_{oij..m}]$ will be defined by

$$[M_{oij..m}] = \begin{bmatrix} P_{oij..m} & U_{oij..m} & X_{oij..l} \\ Q_{oij..m} & V_{oij..m} & Y_{oij..l} \\ R_{oij..m} & W_{oij..m} & Z_{oij..l} \end{bmatrix}. \tag{3.48}$$

The elements of the direction cosine matrix $[M_{ij..m}]$ are invariant with respect to the choice of reference frame although the base $[B_h]$ is employed to yield the simplest expressions. Similarly the elements of the mutual moment matrix $[M_{oij..m}]$ are also invariant.

The mutual moment matrix $[M_{oij..m}]$ is now expressed in terms of the corresponding direction cosine matrix $[M_{ij..m}]$. This is facilitated by introducing the scalar linear operator,

$$\nabla \equiv \sum_{JK=hi}^{nh} a_{JK} \frac{\partial}{\partial \alpha_{JK}} + \sum_{J=h}^{n} S_{JJ} \frac{\partial}{\partial \theta_J} \tag{3.49}$$

with the following properties,

1. When $f=f(\alpha_{JK}, \theta_J)$ then

$$\nabla f = \sum_{JK=hi}^{nh} a_{JK} \frac{\partial f}{\partial \alpha_{JK}} + \sum_{J=h}^{n} S_{JJ} \frac{\partial f}{\partial \theta_J}. \tag{3.50}$$

2. When $f = g+h$ where $g=g(\alpha_{JK}, \theta_J)$ and $h=h(\alpha_{JK}, \theta_J)$ then

$$\nabla f = \nabla(g+h) = \nabla g + \nabla h. \tag{3.51}$$

3. When $f=gh$ then

$$\nabla f = \nabla(gh) = (\nabla g)h + g(\nabla h). \tag{3.52}$$

It will now be demonstrated that the mutual moments are related to the direction cosines by the operator ∇ and,

$$[M_{oij..m}] = \nabla[M_{ij..m}] \tag{3.53}$$

or more explicitly,

$$\begin{bmatrix} P_{oij..m} & U_{oij...m} & X_{oij..l} \\ Q_{oij..m} & V_{oij...m} & Y_{oij..l} \\ R_{oij..m} & W_{oij...m} & Z_{oij..l} \end{bmatrix} = \begin{bmatrix} \nabla P_{ij..m} & \nabla U_{ij..m} & \nabla X_{ij..l} \\ \nabla Q_{ij..m} & \nabla V_{ij..m} & \nabla Y_{ij..l} \\ \nabla R_{ij..m} & \nabla W_{ij..m} & \nabla Z_{ij..l} \end{bmatrix} \tag{3.54}$$

In this way each element in the mutual moment matrix $[M_{oij..m}]$ is directly related to a corresponding element in the direction cosine matrix $[M_{ij..m}]$. Firstly, using the property Eqn. (3.52)

$$\nabla[M_{ij..m}] = \nabla([B_h]^{\mathrm{T}}[B_m])$$

$$= (\nabla[B_h]^{\mathrm{T}})[B_m] + [B_h]^{\mathrm{T}}(\nabla[B_m])$$

$$= [B_h]^{\mathrm{T}}(\nabla[B_m]) \tag{3.55}$$

where $\nabla[B_h]^{\mathrm{T}} = 0$ since the reference frame $[B_h]$ considered fixed and is therefore independent of α or θ. By definition (see Eqn. (3.50)),

$$\nabla[B_m] = a_{hi}\frac{\partial[B_m]}{\partial\alpha_{hi}} + S_{ii}\frac{\partial[B_m]}{\partial\theta_i} + \cdots + a_{lm}\frac{\partial[B_m]}{\partial\alpha_{lm}} + S_{mm}\frac{\partial[B_m]}{\partial\theta_m}. \tag{3.56}$$

However the cross product of a pair of skew vectors, $\boldsymbol{u}\times\boldsymbol{v}$ where $\boldsymbol{v}=\boldsymbol{v}(\lambda)$ can be expressed using vector mechanics (see Ref. [5]) as the partial derivative $\partial\boldsymbol{v}/\partial\lambda$ where λ is an angle measure about $\boldsymbol{u}$. Applying this result to each of the partial derivatives in Eqn. (3.56) yields,

$$\frac{\partial[B_m]}{\partial\alpha_{hi}} = [\boldsymbol{a}_{hi}\times\boldsymbol{a}_{mn} \quad \boldsymbol{a}_{hi}\times\boldsymbol{b}_{mn} \quad \boldsymbol{a}_{hi}\times\boldsymbol{s}_m]$$

$$\frac{\partial[B_m]}{\partial\theta_i} = [\boldsymbol{s}_i\times\boldsymbol{a}_{mn} \quad \boldsymbol{s}_i\times\boldsymbol{b}_{mn} \quad \boldsymbol{s}_i\times\boldsymbol{s}_m]$$

$$\vdots$$

$$\frac{\partial[B_m]}{\partial\theta_m} = [\boldsymbol{s}_m\times\boldsymbol{a}_{mn} \quad \boldsymbol{s}_m\times\boldsymbol{b}_{mn} \quad \boldsymbol{s}_m\times\boldsymbol{s}_m]. \tag{3.57}$$

Substituting Eqn. (3.57) in the right side of (3.56) together with the expression for $\boldsymbol{r}_{hm}$ (Eqn. (3.45)) yields,

$$\nabla[B_m] = [B_{om}] \tag{3.58}$$

where $[B_{om}]$ is defined by Eqn. (3.46). Further, substituting Eqn. (3.58) in (3.55) gives

$$\nabla[M_{ij..m}] = [B_h]^{\mathrm{T}}[B_{om}]. \tag{3.59}$$

Finally substituting Eqn. (3.47) in (3.59) yields the desired relationship between the mutual moment and direction cosine matrices

$$[M_{oij..m}] = \nabla[M_{ij..m}]. \qquad [3.53]$$

Using Eqn. (3.53), tables of mutual moment elements can be constructed directly from the tables of direction cosines with elements (Tables 3.1 - 3.3) with either ascending or descending subscripts in terms of the recursive notation simply by the appropriate introduction of the subscript o. For instance,

TABLE 3.4

TRIPLE PRODUCTS--DEXTRAL BASES $[B_h]^{\mathrm{T}}[B_{om}] = [M_{oi..m}]$

$\cdot$	$\boldsymbol{a}_{omn}$	$\boldsymbol{b}_{omn}$	$\boldsymbol{s}_{om}$
$\boldsymbol{a}_{hi}$	$P_{oi..m}$	$U_{oi..m}$	$X_{oi..l}$
$\boldsymbol{b}_{hi}$	$Q_{oi..m}$	$V_{oi..m}$	$Y_{oi..l}$
$\boldsymbol{s}_h$	$R_{oi..m}$	$W_{oi..m}$	$Z_{oi..l}$

table of mutual moment elements $[B_h]^{\mathrm{T}}[B_{om}]=[M_{oij..m}]$ corresponding to Table 3.1a of direction cosine elements $[B_h]^{\mathrm{T}}[B_m]=[M_{ij..m}]$ can be expressed in the form given by Table 3.4, and, the table of mutual moments and their alternative forms corresponding to Table 3.3 Sets 1a and 1b can be expressed in the form given by Table 3.5.

The expansion of the recursive rotation is now illustrated by an example. In Table 3.3 Set 1 the direction cosine and its alternative expression for $\boldsymbol{s}_1 \cdot \boldsymbol{s}_3$ yields

$$Z_2 = Z_{4567}. \tag{3.60}$$

Using Table 3.1b the single subscript term Z_2 is given by

$$Z_2 \equiv (c_{12}c_{23} - s_{12}s_{23}c_2). \tag{3.61}$$

Using the backward recursion formulae (Table 3.1c), Z_{4567} is expanded as follows

$$Z_{4567} \equiv s_{34}(s_4 X_{567} + c_4 Y_{567}) + c_{34} Z_{567} \tag{3.62}$$

$$X_{567} \equiv c_5 X_{67} - s_5 Y_{67}$$

$$Y_{567} \equiv c_{45}(s_5 X_{67} + c_5 Y_{67}) - s_{45} Z_{67}$$

$$Z_{567} \equiv s_{45}(s_5 X_{67} + c_5 Y_{67}) + c_{45} Z_{67}. \tag{3.63}$$

The remaining terms X_{67}, Y_{67}, Z_{67}, are expanded similarly.

In Table 3.5 Set 1 the mutual moment and its alternative expression for $\boldsymbol{s}_1 \cdot \boldsymbol{s}_{03}$ can be derived directly from Eqn. (3.60) using the linear operator ∇ and,

$$Z_{02} = Z_{04567}. \tag{3.64}$$

or equivalently,

$$\nabla Z_2 = \nabla Z_{4567}\ . \tag{3.65}$$

TABLE 3.5

MUTUAL MOMENTS AND THE ALTERNATIVE FORMS

	SET 1a			*SET* 1b		
	$\boldsymbol{a}_{12}$	$\boldsymbol{b}_{12}$	$\boldsymbol{s}_1$	$\boldsymbol{a}_{12}$	$\boldsymbol{b}_{12}$	$\boldsymbol{s}_1$
$\boldsymbol{a}_{012}$	(0 ,	0 ,	0)	($P_{02345671}$,	$U_{02345671}$,	$X_{0234567}$)
$\boldsymbol{a}_{023}$	(P_{02} ,	Q_{02} ,	R_{02})	($P_{0345671}$,	$U_{0345671}$,	X_{034567})
$\boldsymbol{a}_{034}$	(P_{023} ,	Q_{023} ,	R_{023})	(P_{045671} ,	U_{045671} ,	X_{04567})
$\boldsymbol{a}_{045}$	(P_{0234} ,	Q_{0234} ,	R_{0234})	(P_{05671} ,	U_{05671} ,	X_{0567})
$\boldsymbol{a}_{056}$	(P_{02345} ,	Q_{02345} ,	R_{02345})	(P_{0671} ,	U_{0671} ,	X_{067})
$\boldsymbol{a}_{067}$	(P_{023456} ,	Q_{023456} ,	R_{023456})	(P_{071} ,	U_{071} ,	X_{07})
$\boldsymbol{a}_{071}$	($P_{0234567}$,	$Q_{0234567}$,	$R_{0234567}$)	(P_{01} ,	U_{01} ,	0)
$\boldsymbol{b}_{012}$	(0 ,	0 ,	0)	($Q_{02345671}$,	$V_{02345671}$,	$Y_{0234567}$)
$\boldsymbol{b}_{023}$	(U_{02} ,	V_{02} ,	W_{02})	($Q_{0345671}$,	$V_{0345671}$,	Y_{034567})
$\boldsymbol{b}_{034}$	(U_{023} ,	V_{023} ,	W_{023})	(Q_{045671} ,	V_{045671} ,	Y_{04567})
$\boldsymbol{b}_{045}$	(U_{0234} ,	V_{0234} ,	W_{0234})	(Q_{05671} ,	V_{05671} ,	Y_{0567})
$\boldsymbol{b}_{056}$	(U_{02345} ,	V_{02345} ,	W_{02345})	(Q_{0671} ,	V_{0671} ,	Y_{067})
$\boldsymbol{b}_{067}$	(U_{023456} ,	V_{023456} ,	W_{023456})	(Q_{071} ,	V_{071} ,	Y_{07})
$\boldsymbol{b}_{071}$	($U_{0234567}$,	$V_{0234567}$,	$W_{0234567}$)	(Q_{01} ,	V_{01} ,	$-a_{71}c_{71}$)
$\boldsymbol{s}_{01}$	(0 ,	0 ,	0)	($R_{02345671}$,	$W_{02345671}$,	$Z_{0234567}$)
$\boldsymbol{s}_{02}$	(0 ,	$-a_{12}c_{12}$,	$-a_{12}s_{12}$)	($R_{0345671}$,	$W_{0345671}$,	Z_{034567})
$\boldsymbol{s}_{03}$	(X_{02} ,	Y_{02} ,	Z_{02})	(R_{045671} ,	W_{045671} ,	Z_{04567})
$\boldsymbol{s}_{04}$	(X_{023} ,	Y_{023} ,	Z_{023})	(R_{05671} ,	W_{05671} ,	Z_{0567})
$\boldsymbol{s}_{05}$	(X_{0234} ,	Y_{0234} ,	Z_{0234})	(R_{0671} ,	W_{0671} ,	Z_{067})
$\boldsymbol{s}_{06}$	(X_{02345} ,	Y_{02345} ,	Z_{02345})	(R_{071} ,	W_{071} ,	Z_{07})
$\boldsymbol{s}_{07}$	(X_{023456} ,	Y_{023456} ,	Z_{023456})	(R_{01} ,	W_{01} ,	$-a_{71}s_{71}$)

Using Eqn. (3.61) and ∇ in Eqn. (3.49), Z_{02} is expanded by

$$Z_{02} \equiv \nabla(c_{12}c_{23}-s_{12}s_{23}c_2)$$

$$\equiv -a_{12}(s_{12}c_{23}+c_{12}s_{23}c_2)+S_{22}s_{12}s_{23}s_2 -a_{23}(c_{12}s_{23}+s_{12}c_{23}c_2). \tag{3.66}$$

Furthermore, from Eqn. (3.62),

$$Z_{04567} \equiv \nabla\{s_{34}(s_4X_{567}+c_4Y_{567})+c_{34}Z_{567}\}$$

$$\equiv a_{34}\{c_{34}(s_4X_{567}+c_4Y_{567})-s_{34}Z_{567}\} + S_{44}s_{34}(c_4X_{567}-s_4Y_{567})$$

$$+ s_{34}(s_4X_{0567}+c_4Y_{0567})+c_{34}Z_{0567} \tag{3.67}$$

where the remaining terms are expanded using

$$X_{0567} \equiv \nabla X_{567},\ Y_{0567} \equiv \nabla Y_{567},\ Z_{0567} \equiv \nabla Z_{567} \tag{3.68}$$

etc.

As stated earlier the scalar components of the vector loop equation are special cases of the mutual moment equations. They are in fact the mutual moments of a base with respect to itself. For instance, applying the linear operator to

$$[M_{ij..nh}] = [I] \tag{3.69}$$

yields

$$[M_{oij..nh}] = [B_h]^T[B_{oh}] = [0]. \tag{3.70}$$

By analogy with Eqn. (3.47)

$$[B_h]^{\mathrm{T}}[B_{oh}] = \begin{bmatrix} \boldsymbol{a}_{hi}\cdot\boldsymbol{R}\times\boldsymbol{a}_{hi} & \boldsymbol{a}_{hi}\cdot\boldsymbol{R}\times\boldsymbol{b}_{hi} & \boldsymbol{a}_{hi}\cdot\boldsymbol{R}\times\boldsymbol{s}_h \\ \boldsymbol{b}_{hi}\cdot\boldsymbol{R}\times\boldsymbol{a}_{hi} & \boldsymbol{b}_{hi}\cdot\boldsymbol{R}\times\boldsymbol{b}_{hi} & \boldsymbol{b}_{hi}\cdot\boldsymbol{R}\times\boldsymbol{s}_h \\ \boldsymbol{s}_h\cdot\boldsymbol{R}\times\boldsymbol{a}_{hi} & \boldsymbol{s}_h\cdot\boldsymbol{R}\times\boldsymbol{b}_{hi} & \boldsymbol{s}_h\cdot\boldsymbol{R}\times\boldsymbol{s}_h \end{bmatrix} \tag{3.71}$$

where $\boldsymbol{R}$ is defined by Eqn. (3.44). Simplifying Eqn. (3.71) and substituting it in Eqn. (3.70) yields

$$\begin{bmatrix} 0 & -\boldsymbol{s}_h\cdot\boldsymbol{R} & \boldsymbol{b}_{hi}\cdot\boldsymbol{R} \\ \boldsymbol{s}_h\cdot\boldsymbol{R} & 0 & -\boldsymbol{a}_{hi}\cdot\boldsymbol{R} \\ -\boldsymbol{b}_{hi}\cdot\boldsymbol{R} & \boldsymbol{a}_{hi}\cdot\boldsymbol{R} & 0 \end{bmatrix} = [0]. \tag{3.72}$$

In this way the three scalar components of the vector loop equation, $\boldsymbol{R} = \boldsymbol{0}$ in the directions of the base $[B_h]$ are expressed as the elements of the skew-symmetric matrix Eqn. (3.72).

Summarizing, the recursive notation has a direct vector interpretation which is extremely useful in the analysis of spatial mechanisms and corresponding robot manipulators. Further, the recursive notation facilitates the selection of appropriate equations for analysis since the subscripts labeling direction cosine and mutual moment elements are indicators that they are functions of various joint angles $\theta_h, \theta_i \dots \theta_n$.

4. ON THE SOLUTION OF TRIGONOMETRICAL EQUATIONS

It is important to examine alternative solutions of trigonometrical equations which are frequently used to compute joint displacements of manipulators. In practice it is desirable to utilize methods of solution which are not sensitive to errors contained in the coefficients of such equations. Further, certain indeterminacies may occur in one type of solution making an alternative formulation preferable.

A trigonometric equation which is perhaps most commonly used can be expressed in the form,

$$a \cos\theta + b \sin\theta - d = 0 \tag{4.1}$$

where the coefficients a, b and d are known quantities. Two methods of solution are presented for Eqn. (4.1) and the first is obtained by dividing Eqn. (4.1) by $(a^2+b^2)^{1/2}$ and using the identity

$$\cos(\alpha-\beta) = \cos\alpha \cos\beta + \sin\alpha \sin\beta \tag{4.2}$$

to yield,

$$\cos(\theta-\gamma) = \frac{d}{(a^2+b^2)^{1/2}} \tag{4.3}$$

where

$$\cos\gamma \equiv \frac{a}{(a^2+b^2)^{1/2}}, \quad \sin\gamma \equiv \frac{b}{(a^2+b^2)^{1/2}}. \tag{4.4}$$

Using the trigonometric identity,

$$\cos^2\alpha + \sin^2\alpha = 1 \tag{4.5}$$

with Eqn. (4.3) yields,

$$\sin(\theta-\gamma) = \pm \frac{(a^2+b^2-d^2)^{1/2}}{(a^2+b^2)^{1/2}}. \tag{4.6}$$

Equations (4.3) and (4.6) define a pair of angles which will be denoted by $\psi^{(\pm)}$ where the upper and lower signs correspond to the upper and lower signs in Eqn. (4.6). Since angle γ is single valued and angle ψ is double valued then it follows that

$$\theta^{(\pm)} \equiv \psi^{(\pm)} + \gamma \tag{4.7}$$

and

$$\cos\theta^{(\pm)} = \cos\psi^{(\pm)}\cos\gamma - \sin\psi^{(\pm)}\sin\gamma. \tag{4.8}$$

For Eqn. (4.8), substituting in (4.4) and (4.3), (4.6) using (4.7) yields,

$$\cos\theta^{(\pm)} = \frac{ad \mp b\,(a^2+b^2-d^2)^{1/2}}{(a^2+b^2)} \tag{4.9}$$

and then substituting Eqn. (4.9) in (4.1) yields,

$$\sin\theta^{(\pm)} = \frac{bd \pm a\,(a^2+b^2-d^2)^{1/2}}{(a^2+b^2)}. \tag{4.10}$$

Two values $\theta^{(\pm)}$ can be computed from Eqns. (4.9) and (4.10) using the corresponding upper and lower signs. The relationship between the angles γ, $\psi^{(\pm)}$ and $\theta^{(\pm)}$ is illustrated in Figure 4.1. The line defined by γ is symmetrical with respect to the lines defined by $\theta^{(+)}$ and $\theta^{(-)}$.

The necessary and sufficient condition for Eqns. (4.9) and (4.10) to become indeterminate is $a=b=0$. Simultaneously, Eqn. (4.1) must vanish identically since the condition results in $d=0$. *This indeterminacy flags a geometric singularity for which the manipulator is said to be in an uncertainty configuration.*

In practice the value of θ is computed by a function that uses two arguments to determine the proper quadrant of θ,

$$\theta \equiv \text{ATAN2}\ (N,D). \tag{4.11}$$

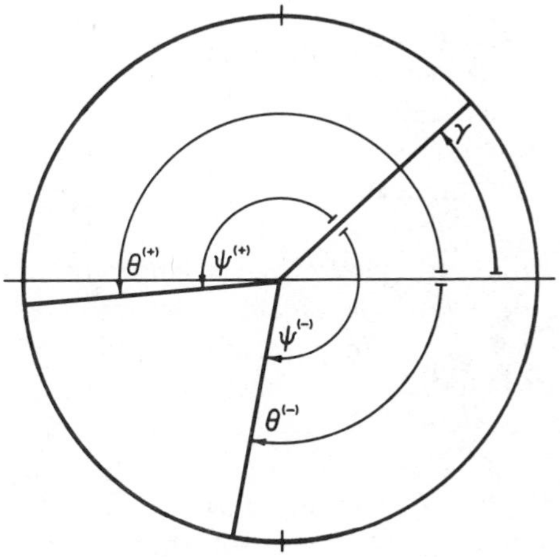

Figure 4.1 The two solutions to a $\cos\theta + b\sin\theta - d = 0$. *The lines defined by* $\theta^{(+)}$ *and* $\theta^{(-)}$ *are symmetrical with the line defined by* γ.

Arguments N and D are respectively $\sin\theta$ and $\cos\theta$ or equivalently the numerators in Eqns. (4.10) and (4.9) since the denominators are positive and equal. When $||N/D|| \leqslant 1$, the ATAN2 function computes the value of θ as,

$$\theta = \arctan(N/D), \quad ||N/D|| \leqslant 1. \tag{4.12}$$

When $||N/D|| > 1$, it is preferable to use the identify,

$$\arctan(N/D) + \operatorname{arccot}(N/D) = \frac{\pi}{2}, \tag{4.13}$$

so that the ATAN2 function can compute the value of θ as,

$$\theta = \frac{\pi}{2} - \arctan(D/N), \quad ||N/D|| > 1. \tag{4.14}$$

This algorithm for the ATAN2 function limits the domain of $\arctan(z)$ to $-1 \leqslant z \leqslant 1$ and avoids a possible division by zero.

The second method of solution of Eqn. (4.1), which at the outset appears preferable to the above method, is obtained by transforming Eqn. (4.1) into an algebraic equation using the trigonometric identities,

$$\cos\theta = \frac{(1-t^2)}{(1+t^2)}, \quad \sin\theta = \frac{2t}{(1+t^2)} \tag{4.15}$$

where $t \equiv \tan(\theta/2)$, is the tan-half angle of θ.

Substitution of Eqn. (4.15) in (4.1) yields the quadratic equation,

$$(a+d)t^2 - (2b)t + (d-a) = 0 \tag{4.16}$$

the solution of which is

$$t = \frac{b \pm (a^2+b^2-d^2)^{1/2}}{(a+d)}. \tag{4.17}$$

The two values for θ are computed directly from Eqn. (4.17) since the arctan-half angle function is single valued in the range $-\pi < \theta \leqslant \pi$. Similar to the ATAN2 function the numerator and denominator of Eqn. (4.17) are used as two arguments in order to avoid the possibility of division by zero. The cosine and sine of θ can be immediately obtained by back substituting t in Eqn. (4.15).

There are, however, four indeterminacies inherent in the tan-half angle solution. The first, $a=b=d=0$ has already been discussed and is a geometric singularity that flags a manipulator uncertainty condition. Table 4.1 summarizes the remaining three tan-half angle indeterminacies along with the values of $\cos\theta$ and $\sin\theta$ which are determined by Eqns. (4.9) and (4.10). Without loss of generality, the assumption $b \geqslant 0$ is made for clarity. The three tabulated cases do not represent geometric singularities since the values of θ can be calculated using the cosine-sine solution and will therefore be referred to as *algebraic indeterminacies* for the tan-half angle solution.

A further undesirable feature occurs when the conditions for an algebraic indeterminacy are approximately satisfied and numerical errors in the

TABLE 4.1

ALGEBRAIC INDETERMINACIES FOR THE TAN-HALF ANGLE SOLUTION EQN. (4.17)

Case	Conditions	Cosine-Sine Solution	Tan-half Angle Solution
i)	$a = -d \neq 0$	$c = -1, \dfrac{-a^2+b^2}{a^2+b^2}$	$t = \dfrac{2b}{0}, \dfrac{0}{0}$
	$b \neq 0$	$s = 0, \dfrac{-2ab}{a^2+b^2}$	
		$\theta = \pi, \text{ATAN2}(\dfrac{-2ab}{a^2+b^2})$	$\theta = \pi, \dfrac{0}{0}$
ii)	$a = -d \neq 0$	$c = -1, -1$	$t = \dfrac{0}{0}, \dfrac{0}{0}$
	$b = 0$	$s = 0, 0$	
		$\theta = \pi, \pi$	$\theta = \dfrac{0}{0}, \dfrac{0}{0}$
iii)	$a = d = 0$	$c = -1, 1$	$t = \dfrac{2b}{0}, \dfrac{0}{0}$
	$b \neq 0$	$s = 0, 0$	
		$\theta = \pi, 0$	$\theta = \pi, \dfrac{0}{0}$

Note: The assumption $b \geqslant 0$ is made for clarity.

coefficients can yield highly unstable results. The solution can be somewhat improved by rationalizing the numerator of Eqn. (4.17) such that,

$$t^{(+)} = \frac{b+(a^2+b^2-d^2)^{1/2}}{a+d} = \frac{d-a}{b-(a^2+b^2-d^2)^{1/2}} \tag{4.18}$$

and

$$t^{(-)} = \frac{b-(a^2+b^2-d^2)^{1/2}}{a+d} = \frac{d-a}{b+(a^2+b^2-d^2)^{1/2}}. \tag{4.19}$$

When $b > 0$ then the tan-half angle solution is best obtained by the first expression in Eqn. (4.18) combined with the second expression in Eqn. (4.19),

$$t^{(+)} = \frac{b+(a^2+b^2-d^2)^{1/2}}{a+d}, \quad t^{(-)} = \frac{d-a}{b+(a^2+b^2-d^2)^{1/2}}, \quad b > 0. \tag{4.20}$$

Eqn. (4.20) yields the correct results in cases i) and iii) in Table 3.4.1 but for case ii) it still yields an indeterminate result. Similarly when $b < 0$ then the tan-half angle solution is best obtained by selecting the second expression in Eqn. (4.18) and the first expression in Eqn. (4.19),

$$t^{(+)} = \frac{d-a}{b-(a^2+b^2-d^2)^{1/2}}, \quad t^{(-)} = \frac{b-(a^2+b^2-d^2)^{1/2}}{a+d}, \quad b < 0. \tag{4.21}$$

The correct result for case ii) is ascertained by first dividing Eqn. (4.16) by t^2 to yield $t^2 = (d-a)/0$ or $\theta = \pi,\pi$. *In general, Eqn. (4.1) is best solved by the cosine-sine solution since algebraic indeterminacies are not present and consequently, alternative formulations are unnecessary.*

The solution of an equation which is of integer order n in $\cos\theta$ and $\sin\theta$ is somewhat more difficult. A polynomial of order $2n$ in t can readily be obtained using the tan-half angle substitutions in Eqn. (4.15). However, it is reasonable to postulate that the solution of such a polynominal with real coefficients will include algebraic indeterminacies since it can always be expressed in terms of quadratic factors with real coefficients.

The cosine-sine solution of Eqn. (4.1) is not readily extendible to higher order equations. It is however, possible to reformulate the solution and then extend it to solve the higher order equations. This is accomplished by writing Eqns. (4.1) and (4.5) in the form,

$$ax + by - d = 0 \tag{4.22}$$

and

$$x^2 + y^2 = 1 \tag{4.23}$$

where $x \equiv \cos\theta$ and $y \equiv \sin\theta$. The simultaneous solution of Eqns. (4.22) and (4.23) yields the points of intersection of a line with the unit circle. There are two intersections which can be either real (distinct or coincident) or imaginary.

An equation of order n in x and y represents a curve which will in general intersect the unit circle in $2n$ points and using Eqn. (4.23) it can always be expressed in the form,

$$(a_n x^n + a_{n-1}x^{n-1} + \cdots + a_o) = y(b_{n-1}x^{n-1} + b_{n-2}x^{n-2} + \cdots + b_o). \tag{4.24}$$

squaring Eqn. (4.24), making the substitution $y^2=1-x^2$ and then regrouping terms yields a polynomial of order $2n$ which can be expressed in the form,

$$f_{2n}\, x^{2n} + f_{2n-1}\, x^{2n-1} + \cdots + f_o = 0. \tag{4.25}$$

Each of the roots of Eqn. (4.25) can be back substituted into Eqn. (4.24) to obtain a unique corresponding value of y. Using this formulation to solve the linear relation Eqn. (4.22) yields the following pair of expressions.

$$x^{(\pm)} = \frac{ad \pm \sigma_b b\,(a^2+b^2-d^2)^{1/2}}{(a^2+b^2)} \tag{4.26}$$

$$y^{(\pm)} = \frac{bd \pm \sigma_b a\,(a^2+b^2-d^2)^{1/2}}{(a^2+b^2)} \tag{4.27}$$

where $\sigma_b = (1, b \geqslant 0; -1, b < 0)$ is the sign of b. These expressions differ from Eqn. (4.9) and (4.10) and their geometrical interpretation is somewhat less direct due to the inclusion of σ_b.

Frequently, a joint variable is determined by the simultaneous solution of a pair of linear equations,

$$a_i x + b_i y - d_i = 0, \; i = 1,2. \tag{4.28}$$

When the coefficients are known quantities then the solution can be expressed in the form,

$$x = |db|/|ab|, \; y = |ad|/|ab| \tag{4.29}$$

where $|ab| \equiv a_1 b_2 - a_2 b_1$, etc., and represents the intersection of a pair of lines. Since this point must lie on the unit circle then substitution of Eqn. (4.29) in (4.23) yields an expression that must be satisfied by coefficients,

$$|ad|^2 + |db|^2 - |ab|^2 = 0. \tag{4.30}$$

In practice the coefficients will inevitably contain numerical and/or measurement errors and Eqn. (4.30) will not be satisfied exactly. When the discrepancy is significant then x and y can be computed by normalizing Eqn. (4.29),

$$x = \frac{\sigma_{ab}|db|}{(|db|^2+|ad|^2)^{1/2}}, \; y = \frac{\sigma_{ab}|ad|}{(|db|^2+|ad|^2)^{1/2}} \tag{4.31}$$

where σ_{ab} is the sign of $|ab|$. (Numerical accuracy can be improved by dividing the numerators and the denominators using the larger absolute magnitude of either $|db|$ or $|ad|$.)

When each of the coefficients of Eqn. (4.28) are linear functions of $u \equiv \cos\lambda$ and $v \equiv \sin\lambda$ where λ is some other joint variable, application of Eqn. (4.30) results in a quartic relation in u and v. Analogous to Eqns. (4.24) and (4.25) this quartic equation leads to an eighth order polynomial in u the roots of which can be used to determine the corresponding values for v. Back substituting each pair of values (u,v) in the coefficients a, b, d of Eqn. (4.29) yields a corresponding pair of values (x,y). Usually the eighth degree polynomial in u reduces to a quartic or quadratic equation when the manipulator has a special geometry (e.g., three co-intersecting axes or three parallel axes).

5. DISPLACEMENT ANALYSIS

It has been demonstrated in Section 2 that when the end-effector location of a six degree-of-freedom manipulator is specified, a corresponding single loop spatial mechanism of mobility one can be formed. The hypothetical angle θ_7 becomes the input parameter of the mechanism and is analogous to the end-effector location being the input parameters for the manipulator. The object of the displacement analysis is to use the input parameter(s) to determine the unknown mechanism joint angles (manipulator actuator angles). The process is sometimes referred to as the reverse or inverse displacement analysis when applied to manipulators. This distinguishes it from the complementary problem referred to as the forward displacement analysis which uses known values of the manipulator actuator angles to locate the end-effector. In comparison to the reverse displacement analysis this problem is simple and direct.

The manipulator to be analyzed is illustrated in Figure 5.1. It is primarily anthropomorphic in design with the second and third joint axes parallel, and with the last three axes co-intersecting at a point to form an equivalent ball and socket joint. The dimensions are listed in Table 5.1. The analysis will be performed with $S_{22} \neq 0$, and $a_{23} \neq S_{44}$, and will be compared with the analysis of the PUMA 600 series robot for which $S_{22} \neq 0$ and $a_{23} = S_{44}$. This comparison together with the analysis of a further special case for which $S_{22} = 0$, provides much useful information in that it demonstrates that the choice of the dimensions (S_{22}, a_{23}, S_{44}) can effect singular configurations. Figure 5.1 illustrates the seven unit vectors $\boldsymbol{s}_i$ and the seven unit vectors $\boldsymbol{a}_{ij}$ that are used to specify the geometry of the spatial mechanism. The general vector loop equation,

$$\boldsymbol{R} \equiv \sum_{i=1}^{7} S_{ii}\boldsymbol{s}_i + \sum_{ij=12}^{71} a_{ij}\boldsymbol{a}_{ij} = \boldsymbol{0} \tag{5.1}$$

reduces to,

$$\boldsymbol{R} = S_{11}\boldsymbol{s}_1 + S_{22}\boldsymbol{s}_2 + a_{23}\boldsymbol{a}_{23} + S_{44}\boldsymbol{s}_4 + S_{66}\boldsymbol{s}_6 + S_{77}\boldsymbol{s}_7 + a_{71}\boldsymbol{a}_{71} = \boldsymbol{0} \tag{5.2}$$

for the mechanism link lengths and offsets listed in Table 5.1. (Note that for the purpose of generality $\boldsymbol{s}_2$ has been selected such that $S_{22} < 0$, see Figure 5.1. If $\boldsymbol{s}_2$ had been chosen in the opposite direction then $S_{22} > 0$, $\alpha_{12} = -\pi/2$ and $\alpha_{23} = \pi$.)

It is first necessary to obtain θ_1 as a function of θ_7. This is known as the input-output function, and it can be obtained by taking the projection of the vector loop equation in the direction of joint axis $\boldsymbol{s}_2$,

$$\boldsymbol{R} \cdot \boldsymbol{s}_2 = 0. \tag{5.3}$$

Using the mechanism twist angles in Table 5.1 this equation reduces to,

$$S_{22}\boldsymbol{s}_2 \cdot \boldsymbol{s}_2 + S_{66}\boldsymbol{s}_6 \cdot \boldsymbol{s}_2 + S_{77}\boldsymbol{s}_7 \cdot \boldsymbol{s}_2 + a_{71}\boldsymbol{a}_{71} \cdot \boldsymbol{s}_2 = 0. \tag{5.4}$$

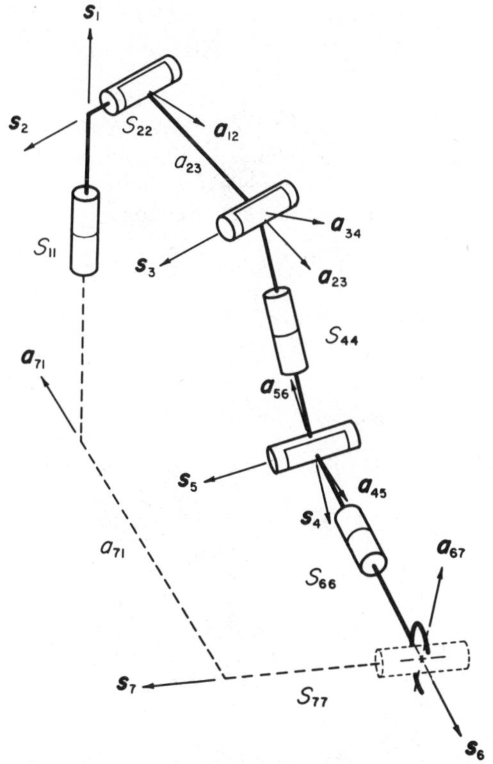

Figure 5.1 An anthropomorphic manipulator. The second and third joint axes are parallel and the last three joint axes co-intersect at a point to form an equivalent ball and socket joint (wrist). This simple geometry leads to quadratic expressions for angles θ_1, θ_3, θ_5 and linear expressions for angles θ_2, θ_4, θ_6. (Note that the direction of $\mathbf{s}_2$ has been chosen such that $S_{22} < 0$).

TABLE 5.1

MECHANISM DIMENSIONS

$S_{11} = *$	$a_{12} = 0$	$\alpha_{12} = \pi/2$
$S_{22} = \dagger$	$a_{23} = \dagger$	$\alpha_{23} = 0$
$S_{33} = 0$	$a_{34} = 0$	$\alpha_{34} = \pi/2$
$S_{44} = \dagger$	$a_{45} = 0$	$\alpha_{45} = -\pi/2$
$S_{55} = 0$	$a_{56} = 0$	$\alpha_{56} = \pi/2$
$S_{66} = \dagger$	$a_{67} = 0$	$\alpha_{67} = -\pi/2$
$S_{77} = *$	$a_{71} = *$	$\alpha_{71} = *$

† Non-zero dimension.

* Determined by completion of the spatial loop.

Equation (5.4) can now be expressed in terms of θ_7 and θ_1 by using the direction cosines listed in Table 3.3, Set 14 (reference frame at $[\boldsymbol{a}_{12}\ \boldsymbol{d}_{12}\ \boldsymbol{s}_2]$),

$$S_{22} + S_{66}\, Z_{17} + S_{77}\, \bar{Z}_1 + a_{71}\, \bar{R}_1 = 0. \tag{5.5}$$

The definitions of these terms together with their simplifications using the twist angle values are tabulated in Table 5.2. Substituting these values into Eqn. (5.5) gives the required input-output function,

$$A_7\, c_1 + B_7\, s_1 + S_{22} = 0 \tag{5.6}$$

where,

$$A_7 = S_{66}\, \bar{Y}_7 - S_{77}\, s_{71} \tag{5.7}$$

$$B_7 = S_{66}\, \bar{X}_7 + a_{71}. \tag{5.8}$$

The explicit solution for θ_1 is obtained by using Eqns. (4.9) and (4.10),

$$c_1 = \frac{-A_7 S_{22} \mp B_7 (A_7^2 + B_7^2 - S_{22}^2)^{1/2}}{(A_7^2 + B_7^2)} \tag{5.9}$$

$$s_1 = \frac{-B_7 S_{22} \pm A_7 (A_7^2 + B_7^2 - S_{22}^2)^{1/2}}{(A_7^2 + B_7^2)}. \tag{5.10}$$

TABLE 5.2

DEFINITIONS OF THE DIRECTION COSINE NOTATION AND SIMPLIFICATIONS USING THE TWIST ANGLE VALUES α_{ij} FROM TABLE 5.1

$X_{3217} = (c_{2+3} X_{17} - s_{2+3} Y_{17})$ Note: $c_{2+3} = \cos(\theta_2+\theta_3)$, $s_{2+3} = \sin(\theta_2+\theta_3)$

$Y_{3217} = c_{34}(s_{2+3} X_{17} + c_{2+3} Y_{17}) - s_{34} Z_{17} = -Z_{17}$

$Z_{3217} = s_{34}(s_{2+3} X_{17} + c_{2+3} Y_{17}) + c_{34} Z_{17} = (s_{2+3} X_{17} + c_{2+3} Y_{17})$

$X_{17} = (c_1 \bar{X}_7 - s_1 \bar{Y}_7)$

$Y_{17} = c_{12}(s_1 \bar{X}_7 + c_1 \bar{Y}_7) - s_{12} \bar{Z}_7 = -\bar{Z}_7$

$Z_{17} = s_{12}(s_1 \bar{X}_7 + c_1 \bar{Y}_7) + c_{12} \bar{Z}_7 = (s_1 \bar{X}_7 + c_1 \bar{Y}_7)$

TABLE 5.2 (cont'd)

$$\bar{X}_7 = s_{67}s_7 = -s_7$$

$$\bar{Y}_7 = -(s_{71}c_{67} + c_{71}s_{67}c_7) = c_{71}c_7$$

$$\bar{Z}_7 = c_{71}c_{67} - s_{71}s_{67}c_7 = s_{71}c_7$$

$$X_{54321} = \quad (c_5X_{4321} - s_5Y_{4321})$$

$$Y_{54321} = c_{56}(s_5X_{4321} + c_5Y_{4321}) - s_{56}Z_{4321} = -Z_{4321}$$

$$X_{4321} = \quad (c_4X_{321} - s_4Y_{321})$$

$$Y_{4321} = c_{45}(s_4X_{321} + c_4Y_{321}) - s_{45}Z_{321} = Z_{321}$$

$$Z_{4321} = s_{45}(s_4X_{321} + c_4Y_{321}) + c_{45}Z_{321} = -(s_4X_{321} + c_4Y_{321})$$

$$X_{321} = \quad (c_{2+3}\bar{X}_1 - s_{2+3}\bar{Y}_1)$$

$$Y_{321} = c_{34}(s_{2+3}\bar{X}_1 + c_{2+3}\bar{Y}_1) - s_{34}\bar{Z}_1 = -\bar{Z}_1$$

$$Z_{321} = s_{34}(s_{2+3}\bar{X}_1 + c_{2+3}\bar{Y}_1) + c_{34}\bar{Z}_1 = (s_{2+3}\bar{X}_1 + c_{2+3}\bar{Y}_1)$$

$$\bar{X}_1 = s_{12}s_1 = s_1$$

$$\bar{Y}_1 = -(s_{71}c_{12} + c_{71}s_{12}c_1) = -c_{71}c_1$$

$$\bar{Z}_1 = c_{71}c_{12} - s_{71}s_{12}c_1 = -s_{71}c_1$$

$$R_{32} = s_{34}(s_3\bar{P}_2 + c_3\bar{Q}_2) + c_{34}\bar{R}_2 = s_3c_2 + c_3s_2 = s_{2+3}$$

$$\bar{P}_2 = c_2$$

$$\bar{Q}_2 = c_{23}s_2 = s_2$$

$$\bar{R}_2 = s_{23}s_2 = 0$$

TABLE 5.2 (cont'd)

$$W_{32} = s_{34}(s_3\bar{U}_2 + c_3\bar{V}_2) + c_{34}\bar{W}_2 = c_2c_3 - s_2s_3 = c_{2+3}$$

$$\bar{U}_2 = -s_2$$

$$\bar{V}_2 = c_{23}c_2 = c_2$$

$$\bar{W}_2 = s_{23}c_2 = 0$$

$$R_{54} = s_{56}(s_5\bar{P}_4 + c_5\bar{Q}_4) - c_{56}\bar{R}_4 = -s_5c_4$$

$$\bar{P}_4 = \quad c_4$$

$$\bar{Q}_4 = c_{45}s_4 = 0$$

$$\bar{R}_4 = s_{45}s_4 = -s_4$$

$$W_{54} = s_{56}(s_5\bar{U}_4 + c_5\bar{V}_4) - c_{56}\bar{W}_4 = -s_5s_4$$

$$\bar{U}_4 = -s_4$$

$$\bar{V}_4 = c_{45}c_4 = 0$$

$$\bar{W}_4 = s_{45}c_4 = -c_4$$

$$\bar{P}_1 = \quad c_1$$

$$\bar{Q}_1 = c_{12}s_1 = 0$$

$$\bar{R}_1 = s_{12}s_1 = s_1$$

$$\bar{Z}_5 = c_{56}c_{45} - s_{56}s_{45}c_5 = c_5$$

$$\bar{R}_6 = s_{67}s_6 = -s_6$$

$$\bar{W}_6 = s_{67}c_6 = -c_6$$

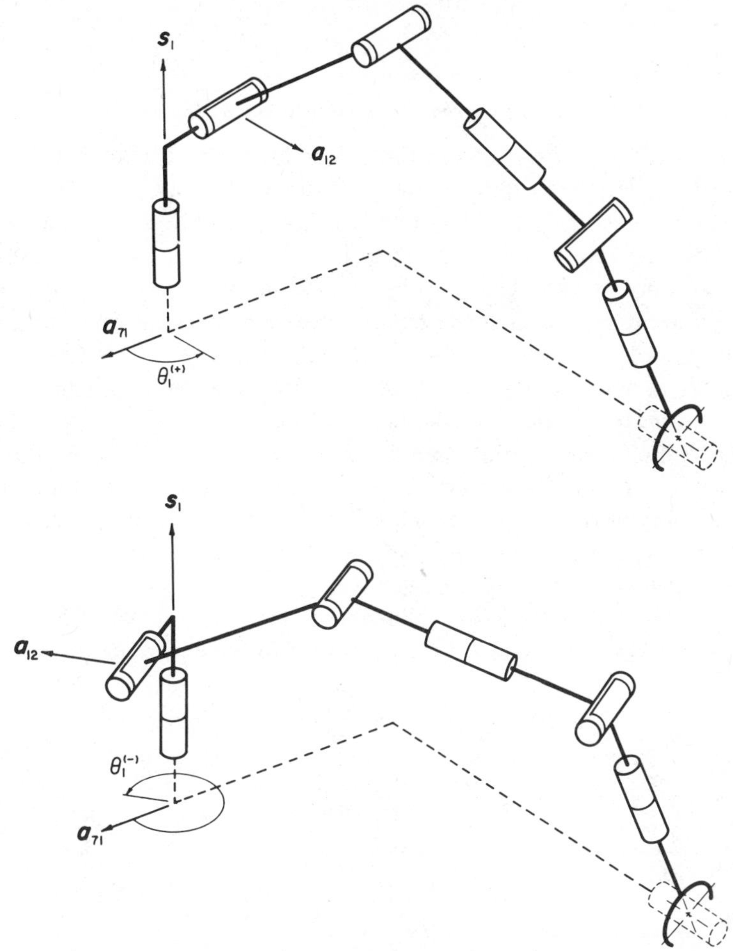

Figure 5.2 The two values of θ_1 yield distinct configurations corresponding to selection of a) the upper signs and b) the lower signs of Eqns. (5.9) and (5.10).

The two mechanism configurations corresponding to Eqns. (5.9) and (5.10) are illustrated in Figure 5.2. The necessary and sufficient condition for θ_1, to be indeterminate is $A_7=B_7=0$ (see Section 4). However, from Eqn. (5.6), $A_7=B_7=0$ implies the contradictory condition $S_{22}=0$ and therefore θ_1 is never indeterminate for the PUMA robot. It is interesting to examine the geometry when $S_{22}=0$ since this models a number of industrial robots such as the Cincinnati Milacron T3-R3 robot. The physical interpretation of the indeterminancy is revealed by expressing Eqns. (5.7) and (5.8) in vector form using the direction cosines listed in Table 3.3, Set 8 (reference frame $[\boldsymbol{a}_{71}\ \boldsymbol{d}_{71}\ \boldsymbol{s}_1]$),

$$A_7 = \boldsymbol{u} \cdot \boldsymbol{d}_{71} \tag{5.11}$$

$$B_7 = \boldsymbol{u} \cdot \boldsymbol{a}_{71} \tag{5.12}$$

where,

$$\boldsymbol{u} \equiv S_{66}\boldsymbol{s}_6 + S_{77}\boldsymbol{s}_7 + a_{71}\boldsymbol{a}_{71}. \tag{5.13}$$

The position vector $\boldsymbol{u}$ points from the center of the equivalent ball joint (the point where the last three axes co-intersect) to the frame $[\boldsymbol{a}_{71}\ \boldsymbol{d}_{71}\ \boldsymbol{s}_1]$.

Two cases arise when both A_7 and B_7 are zero. The first occurs when $\boldsymbol{u}$ vanishes and the second occurs when $\boldsymbol{u}$ is collinear with $\boldsymbol{s}_1$ (see Figure 5.3). Both cases imply that the ball joint center lies on the $\boldsymbol{s}_1$ axis such that $\boldsymbol{s}_1$, $\boldsymbol{s}_4$, $\boldsymbol{s}_5$, and $\boldsymbol{s}_6$ co-intersect at a single point to create a spherical four bar mechanism which has mobility one see Eqn. (2.1) with $(m, n, j, f_i) = (3, 4, 4, 1)$.

Angle θ_1 is now geometrically indeterminate since link a_{23} and offset S_{44} together can lie in any plane of the pencil of planes through the axis of $\boldsymbol{s}_1$ for the given end-effector location (see Figure 5.3). The direction of the vector $\boldsymbol{a}_{71}$ is fixed by the specification of the end-effector location; therefore, angle θ_1 can assume any value within the range 0 to 2π since it is measured from $\boldsymbol{a}_{71}$ to $\boldsymbol{a}_{12}$ about $\boldsymbol{s}_1$. *Because of this indeterminancy, the manipulator is said to be in an uncertainty configuration.*

The subsequent analysis is valid for a manipulator with $S_{22} = 0$ or $S_{22} \neq 0$ since this joint offset does not appear in the equations.

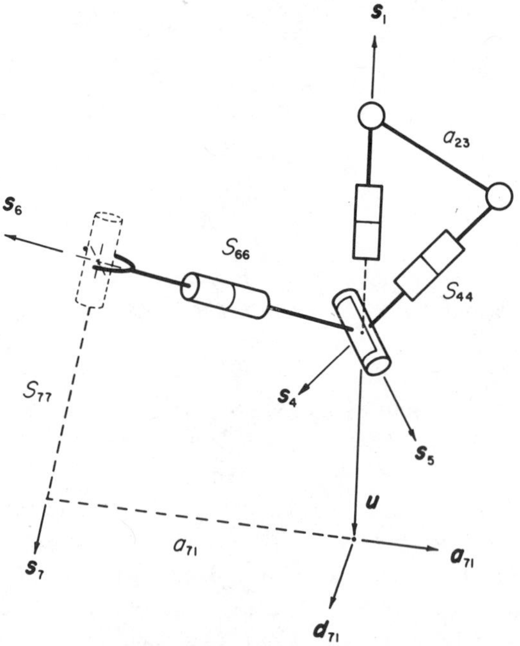

Figure 5.3 Uncertainty configuration for which the expressions for θ_1 become indeterminate. Axes $\boldsymbol{s}_1$, $\boldsymbol{s}_4$, $\boldsymbol{s}_5$, $\boldsymbol{s}_6$ co-intersect at a point to form a spherical four bar mechanism.

In order to position the manipulator, it is necessary to determine the actuator angle ϕ_1 which is measured from the fixed $\boldsymbol{i}$ axis to $\boldsymbol{a}_{12}$ about $\boldsymbol{s}_1$. This can be determined directly using Eqns. (2.16), (2.17), (5.9) and (5.10) in the trigometrical identities,

$$\cos\phi_1 = \cos\theta_1 \cos(\theta_1 - \phi_1) + \sin\theta_1 \sin(\theta_1 - \phi_1) \tag{5.14}$$

$$\sin\phi_1 = \sin\theta_1 \cos(\theta_1 - \phi_1) - \cos\theta_1 \sin(\theta_1 - \phi_1). \tag{5.15}$$

The second and third joint angles are determined by a pair of equations representing the projection of the vector loop equation in the directions of $\boldsymbol{a}_{12}$ and $\boldsymbol{d}_{12}$,

$$\boldsymbol{a}_{12} \cdot \boldsymbol{R} = 0 \tag{5.16}$$

$$\boldsymbol{d}_{12} \cdot \boldsymbol{R} = 0. \tag{5.17}$$

These relations can be expressed in terms of θ_7, θ_1, θ_2 and θ_3 using Table 3.3, Set 14 (reference frame at $[\boldsymbol{a}_{12}\ \boldsymbol{d}_{12}\ \boldsymbol{s}_2]$),

$$a_{23}(\bar{P}_2) + S_{44}(R_{32}) = A_{17} \tag{5.18}$$

$$a_{23}(\bar{U}_2) + S_{44}(W_{32}) = B_{17} \tag{5.19}$$

where,

$$A_{17} \equiv -(S_{66}X_{17}+S_{77}\bar{X}_1+a_{71}\bar{P}_1) = \boldsymbol{v}\cdot\boldsymbol{a}_{12} \tag{5.20}$$

$$B_{17} \equiv -(S_{66}Y_{17}+S_{77}\bar{Y}_1+a_{71}\bar{Q}_1) = \boldsymbol{v}\cdot\boldsymbol{d}_{12} \tag{5.21}$$

$$\boldsymbol{v} \equiv -(S_{66}\boldsymbol{s}_6+S_{77}\boldsymbol{s}_7+a_{71}\boldsymbol{a}_{71}+S_{11}\boldsymbol{s}_1+S_{22}\boldsymbol{s}_2). \tag{5.22}$$

The definitions of the direction cosine terms together with their simplifications using the twist angle values are listed in Table 5.2. Substitution of these terms into the left sides of Eqns. (5.18) and (5.19) yields,

$$(a_{23} + S_{44}\, s_3)c_2 + (S_{44}\, c_3)s_2 = A_{17} \tag{5.23}$$

$$(S_{44}\, c_3)c_2 - (a_{23} + S_{44}\, s_3)s_2 = B_{17}. \tag{5.24}$$

Using Eqn. (4.30) to eliminate angle θ_2 between the pair of equations should yield a quartic relation in c_3 and s_3. However, due to the simple geometry of the manipulator this relation is reduced to an equation that is linear in s_3 only,

$$a_{23}^2 + S_{44}^2 + 2a_{23}\, S_{44}\, s_3 = v^2 \tag{5.25}$$

where,

$$v^2 = A_{17}^2 + B_{17}^2. \tag{5.26}$$

It is noted that Eqn. (5.25) can also be obtained by adding the squares of Eqns. (5.23) and (5.24). Solving Eqn. (5.25) for angle θ_3 yields,

$$s_3 = \frac{v^2 - a_{23}^2 - S_{44}^2}{2a_{23}S_{44}} \tag{5.27}$$

$$c_3 = \pm\,(1 - s_3^2)^{1/2}. \tag{5.28}$$

The two mechanism configurations corresponding to Eqns. (5.27) and (5.28) are shown in Figure 5.4. The vector $\boldsymbol{v}$ lies in the plane of a_{23} and S_{44} and points from the reference frame $[\boldsymbol{a}_{12}\ \boldsymbol{d}_{12}\ \boldsymbol{s}_2]$ to the position of the ball joint center which is defined by the specification of the manipulator end-effector. By making the definition,

$$\theta_{3'} \equiv \theta_3 - \pi/2, \tag{5.29}$$

it is interesting to note that Eqn. (5.25) can be expressed as the cosine law for a planar triangle (see Figure 5.4),

$$a_{23}^2 + S_{44}^2 - 2a_{23}S_{44}c_{3'} = v^2. \tag{5.30}$$

Since the denominator in Eqn. (5.27) is a non-zero constant, angle θ_3 is never indeterminate. Angle θ_3 will yield real values when

$$-1 \leqslant s_3 \leqslant 1. \tag{5.31}$$

By substituting Eqn. (5.27) in (5.31) this condition can be expressed in the form,

$$(a_{23} - S_{44})^2 \leqslant v^2 \leqslant (a_{23} + S_{44})^2. \tag{5.32}$$

Alternatively, this result can be deduced when Eqns. (5.25) and (4.9) are used to express c_3 in the form,

$$c_3 = \frac{\pm([(a_{23} + S_{44})^2 - v^2]\,[v^2 - (S_{44} - a_{23})^2])^{1/2}}{2a_{23}S_{44}}. \tag{5.33}$$

Equation (5.33) yields the conditions for imaginary values of the angle θ_3. When angle θ_3 is imaginary, then the specified ball joint location $\boldsymbol{v}$ requires that the manipulator must either overextend $(v^2 > (a_{23} + S_{44})^2)$ or underextend $(v^2 < (a_{23} - S_{44})^2)$ outside its physical range. Under these conditions it is sometimes useful to allow the manipulator to achieve the specification as closely as possible. This is accomplished by setting the radical in Eqn. (5.33) (or alternatively (5.28)) to zero so that $c_3 = 0$ and $s_3 = 1$ or $s_3 = -1$ for the overextended or underextended cases respectively. Therefore, for the overextended case, $\boldsymbol{a}_{23}$ and $\boldsymbol{s}_4$ become parallel and the distance from $\boldsymbol{s}_2$ to the ball joint is maximized. For the underextended case, $\boldsymbol{a}_{23}$ and $\boldsymbol{s}_4$ are antiparallel. The arm now folds back on itself and the distance from $\boldsymbol{s}_2$ to the ball joint is minimized. Setting the radical of Eqn. (5.33) to zero is equivalent to selecting $\theta_3^{(\pm)} = \gamma$ in Figure 4.1.

The PUMA robot has the special dimensions $a_{23} = S_{44}$ for which Eqns. (5.32) and (5.33) simplify and imaginary values for θ_3 occur only for the overextended case $(v^2 > (a_{23} + S_{44})^2)$.

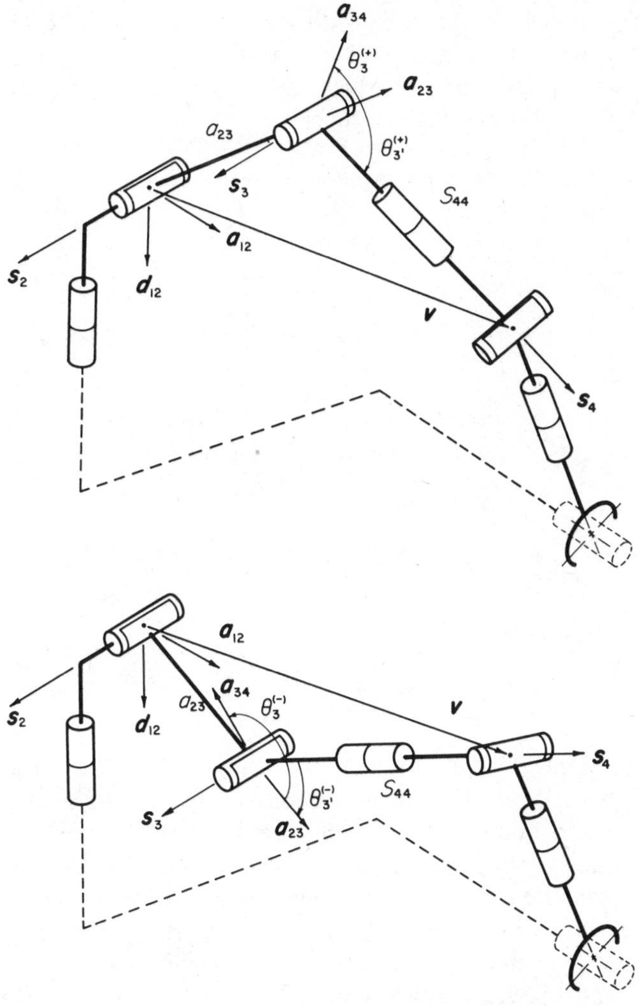

Figure 5.4 The two values of θ_3 yield distinct configurations corresponding to selection of (a) the upper sign and (b) the lower sign in Eqns. (5.28) or (5.33).

The second joint angle is determined by back substituting the values of c_3 and s_3 Eqns. (5.23) and (5.24) to yield,

$$c_2 = \frac{A_{17}(a_{23} + S_{44}\, s_3) + B_{17}(S_{44}\, c_3)}{a_{23}^2 + S_{44}^2 + 2a_{23}\, S_{44}\, s_3} \tag{5.34}$$

$$s_2 = \frac{A_{17}(S_{44}\, c_3) - B_{17}(a_{23} + S_{44}\, s_3)}{a_{23}^2 + S_{44}^2 + 2a_{23}\, S_{44}\, s_3}. \tag{5.35}$$

A necessary condition for θ_2 to be indeterminate is that the denominators of Eqns. (5.34) and (5.35) must vanish. However, using Eqns. (5.25), (5.32) and the preceding analysis for angle θ_3 with $a_{23} \neq S_{44}$, it can be shown that θ_3 must be imaginary and therefore real (physical) indeterminacy is precluded.

In the underextended case for which the assignments $c_3 = 0$ and $s_3 = -1$ are made, Eqns. (5.23) and (5.24) reduce to,

$$c_2 = \frac{A_{17}}{(a_{23} - S_{44})} \tag{5.36}$$

and

$$s_2 = \frac{-B_{17}}{(a_{23} - S_{44})}. \tag{5.37}$$

Since the constraint $c_2^2 + s_2^2 = 1$ must be satisfied, these two relations must be normalized (see Section 4, Eqns. (4.28) - (4.31)). Here it will be assumed that $a_{23} - S_{44} < 0$ so that the normalization of Eqns. (5.36) and (5.37) yields

$$c_2 = \frac{-A_{17}}{(A_{17}^2 + B_{17}^2)^{1/2}} \tag{5.38}$$

and

$$s_2 = \frac{B_{17}}{(A_{17}^2 + B_{17}^2)^{1/2}}. \tag{5.39}$$

Theoretically, Eqns. (5.38) and (5.39) can become indeterminate when $A_{17} = B_{17} = 0$ ($\boldsymbol{v} = \boldsymbol{0}$). However, physically this cannot be achieved because of the dimensions of the manipulator ($a_{23} \neq S_{44}$) and it results from the under-extended assignments. For this reason such an indeterminacy is imaginary. *The condition* $\boldsymbol{v} = \boldsymbol{0}$ *flags what will be called an imaginary uncertainty configuration.* For the overextended case ($c_3 = 0$ and $s_3 = 1$) the expressions for c_2 and s_2 are negative the values on the right side of Eqns. (5.38) and (5.39). Since $\boldsymbol{v}$ does not vanish for this case, the imaginary uncertainty configuration does not occur.

However, for the PUMA robot ($a_{23} = S_{44}$), the indeterminacy for the underextended case is real. This is because when the center of the ball joint lies on the second joint axis, $\boldsymbol{v} = \boldsymbol{0}$, $c_3 = 0$ and $s_3 = -1$. All four joint axes $\boldsymbol{s}_2$, $\boldsymbol{s}_4$, $\boldsymbol{s}_5$ and $\boldsymbol{s}_6$ now co-intersect at a point and create a spherical four bar mechanism which has mobility one (see Eqn. (2.1) with $(m, n, j, f_i) = (3, 4, 4, 1)$).

Joint angle θ_5 is determined from a spherical cosine law using Table 3.3 Set 12 (reference frame at $[\boldsymbol{a}_{34}\ \boldsymbol{d}_{34}\ \boldsymbol{s}_4]$),

$$\bar{Z}_5 = Z_{3217}. \tag{5.40}$$

Using the simplified expression for $\bar{Z}_5$ in Table 5.2 yields,

$$c_5 = Z_{3217} \tag{5.41}$$

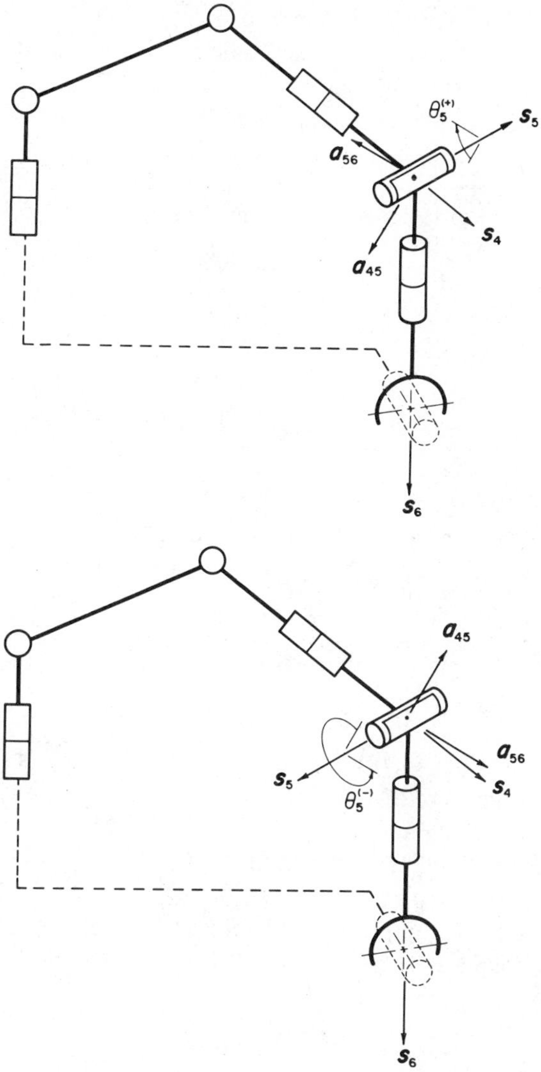

Figure 5.5 The two values of θ_5 yield similar yet distinct configurations corresponding to selection of (a) the upper sign and (b) the lower sign in Eqn. (5.42).

and therefore,

$$s_5 = \pm\ (1 - Z_{3217}^2)^{1/2}. \tag{5.42}$$

The two mechanism configurations corresponding to Eqns. (5.41) and (5.42) are shown in Figure 5.5. Angle θ_5 can never become imaginary because $-1 \leqslant Z_{3217} \leqslant 1$. Since c_5 and s_5 are not expressed as ratios, it can also be concluded that angle θ_5 is never indeterminate.

Joint angle θ_4 is determined from a pair of spherical sine and sine-cosine laws using Table 3.3, Set 12 (reference frame at $[\boldsymbol{a}_{34}\ \boldsymbol{d}_{34}\ \boldsymbol{s}_4]$),

$$R_{54} = X_{3217} \tag{5.43}$$

$$W_{54} = Y_{3217}. \tag{5.44}$$

Using the simplified expressions for R_{54} and W_{54} in Table 5.2 yields,

$$c_4 = \frac{-X_{3217}}{s_5} \tag{5.45}$$

and

$$s_4 = \frac{-Y_{3217}}{s_5}. \tag{5.46}$$

It is now shown that when $\theta_5 = 0$ or π both Eqns. (5.45) and (5.46) become indeterminate by utilizing the vectoral definitions,

$$X_{3217} \equiv \boldsymbol{a}_{34} \cdot \boldsymbol{s}_6 \tag{5.47}$$

$$Y_{3217} \equiv \boldsymbol{d}_{34} \cdot \boldsymbol{s}_6 \tag{5.48}$$

$$s_5 \equiv \boldsymbol{s}_5 \cdot \boldsymbol{a}_{45} \times \boldsymbol{a}_{56} = \boldsymbol{s}_5 \cdot \boldsymbol{s}_4 \times \boldsymbol{s}_6. \tag{5.49}$$

The right side of Eqn. (5.49) is obtained by expanding $\boldsymbol{s}_5 \cdot \boldsymbol{a}_{45} \times \boldsymbol{a}_{56}$ using $\boldsymbol{a}_{45} = \boldsymbol{s}_5 \times \boldsymbol{s}_4$ ($\alpha_{45} = -\pi/2$) and $\boldsymbol{a}_{56} = \boldsymbol{s}_5 \times \boldsymbol{s}_6$ ($\alpha_{56} = \pi/2$) and it vanishes only when the axes of $\boldsymbol{s}_4$ and $\boldsymbol{s}_6$ are collinear since both vectors are normal to $\boldsymbol{s}_5$. It follows immediately that the right sides of Eqns. (5.47) and (5.48) must also vanish since $\boldsymbol{a}_{34} \cdot \boldsymbol{s}_4 = \boldsymbol{d}_{34} \cdot \boldsymbol{s}_4 = 0$.

The remaining joint angle θ_6 is determined from the spherical sine and sine-cosine laws found in Table 3.3, Set 10 (reference frame at $[\boldsymbol{a}_{56}\ \boldsymbol{d}_{56}\ \boldsymbol{s}_6]$),

$$\bar{R}_6 = X_{54321} \tag{5.50}$$

$$\bar{W}_6 = Y_{54321}. \tag{5.51}$$

Using the simplifications in Table 5.2 yields the explicit solution,

$$s_6 = -X_{54321} \tag{5.52}$$

$$c_6 = -Y_{54321}. \tag{5.53}$$

Also from Table 3.3, Set 10 it is determined that $Z_{54321} = 0$ ($\alpha_{67} = -\pi/2$) and since

$$X_{54321}^2 + Y_{54321}^2 + Z_{54321}^2 = 1 \tag{5.54}$$

it is clear that Eqns. (5.52) and (5.53) cannot vanish simultaneously or represent imaginary angles.

The reverse displacement analysis for the manipulator has been accomplished by solving for the joint angles of the corresponding spatial mechanism. *This process has been facilitated by the utilization of the recursive notation developed in Section 3 and has a major advantage in that it has enabled the introduction of vector expressions to identify the physical significance of indeterminate and imaginary joint angles.* Further, this method of analysis can be employed to examine the role of the robot dimensions which produce and avoid singularities, and as such it also provides a useful tool for the design of manipulators.

An algorithm based on this method of analysis has been successfully employed to control a robot Ref. [12]. However, it is recognized that for real-time computation this method has the disadvantage of necessitating additional calculations since it requires completion of the spatial loop (see Section 2) and the determination of the mechanism angle θ_1. Also, the method does not exploit the direct inclusion of the parameters that specify the end-effector location in the joint angle solutions which can also reduce the required computation. However, since the reverse solution has been accomplished using vector expressions, the problem can be easily reformulated for a more computationally efficient result. The physical interpretation of the solution will remain unaltered.

It is noted that the reverse displacement analysis mathematically assumes that all joint angles can turn through a full revolution. In practice this assumption is usually not valid and it must be considered in the implementation of the solution.

Finally, the forward displacement analysis can be easily determined using Table 3.3, Set 8b. The associated dextral frame $[\boldsymbol{a}_{71}\ -\boldsymbol{d}_{71}\ \boldsymbol{s}_1]$ is aligned with the origin of the fixed reference frame $[\boldsymbol{i}\ \boldsymbol{j}\ \boldsymbol{k}]$ on the $\boldsymbol{s}_1$ axis. The subscript 1 in the elements of Set 8b now represents the actuator angle ϕ_1 rather than the mechanism angle θ_1 (see Figure 2.2). The definitions are most easily expressed using the forward recursion formulas (Table 3.2d). It is noted that none of the recursive elements in Set 8b contain the hypothetical mechanism dimensions which complete the spatial loop (with the obvious exclusion of the elements for the reference frame). Once the expressions for the unit vectors $\boldsymbol{s}_i$ and $\boldsymbol{a}_{ij}$ are determined, the orientation of the end-effector is given by $\boldsymbol{a}_{67}$ and $\boldsymbol{s}_7$ (also $-\boldsymbol{d}_{67} = \boldsymbol{s}_7 \times \boldsymbol{a}_{67}$) and the target point of the end-effector is determined by (see Section 2),

$$\boldsymbol{r} = S_{11}\boldsymbol{s}_1 + S_{22}\boldsymbol{s}_2 + a_{23}\boldsymbol{a}_{23} + S_{44}\boldsymbol{s}_4 + S_{66}\boldsymbol{s}_6. \tag{5.55}$$

The forward displacement analysis will now be incorporated in the linearized singularity analysis.

6. LINEARIZED SINGULARITY ANALYSIS

In the preceding section expressions for the joint angle displacements are examined for geometrical indeterminacies. When considering instantaneous motion, the joint displacements can be formulated using a linear algebra approach. The Jacobian is a linear transformation that maps manipulator joint velocities into the instantaneous end-effector motion. This section deals with the examination of corresponding manipulator configurations when the Jacobian becomes singular (see Refs. [13,14]).

Using screw coordinates (Ref. [14]) the instantaneous end-effector motion for a six degree-of-freedom manipulator with revolute joints can be written as,

$$\begin{bmatrix} \boldsymbol{\omega} \\ \\ \boldsymbol{v} \end{bmatrix} = [J] \begin{bmatrix} \omega_1 \\ \vdots \\ \omega_6 \end{bmatrix}, \tag{6.1}$$

where,

$\omega_1 \cdots \omega_6 =$ the six actuator angular velocities

$\boldsymbol{\omega} =$ the end-effector angular velocity vector

$\boldsymbol{v} =$ the end-effector translational velocity vector at a point coincident with the coordinate origin

$[J] =$ the Jacobian defined by the 6×6 matrix,

$$[J] = \begin{bmatrix} \boldsymbol{s}_1 & \boldsymbol{s}_2 & \boldsymbol{s}_3 & \boldsymbol{s}_4 & \boldsymbol{s}_5 & \boldsymbol{s}_6 \\ \boldsymbol{r}_1 \times \boldsymbol{s}_1 & \boldsymbol{r}_2 \times \boldsymbol{s}_2 & \boldsymbol{r}_3 \times \boldsymbol{s}_3 & \boldsymbol{r}_4 \times \boldsymbol{s}_4 & \boldsymbol{r}_5 \times \boldsymbol{s}_5 & \boldsymbol{r}_6 \times \boldsymbol{s}_6 \end{bmatrix} \tag{6.2}$$

and where,
$\boldsymbol{s}_i$ = the unit vector along the joint axis i
$\boldsymbol{r}_i$ = the position vector from the origin to any point on joint axis i.

(For a prismatic joint, the first and last three rows of the ith column are $\mathbf{0}$ and $\boldsymbol{s}_i$ respectively.)

The reader not familiar with the theory of screws should consult Refs. [14-17]. *It should be noted that the projection of the vectors $\boldsymbol{s}_i$ and $\boldsymbol{r}_i \times \boldsymbol{s}_i$ in Eqn. (6.2) on the coordinate base $[\boldsymbol{i}\ \boldsymbol{j}\ \boldsymbol{k}]$ are respectively direction cosines and mutual moments.*

The most general instantaneous motion of a rigid body in space can be described as a twist upon a screw. That is, the motion is equivalent to a rotation about a unique line together with a translation parallel to that line. For instantaneous motion of the end-effector, the pitch of the screw is unique and is determined by the ratio,

$$h = \frac{\boldsymbol{v} \cdot \boldsymbol{\omega}}{\boldsymbol{\omega} \cdot \boldsymbol{\omega}}. \tag{6.3}$$

The perpendicular distance from the origin to the axis of the screw is determined by

$$\boldsymbol{r} = \frac{\boldsymbol{v} \times \boldsymbol{\omega}}{\boldsymbol{\omega} \cdot \boldsymbol{\omega}}. \tag{6.4}$$

There are two special screws; pure rotations and pure translations are screws of zero and infinite pitch respectively. Revolute and prismatic joints permit these special relative motions.

When the manipulator joint angles $(\phi_1, \theta_2, \cdots, \theta_6)$ are known then the Jacobian, $[J]$ (Eqn. 6.2), can be calculated using the forward displacement analysis (Section 5). When the instantaneous end-effector motion is specified, then the actuator joint velocities ω_i i=1...6 can be determined by inverting $[J]$ and solving Eqn. (6.1). Further, when $[J]$ is singular the manipulator is said to be in a special configuration (Ref. [13]). The columns of $[J]$ no longer span a six dimensional space but instead span a space of a lower dimension given by the rank of $[J]$. For the PUMA robot, the form of the Jacobian can be simplified by selecting the origin of the coordinate system to coincide with the point where the last three joint axes co-intersect. The $\boldsymbol{i}$ direction is taken along $\boldsymbol{s}_4$ and the $\boldsymbol{k}$ direction is taken along $\boldsymbol{s}_5$ (Figure 6.1). The Jacobian can now be expressed in the simplified form,

$$[J] = \begin{bmatrix} \boldsymbol{s}_1 & \boldsymbol{s}_2 & \boldsymbol{s}_3 & \boldsymbol{i} & \boldsymbol{k} & \boldsymbol{s}_6 \\ \boldsymbol{r}_1 \times \boldsymbol{s}_1 & \boldsymbol{r}_2 \times \boldsymbol{s}_2 & \boldsymbol{r}_3 \times \boldsymbol{s}_3 & \mathbf{0} & \mathbf{0} & \mathbf{0} \end{bmatrix}. \tag{6.5}$$

A necessary and sufficient condition for J to be singular is that its determinant vanishes and for Eqn. (6.5) this relation can be expressed in the simplified form,

$$|J| = |\boldsymbol{r}_1 \times \boldsymbol{s}_1 \;\; \boldsymbol{r}_2 \times \boldsymbol{s}_2 \;\; \boldsymbol{r}_3 \times \boldsymbol{s}_3| \, (\boldsymbol{s}_6 \cdot \boldsymbol{j}) = 0. \tag{6.6}$$

Expanding the 3×3 determinant in Eqn. (6.6) and making the substitution $|\boldsymbol{s}_2 \; \boldsymbol{r}_3 \; \boldsymbol{s}_3| = 0$ ($\boldsymbol{s}_2$ and $\boldsymbol{s}_3$ are parallel) yields the product

$$|J| = -\,|\boldsymbol{r}_1 \; \boldsymbol{s}_1 \; \boldsymbol{s}_2| \; |\boldsymbol{r}_2 \; \boldsymbol{r}_3 \; \boldsymbol{s}_3| \; |\boldsymbol{s}_4 \; \boldsymbol{s}_5 \; \boldsymbol{s}_6| = 0 \tag{6.7}$$

where additionally, the substitution $\boldsymbol{j} = \boldsymbol{k} \times \boldsymbol{i} = \boldsymbol{s}_5 \times \boldsymbol{s}_4$ has been introduced.

The conditions for $|J| = 0$ will now be considered in detail and the results will be compared with those derived using the displacement analysis in Section 5. It should be noted that there are two distinct types of singularities for which $|J| = 0$. Uncertainty configurations, for which indeterminacies occur

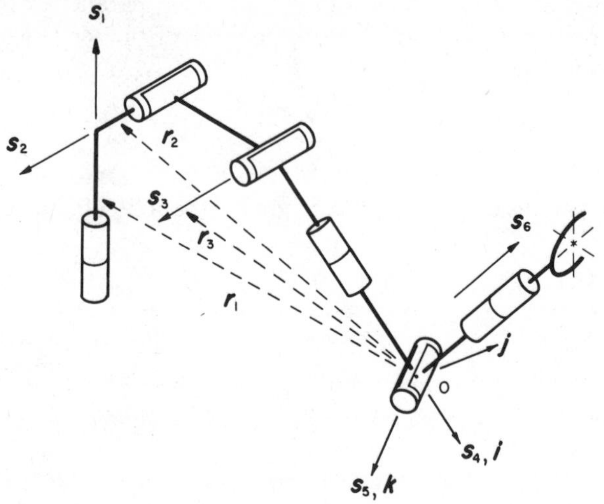

Figure 6.1 Selecting the coordinate origin at the point where the last three joint axes co-intersect simplifies the form of the Jacobian.

in the calculation of joint angles have been identified in Section 5. *Here a second type of singularity, namely stationary configurations, are identified for which it is possible to compute the joint angles.* Various cases for the vanishing of a single factor on the right side of Eqn. (6.7) will now be examined, the first of which is summarized in Table 6.1 where the columns of $[J]$ are denoted by the screw notation $\$_i$, $i = 1 \cdots 6$.

TABLE 6.1

Case (i): $|\boldsymbol{r}_1\, \boldsymbol{s}_1\, \boldsymbol{s}_2| = 0$, $|\boldsymbol{r}_2\, \boldsymbol{r}_3\, \boldsymbol{s}_3| \neq 0$, $|\boldsymbol{s}_4\, \boldsymbol{s}_5\, \boldsymbol{s}_6| \neq 0$.

	$a_{23} \neq S_{44}$	$a_{23} = S_{44}$
$S_{22} \neq 0$	$\boldsymbol{r}_1, \boldsymbol{s}_1, \boldsymbol{s}_2$ coplanar $[\$_1\ \$_2\ \$_4\ \$_5\ \$_6]_4$ Stationary	$\boldsymbol{r}_1, \boldsymbol{s}_1 \boldsymbol{s}_2$ coplanar $[\$_1\ \$_2\ \$_4\ \$_5\ \$_6]_4$ Stationary
$S_{22} = 0$	$\boldsymbol{r}_1 = \boldsymbol{0}$ $[\$_1\ \$_4\ \$_5\ \$_6]_3$ Uncertainty	$\boldsymbol{r}_1 = \boldsymbol{0}$ and $\boldsymbol{r}_2 \neq \boldsymbol{0}$ $[\$_1\ \$_4\ \$_5\ \$_6]_3$ Uncertainty

When $S_{22} \neq 0$, and either $a_{23} \neq S_{44}$ or $a_{23} = S_{44}$ then the vectors $\boldsymbol{r}_1$, $\boldsymbol{s}_1$ and $\boldsymbol{s}_2$ are coplanar, and the center of the ball and socket joint lies in the plane defined by the axes $\boldsymbol{s}_1$ and $\boldsymbol{s}_2$ as illustrated in Figure 6.2. The robot is said to be in a stationary configuration, the rank of $[J]$ is 5 because the rank of the submatrix $[\$_1\ \$_2\ \$_4\ \$_5\ \$_6]_4$ is 4. The linear dependence of this submatrix occurs because the equivalent ball and socket joint (composed of $\boldsymbol{s}_4$, $\boldsymbol{s}_5$, $\boldsymbol{s}_6$) is capable of producing a rotation about an axis $\boldsymbol{s}_{456}$ which is coplanar with and co-intersects the axes of $\boldsymbol{s}_1$ and $\boldsymbol{s}_2$.

When $S_{22}=0$, and either $a_{23} \neq S_{44}$ or $a_{23} = S_{44}$ then $\boldsymbol{r}_1 = \boldsymbol{0}$, and the center of the equivalent ball and socket joint lies on the first joint axis $\boldsymbol{s}_1$. This is the uncertainty configuration which has already been illustrated in Figure 5.3. The rank of the submatrix $[\$_1\ \$_4\ \$_5\ \$_6]_3$ is 3 and therefore the rank of $[J]$ is 5. The linear dependence of the submatrix occurs because all the four joint axes co-intersect at a point and form a spherical four bar mechanism with mobility one.

TABLE 6.2

Case (ii): $|\boldsymbol{r}_1\ \boldsymbol{s}_1\ \boldsymbol{s}_2| \neq 0$, $|\boldsymbol{r}_2\ \boldsymbol{r}_3\ \boldsymbol{s}_3| = 0$, $|\boldsymbol{s}_4\ \boldsymbol{s}_5\ \boldsymbol{s}_6| \neq 0$.

	$a_{23} \neq S_{44}$	$a_{23} = S_{44}$
$S_{22} \neq 0$	$\boldsymbol{r}_2$, $\boldsymbol{r}_3$, $\boldsymbol{s}_3$ coplanar $[\$_2\ \$_3\ \$_4\ \$_5\ \$_6]_4$ Stationary	$\boldsymbol{r}_2$, $\boldsymbol{r}_3$, $\boldsymbol{s}_3$ coplanar and $\boldsymbol{r}_2 \neq \boldsymbol{0}$ $[\$_2\ \$_3\ \$_4\ \$_5\ \$_6]_4$ Stationary
		$\boldsymbol{r}_2 = \boldsymbol{0}$ $[\$_2\ \$_4\ \$_5\ \$_6]_3$ Uncertainty
$S_{22}=0$	$\boldsymbol{r}_2$, $\boldsymbol{r}_3$, $\boldsymbol{s}_3$ coplanar and $\boldsymbol{r}_1 \neq \boldsymbol{0}$ $[\$_2\ \$_3\ \$_4\ \$_5\ \$_6]_4$ Stationary	$\boldsymbol{r}_2$, $\boldsymbol{r}_3$, $\boldsymbol{s}_3$ coplanar and $\boldsymbol{r}_2 \neq \boldsymbol{0}$ $[\$_2\ \$_3\ \$_4\ \$_5\ \$_6]_4$ Stationary

When $S_{22} \neq 0$ and $a_{23} \neq S_{44}$ then the center of the equivalent ball and socket joint lies in the plane defined by the parallel joint axes of $\boldsymbol{s}_2$ and $\boldsymbol{s}_3$ as illustrated by Figure 6.3. The linear dependence of the submatrix occurs because the equivalent ball and socket joint is capable of producing a rotation about an axis $\boldsymbol{s}_{456}$ which is parallel and coplanar with the axes of $\boldsymbol{s}_2$ and $\boldsymbol{s}_3$.

This stationary configuration corresponds to both the overextended case (Figure 6.3) and the underextended case described in Section 5.

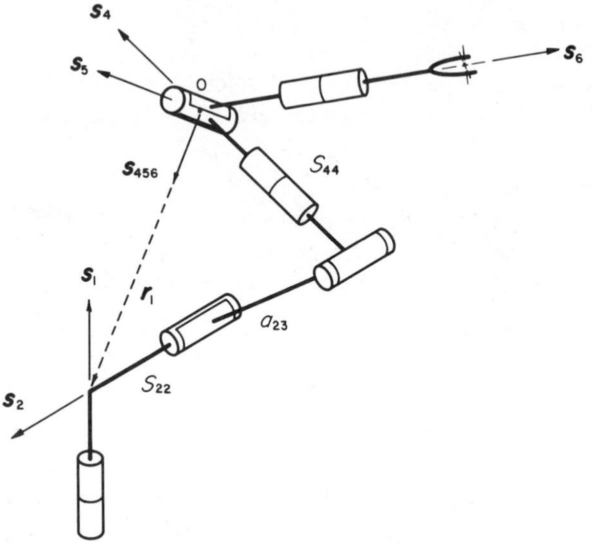

Figure 6.2 For $S_{22} \neq 0$, a stationary configuration occurs when the origin 0 lies in the plane of the first two joint axes. The last three axes are capable of producing a rotation about an axis $\mathbf{s}_{456}$ which is coplanar with an co-intersects the axes of $\mathbf{s}_1$ and $\mathbf{s}_2$.

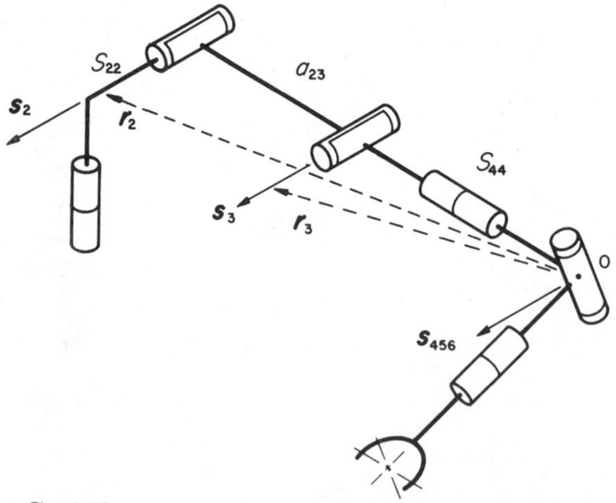

Figure 6.3 For $S_{22} \neq 0$ and $a_{23} \neq S_{44}$, a stationary configuration occurs when the origin 0 lies in the plane of the second and third joint axes which is illustrated for the over-extended case. The last three joint axes are capable of producing a rotation about an axis $\mathbf{s}_{456}$ which is coplanar with and parallel to the axes of $\mathbf{s}_2$ and $\mathbf{s}_3$.

Analogously when $S_{22} \neq 0$, $a_{23}=S_{44}$ and $\boldsymbol{r}_2 \neq \boldsymbol{0}$ the robot is in a stationary configuration which corresponds to the overextended case. However, when $\boldsymbol{r}_2=\boldsymbol{0}$ the robot is in an uncertainty configuration (the underextended case) for which the joint axes of $\boldsymbol{s}_2$, $\boldsymbol{s}_4$, $\boldsymbol{s}_5$ and $\boldsymbol{s}_6$ co-intersect at a point to form a spherical four bar mechanism.

For $S_{22}=0$ and $a_{23} \neq S_{44}$ the robot is in a stationary configuration corresponding to the overextended and underextended cases. For $S_{22}=0$ and $a_{23}=S_{44}$, the stationary configuration is represented by the overextended case.

Case (iii): $|\boldsymbol{r}_1\ \boldsymbol{s}_1\ \boldsymbol{s}_2| \neq 0$, $|\boldsymbol{r}_2\ \boldsymbol{r}_3\ \boldsymbol{s}_3| \neq 0$, $|\boldsymbol{s}_4\ \boldsymbol{s}_5\ \boldsymbol{s}_6| = 0$.

For this case the axes of $\boldsymbol{s}_4$ and $\boldsymbol{s}_6$ are collinear. The submatrix $[\$_4\ \$_6]_1$ has rank 1 and the robot is in an uncertainty configuration as previously described in Section 5.

Numerous special singularities occur since two or more screws in a submatrix can themselves become linearly dependent. Further, two or all three of the factors of $|J|$ can vanish simultaneously. All such special cases have been described in detail for a similar manipulator geometry in Ref. [12].

Introducing the terminology of linear algebra, when the Jacobian spans a space of rank $1 \leqslant n \leqslant 5$ there exists a complementary orthogonal space of rank $6-n$. The inner product of any screw $\$_c$ in the orthogonal space with any screw $\$_i$ $(i = 1,2...6)$ of the Jacobian vanishes,

$$\$_c^{\mathrm{T}} \$_i = 0 \tag{6.8}$$

and the end-effector cannot have motion about any screw with a non-zero projection in the orthogonal space (see Refs. [17-20]). The effects of a stationary or uncertainty configuration on the performance of a robot will now be examined using orthogonal screws.

Generally when a robot is in a special configuration the specified end-effector motion will not be spanned by the columns of $[J]$ and the actuator velocities of the linearly dependent joints will mathematically become infinite. This implies that excessive torque demands will develop for these joint actuators in the neighborhood of the singularity. As one or more of the actuators saturate, the end-effector will deviate from its prescribed trajectory.

Consider for example that the end-effector of the PUMA robot is traveling on a path and that at some instant it approaches the stationary configuration illustrated in Figure 6.3 (see Table 6.2) for which the joint screws $\$_2, \$_3 \cdots \$_6$ are linearly dependent. In order to determine the angular velocity ω_2 of the second joint actuator, it is useful to define

$$\Delta_{23456} \equiv \$_2+\lambda_3\$_3+\lambda_4\$_4+\lambda_5\$_5+\lambda_6\$_6 \tag{6.9}$$

where the coefficients λ_i are constants which are selected such that Δ_{23456} vanishes in the stationary configuration. Equation (6.1) can now be written in the form

$$\omega\$ = \omega_1\$_1+\omega_2\Delta_{23456}+(\omega_3-\lambda_3)\$_3+(\omega_4-\lambda_4)\$_4$$

$$+(\omega_5-\lambda_5)\$_5+(\omega_6-\lambda_6)\$_6 \tag{6.10}$$

where $\omega\$$ is the prescribed end-effector motion. It is always possible to determine a screw $\$_2'$ which is orthogonal to each of the five screws in $[J]$ where $\$_2$ has been deleted. Forming the inner product of Eqn. (6.10) with $\$_2'$ and solving for ω_2 yields

$$\omega_2 = \frac{(\$_2')^{\mathrm{T}}\ \$}{(\$_2')^{\mathrm{T}}\ \Delta_{23456}}. \tag{6.11}$$

As the manipulator approaches the stationary configuration, the numerator will approach a non-zero value since the prescribed end-effector motion is by assumption not spanned by $\$_1,\$_3 \cdots \$_6$. However, the denominator will vanish since $\Delta_{23456} \rightarrow \mathbf{0}$ and therefore ω_2 will be become unbounded. Similarly it can be shown that the angular velocities of the actuators for the remaining dependent joints become unbounded.

Finally, the end-effector can be driven through a special configuration with finite actuator velocities when the end-effector motion is selected such that it is spanned by the columns of $[J]$. This can be accomplished by imposing one or more additional constraints to solve Eqn. (6.1) uniquely (see Ref. [21] for example).

ACKNOWLEDGMENTS

The authors gratefully acknowledge the financial support of the National Science Foundation (Grant No. ENG78-20112), the Department of Energy (Contract No. DE-AC05-79ER10013), the Westinghouse Research and Development Center (Pittsburgh, PA), and the State of Florida Center of Excellence Fund granted to the Center for Intelligent Machines and Robotics (CIMAR) of the University of Florida.

Illustrations were prepared by Jeffery Hudgens, Department of Mechanical Engineering, University of Florida.

REFERENCES

[1] D. L. Pieper, and B. Roth, "The Kinematics of Manipulators Under Computer Control," *Proceedings 2nd International Congress on the Theory of Machines and Mechanisms,* Vol. 2, 1969.

[2] J. Duffy, *Analysis of Mechanisms and Robot Manipulators,* John Wiley and Sons, Inc., New York, 1980.

[3] A. T. Yang and F. Freudenstein, "Application of Dual-Number Quaternion Algebra to the

Analysis of Spatial Mechanisms." *J. Applied Mechanics (ASME),* V. 31, pp. 300-308, Series E (June, 1964).

[4] F. M. Dimentberg, *The Screw Calculus and its Applications to Mechanics,* Moscow, 1965. English Translation, U. S. Dept. of Commerce, (N.T.I.S.), No. AD 680 993, 1969.

[5] L. Brand, *Vector and Tensor Analysis,* John Wiley and Sons, Inc., New York, 1948.

[6] J. Rooney, *A Unified Theory for the Analysis of Spatial Mechanisms Based on Spherical Trigonometry.* Ph.D. Thesis, Liverpool Polytechnic, 1974.

[7] J. Duffy and J. Rooney, "A Foundation for a Unified Theory of Analysis of Spatial Mechanisms," Trans. of the ASME, *Journal of Engineering for Industry,* Vol. 97, Series B, No. 4, 1975.

[8] R. P. Paul, B. Shimano, and G. E. Mayer, "Kinematic Control Equations for Simple Manipulators," *IEEE Trans. on Systems, Man, and Cybernetics,* Vol. SMC-11, No. 6 June, 1981.

[9] G. S. G. Lee and M. Ziegler, "A Geometric Approach in Solving Inverse Kinematics of PUMA Robots," *Proceedings 13th International Symposium on Industrial Robots and Robots 7,* Vol. 2, 1983.

[10] D. R. Keen and J. Duffy, "A Unified Theory for the Analysis of Single Loop Spatial Mechanisms," *Proceedings of IFToMM Fourth World Congress on the Theory of Machines and Mechanisms,* Newcastle upon Tyne, U.K., 1975.

[11] G. E. Hay, *Vector and Tensor Analysis,* Dover, 1952.

[12] H. Lipkin, J. Duffy, and D. Tesar, *Kinematic Control of a Robotic Manipulator With a Unilateral Manual Controller,* DOE Grant No. DE-AC05-79ER10013, 1983.

[13] K. Sugimoto, J. Duffy, and K. H. Hunt, "Special Configurations of Spatial Mechanisms and Robot Arms," *Mechanism and Machine Theory,* Vol. 17, No. 2, 1982.

[14] K. H. Hunt, *Kinematic Geometry of Mechanisms,* Clarendon Press, Oxford, 1978.

[15] R. S. Ball, *A Treatise on the Theory of Screws,* Cambridge University Press, 1900.

[16] O. Bottema and B. Roth, *Theoretical Kinematics,* North-Holland Publishing Co., Amsterdam, 1979.

[17] H. Lipkin and J. Duffy, "Analysis of Industrial Robots via the Theory of Screws," *Proceedings 12th International Symposium on Industrial Robots,* Paris, 1982.

[18] J. Duffy and K. Sugimoto, "Application of Linear Algebra to Screw Systems," *Mechanism and Machine Theory,* Vol. 17, No. 1, 1982.

[19] H. Lipkin and J. Duffy, "On the Geometry of Orthogonal and Reciprocal Screws," 5th CISM-IFToMM Symposium on Theory and Practice of Robots and Manipulators Udine, Italy, June 1984.

[20] H. Lipkin and J. Duffy, "The Elliptic Polarity of Screws," to appear in, J. of Mechanisms, Transmissions and Automation in Design, *Trans. of the ASME.*

[21] D. E. Whitney, "The Mathematics of Coordinated Control of Prosthetic Arms and Manipulators," J. of Dynamic Systems, Measurement and Control, *Trans. of the ASME,* December, 1972.

6

Six-Legged Walking Robots

J. J. KESSIS, J. P. RAMBAUT, J. PENNE, R. WOOD

1. INTRODUCTION

1.1 Potential Usefulness of Walking Robots

At the present time, stationary manipulator type robots are used mainly in production tasks in factories. Yet, both inside and outside of industry, many more tasks could be performed by a mobile robot. Among these tasks are spatial and terrestrial exploration, mineral prospecting, maintenance or repair in difficult environments (such as nuclear plants), and farming. Current research on mobile robots [1,2,3] has the potential to open such tasks to robots if a wide range of robotics problems can be solved in the fields of microcomputer control, sensors, perception, world modeling, and decision making.

Many tasks require that the mobile robot have off-road capabilities similar to those of terrestrian animals; that is, the robot must have better mobility than conventional wheeled or even on-tracks robots. *Legged locomotion* appears to be the best way to traverse virtually any terrain, both because the legged device causes the least ecological damage and because it has good body stability. Currently, however, only a few researchers are investigating walking machines [4,5,6], probably due to inherent complexity of the matter and its especially multidisciplinary character (involving such different subjects as Biology and Computer Science). In our opinion, walkers have to be, from the outset, fully robotic (i.e., designed with high Artificial Intelligence software levels). The early General Electric Machine [7], although quite impressive, showed the inefficiency of pure mechanics in the domain.

1.2 Position of Walkers in Robotic Research

1.2.1 Shared Problems

The "psychologic" similarities with other *Mobile Robots* are obvious: in an unknown environment, world sensing and understanding capabilities are essential, concurrently with a goal-oriented decision system. Mobile Robots, including walkers, thus offer a concrete support for experimenting with A.I. concepts.

Walking Robots, somewhat unexpectedly, show many common features with *manipulators*, particularly with respect to mechanics and control. First, the Walker's flexible "anatomy", with many degrees of freedom, is similar to that of the industrial robots. Therefore, many development problems and implementation methods are common to both walkers and manipulators. For instance, given a locomotion task in *task coordinates,* one has to find a corresponding set of the numerous *articular coordinates* (not necessarily by implementing a real coordinate transformer). This is the so-called "resolvability problem" [8], whose wide indefiniteness is reduced in walkers by introducing the concept of gait. In addition, there is a need for high-level computer languages specifying tasks rather than tedious effector moves. Again it is the same for manipulators, contrary to wheeled mobile Robots whose rather trivial mechanics is easy to control.

1.2.2 Specific Features

Of course, Walkers exhibit specific features, concepts and problems. First, Walker's legs have to be mechanically well-designed according to walking and control needs and efficiency. Common robotic concepts, such as passive or active compliance, are pertinent in the domain. We evoked above the essential concept of *gait*; possibly in the future the involved ideas may apply to multi-manipulator control.

The *terrain adaptability* is perhaps the more specific problem of Walkers. In addition to modeling "world" as other Mobile Robots, it is necessary to model *terrain* (closer and beneath world) and generate decisions according to both models. We shall see below that a part of the task may be made at a lower level.

The reader will find a review of the Mobile Robot Research in [9], and a reference list about Walking Machines in [10]. By comparison with these current projects which seem at the moment to emphasize "Supervisory Control" (i.e., man-driven vehicles) our approach is characterized by the particular attention we pay to autonomous behavior and plan-language generation.

1.3 The Hexapod Robots H1 and H2

Hereafter we shall describe the current state of research work in our Laboratory. We have designed and built two Hexapod (six-legged) Robots, sharing the same general logic and control architecture.

The first, H1, is operating. The second, H2, which is heavier, is being finished; its 40 Kg payload will allow various sensor equipment and possible battery operation for autonomy, which is not so easy as for wheeled Robots, due to the added task of bearing itself on reversible legs.

Both robots have at the present a rotating ultrasonic range finder, operating under the control of the Plan Language, and an equilibrium sensor. A joystick

Figure 1 Hexapod H1 with pantograph-type leg mechanisms.

Figure 2 Hexapod H2 (40 kg payload)

treated by the Plan Level as an extra sensor, allows if desired (under Plan control) the hand-driving of the Robot. Although anti-robotic, the capability was easy to implement and may be useful, for maintenance tasks for instance.

2. MULTI-LEVEL CONTROL ARCHITECTURE

At the outset we devised a multi-level logical architecture [11,12] for modularity and flexibility.

As experiments on the Robot progressed, the basic concepts were expanded and refined without changing the original architecture. The only additional level is the 1.5 (Tonus). Sensory levels, non-hierarchical, are outside of this scheme. We attempted to apply a *least level principle:* that is, to accomplish a task at the lowest level able to manage it in a natural way. For instance, equilibrium control, which was tried at level 3 for convenience, will be lowered to 1.5 level when operational. A part of terrain adaptability may be situated at this "reflex" level.

Table I shows the present implementation of the control levels: leg, tonus, gait, plan and intelligence.

3. THE PHYSICAL CONTROL LEVELS

3.1 The Leg Level

This mechanical and servo level was designed to simplify control. So each leg is a cartesian (i.e., the two servo motors control respectively x and y coordinates) planar reversible mechanism. This design allows an intrinsic limited terrain-adaptability; e.g., the robot walks on small blocks without limping. On H1, the leg mechanism is an original triangle and pantograph system; H2 has a direct cartesian mechanism. It should be noted that the Robots have only 12 leg motors, instead of 18 in a nonplanar configuration generally seen. Each leg exhibits a lateral compliance instead of a third motor. This design proved to be efficient and allows turn-in-place moves.

3.2 The Tonus Level

This level was added for tonicity: a separate microprocessor was shown to be necessary for continuously sending orders to servo motors, simultaneously with gait calculation. This is not required on H2 motors, but as stated above, this 1.5 level is conceptually useful for implementing "reflex" adaptability.

4. THE GAIT CONCEPT

4.1 History

The observation of moving animals is immemorial: Muybridge [13] points out a correct positioning of legs in cave drawings, the well-known religious

TABLE 1

MULTI-LEVEL ARCHITECTURE

Level		
1	*LEG:*	Mechanic level. Intrinsic Terrain adaptability (by reversible gearing of the Servomotors).
1.5	*TONUS:*	– Generates continuous motor control. – Future equilibrium control and uneven terrain adaptation.
2	*GAIT:*	Generalized gait generator (including turning gaits, etc.). Gaits are primitives for LP 4.5.
3	*PLANS:*	Plan Language LP 4.5 Interpreter. LP 4.5 Plans are conditional allowing autonomous behavior.
4	*INTELLIGENCE* *Perception:* *Decision:*	(in development) World modeling in an unknown environment. Path finding, Plan generation.

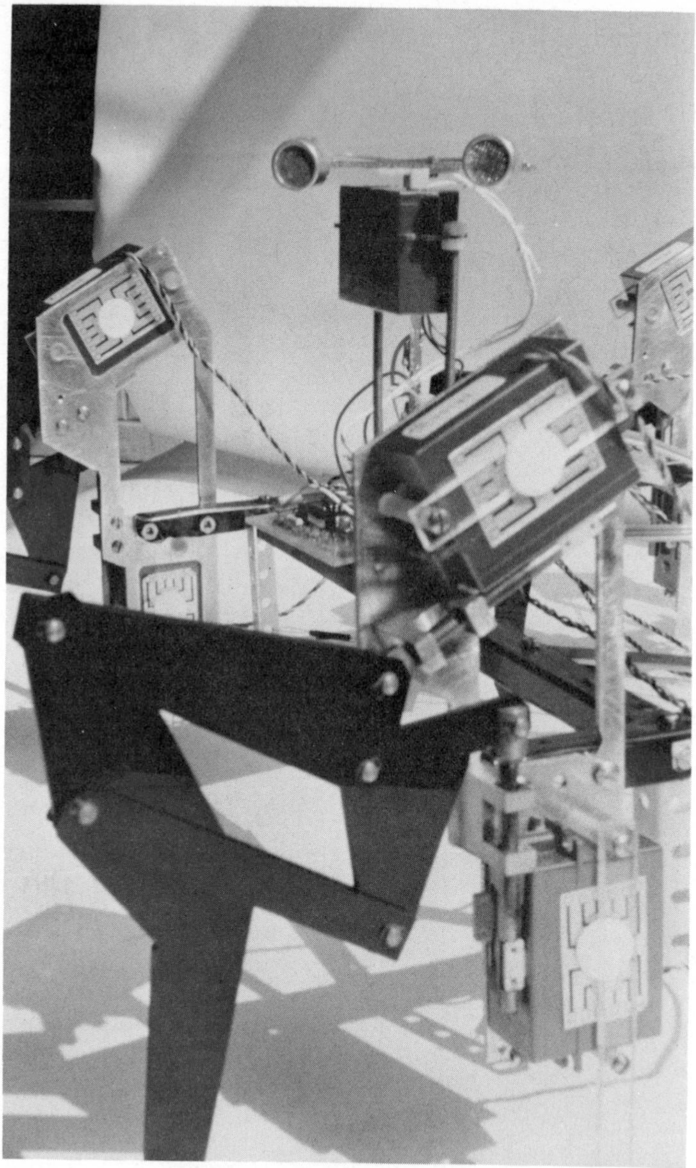

Figure 3 Closer view of H1 showing leg mechanism and ultrasonic range finder.

stereotyping of human and animal postures in Aegyptian art, and various errors and prejudices in more recent oil paintings. The gait idea, too, seems to be very ancient, thanks to the rich variety exhibited by horses. About the 1880's some scientists, Muybridge and Marey principally, followed artists and horsemen on the subject, applying photographic techniques, mainly to mammals. Studies on insects followed, see [14,15].

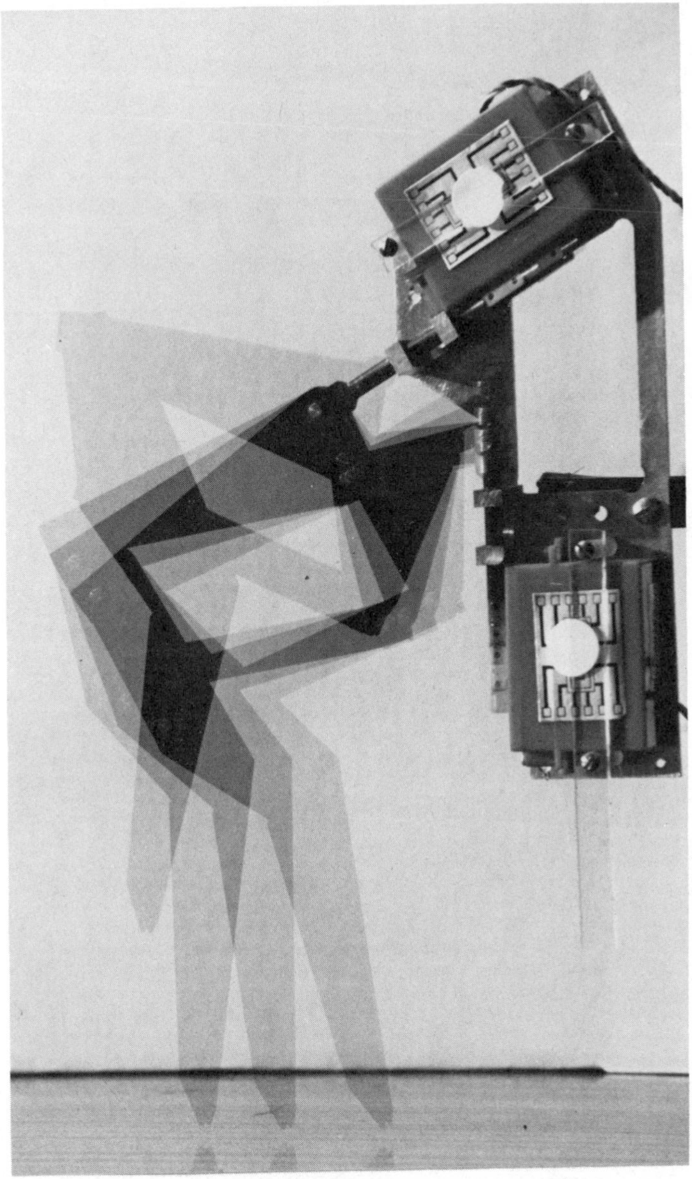

Figure 4 Four-times leg cycle $(T = 4)$.

In the 1960s a new generation of scientists payed attention to the theory of gaits, first as finite state automata (McGhee, 16), next studying static stability (Bessonov, 17). For hexapods - our subject - it was found that among the 1030 statically possible gaits the better were in the insect-used "wavy symmetric" class. Stability decreases with the *duty factor* of a leg (support time/total leg cycle time).

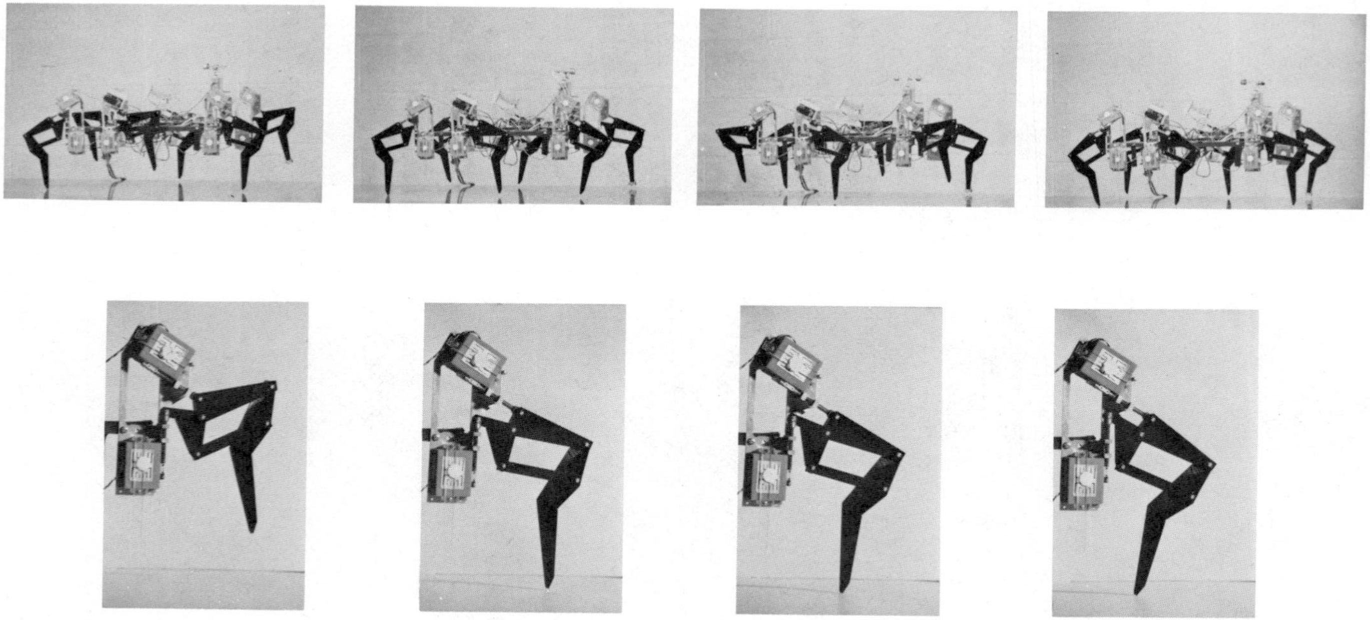

Figure 5 Four-times "tripled" gait.

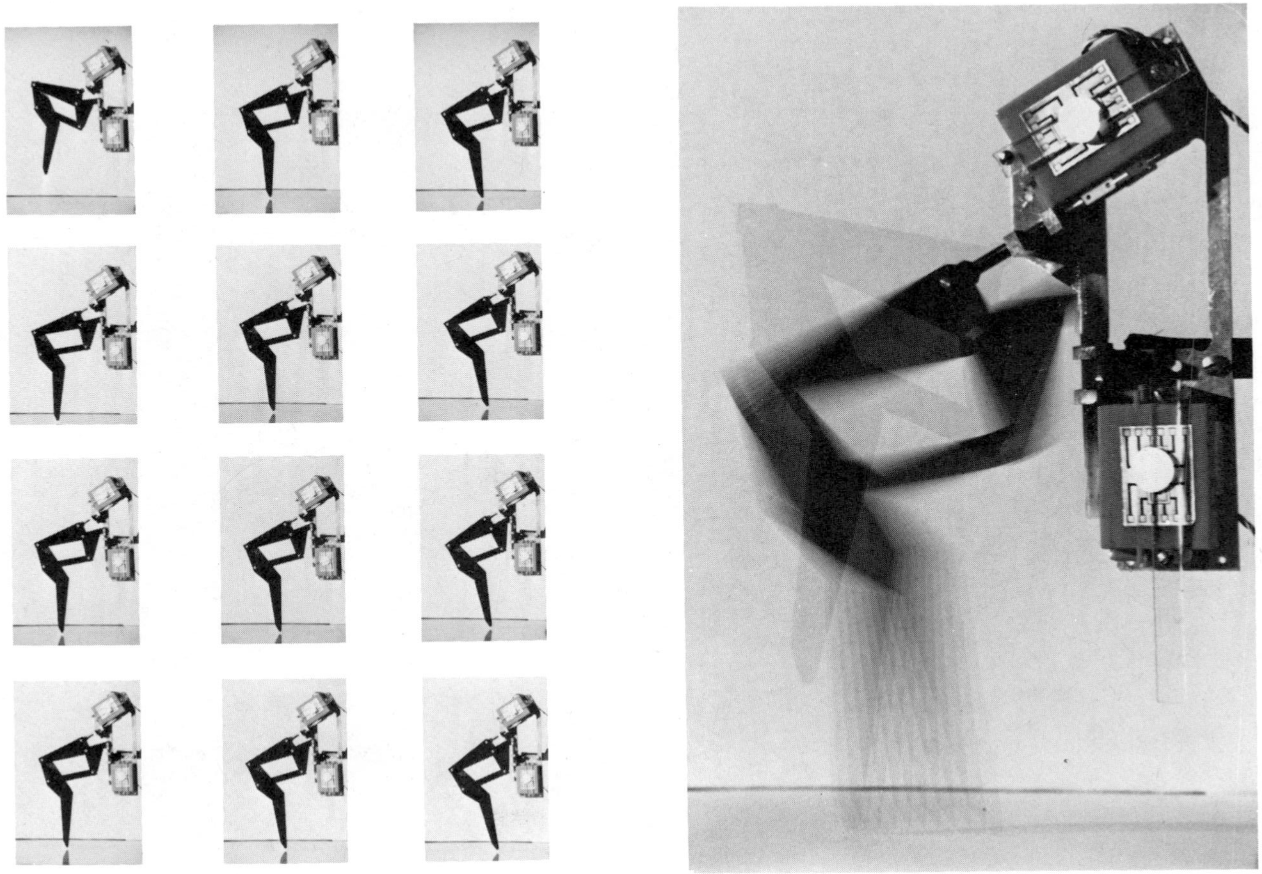

Figure 6 Twelve-times leg cycle (T = 12).

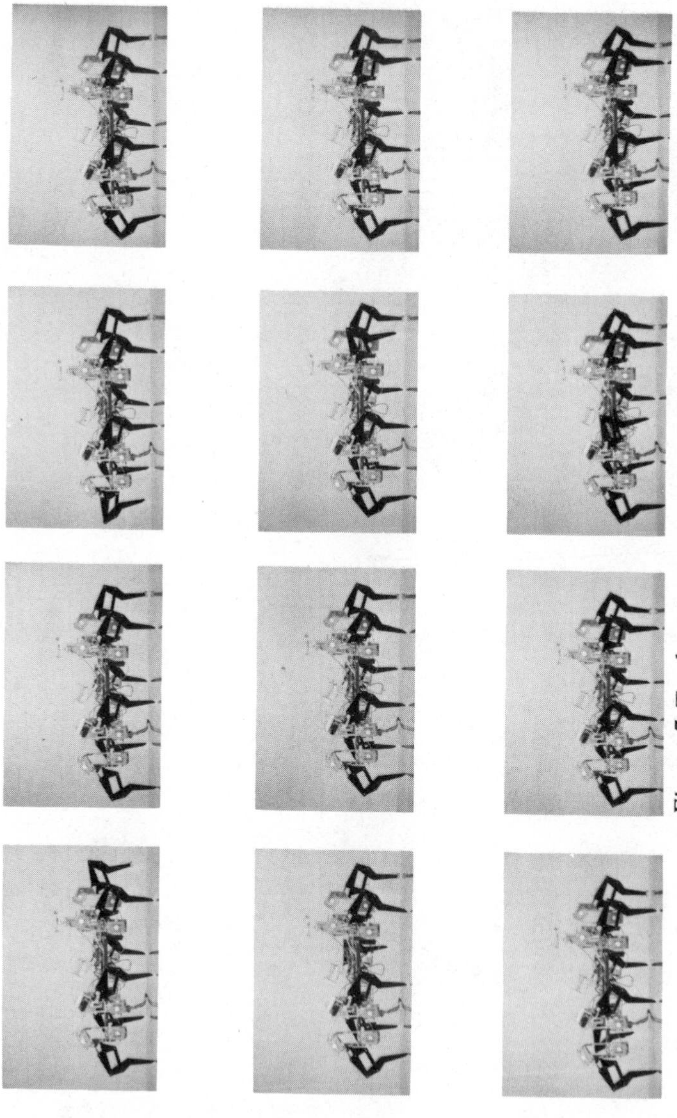

Figure 7 Twelve-times wavy symmetric gait.

4.2 The Gait Generator

The problem is to design a *gait generator* that gives a continuous *control pattern* for all degrees of freedom (DOF) of the robot. The following detailed description of the current status of the model shows *generalized gaits* (this approach is much more versatile than previously proposed non-parameterized models).

Let n be the number of legs, each with p DOF. The total number of DOF is np (all variables considered hereafter are integers). The unit of elementary time (e.t.) is assumed to be long enough for execution of any elementary command. In our basic model all legs have the same cycle (regular gaits). A gait is then only a *vector of phase shifts* ϕ_i between legs.

Let T be the duration of a *leg cycle*. T is a multiple of the number of independent legs. For example, in a hexapod "tripled" gait, there are only 2 independent legs; T may be as short as 4, because 2 e.t. are required for the return stroke, and 2 more for power stroke (otherwise the duty factor is under 1/2, and 4 legs are up together at times, a violation of the quasistatic model).

Since gait is cyclic, it is of interest to consider a *cyclical time:*

$$\tau = t \pmod{T}.$$

So, the control sequence of the *reference leg* (leg 1) is a vector function of cyclical time:

$$L_k(\tau) \quad (k = 1,p).$$

Then T different L_k form a $p \times T$ matrix, which describes the leg cycle.

Given the $L_k(\tau)$ and the ϕ_i, one may generate the whole control sequence. It is convenient to introduce a *leg cyclical time,* θ_i:

$$\theta_i = (\tau + \phi_i) \pmod{T}$$

with $\theta_i = \tau$ for the reference leg 1.

The control sequence of leg i is then expressed as a function of its own cyclical time:

$$C_{ik} = S_{ik} A_{ik} L_{ik}(\theta_i) \quad (k = 1, \ldots, p).$$

The robot's control pattern (elementary command) at time t is the $p \times n$ matrix C_{ik}. The A_{ik} and S_{ik} parameters modulate every DOF amplitude and sign, respectively, in order to solve turn problems, for instance. Centering parameters are being introduced for terrain adjusting.

The gait generation problem now is reduced to finding an algorithm which gives the ϕ_i. For instance, a gait is *symmetrical* when:

$$\phi_{n-i+1} = (\phi_i + T/2) \pmod{T} \quad (i = 1, \ldots, n/2).$$

It is *wavy* with a propagation time w when

$$\phi_i = (\phi_{i-1} - w) \pmod{T} \quad (i = 1, \ldots, n/2).$$

Thus, as is well known, a *wavy symmetrical* gait depends only on parameter w which may vary between 1 and $T-1$. For the hexapod, $T = 12$ and $w = 8$ gives the maximal stability 142635 gait, while $T = 12$ and $w = 6$ produce the "tripled" gait (or, according to the above discussion, $T = 4$ and $w = 2$).

In short, our system generates gaits as a function of duration of a leg cycle T, the control sequence of a reference leg $L_k(\tau)$, amplitudes A_{ik} and signs S_{ik}, as well as a phase shift rule giving the θ_i. The function may be arbitrary ($n-1$ parameters) or structured according to classical descriptions (e.g., wavy). Every gait is referenced in the system by a *gait number* which is mapped to a *gait table.* The gait table includes all the gait parameters as described above. The gait generator uses the gait table to produce the robot's control pattern.

5. THE PLAN LEVEL AND LP 4.5 LANGUAGE

Our aim is to obtain autonomous behavior by generating plans according to goals, world models, and terrain. The third level is a plan interpreter of a plan language, called LP 4.5, which has as its fundamental statement a pair (gait and duration). Gaits, as defined by level 2, are thus among the primitives of level 3. Currently, the plan language is enriched with conditionals, sensor tests, and linkage to special purpose machine language calls.

Figure 8 Hexapod H1: escaping behavior.

Several authors have distinguished between low level and higher level robot control languages. For instance, Latombe [18] describes motor, effector, object, and goal languages. It is hard to characterize LP 4.5, the first walking robot language, within such a framework. LP 4.5 is a high enough level language to relieve the programmer of the task of taking care of "effector" leg control. This capability characterizes LP 4.5 as an object type language for manipulators according to Latombe's classification. From a computer science point of view, however, LP 4.5 is similar internally to an assembly language, including conditionals. The source code is "assembled" to generate *evolutive plans*. The plans are then interpreted to control the robot.

Table 2 shows the current instruction set of LP 4.5. Although LP 4.5 is designed for generation by level 4, handwritten LP 4.5 "plans" allow such autonomous behaviors as edge following, escaping, avoiding obstacles, etc.; LP

TABLE 2

INSTRUCTION SET FOR LP 4.5

Instruction	Description
ALLU n,t	Select gait n and execute that gait for t e.t.'s.
ALLC t	Execute the current gait for t e.t.'s.
CHBA a	Switch to the ath gait bank.
REPO t,n	Initialization command. The robot retracts its legs (the robot rests on a pedestal), resets the gait cycle counter, and initializes for gait n, and wait t e.t.'s.
DEBT	Stand up command (executed after a REPO)
TMPO n	Sets the value of e.t. to n
CNTR p,v	Equilibrium adjustment. Sets leg p's vertical centering to v.
BRAN 1	Unconditional branch to label 1.
BEVE c, 1	Branch on condition code c to label 1.
BARI p, 11, 12, 13	Arithmetic branch on contents of register p.
CONS vl,vd	Set sensor register to range vl $\pm$ vd.
RADF p	Set telemeter to angular position p.
RADM p	Set telemeter to p and rotate 360 degrees.
EXAM p,n	Call routine p, passing argument n.
MONI n	Branch to monitor entry point n.
EXIT	Return.

4.5 is also useful for testing capability such as equilibrium recovery before implementing at a lower level.

6. INTELLIGENCE LEVEL

The intelligence level controls the overall behavior of the robot by generating LP 4.5 programs for the Plan Level. These programs are then transferred to the Plan Level for execution. The Intelligence Level constructs the programs by selecting a strategy for the robot to follow to achieve a goal, for example moving the robot from room to room. Goals may be externally specified by a user, or may be generated as sub-goals during the solution of another goal. Each strategy contains a prototype LP 4.5 program segment which is instantiated with data corresponding to the robot's current environment. In this way, general plans for traversing rooms or following passages that are encoded as plan segments can be adapted to specific situations. Furthermore, several plan segments can be joined together to achieve more complicated goals for which a single strategy is not sufficient.

Intelligent strategy selection is the ability to choose an appropriate method to achieve the goal given the robot's current environment. The robot internally represents his environment in a *World Map*. This map is a hierarchical symbolic description of spaces, objects, and their connections. This map is incrementally constructed as the robot explores and encounters new situations. The robot's ability to act intelligently in his environment depends on the quality of the World Map. With a complete and current World Map, the robot can discriminate among competing strategies. With an incomplete or erroneous World Map, the robot cannot make appropriate selections.

In the following sections, the system structure will be defined and the issues of map construction and strategy selection will be examined in greater detail. To accomplish both these tasks the Intelligence Level uses a single structure, based on the Hearsay-II system [19]. In this structure multiple "experts" monitor the robot's receipt of data and interpretation of this data. When a particular expert recognizes an appropriate situation, this expert is executed and generates new information. This information may, in turn, cause other experts to run, each analyzing earlier information and possibly contributing new information to the robot.

The experts are arranged in several levels which correspond to different levels of description of the robot's environment and planning. This arrangement is parallel to the relationship of the Intelligence Level in the lower Plan, Gait and Leg Levels of the robot itself. In the robot's overall structure the Intelligence Level accepts information from the Plan Level and generates directions for the Plan Level to follow. Similarly, within the Intelligence Level structure, lower level experts generate information (i.e., interpretations) for higher level experts, which ultimately select a strategy.

6.1 System Structure within the Intelligence Level

The system consists of a data base of symbolic descriptions, a set of experts, and a scheduler. The data base contains descriptions of facts in the robot's environment. Each fact is encoded as a LISP list and stored into the data base. Individual facts, or sets of facts corresponding to a complex description can be located by pattern-directed retrieval (e.g., similar to the data bases of Conniver [20] and Planner [21]. For example, the fact that two rooms are connected by a passageway is represented as (CONNECTED ROOM-1 ROOM-2 PASSAGE-1) where ROOM-1, ROOM-2 and PASSAGE-1 are taken representing more complete description of those objects.

The data base of symbolic descriptions is divided into several levels each of which corresponds to a different level of descriptive abstraction. The lowest level, Sensor, contains descriptions of individual point sensor readings by the robot. Sensor level descriptions are aggregated together to form Segment level descriptions which correspond to contiguous boundaries. Segment descriptions are further combined to form Area descriptions, which correspond, for example, to corners or objects. All these descriptions are used by the Room level to describe the possible passable spaces for the robot. Currently, the Room level is the most abstract descriptive level in the World Map. Thus, the addition of a single new point's description at the sensor level may cause the propagation of new description to higher levels.

The actual transfer of information among levels is controlled by the experts. Each expert consists of a triggering pattern which describes the necessary conditions for the expert to be executed and a function body defining what the expert will do if it is triggered. These experts are arranged in four different levels which correspond to the levels of the data base and monitor the addition and modification of information to each specific level. When an expert is triggered, it is added to a list of runnable experts.

The scheduler controls which expert is to be executed by examining the set of runnable experts and selecting a single expert to be run. Currently, the scheduler uses a priority queue, in which experts are assigned a set priority based on the level of the data base which is monitored. Because lower level experts are given more priority than higher level experts, the system operates primarily in a data driven mode.

6.2 Map Construction

The robot constructs the World Map by receiving sensor data from the Plan Level and constructing an abstract interpretation of that data. Because only a single sonar sensor currently provides this information, many more lower level experts are needed to interpret this information. The sonar sensor generates a pair (Q, D) which describes the detection of an object at approximate distance D in sector Q.

The robot-centric information is translated to a global cartesian coordinate set and entered in the Sensor level of the data base. Experts at the level compare the new description with existing map information (for detecting false or previously false readings) and other neighboring sensor points. Neighboring points are joined to form primitive edges, whose descriptions are fed to the next higher Segment level. Some Segment level experts classify those edges to determine if they are part of existing segments or form the beginning of a new segment. Other segment level experts join two segment descriptions together to form a larger segment (in the case of co-linear segments), corners or points (in the case of common endpoints), or split a single segment into two segments (in the case of a new segment intersecting an existing one). Segments joined in corners are the basis for area descriptions, while segments joined in points can be part of objects or boundaries between rooms. The Room level experts record connectivity relationships between passable spaces (i.e., rooms and passages).

6.3 Strategy Selection and Plan Generation

Strategy selection is controlled by other experts assigned to levels higher even than the map construction experts. These experts react as *goal* descriptions are added to the data base. A goal description describes a set of conditions the robot must achieve. For example, the description (GOAL (LOC ROBOT ROOM-1)) is a request for the robot to move to ROOM-1 (if ROOM-1 is known) or an exploration to attempt to locate ROOM-1. In the former case, an expert which generates routes following the connectivity relationships of the Room level is triggered. The set of all possible (non-cyclic) routes to the goal location is generated and the route with minimum cost (currently shortest distance) is selected. In the later case, when ROOM-1 is an unknown location, an expert is triggered which adds to new goal (AND (LOC ROBOT ?ROOM) (GOAL (EXPLORE ?ROOM))). (This new goal can be translated as, "Explore the location where the robot currently is".) Explore experts look at the current state of the descriptions in the data base and select a candidate for further examination. For example, if the robot is in a room with three exits, two of which lead to known areas, the explore expert would indicate that the robot move through the unknown exit. Movement into an unknown area will augment the World Map (through the interpretation of new sensor data) and hopefully will add the missing information.

Thus, the robot first tries to directly achieve the specified goal, if possible. If the robot lacks sufficient information to achieve the goal, however, the robot suspends the achievement of the goal and enters an exploration phase in an attempt to augment its knowledge. It then returns to the suspended goal after a period of exploration and reattempts the goal.

Each strategy contains a set of actions for the robot to perform. These actions are represented as sequences of pseudo LP 4.5 instructions called prototype plans. Pseudo LP 4.5 instructions contain variable references which are replaced by actual values before being sent to the Plan Level. For example, the strategy for approaching a wall contains a variable for the distance to advance. After the distance is calculated from the World Map, the actual LP 4.5 instruction replaces the pseudo instruction. In this way, prototype plans can be shared among many strategies. The instantiated prototype plan is then transferred to the Plan Level for execution.

7. CONCLUSIONS

In this research, two prototype walking robots have been built and a walking robot language has been developed, LP 4.5. When the robot is equipped with plans that are controlled by the intelligence level, it can avoid obstacles, walk along walls, and escape from partial confinement. Future work will be focussed on further development of the walking robot language, automatic generation of such languages using artificial intelligence techniques, and testing prototype robots.

Currently, the intelligence level is interfaced with a simulation of the actual robot. Actions performed by the robot are reflected in an independent model of the world. Sensory information is entered as symbolic descriptions directly into the data base. The interface to the prototype robots is achieved. This prototype will allow better measurement of how the robot performs when the difficulties of noisy data and imprecise execution of actions are introduced.

ACKNOWLEDGMENTS

We thank for support the Centre National de la Recherche Scientifique (CNRS) and the Agence de l'Informatique, and the Franco-American Exchange Commission for allowing Dr. Richard Wood to teach and work at our Laboratory as a Fullbright "Junior Lecturer".

REFERENCES

[1] G. Giralt and R. Sobek, "A multi-level planning and navigation system for a mobile robot: a first approach to Hilare", *Proc. of 6th IJCAI,* Tokyo, 1979, p. 335.

[2] L. Marce, M. Jullière, H. Place, and H. Perichot, "A semiautonomous remote controlled mobile robot", *Industrial Robot, 7,* (4), 1980, 232.

[3] R. Chatila, "Système de Navigation pour un robot mobile autonome", *Thesis,* Toulouse, 1981.

[4] R. B. McGhee, C. G. Chao, V. C. Jaswa, and D. E. Orin, "Real time computer control of a hexapod vehicle", *Third Int. Symp. on Robots and Manipulators,* Udine, 1978, p. 323.

[5] D. E. Okhotsimsky, V. S. Gurfinkel, E. A. Devyanin, and A. K. Platonov, "Integrated walking robot", *Machine Intelligence 9,* J. Wiley and Sons, New York, 1979, p. 313.

[6] S. Hirose and Y. Umetani, "Some consideration of a feasible walking mechanism", *Third Int. Symp. on Robots and Manipulators,* Udine, 1978, p. 357.

[7] R. S. Mosher, "Exploring the potential of a quadruped", *Int. Automative Engineering Conf.,* SAE Paper n°690191, Detroit, 1969.

[8] P. Coiffet, "Les Robots", *Hermès, Paris* Vol. 1, 1981.

[9] H. Place, M. Jullière, and L. Marcé, "Qu'en est-il des Robots Mobiles", *Le Nouvel Automatisme,* 35, 1983, pp. 31-39.

[10] D. E. Orin, "Supervisory control of a multilegged robot", *Int. J. Robotics Research,* 1, 1, 1982, pp. 79-91.

[11] J. J. Kessis, J. P. Rambaut, and J. Penné, "Walking robot multi-level architecture and implementation", *Fourth Symp. on Theory and Practice of Robots and Manipulators, Romansy'81,* Proceedings, PWN, Warsaw, 1983, p. 297.

[12] J. J. Kessis, J. P. Rambaut, and J. Penné, "Un robot mobile hexapode - Architecture et langage de plans", *Etat de la Robotique en France,* Tome 1: Recherche, Hermès, Paris, 1982, p. 3.

[13] Muybridge, "Animals in motion", Dover, New York, 1957 (lst published, 1898).

[14] D. H. Wilson, "Insect walking", *Annual Review of Entomology, 11, 1966, pp. 103-121.*

[15] R. F. Chapman, "The insects, Structure and Function", The English University Press, London, 1972, p. 142.

[16] R. B. McGhee, "Finite state control of quadruped locomotion", *Simulation,* 1967, pp. 135-140.

[17] A. P. Bessonov and N. V. Umnov, "The analysis of gaits in six-legged vehicles according to their static stability", *First Symp. on Robots and Manipulators,* Udine, 1974, pp. 1-10.

[18] J. C. Latombe, in *Journées ARA-Pole Robotique Générale,* Poitiers, 1982.

[19] L. D. Erman, R. Hayes-Roth, V. R. Lesser and R. Reddy, "The Hearsay-II speech-understanding system: integrating knowledge to resolve uncertainty", *ACM Computing Surveys,* Vol. 12 (2), 1980, pp. 213-253.

[20] D. McDermott and G. J. Sussman, "The Conniver reference manual", *MIT A.I. Laboratory,* 1974, AI Memo 259a.

[21] G. J. Sussman, T. Winograd, and E. Charniak, "Micro-planner reference Manual", *MIT A.I. Laboratory,* 1971, A.I. Memo 203a.

PART III SENSORS

7

Three-Dimensional Robot Vision Techniques

CHARLES A. MC PHERSON
AND
ERNEST L. HALL

1. INTRODUCTION

Computer vision systems have greatly enhanced the application of computers to areas which require a knowledge of the external environment such as the field of robotics. Many robotics systems are trained to perform specific tasks and require no human supervision once the training is complete. However, the training process is sufficient to accomplish the task only if the task is well defined and invariant each time the task is performed. This requires that the environment be well controlled and synchronized with the movements of the robot manipulator. If, for example, a robot is trained to assemble a machine part, each piece must be positioned in the proper place so that the robot can locate it. Computer vision systems could eliminate this constraint by allowing the robot manipulator to locate each piece and identify them before assembly. This enhancement to robotics provides a feedback path to adjust a trained procedure in accordance with the environment and the robot manipulator becomes capable of performing more "intelligent" tasks.

Most early work in object location dealt with the interpretation of polygonal surfaces [1,2]. Objects of this type have distinguishable features which can be identified by segmentation methods [3,4]. These features may be used as reference points of the surface for object location [5]. However, featureless curved objects that have no unique features on the surface are difficult to measure from a two-dimensional image. Unfortunately, this class of objects encompasses many important man-made or natural structures. For example, a major task in radiotherapy treatment is to obtain a description of the surface of the patient's body so that a uniform dose can be delivered to a specific region.

Renner et al. [6] gives an excellent description of an interactive technique that uses a single image to measure the body contours and obtains the three-dimensional measurements through computer graphics. Parke [2] used a two-dimensional tablet to measure the three-dimensional position of points on a surface. Estimation of location and orientation of three-dimensional surfaces using a two-dimensional image is described by Cernuschi-Frias et al. [7]. The use of a robot system consisting of a vacuum cup to hold the object for presentation to a camera and computation of its orientation was developed by Kelley et al. at the University of Rhode Island [8].

Many different approaches provide the three-dimensional solution necessary for curved object location, recognition and manipulation. The region method of representing polyhedra used by Shirai et al. [9], incorporates an approach called "active imaging," which involves a projection system and a camera system. The projection system is used to impose features on the surface of the object, and the many resultant patterns can be used to obtain different types of images. The camera system acquires the object image with the imposed features and necessary calculations are done to interpret the image for analysis. Active imaging techniques are ideally suited for industrial applications. One advantage of the active approach is that it provides a high degree of control over the environment by providing special illumination or markers for accurate location. The accuracy of the object location derived from a projected grid system can be controlled by the grid spacing. "Stereo" imaging with a single image may appear difficult at first. However, if one considers the projected pattern as the first image and the received image as the second, a standard stereo solution results. The use of cooperative algorithms to determine the orientation from a single view is given by Woodham [10]. Using a single camera to recognize overlapping parts is described by Berman [11]. The use of geometric and relational reasoning to recognize a three-dimensional object from a single view is described by Mulgaonkar [12].

Successful developments in active imaging techniques include the Consight system developed by General Motors Technical Center [13,14] and the system developed by Albus [15] at the National Bureau of Standards. The Consight system uses a linear array camera and two projected light lines focused as one line on a belt. The system developed by Albus uses a plane of light to determine the position and orientation of parts on a table. The line projection technique works quite well for locating the edges of an object but requires scanning to locate object surface points.

The use of grid coding for image segmentation based on the spatial frequencies of the projected image is described by Will and Pennington [16], who traced the history of the grid patterns from the 1850s.

The computational techniques associated with a recently developed laser shutter/space encoding system are described by Posdamer and Altschuler [17]. This system employs time and space coding of dots to measure surface coordinates. Paul [18] gives an introduction to robot manipulators and their computer control.

The human visual system receives two perspective views of objects from which distance or depth information is derived [19]. Similarly, the computer vision system is capable of obtaining depth information from two cameras viewing an object from two different perspectives. The human visual system also has the ability to detect depth changes from shading information. Depth information can also be derived from images obtained by a computer vision system through an analysis of the shading of the surface along with a knowledge of lighting conditions. A unique surface description exists in the form of a mathematical model if a set of sample surface points, for which their absolute or relative three-dimensional coordinates are known, satisfy the Nyquist Criterion. This model may be obtained, given a format for the model, using numerical or analytic techniques of least-squares curve fitting. Furthermore, once a model is obtained, the mathematical model may be decomposed into a qualitative description of the surface [20].

The study of reflectance and shading is important in a variety of fields [21]. In radar systems, the description of the power of the return signal as a function of the incidence angle between the surface normal and the transmitted signal vector has been recognized as an important step in predicting radar system performance [22]. Since a radar signal is coherent, the phase of the reflected signal is of great concern in radar systems. This further complication is not presented under normal lighting because the light is noncoherent or randomly phased. The noncoherent lighting condition is the case presented in this chapter.

A concern for characterizing the reflection of light on surfaces having different orientations exists in computer graphics for image synthesis. Numerous computer aided design (CAD) systems, such as those marketed by Computer Vision Corporation and Applicon Systems, have the capability of simulating the appearance of a mechanical part or architectural structure through graphical reconstruction. In order for these reconstructions to appear natural, an understanding of the reflection characteristics of surfaces is required. Numerous publications on surface shading for image synthesis exist and suggest numerous techniques according to the type of surface being viewed [23,24,25,26,27].

An accurate representation of arbitrary surfaces is impossible on existing computer systems because of the extreme complexity of even the simplest object. An interesting argument related to scene representation has been described by Carl Sagan in his book, *Broca's Brain* [28]. Sagan states that the storage capacity of neither the largest computers nor the human brain is sufficient to describe even a barely visible grain of salt since an accurate description must be made on the molecular level. The choice of a method for describing arbitrary surfaces should be based upon whether the description fits a particular need.

Polygonal surfaces are among the simplest to describe as is evidenced by the simple techniques used in their reconstruction on graphical displays. The popular computer program, MOVIE.BYU, developed by Dr. Hank Christenson and Dr. Mike Stevenson at Brigham-Young University for the reconstruction of polygonal surfaces on graphical display devices, used a vertex list and a vertex connection table to describe a surface. This representation provides sufficient information to reconstruct the projection of the surface onto a plane positioned anywhere in space and oriented in any way relative to a global coordinate system. Other data structures for describing polygonal surfaces are presented by Giloi [29] which also contains node or vertex locations with a variety of schemes for describing the connection of points in graphical reproduction. However, similar techniques cannot be applied to curved surface representation since there exist no vertices, edges or sides on a curved surface such as a sphere.

Polygonal surfaces are constructed of multiple intersecting planes, each of which may be described by a linear equation. The line defined by the intersection of two or more planes is considered an edge and a vertex is the point of intersection of two or more edges. The edges of a polygonal object are discontinuities in the surface description. Since an object of this type is constructed of bounded planes which form the polygons, the edges are line segments at which one element of the surface ends and another begins. Curved surfaces may also have edges formed by discontinuities if the surface is formed by multiple bounded curved surfaces where each of the elemental curved surfaces describes only a bounded region of the entire surface. Unlike the polygonal surfaces, where bounded planes are describable by first order equations, each of the elements are bounded curved surfaces which may be described by higher order equations. With no restriction on the order of the equations required to describe a general surface, the complexity of the mathematical model could approach that of the molecular level description. Again, the search for an accurate description of a general surface becomes futile. Therefore, the methods employed in the representation of arbitrary curved surfaces must rely on approximation techniques. Many man-made objects and structures can often be described by low order equations for which a mathematical model can be derived. In such cases, a mathematical model would be the preferred method of representing the surface, since a single equation may be used to obtain any point on the surface. Numerous methods of describing curved surfaces, both of mathematical models and by approximations, are presented in this chapter.

The data generated by the processes discussed in the following chapters is a set of discrete surface points of the object. If the data obtained through these methods do not cover the entire surface, extrapolation and interpolation techniques must be used to estimate the size and shape from the available data. Methods of constructing mathematical models from discrete data techniques are also presented in this chapter.

2. SURFACE MEASUREMENT USING SHADING INFORMATION

A surface recognition technique requires a sufficient sample of surface points, as required by the sampling theorem, for adequate surface description. Stereo vision has long been accepted as a valid method of surface measurement; however, since the stereo vision technique requires the identification of corresponding points in the two images, this method is difficult to apply to curved surfaces. The vertices of polygonal solids are the features most commonly used to match two points in a pair of images. Curved surfaces may not contain vertices and still be closed surfaces. If a featureless curved surface, such as a sphere, is viewed from two different positions, the two images may appear identical under uniform lighting. One solution of curved surface measurement is suggested by B. K. P. Horn [21] which utilizes the characteristics of the change in surface reflectivity as the illumination source is moved from one position to another.

The models that are applied to image synthesis suggest the ability to recognize surface shapes from the shading information. The next section concentrates on surface shading from the point of view of image synthesis and the models applied to various surface types. Then the application of these shading models to the problem of measuring surface shape is considered.

2.1 Surface Shading

An important feature of an image synthesis system is surface shading prediction. The process of surface shading involves the assignment of an intensity or color to every picture element in the image which accurately simulates the viewing situation. The shading of a surface point depends upon the surface reflection characteristics, the surface geometry and the lighting conditions. Each of these properties must be considered in the development of a surface shading model.

One important property in describing the contribution of the surface geometry to surface shading is the surface normal vector at the point of interest. This normal vector is identically the gradient of the surface at the point, or

$$\boldsymbol{n} = \frac{\partial f}{\partial x}\boldsymbol{i} + \frac{\partial f}{\partial y}\boldsymbol{j} + \frac{\partial f}{\partial z}\boldsymbol{k} \tag{1}$$

where $\boldsymbol{i}$, $\boldsymbol{j}$, and $\boldsymbol{k}$ are the unit vectors in three-space. If the surface is a plane represented by the equation

$$f(x, y, z) = ax + by + cz + d = 0, \tag{2}$$

the normal vector to the surface at any point is

$$\boldsymbol{n} = a\boldsymbol{i} + b\boldsymbol{j} + c\boldsymbol{k}.$$

For curved surfaces, however, the normal vector is dependent upon the surface point that is considered and is not constant over the surface.

A second characteristic of surfaces which must be considered is the reflection property corresponding to a given surface type. These surface properties include spectral reflectance, which determines how the surface reflects light of specific wavelengths, the surface texture, which determines the diffusivity and specularity components of reflection, and the surface transparency, which determines the amount of light that is refracted by the surface rather than reflected. These characteristics of surfaces identify the shading model required for proper surface simulation.

One model for simulating an ideal diffuse reflector is based upon Lambert's law [23], which states that a surface will diffuse incident light equally to all directions. For a Lambertian surface, the quantity of reflected light is proportional to the amount of incident light that is intercepted by the surface. The intensity of a surface point, I, may, therefore, be modeled by the cosine of the angle between the normal vector, $\boldsymbol{n}$, to the surface point and the light source vector, $\boldsymbol{\ell}$, as shown in Figure 1(a). The cosine of this angle, θ, may be computed by the inner product of the two vectors. If the cosine is negative, the surface point is hidden from the light source and the intensity is zero. This is a condition known as "self shadowing." The Lambertian model is, therefore,

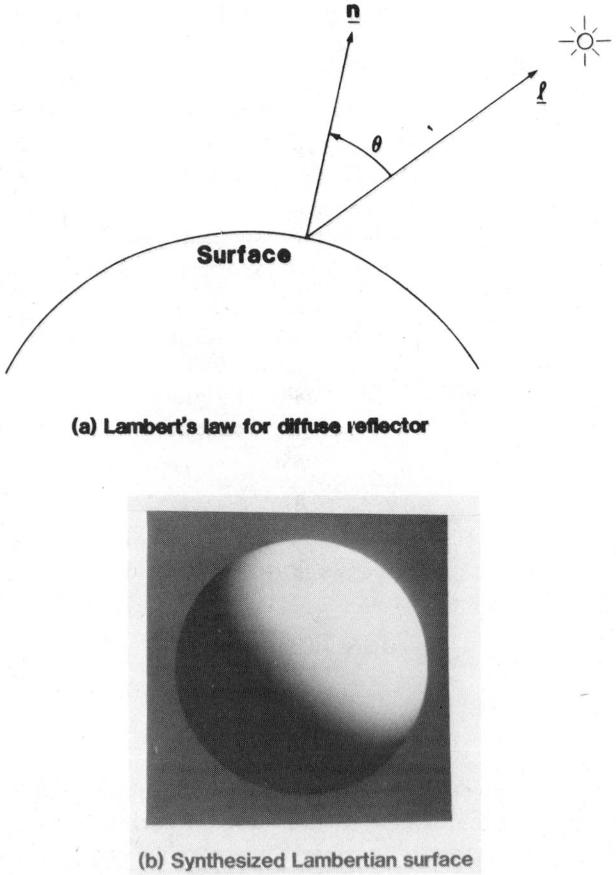

Figure 1 The Lambertian reflection model. a. Lamber's law for diffuse reflector. b. Synthesized Lambertian surface.

$$I = \begin{cases} \cos\theta, & -\pi/2 < \theta < \pi/2 \\ 0, & \text{otherwise.} \end{cases} \tag{3}$$

Figure 1(b) is a synthesized image of a sphere which is modeled by the Lambertian reflection model.

A model for simulating the reflection characteristics of a specular surface was introduced by Phong [24] which suggests that more light is reflected in the direction forming an equal but opposite angle to the normal vector from the light source vector as shown in Figure 2(a). This reflection vector may be determined by solving

$$\boldsymbol{r} = \sqrt{2 + 2\cos(2\theta)}\,\boldsymbol{n} - \boldsymbol{\ell} \tag{4}$$

where θ is the angle between $\boldsymbol{\ell}$ and $\boldsymbol{n}$ and the light source vector, $\boldsymbol{\ell}$, and the normal vector, $\boldsymbol{n}$ are normalized. Unlike an ideal Lambertian surface, the intensity of a point on a specular surface is dependent on the viewing geometry. For a mirror type surface the intensity, as seen by the camera, would be zero everywhere except for the points which reflect the light exactly in the direction

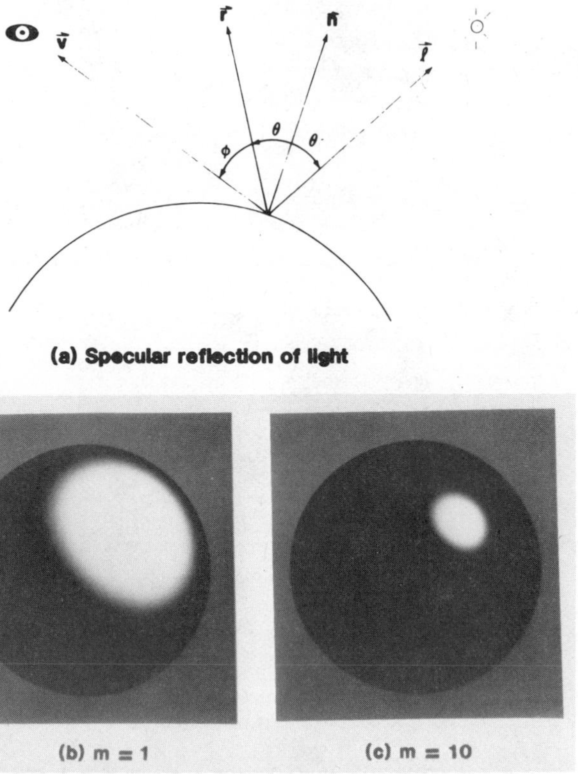

Figure 2 The specular reflection model and synthesized specular surfaces. a. Specular reflection of light. b. m = 1 c. m = 10

of the camera. For a duller surface, the intensity falls off as the angle between the viewing vector and reflection vector increases. The relationship between the intensity and the angle ϕ, between the view vector, $\boldsymbol{v}$, and $\boldsymbol{r}$, may be modeled by

$$I = \begin{cases} \cos^m \phi, & -\pi/2 < \phi < \pi/2 \\ 0, & \text{otherwise} \end{cases} \tag{5}$$

where the coefficient m is a measure of the shininess of the surface. A mirrorlike surface would correspond to a large value of m (infinite in the ideal case) and a duller surface would exhibit a value near unity. Typical values for such surfaces as nickel would be between 10 and 100. Figures 2(b) and 2(c)

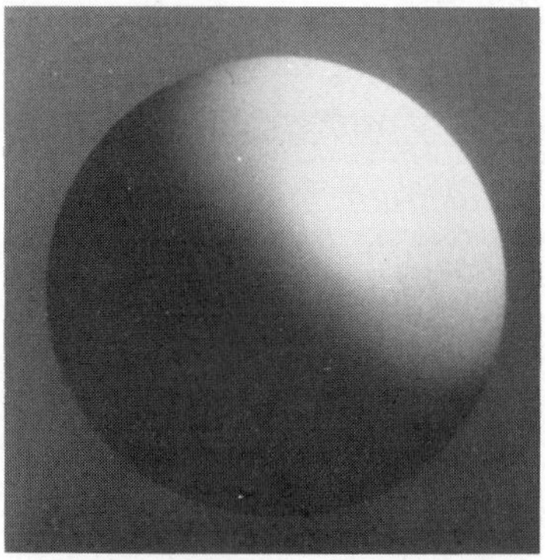

(a) A primarily diffuse reflector with a = 0.7, b = 0.3, c = 0.0 and m = 10

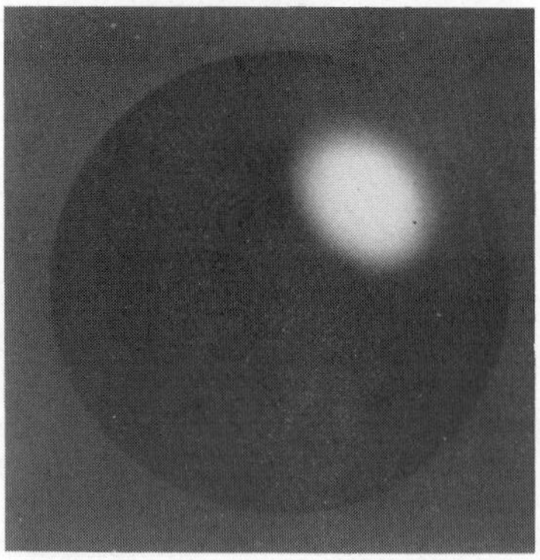

(b) A primarily specular reflector with a = 0.4, b = 0.6, c = 0.0 and m = 10

Figure 3 Synthesized images of a sphere using both specular and diffuse components of shading. a. A primarily diffuse reflector with $a = 0.7$, $b = 0.3$, $c = 0.0$, *and* $m = 10$. *b. A primarily specular reflector with* $a = 0.4$, $b = 0.6$, $c = 0.0$, *and* $m = 10$.

are synthesized images of a sphere using the specular model for specular reflectors with $m = 1$ and $m = 10$ respectively. Other models for specular reflectors, such as the Torrance-Sparrow model [25], are also widely used.

Most real surfaces are neither ideal diffuse reflectors nor ideal specular reflectors [26]. A model which combines the effects of diffuse and specular reflectance as well as ambient light produces synthesized images which appear realistic for certain types of materials. The ambient component is light which is assumed to be uniformly distributed over the surface. The combination of these components yields the reflection model.

$$I = a \cos(\theta) + b \cos^m(\phi) + c \tag{6}$$

by Equations 3 and 5. The terms, a, b, and m, depend upon the materials and surface roughness and the received intensity depends upon these parameters as well as the surface and viewing geometries. If the intensity, I, is defined to be in a range of zero to one, then

$$a + b + c = 1. \tag{7}$$

The specular component represents light that is reflected from the surface of the material. The diffuse component characterizes internal scattering of light that penetrates beneath the surface of the material [26]. Most metals, such as copper or bronze, exhibit a large specular component ($b = 0.7$ to $b = 1.0$) and a small diffuse component ($a = 0.3$ to $a = 0.0$). Rough non-metallic surfaces, such as carbon or rubber, are characterized by a large diffuse component and a small specular component. The synthetic images of Figure 3 were produced using the model of Equation 6 for a primarily diffuse reflector and a primarily specular reflector.

Numerous other models consisting of both diffuse and specular components have been proposed by Blinn [23], Whitted [27], Cook and Torrance [26], and others which utilize different models of specular or diffuse reflectors. These models have achieved reasonable success in the synthesis of various surface materials. This suggests that such models are much more applicable to the representation of surface reflectivity of natural surfaces than models consisting of only the diffuse type or the specular type. The three models described by Equations 3, 5, and 6 will be utilized in the following section to obtain three-dimensional information that represents the surface shape by examing the observed intensity of a set of surface points.

2.2 Surface Shape Measurement from Shading

Since the shading of surfaces in image synthesis is modeled by functions of the surface geometry, it is apparent that the surface geometry could be derived from the shading information. Horn [21] suggests that the intensity of a surface point identifies the solution space for the normal vector to the surface at that point if the viewing geometry and lighting conditions are known. The process of determining the surface normal at a point consists of performing an inverse process of image synthesis. This process was applied to polyhedral objects by Macworth [30] and was shown to yield good results for surfaces in which the reflection characteristics are known. This section utilizes the surface shading models discussed in Section 2.1 to obtain the surface geometry of an object from the shading of the surface in an image.

The Lambertian model for diffuse surfaces, Equation 3, describes the intensity of a surface point in the image in relationship with the quantity of light per unit area received by the surface at that point. This relationship is

$$I = \cos(\theta), \frac{-\pi}{2} < \theta < \frac{\pi}{2} \tag{8}$$

where θ is the angle between the light source vector and the normal vector to the surface at that point. Using a two-space example shown in Figure 4(a), the direction of the normal vector is defined as the angle β and the direction of the light source vector is defined as the angle α. With these definitions,

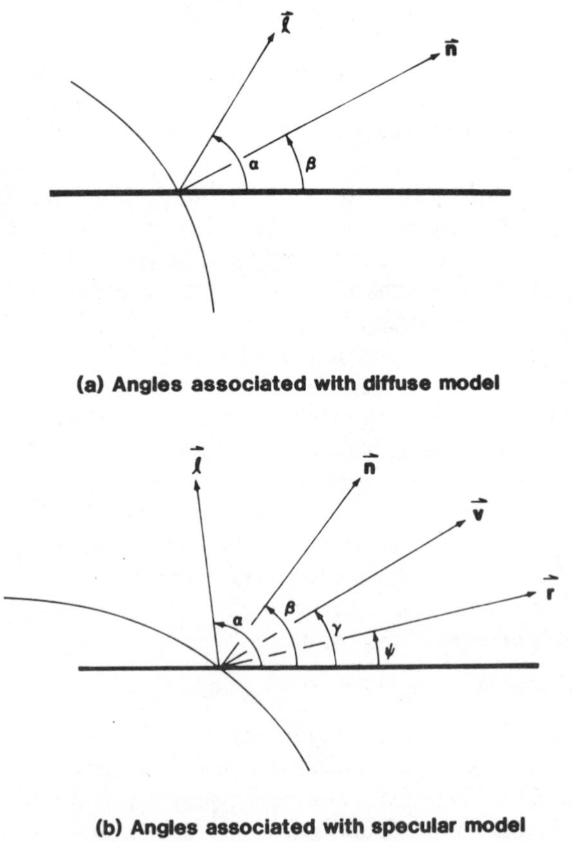

(a) Angles associated with diffuse model

(b) Angles associated with specular model

Figure 4 Definition of angles used in measuring surface shape from shading. a. Angles associated with diffuse model. b. Angles associated with specular model.

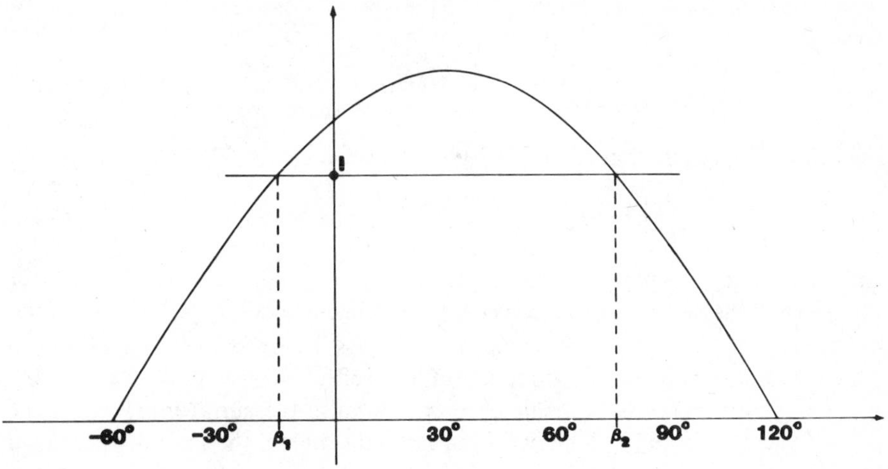

Figure 5 Example showing the two solutions, β_1 and β_2, given an intensity for a diffuse reflector ($\alpha = 30°$).

Equation 8 may be rewritten as

$$I = \cos(\alpha - \beta). \tag{9}$$

For any intensity, I, there exist two solutions for the angle θ and, if the direction of the light source, α, is known, two solutions for the direction of the normal vector, β, exist as shown in Figure 5. If a second image is obtained with the light source in another position, two solutions will again be obtained for β. However, if the two solutions $\alpha_1 \neq \alpha_2$, the true solution of the direction of the normal vector is the solution that is obtained in both cases.

The specular model discussed in the previous section models the surface shading as a function of the reflection vector, or

$$I = \cos^m(\phi), \quad \frac{-\pi}{2} < \phi < \frac{\pi}{2}, \tag{10}$$

where ϕ is the angle between the view vector, $\boldsymbol{v}$, and the reflection vector, $\boldsymbol{r}$, as shown in Figure 2. If the direction of the view vector is represented by the angle γ and the direction of the reflection vector by the angle ψ, Equation 10 becomes

$$I = \cos^m(\psi - \gamma). \tag{11}$$

Since the direction of the normal vector, β is half-way between the reflection vector, ψ, and the light source vector, α,

$$\beta = \frac{1}{2}(\psi + \alpha)$$

or

$$\psi = 2\beta - \alpha. \tag{12}$$

Using the relationship of Equation 12, the specular model of surface shading becomes

$$I = \cos^m(2\beta - \alpha - \gamma). \tag{13}$$

Equation 13 suggests that the intensity is positive for

$$\frac{1}{2}(\alpha + \gamma) - \frac{\pi}{4} < \beta < \frac{1}{2}(\alpha + \gamma) + \frac{\pi}{4} \tag{14a}$$

and is zero everywhere outside of this range. Note that the function of Equation 13 has two positive peaks in the range $-\pi \leqslant \beta \leqslant \pi$, unlike the diffuse model in Equation 9. One peak occurs half-way between α and γ which was shown to be the direction of the reflection vector, ψ. The second positive peak occurs at the angle $\psi - \pi$. Because the surface, for which the normal vector is $\psi - \pi$, is hidden from the light source, the intensity is zero in the range

$$-\frac{1}{2}(\alpha + \gamma) - \frac{\pi}{4} < \beta < -\frac{1}{2}(\alpha + \gamma) + \frac{\pi}{4}. \tag{14b}$$

This condition yields, like the diffuse shading model, two possible solutions for the direction of the normal vector, β, given an intensity of the point in two images with the light source in two different positions for the two-space geometry.

For the specular model and the diffuse model, the direction of the normal vector may be determined analytically. Using the diffuse shading model of Equation 9, the normal vector direction is

$$\beta = \alpha \pm \cos^{-1}(I). \tag{15}$$

Similarly, the specular shading model may be solved to yield

$$\beta = \frac{1}{2}[\alpha + \gamma \pm \cos^{-1}(I^{1/m})]. \tag{16}$$

As an example, consider the two-dimensional imaging situation where an image pixel and the image center form a vector

$$\boldsymbol{v} = 0.866\,\boldsymbol{i} + 0.5\,\boldsymbol{j} \tag{17}$$

or $\gamma = 30°$. The light source is illuminating the surface from the direction of

$$\boldsymbol{\ell} = -0.259\,\boldsymbol{i} + 0.966\,\boldsymbol{j} \tag{18}$$

for which $\alpha = 105°$. The intensity of the surface point is 0.841 as seen by the pixel. The surface material is known to have a coefficient of specularity, $m = 5$. Using Equation 15,

$$\beta = \frac{1}{2}[105° + 30° \pm \cos^{-1}(0.841^{1/5})]$$

$$\beta = \frac{1}{2}[135° \pm \cos^{-1}(0.966)] \tag{19}$$

which yields the two solutions $\beta_1 = 75°$. or $\beta_2 = 60°$. If the light source is then moved to

$$\boldsymbol{\ell} = -0.5\, \boldsymbol{i} + 0.866\, \boldsymbol{j}, \tag{20}$$

$\alpha = 120°$, the intensity of the pixel becomes 0.487. Again, Equation 15 is applied to yield the solutions $\beta_1 = 90°$ and $\beta_2 = 60°$. Because $\beta = 60°$ is a solution to the normal vector in both cases, it is the true solution.

An analytical solution does not exist in the case of the combined surface shading model. The composite shading model of Equation 6 may be expressed as

$$I = a \cos(\alpha - \beta) + b \cos^m(2\beta - \alpha - \gamma) + c \tag{21}$$

by incorporating Equations 9 and 13 for the diffuse and specular components of the shading model. Like the diffuse model and the specular model, there exist two possible solutions for the normal vector given an intensity for Equation 21. The normal vector may be determined using numerical techniques alone. Newton's method for solving non-linear equations is one accepted technique that may be applied (31). In order to obtain both solutions for the normal vector, Newton's method must be applied twice using two initial values that would force convergence to each solution. If the two solutions are sufficiently separated, one method that may be applied is to obtain a coarse estimate of each solution for use as the initial estimate for Newton's method.

The graph of Figure 6 shows a two-space imaging situation for Equation 21 as a function of β. In this example the view vector direction, γ, is 0° and the light source direction, α, is 60°. The coefficients are $a = 0.5$, $b = 0.5$, $c = 0.0$ and $m = 1$. Given an intensity, the solutions β_1 and β_2 to Equation 21 are an opposite sides of the maximum. The maximum can be found by solving

$$0 = a \sin(\alpha - \beta) - 2mb \cos^{m-1}(2\beta - \alpha - \gamma) \sin(2\beta - \alpha - \gamma) \tag{22}$$

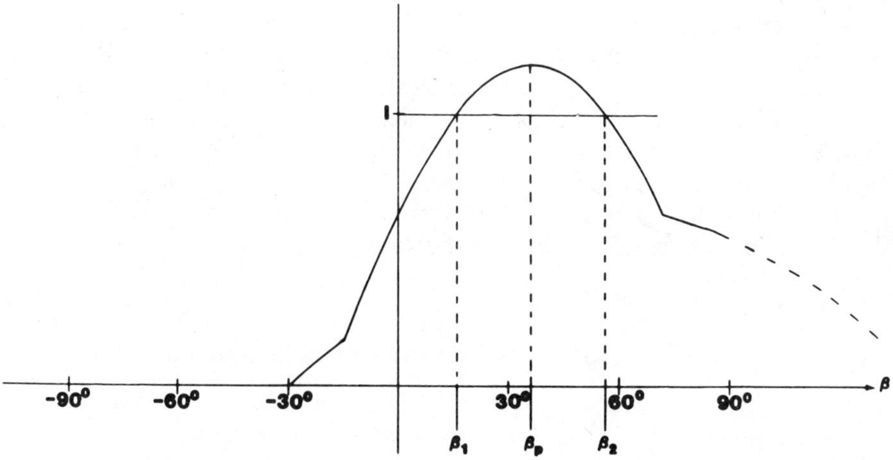

Figure 6 Graph of Equation 21 for sample parameters.

for β. Because there exist two peaks and two minima to Equation 21, a coarse estimate of the desired maximum in the range $(\alpha - \pi/2) \leqslant \beta \leqslant (\alpha + \pi/2)$ must be obtained as an initial estimate for Newton's method. This coarse estimate may be obtained by searching the value of Equation 21 at intervals of β for a maximum. Using this value of β as an initial value, Newton's method may be applied to Equation 22 to refine this estimate and obtain β_p. Since each of the two solutions β_1 and β_2 to Equation 21 are on opposite sides of β_p, another coarse search may be applied to Equation 21 for the given intensity using β_p as a limit to the search space for each solution. Newton's method may then be applied to obtain the more accurate approximation for the solutions.

The reflection model which combines the specular and diffuse model is more applicable to the recognition of "real" surfaces as is evidenced by the realistic appearance of synthetic images of objects for which this model is applied [26]. The increased complexity of the shading model is reflected in the computation required for surface measurement. The increased computation time over that of simpler shading models is, however, warranted since the accuracy of the measurement is substantially improved.

2.3 Summary

It has been demonstrated that the models used in image synthesis for the shading of the surface of an object can be applied to image analysis to obtain the surface shape from the shading information. However, the ability to perform this task with sufficient accuracy depends upon the ability to select an appropriate shading model that closely approximates the reflection characteristics of the surface material. The problem of selecting an appropriate shading model is a topic which requires continued research. Present techniques of calibration are difficult to apply to automated systems since a new set of calibration parameters must be obtained for every surface that is to be examined.

If an appropriate model of the surface shading is obtained, a measurement of the surface normal may be obtained for every pixel of the image that views the object. This is advantageous since this technique of surface measurement would not require a high resolution image in order to obtain a sufficient number of sample points for surface modeling. A limitation of the analysis of shading to obtain depth perception exists, however, in that this procedure alone can only yield an objects shape. Other types of analysis and procedures must be applied in order to derive absolute size, orientation, and location.

3. RECOGNITION BY STEREO IMAGING

In the previous section, a method of curved surface measurement was discussed for which image grey-levels were employed. This method does, however, require that certain luminous characteristics of the surface material be known.

These characteristics are often difficult to obtain through experiment. The problem is further complicated if these luminous characteristics vary over the surface. For this reason, a method of surface measurement may be required which has no dependence upon the characteristics of the surface reflectance. The method described in this section is based upon the stereo vision system approach which utilizes coincident surface points from two views of the surface as a means of surface measurement.

The method of triangulation is widely used in computing the location of a three-dimensional object from a pair of corresponding points in two images. Any point in one image defines a line in space, and the corresponding point in another image defines a second line in space. The method of triangulation involves determining the intersection of the two lines using geometric and trigonometric relationships. The point in space where the two lines intersect is the three-dimensional location of that feature point of the object. The following section discusses the technique of triangulation as a means of locating object points of polygonal surfaces. The expansion of this technique into the realm of curved surfaces is presented in Section 3.2. A final discussion of three-dimensional curved surface measurement using the stereo vision principles with only a single view of the surface is presented in Section 3.3.

3.1 The Camera Model

The task of three-dimensional object location and surface measurement involves a modeling of the imaging process and an inverse transformation (32) to determine the line equations and search for the corresponding image points. These points are required for computing the three-dimensional object points using the triangulation method which describes the scene. The general steps required for the three-dimensional object location using stereo images are:

1. Modeling the cameras.
2. Calibrating the cameras.
3. Computing the perspective transformation using camera location and orientation information along with calibration results.
4. Pairing the corresponding points of the two images.
5. Computing the three-dimensional object surface points corresponding to the image point pairs using the camera model parameters and the imaging geometry.

Camera models may vary in complexity from the simple pin-hole model to such models which include the addition of depth of field due to finite aperture camera, spatial aberration, geometric distortions, spectral aberration, or other non-ideal characteristics of physical lenses [33]. The pin-hole camera model is widely accepted as a reasonable approximation of a true camera system and, due to its simple mathematical representation, is used to characterize the imaging process in this chapter. The geometry of the pin-hole camera model is illustrated in Figure 7.

The perspective projection transformation in three space contains the components of projection, perceptive, rotation and translation transformations.

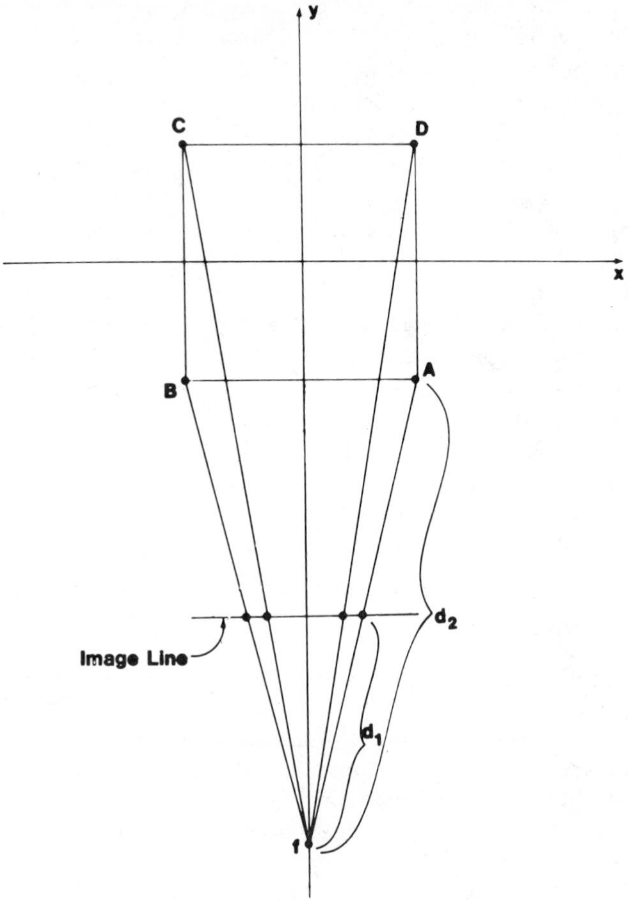

Figure 7 Two-dimensional perspective projection.

The projection transformation matrix has the form and is a model of the film plane,

$$A_p = \begin{bmatrix} 1 & 0 & 0 & 0 \\ 0 & 1 & 0 & 0 \\ 0 & 0 & 0 & 0 \\ 0 & 0 & 0 & 1 \end{bmatrix}, \tag{23}$$

which removes the third dimension information from the image. The perspective transformation is a model of the camera lens and is of the form

$$A_f = \begin{bmatrix} 1 & 0 & 0 & 0 \\ 0 & 1 & 0 & 0 \\ 0 & 0 & 1 & 0 \\ 0 & 0 & \dfrac{1}{f} & 1 \end{bmatrix}. \tag{24}$$

Similarly, the translation transformation matrix has the form

$$A_t = \begin{bmatrix} 1 & 0 & 0 & -N_x \\ 0 & 1 & 0 & -N_y \\ 0 & 0 & 1 & -N_z \\ 0 & 0 & 0 & 1 \end{bmatrix}. \tag{25}$$

The rotation transformation is composed of three rotation transformation matrices in the three Euler angles. The three rotation transformations are

$$A_\alpha = \begin{bmatrix} \cos\alpha & \sin\alpha & 0 & 0 \\ -\sin\alpha & \cos\alpha & 0 & 0 \\ 0 & 0 & 1 & 0 \\ 0 & 0 & 0 & 1 \end{bmatrix}, \tag{26}$$

$$A_\theta = \begin{bmatrix} \cos\theta & 0 & -\sin\theta & 0 \\ 0 & 1 & 0 & 0 \\ \sin\theta & 0 & \cos\theta & 0 \\ 0 & 0 & 0 & 1 \end{bmatrix}, \tag{27}$$

$$A_\beta = \begin{bmatrix} 1 & 0 & 0 & 0 \\ 0 & \cos\beta & \sin\beta & 0 \\ 0 & -\sin\beta & \cos\beta & 0 \\ 0 & 0 & 0 & 1 \end{bmatrix}, \tag{28}$$

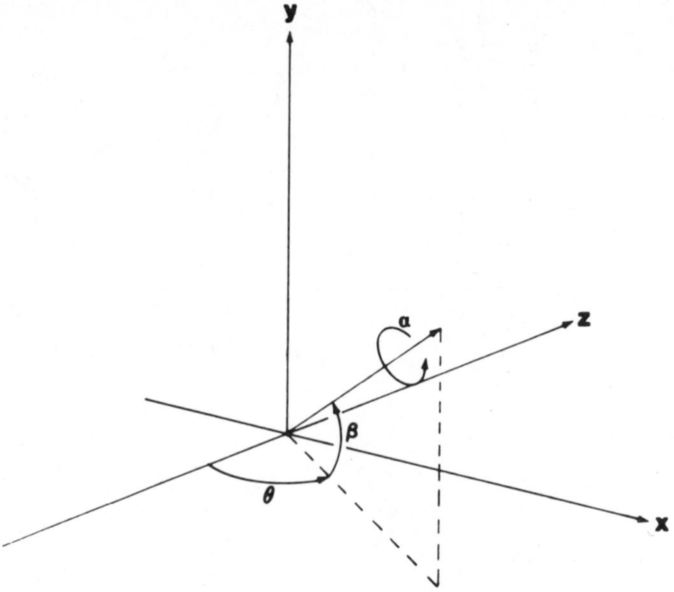

Figure 8 Euler angles in three-space.

with the angles defined as shown in Figure 8. The rotation transformation in three space is the product of the three Euler angle transformations, or

$$A_r = A_\alpha A_\theta A_\beta,$$

$$A_r = \begin{bmatrix} r_{11} & r_{12} & r_{13} & 0 \\ r_{21} & r_{22} & r_{23} & 0 \\ r_{31} & r_{32} & r_{33} & 0 \\ 0 & 0 & 0 & 1 \end{bmatrix}, \tag{29}$$

where
$$\begin{aligned}
r_{11} &= \cos\alpha\cos\theta\sin\alpha\sin\beta\sin\theta + \cos\alpha\cos\theta \\
r_{12} &= -\sin\alpha\cos\beta - \sin\theta\sin\beta\cos\alpha\sin\alpha\cos\beta \\
r_{13} &= -\sin\alpha\sin\beta + \cos\theta\sin\theta\cos\beta\sin\alpha\sin\beta\cos\theta \\
r_{12} &= -\cos\alpha\sin\theta \\
r_{21} &= -\sin\alpha\cos\theta\cos\alpha\sin\beta\sin\theta - \sin\alpha\cos\theta \\
r_{22} &= -\cos\alpha\cos\beta - \sin\alpha\sin\theta\sin\beta\cos\alpha\cos\beta \\
r_{23} &= -\cos\alpha\sin\beta - \sin\alpha\sin\theta\cos\beta\sin\alpha\sin\theta \\
&\quad + \cos\alpha\sin\beta\cos\theta \\
r_{31} &= \sin\theta\cos\beta\sin\theta \\
r_{32} &= \cos\theta\sin\beta - \sin\beta \\
r_{33} &= -\cos\theta\cos\beta\cos\beta\cos\theta
\end{aligned}$$

Using this definition of the tri-angle rotation transformation matrix the perspective projection transformation matrix is

$$A = A_p A_f A_r A_t. \tag{30}$$

From Equation 30 the perspective projection transformation, as derived from the camera model parameters, is

$$A = \begin{bmatrix} r_{11} & r_{12} & r_{13} & (-N_x r_{11} - N_y r_{12} - N_z r_{13}) \\ r_{21} & r_{22} & r_{23} & (-N_x r_{21} - N_y r_{22} - N_z r_{23}) \\ r_{31}/f & r_{32}/f & r_{33}/f & 1+(-N_x r_{31} - N_y r_{32} - N_z r_{33})f \end{bmatrix}. \tag{31}$$

The method of camera calibration in the three space model follows a similar procedure as in the two space example. The perspective projection transformation equation yields the three equations

$$wx' = a_{11}x + a_{12}y + a_{13}z + a_{14}, \tag{32}$$

$$wy' = a_{21}x + a_{22}y + a_{23}z + a_{24}, \tag{33}$$

$$w = a_{31}x + a_{32}y + a_{33}z + a_{34}. \tag{34}$$

The substitution of Equation 34 for the term w in Equations 32 and 33 gives the pair of equations

$$a_{11}x + a_{12}y + a_{13}z + a_{14} - a_{31}xx' - a_{32}yx' - a_{33}zx' - a_{34}x' = 0, \tag{35}$$

$$a_{31}x + a_{22}y + a_{23}z + a_{24} - a_{31}xy' - a_{32}yy' - a_{33}zy' - a_{34}y' = 0, \tag{36}$$

where the scaling term, w, is eliminated. If the coefficients of the perspective projection transformation matrix are arranged in a column vector, B, where

$$B = [a_{11}\ a_{12}\ a_{13}\ a_{14}\ a_{21}\ a_{22}\ a_{23}\ a_{24}\ a_{31}\ a_{32}\ a_{33}\ a_{34}]', \tag{37}$$

then it is evident that the coefficients may be determined by the equation

$$QB = 0, \tag{38}$$

where the row vectors of the matrix, Q, are

$$q_i = [0\ 0\ 0\ 0\ x_k\ y_k\ z_k\ 1\ x_k y_k'\ y_k y_k'\ z_k y_k'\ y_k'] \tag{39}$$

and

$$q_{i-1} = [x_k\ y_k\ z_k\ 1\ 0\ 0\ 0\ 0\ x_k x_k'\ y_k x_k'\ z_k x_k'\ x_k'] \tag{40}$$

for i even and $k = i/2$. Given six known surface points and their corresponding image points, the perspective projection transformation matrix can be derived using Equation 31. Since Equation 38 is homogeneous, one of the coefficients may be chosen arbitrarily. If a_{34} is chosen to have a value of unity and the last column of Q is moved to the right-hand side of Equation 38, the solution may be found by the least squares regression method.

Whether the perspective projection transformation matrix for the camera model is found through a knowledge of the camera parameters or by the camera calibration method, the resulting transformation describes the imaging geometry of the camera system. From the transformation the projection of any point in space onto the image plane may be determined. Using two cameras having known geometries provides the corresponding inverse transformation for three-dimensional surface measurement.

3.2 Stereo Vision Principles

Stereo vision has long been accepted as a valid technique of three-dimensional data acquisition and surface description. For centuries, engineers and architects have used various views of an object or structure to describe it in three-dimensional space. The most notable example of the use of stereo vision for three-dimensional reconstruction or surface measurement is the human visual system. This section describes the principles of stereo vision for three-dimensional data acquisition.

The technique of stereo vision for surface measurement is based on the triangulation principle. A point in an image defines a line in space along which the corresponding surface must lie. The corresponding point in a second image also defines another line in space along which the surface point must lie. Given two images from cameras having known geometries, the surface point corresponding to the image points in the two views is at the intersection of the two defined lines in space as illustrated in Figure 9 for a two-dimensional situation.

If the perspective projection transformation matrix for the two images are known, a set of simultaneous equations is defined for each surface point. If one image in a two space geometry has the transformation of surface points to image points.

$$\begin{bmatrix} wx' \\ w \end{bmatrix} = \begin{bmatrix} a_{11} & a_{12} & a_{13} \\ a_{21} & a_{22} & a_{23} \end{bmatrix} \begin{bmatrix} x \\ y \\ 1 \end{bmatrix} \tag{41}$$

in a two space geometry and the transformation for the second image is

$$\begin{bmatrix} wx'' \\ w \end{bmatrix} = \begin{bmatrix} b_{11} & b_{12} & b_{13} \\ b_{21} & b_{22} & b_{23} \end{bmatrix} \begin{bmatrix} x \\ y \\ 1 \end{bmatrix}, \tag{42}$$

a single solution exists and may be derived. Equation 41 defines the equations

$$wx' = a_{11}x + a_{12}y + a_{13}, \tag{43}$$

$$w = a_{21}x + a_{22}y + a_{23}, \tag{44}$$

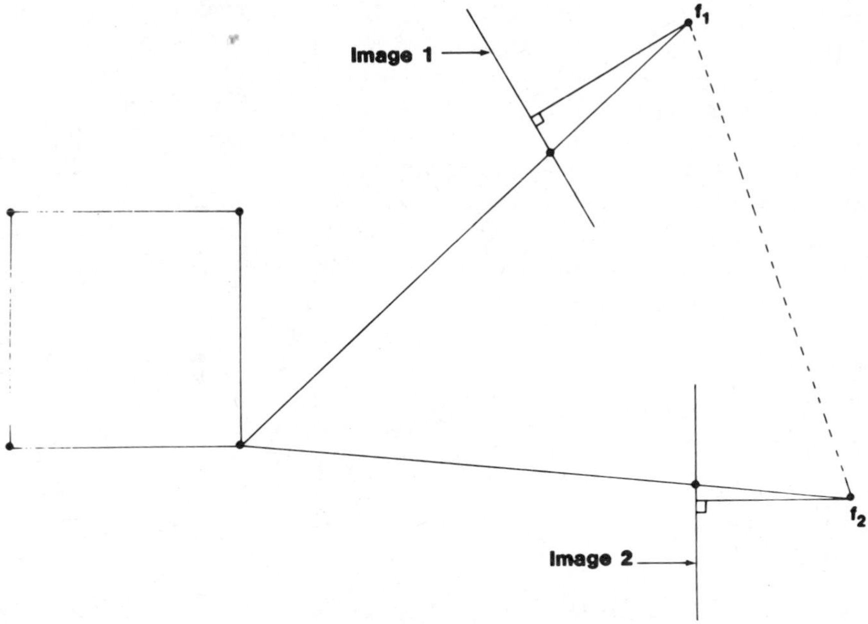

Figure 9 Triangulation of an object point.

which, by substituting the equivalence relationship of Equation 44 for the term, w, in Equation 43, becomes

$$(a_{11}-a_{21}x')x + (a_{12}-a_{22}x')y + (a_{13}-a_{23}x') = 0, \tag{45}$$

for which the only unknown terms are x and y. Similarly, from Equation 42 a second relationship is obtained

$$(b_{11}-b_{21}x'')x + (b_{12}-b_{22}x'')y + (b_{13}-b_{23}x'') = 0. \tag{46}$$

Equations 45 and 46 are two linearly independent equations as long as the lines defined in space by each of the corresponding image points are not coincident. Arranging these equations in the matrix form yields

$$\begin{bmatrix} a_{11}-a_{21}x' & a_{12}-a_{22}x' \\ b_{11}-b_{21}x & b_{12}-b_{22}x'' \end{bmatrix} \begin{bmatrix} x \\ y \end{bmatrix} = \begin{bmatrix} a_{23}x'-a_{13} \\ b_{23}x''-b_{13} \end{bmatrix}. \tag{47}$$

The object surface point in the global coordinate system (x,y) may be found from

$$\begin{bmatrix} x \\ y \end{bmatrix} = \begin{bmatrix} a_{11}-a_{12}x' & a_{12}-a_{22}x' \\ b_{11}-b_{21}x'' & b_{12}-b_{22}x'' \end{bmatrix}^{-1} \begin{bmatrix} a_{23}x'-a_{13} \\ b_{23}x''-b_{13} \end{bmatrix}. \tag{48}$$

Given two images for which the transformations are known and the image coordinates of the projections of an object surface point on each image, Equation 48 yields the global coordinates of the surface point in space.

The expansion of the stereo vision technique of point location into a three space geometry provides one additional unknown, a third dimension, to be measured for each point in space. A parallel to the procedure developed for point location in two space may be adopted for obtaining the three-dimensional location of a point using the projections of that point on a pair of two-dimensional images. The previous section described methods to determine the perspective projection transformation matrix for a given viewing geometry which defines the set of equations

$$wx' = a_{11}x + a_{12}y + a_{13}z + a_{14}, \tag{49}$$

$$wy' = a_{21}x + a_{22}y + a_{23}z + a_{24}, \tag{50}$$

$$w = a_{31}x + a_{32}y + a_{33}z + a_{34}. \tag{51}$$

By substituting Equation 51 for the term, w, in Equations 49 and 50, the pair of equations

$$(a_{11}-a_{31}x')x + (a_{12}-a_{32}x')y + (a_{13}-a_{33}x')z = (a_{34}x'-a_{14}), \tag{52}$$

$$(a_{21}-a_{31}y')x + (a_{22}-a_{32}y')y + (a_{23}-a_{33}y')z = (a_{34}y'-a_{24}) \tag{53}$$

is obtained. Similarly, a second image defines another pair of equations

$$(b_{11}-b_{31}x'')x + (b_{12}-b_{32}x'')y + (b_{13}-b_{33}x'')z = (b_{34}x''-b_{14}), \tag{54}$$

$$(b_{21}-b_{31}y'')x + (b_{22}-b_{32}y'')y + (b_{23}-b_{33}y'')z = (b_{34}y''-b_{24}) \tag{55}$$

for which the coefficients b_{ij} are the coefficients of the transformation matrix for the second image. Arranging Equations 52, 53, 54, and 55 in matrix form yields

$$\begin{bmatrix} a_{11}-a_{31}x' & a_{12}-a_{32}x' & a_{13}-a_{33}x' \\ a_{21}-a_{31}y' & a_{22}-a_{32}y' & a_{23}-a_{33}y' \\ b_{11}-b_{31}x'' & b_{12}-b_{32}x'' & b_{13}-b_{33}x'' \\ b_{21}-b_{31}y'' & b_{22}-b_{32}y'' & b_{23}-b_{33}y'' \end{bmatrix} \begin{bmatrix} x \\ y \\ z \end{bmatrix} \begin{bmatrix} a_{34}x'-a_{14} \\ a_{34}y'-a_{24} \\ b_{34}x''-b_{14} \\ b_{34}y''-b_{24} \end{bmatrix}, \tag{56}$$

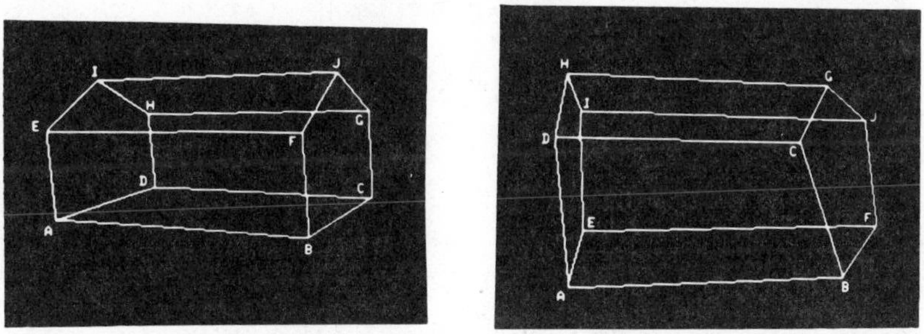

(a) Image #1 (b) Image #2

Figure 10 a. Image ≠ 1 b. Image ≠ 2 Two vies of example polygonal object.

which may be solved to obtain the three-dimensional coordinates (x,y,z). Since Equation 56 consists of four linearly independent equations and there exist only three unknowns, least-squares regression may be applied to yield a minimum mean-square error solution. The primary source of error would be in the measurement of the image coordinates of a point due to the limited resolution of the image in digital format.

Two views of the edges of an object are shown in Figure 10. The image and global coordinates of six calibration points are shown in Table 1 for both images. Using the calibration points and Equation 38, the perspective projection transformation matrices of Table 2 are obtained. Note that once the cameras are calibrated, any object anywhere in space which appears in both views may be located using the same transformation matrices as long as the cameras remain stationary. Table 3 lists the image coordinates for the corresponding vies of each point. From Equation 56 the global coordinates of Table 4 for each object surface point are obtained. The coordinates of these points, or vertices, and their interconnections, which are identical to the corresponding interconnections in the images, can yield sufficient information for the description and reconstruction of the coincidently visible areas of the polygonal solid as well as the object location.

3.3 Curved Surface Measurement and Location

Polygonal solids are constructed of multiple intersecting planes. The intersection of two planar surfaces in polygonal solid creates discontinuities known as edges. Furthermore, the intersection of two or more edges creates a vertex. The vertices, being point features, are often the best features to use for object location and object description since they are more easily correlated in two images than edges or planar surfaces. One inherent problem with curved objects is the possible lack of vertices or edges on the surface which prevents the location of corresponding points in the images.

The technique of measuring the location of surface points on curved surfaces using active stereo vision is identical to that for polygonal solids. The

TABLE 1

CALIBRATION POINTS FOR POLYGONAL OBJECT

	Image #1		*Image #2*		*Object Coordinates*		
	X'	Y'	X''	Y''	X	Y	Z
A	148.0	309.0	152.0	347.0	-3.0	0.0	-2.0
B	323.0	320.0	343.0	338.0	3.0	0.0	-2.0
C	369.0	291.0	310.0	243.0	3.0	0.0	2.0
J	344.0	203.0	358.0	227.0	3.0	3.0	0.0
F	319.0	246.0	367.0	301.0	3.0	2.0	-2.0
E	142.0	247.0	161.0	307.0	-3.0	2.0	-2.0

only difference is that for curved surfaces, the measured points are not vertices of the object but are simply sample points on the surface. Active imaging by grid projection solves the problem of adding identifiable featuresless surface; however, this technique has advantages which extend beyond the more apparent capability. Two recent publications [34,35] have presented an additional advantage of this technique which is referred to as single image stereo vision. These publications describe the ability to perform surface measurement using a single view of the object with a projected grid pattern. When a rectangular grid pattern is projected onto a curved surface from one vantage point and is

TABLE 2

THE PERSEPCTIVE PROJECTION TRANSFORMATION MATRICES OF THE TWO VIEWS

	Perspective Projection of Image #1		
23.531081	-3.106923	20.781372	132.914931
3.328802	31.578038	-1.570048	-171.880776
-0.026788	-0.008146	0.046454	1.000000
	Perspective Projection of Image #2		
32.780992	-2.088013	2.811507	239.581265
3.027502	-29.340755	-14.589663	289.343900
0.012865	-0.037429	0.034846	1.000000

TABLE 3

SAMPLE IMAGE POINTS OF THE POLYGONAL SOLID EXAMPLE

	Image #1		*Image #2*	
	X'	Y'	X''	Y''
A	148.0	309.0	152.0	347.0
B	323.0	320.0	343.0	338.0
C	369.0	291.0	310.0	243.0
D	215.0	286.0	141.0	242.0
E	142.0	247.0	161.0	307.0
F	319.0	246.0	367.0	301.0
G	367.0	230.0	329.0	203.0
H	211.0	234.0	148.0	198.0
I	177.0	211.0	158.0	224.0
J	344.0	203.0	358.0	227.0

TABLE 4

COMPUTED THREE-DIMENSIONAL COORDINATES OF THE OBJECT VERTICES

X	Y	Z
-3.002173	-0.000566	-1.998774
3.002010	0.000216	-1.996167
2.999814	0.000264	1.996891
-3.052032	-0.015057	2.063887
-2.997853	2.000350	-2.001135
2.997871	1.999445	-2.006467
3.012070	2.014838	2.011220
-3.054216	1.990594	2.107293
-3.035907	2.996249	0.068029
3.000348	3.000281	0.005637

viewed from another direction, the grid pattern appears distorted in the image. This geometric distortion of the grid pattern characterizes the shape of the surface. The relationship between this technique of surface measurement and the stereo vision technique is apparent if the relationship between the imaging process and the projection process are considered. Since the geometry of the two processes are identical, the projected grid may be considered a second view for the stereo imaging surface measurement technique. A camera having the same model parameters as the projector would obtain an image that is identical to the projected grid. By using the image as one view of the surface and by treating the projected grid as the second view, the stereo imaging technique may be applied to measure sample points of the surface.

The calibration of the projector is similar to that of the calibration procedure for the camera except that only the grid points may be used as calibration points since these are the only points which may be related to the projected image coordinate. The grid spacing on the projection grid may be assumed to be any constant since an error in the grid spacing would not change the location of the focal point but would only translate the projection grid relative to the focal point as shown in Figure 11. Therefore, for the calibration of the projector, only the correspondence between the projection grid origin and its surface point along with the knowledge that the grid is a square grid are necessary.

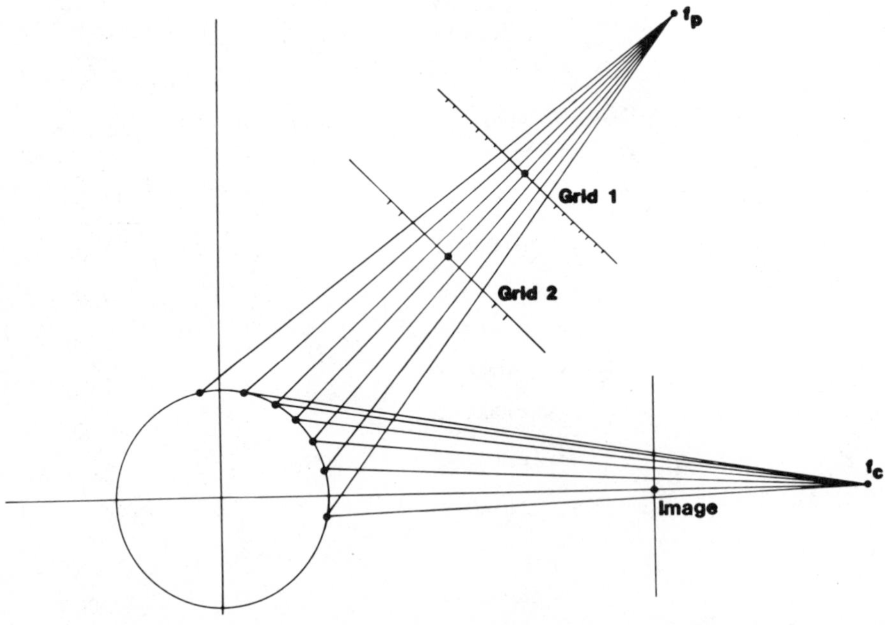

Figure 11 Surface measurement by grid projection. The two grids shown illustrate the resulting location of the grid plane from camera calibration for two assumed grid spacings.

As an example, an aluminum brick was used for calibrating the camera and projector due to the simplicity of obtaining calibration data from such a surface. One face of the brick was defined as the $x-y$ plane and another as the $y-z$ plane as illustrated in Figure 12(a). The grid is projected onto the brick and the camera image is digitized resulting in the image shown in Figure 12(b). Several points of the intersecting grid lines are measured in three-dimensional space using the defined coordinate system and their corresponding image coordinates and grid coordinates are obtained to yield the calibration data of Table 5. Using these points, the perspective projection transformation matrices of Table 6 was obtained for the camera and projector by the application of Equation 38. The image grid coordinates of all grid points as projected onto the surface that are visible in the image are listed in Table 7. Applying Equation 56 to these sample points yields the global coordinates of the corresponding surface points shown in Table 8. Various views of the brick obtained by graphical reconstruction from the computed surface points are shown in Figure 13. The perspective projection transformation matrices may now be applied to measure the surface geometry of any object that is placed within view of the projector and camera.

Other examples of a circular cylinder and a mug were placed in view of both camera and projector yielding the images of Figure 14. Tables 9 and 10 show the image points and the corresponding grid points of the two surfaces. Again, Equation 56 was applied to obtain the sample surface points shown in Tables 11 and 12. The graphical reconstruction of the surfaces are shown in Figure 15 and were obtained using the derived surface points.

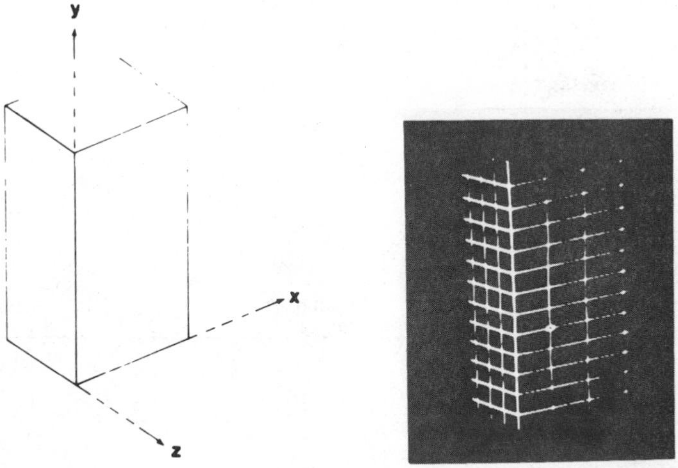

(a) Coordinate system definition **(b) Digital image of calibration surface**

Figure 12 Camera and projector calibration procedure. a. Coordinate system definition. b. Digital image of calibration surface.

TABLE 5

CALIBRATION POINTS FOR CALIBRATION BLOCK

	Image		Grid		Object Coordinates		
	X'	Y'	X''	Y''	X	Y	Z
A	131.0	152.0	0.0	0.0	0.00	2.50	-0.95
B	100.0	158.0	1.0	0.0	0.00	2.44	0.00
C	132.0	186.0	0.0	-2.0	0.00	1.54	-0.95
D	164.0	197.0	-1.0	-3.0	0.00	1.06	-2.10
E	89.0	172.0	2.0	-1.0	0.57	1.97	0.00
F	80.0	202.0	3.0	-3.0	1.11	1.03	0.00
G	103.0	227.0	1.0	-4.0	0.00	0.53	0.00

TABLE 6

PERSPECTIVE PROJECTION MATRICES OF THE CAMERA AND PROJECTOR
AS COMPUTED FOR THE CALIBRATION SURFACE
(THE ALUMINUM BRICK)

	Perspective Projection of Image		
-20.876596	-0.730895	-30.479003	-23.189201
4.491177	36.586780	-1.902322	-116.883184
0.028971	-0.028486	-0.051452	1.000000
	Perspective Project of Grid Pattern		
1.653097	-0.035870	0.963726	1.000816
0.087336	2.065337	0.140802	-5.036822
-0.031991	-0.027664	-0.037502	1.000000

TABLE 7

THE SAMPLE IMAGE AND GRID COORDINATES FOR THE CALIBRATION BLOCK

Image		Grid	
X'	Y'	X''	Y''
67.0	148.0	4.0	0.0
78.0	151.0	3.0	0.0
89.0	155.0	2.0	0.0
100.0	158.0	1.0	0.0
131.0	152.0	0.0	0.0
163.0	145.0	-1.0	0.0
195.0	138.0	-2.0	0.0
67.0	164.0	4.0	-1.0
78.0	168.0	3.0	-1.0
89.0	172.0	2.0	-1.0
101.0	176.0	1.0	-1.0
13.0	169.0	0.0	-1.0
163.0	162.0	-1.0	-1.0
196.0	155.0	-2.0	-1.0
68.0	181.0	4.0	-2.0
79.0	185.0	3.0	-2.0
90.0	189.0	2.0	-2.0
102.0	193.0	1.0	-2.0
132.0	186.0	0.0	-2.0
163.0	180.0	-1.0	-2.0
196.0	173.0	-2.0	-2.0
69.0	198.0	4.0	-3.0
80.0	202.0	3.0	-3.0
91.0	206.0	2.0	-3.0
102.0	209.0	1.0	-3.0
133.0	203.0	0.0	-3.0
164.0	197.0	-1.0	-3.0
197.0	189.0	-2.0	-3.0
69.0	215.0	4.0	-4.0
80.0	219.0	3.0	-4.0
91.0	223.0	2.0	-4.0
103.0	227.0	1.0	-4.0
134.0	220.0	0.0	-4.0
165.0	214.0	-1.0	-4.0
198.0	206.0	-2.0	-4.0

TABLE 8

COMPUTER THREE-DIMENSIONAL COORDINATES OF THE SAMPLE DATA POINTS OF TABLE 7*

X	*Y*	*Z*
1.618133	2.461418	-0.004985
1.122805	2.448250	-0.037719
0.574748	2.422142	-0.029220
0.006982	2.430563	0.000784
0.003661	2.502265	-0.951683
0.076392	2.594568	-2.099441
0.200393	2.689498	-3.429356
1.622374	2.024281	0.024368
1.109095	1.992002	-0.000219
0.559813	1.971950	0.005054
-0.005700	1.957042	0.001475
-0.011841	2.038809	-0.931485
0.053789	2.104670	-2.087185
0.192068	2.160667	-3.487554
1.629867	1.544346	0.025175
1.113674	1.516728	-0.002235
0.564406	1.502009	-0.001478
0.001065	1.491823	-0.011406
-0.005632	1.552337	-0.959074
0.002353	1.580088	-2.053165
0.118050	1.604585	-3.450110
1.631567	1.052423	0.029340
1.114484	1.028955	-0.002061
0.567328	1.018841	-0.007121
0.016163	1.043390	-0.000678
-0.000239	1.050741	-0.986837
0.000205	1.053157	-2.094578
0.139604	1.067063	-3.535990
1.609394	0.559728	0.077183
1.094505	0.539815	0.040584
0.552602	0.533444	0.028279
0.000678	0.529777	0.001506
0.004205	0.533436	-1.014710
-0.003814	0.509606	-2.135949
0.120739	0.489743	-3.590691

* Note that the points are listed in the same order as those in Table 7.

TABLE 9

THE SAMPLE IMAGE AND GRID COORDINATES FOR THE CIRCULAR CYLINDER

Image		*Grid*	
X'	*Y'*	*X"*	*Y"*
86.0	83.0	2.0	4.0
99.0	85.0	1.0	4.0
115.0	86.0	0.0	4.0
134.0	85.0	-1.0	4.0
160.0	81.0	-2.0	4.0
87.0	100.0	2.0	3.0
99.0	103.0	1.0	3.0
115.0	104.0	0.0	3.0
135.0	103.0	-1.0	3.0
161.0	99.0	-2.0	3.0
88.0	118.0	2.0	2.0
100.0	120.0	1.0	2.0
116.0	121.0	0.0	2.0
135.0	121.0	-1.0	2.0
161.0	117.0	-2.0	2.0
88.0	135.0	2.0	1.0
101.0	138.0	1.0	1.0
116.0	139.0	0.0	1.0
135.0	139.0	-1.0	1.0
161.0	135.0	-2.0	1.0
90.0	153.0	2.0	0.0
102.0	155.0	1.0	0.0
118.0	157.0	0.0	0.0
137.0	156.0	-1.0	0.0
163.0	153.0	-2.0	0.0
90.0	170.0	2.0	-1.0
103.0	173.0	1.0	-1.0
118.0	174.0	0.0	-1.0
137.0	173.0	-1.0	-1.0
163.0	170.0	-2.0	-1.0
91.0	187.0	2.0	-2.0
103.0	190.0	1.0	-2.0
119.0	191.0	0.0	-2.0
138.0	190.0	-1.0	-2.0
164.0	187.0	-2.0	-2.0
92.0	204.0	2.0	-3.0
104.0	207.0	1.0	-3.0
119.0	209.0	0.0	-3.0
138.0	208.0	-1.0	-3.0
164.0	205.0	-2.0	-3.0
92.0	221.0	2.0	-4.0
105.0	224.0	1.0	-4.0
120.0	226.0	0.0	-4.0
139.0	226.0	-1.0	-4.0
164.0	222.0	-2.0	-4.0

TABLE 10

COMPUTER THREE-DIMENSIONAL COORDINATES OF THE SAMPLE DATA POINTS FOR THE CYLINDER*

X	*Y*	*Z*	*X*	*Y*	*Z*
0.584329	4.225848	-0.052818	-0.775278	2.542439	-0.597895
0.098042	4.220715	-0.113083	-1.013723	2.596099	-1.264748
-0.350956	4.237379	-0.274320	0.618360	2.008879	-0.062480
-0.750874	4.297401	-0.552649	0.094879	2.005577	-0.124301
-0.970071	4.417300	-1.184336	-0.383257	2.040045	-0.268373
0.607995	3.814390	-0.071159	-0.773875	2.092145	-0.592224
0.078933	3.794478	-0.080658	-1.010817	2.132735	-1.271812
-0.378121	3.814164	-0.241789	0.625641	1.536959	-0.071111
-0.757128	3.870095	-0.570536	0.086440	1.546053	-0.098459
-0.975401	3.981093	-1.215464	-0.367020	1.571538	-0.295288
0.617951	3.356511	-0.079950	-0.750081	1.618599	-0.634227
0.092848	3.377320	-0.097999	-0.984668	1.644533	-1.331604
-0.367560	3.399361	-0.263611	0.630739	1.051738	-0.078557
-0.78661	3.436593	-0.542548	0.097010	1.064122	-0.114547
-1.005517	3.535955	-1.196249	-0.391399	1.078009	-0.263688
0.611005	2.937530	-0.051359	-0.767933	1.119366	-0.614640
0.091470	2.925136	-0.104847	-1.003895	1.130018	-1.322689
-0.393054	2.955280	-0.231797	0.617298	0.564263	-0.044196
-0.813429	2.991699	-0.516330	0.106983	0.567410	-0.130557
-1.032207	3.079500	-1.179740	-0.374171	0.579774	-0.291738
0.630497	2.461002	-0.094819	-0.771108	0.594580	-0.638285
0.103513	2.482927	-0.121520	-0.990250	0.619043	-1.339032
-0.377935	2.486680	-0.286128			

* Note that the points are listed in the same order as those in Table 9.

TABLE 11

THE SAMPLE IMAGE AND GRID COORDINATES FOR THE MUG

Image		*Grid*	
X'	*Y'*	*X''*	*Y''*
87.0	127.0	3.0	1.0
98.0	130.0	2.0	1.0
112.0	132.0	1.0	1.0
129.0	133.0	0.0	1.0
148.0	132.0	-1.0	1.0
173.0	129.0	-2.0	1.0
89.0	144.0	3.0	0.0
100.0	147.0	2.0	0.0
114.0	149.0	1.0	0.0
130.0	150.0	0.0	0.0
150.0	159.0	-1.0	0.0
175.0	146.0	-2.0	0.0
90.0	160.0	3.0	-1.0
101.0	164.0	2.0	-1.0
115.0	166.0	1.0	-1.0
131.0	167.0	0.0	-1.0
151.0	166.0	-1.0	-1.0
177.0	162.0	-2.0	-1.0
91.0	176.0	3.0	-2.0
102.0	181.0	2.0	-2.0
115.0	183.0	1.0	-2.0
132.0	184.0	0.0	-2.0
152.0	183.0	-1.0	-2.0
178.0	179.0	-2.0	-2.0
92.0	193.0	3.0	-3.0
103.0	197.0	2.0	-3.0
116.0	200.0	1.0	-3.0
132.0	201.0	0.0	-3.0
152.0	200.0	-1.0	-3.0
179.0	196.0	-2.0	-3.0
93.0	210.0	3.0	-4.0
103.0	214.0	2.0	-4.0
117.0	217.0	1.0	-4.0
133.0	218.0	0.0	-4.0
153.0	217.0	-1.0	-4.0
180.0	213.0	-2.0	-4.0

TABLE 12

COMPUTED THREE-DIMNESIONAL COORDINATES OF THE SAMPLE DATA POINTS FOR THE MUG*

X	*Y*	*Z*
1.445791	3.015824	-0.533049
0.920245	3.000210	-0.541231
0.442412	3.000526	-0.662759
0.011331	3.014786	-0.904018
-0.374765	3.070285	-1.243175
-0.601623	3.141338	-1.921236
1.491432	2.542334	-0.595014
0.960539	2.532520	-0.605418
0.479472	2.535404	-0.732426
0.023925	2.556114	-0.934661
-0.335589	2.600847	-1.336702
-0.560000	2.659099	-2.036122
1.521471	2.086291	-0.619173
0.970751	2.057615	-0.620909
0.489178	2.062638	-0.754011
0.035549	2.083213	-0.965330
-0.320779	2.122118	-1.380554
-0.492425	2.179494	-2.174731
1.548987	1.617113	-0.642001
0.976178	1.569716	-0.633687
0.475794	1.585756	-0.728732
0.045928	1.595601	-0.995863
-0.306420	1.627522	-1.425180
-0.477168	1.667757	-2.237435
1.551974	1.110680	-0.648649
1.001099	1.091245	-0.660253
0.481491	1.086315	-0.748101
0.036971	1.103720	-0.980082
-0.310604	1.127675	-1.421806
-0.463550	1.138354	-2.300606
1.548925	0.591330	-0.651803
0.981804	0.587793	-0.625505
0.484749	0.572225	-0.766359
0.046018	0.586300	-1.010604
-0.296716	0.600813	-1.467723
-0.451918	0.5960678	-2.364033

* Note that the points are list in the same order as those in Table 11.

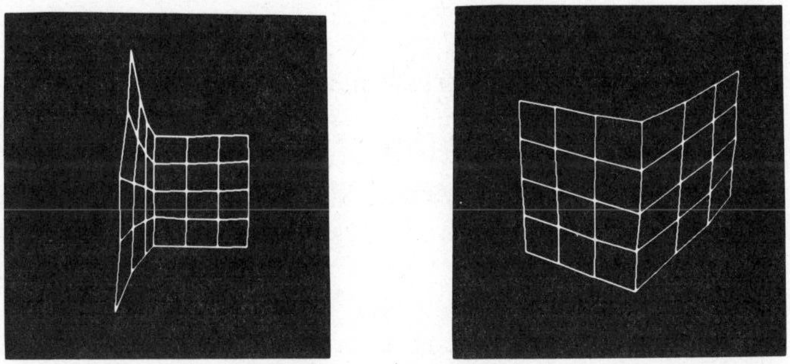

Figure 13 Two views of the calibration surface generated.

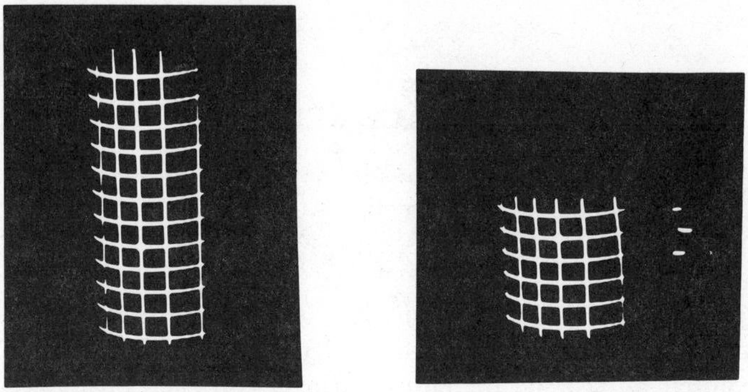

(a) A circular cylinder **(b) Mug**

Figure 14 Two curved surface examples. a. A circular cylinder. b. Mug

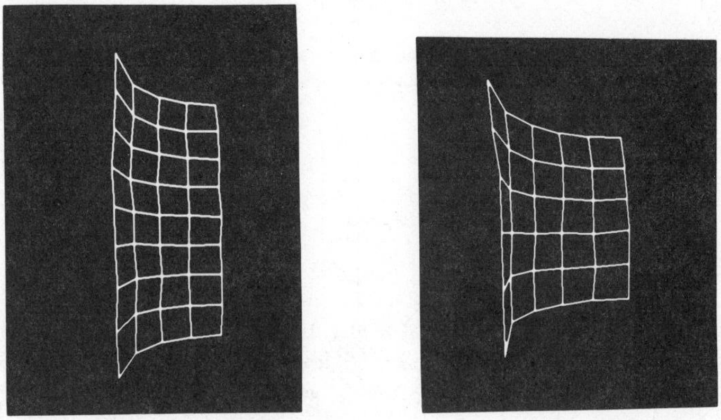

(a) Circular cylinder from computed surface points. **(b) Mug**

Figure 15 Graphical reconstructions of example curved surfaces. a. Circular cylinder from computed surface points. b. Mug

3.4 Summary

The technique of stereo vision for surface measurement is well known and has long been applied to three-dimensional data acquisition and object description. The stereo vision process requires that the camera models be known for each view of the object. Through a knowledge of the camera model parameters of location, orientation and focal length, the perspective projection transformation matrix for each camera can be determined from Equation 31. Since it is often difficult to obtain the model parameters directly, the camera calibration procedure may be applied using six points for which the global coordinates and image coordinates are known and Equation 38. Once the transformation matrices are known, the global coordinates of a surface point may be computed from Equation 56 accompanied by the image coordinates of the corresponding image points.

4. CURVED SURFACE REPRESENTATION

A mathematical model of a surface is an exact function describing the surface of an object for all points. Such a function would be of the form

$$f(x,y,z) = 0, \tag{57}$$

where all points which satisfy the equation lie on the surface of the object. In cases where discontinuities exist on the surface, multiple functions of this form may be required to describe the entire surface with each function being valid only over a certain region of the surface.

The description of a polygonal surface that is constructed of planes would be modeled by a set of linear equations of the form

$$ax + by + cz + d = 0, \tag{58}$$

where each equation represents one side or face of the object. A matrix representation of Equation 57 would be

$$[x\ y\ z\ 1]\begin{bmatrix} a \\ b \\ c \\ d \end{bmatrix} = 0. \tag{59}$$

It is not always possible to represent a surface in this manner since all surface descriptions may not be expressed by low order polynomials. One such example is the ripples in water after a stone is dropped into it. Surface representations of the more complex forms are much more difficult to derive for a given surface.

Polynomial surface representations can, with unlimited numerical capabilities, describe any continuous surface. The coefficient matrix becomes increasingly large as the order of the polynomial increases; in fact, the number of $k-th$ order coefficients is

$$N = \sum_{i=1}^{k+1} i = \frac{(k+1)(k+2)}{2}. \tag{60}$$

Therefore a quadratic equation would have six second order coefficients. in order to include translation of object coordinates, however, the first order terms must also be included in the mathematical model. In general, a model of order k should include all terms of order n such that $n = 0,1,2,...,k$. The total number of coefficients required to model a surface of order k would be

$$T = \sum_{n=1}^{k+1} N(n) = \frac{(k+1)(k+2)[2(k+1)+1]}{6} + \frac{(k+1)(k+2)}{4}, \tag{61}$$

if the model is to include all terms less than or equal to k, which provides the position variant terms necessary in many applications. Note that the number of coefficients required grows exponentially with the order of the polynomial. Since each coefficient requires at least one measurement for its determination, the number of measurements required would also increase exponentially. The large effort is one of the major reasons that low order polynomials are preferable. Another reason is that generally the simpler a surface the easier it is to work with. Texture surfaces are an exception, of course. However, many man made objects have rather "smooth or simple" surfaces. The first and second order surfaces fit a large number of applications. A complete representation of a quadric surface with $k = 2$ would require ten coefficients from Equation 61.

4.1 Quadric Polynomial Representations

Numerous methods, both exact and approximate, exist for obtaining the coefficients of a polynomial from known points. An exact method of determining the coefficient matrix is by the application of direct methods for solving linear systems of equations. Higher order polynomials are nonlinear equations; however, the polynomials are nonlinear in the variables x, y, and z. The equation is linear with respect to the coefficients which are the unknown quantities. Let the coefficient matrix A be defined by

$$A = [a_1, a_2, \dots, a_n]', \tag{62}$$

where A' denotes the transpose of the matrix A and a_i for $i = 1, 2, \dots, n$ are each of the unknown coefficients of the polynomial. The matrix X is comprised of the data points contained on the curved surface in the polynomial form. For example, the general quadratic polynomial would yield the matrix

$$X = \begin{bmatrix} x_1^2 & y_{1_2} & z_1^2 & x_1y_1 & x_1z_1 & y_1z_1 & x_1 & y_1 & z_1 \\ x_2^2 & y_2^2 & z_2^2 & x_2y_2 & x_2z_2 & y_2z_2 & x_2 & y_2 & z_2 \\ & & & & \vdots & & & & \\ x_9^2 & y_9^2 & z_9^2 & x_9y_9 & x_9z_9 & y_9z_9 & x_9 & y_9 & z_9 \end{bmatrix}, \tag{63}$$

where (x_i,y_i,z_i) are each of the nine known data points on the quadratic surface. If X^{-1} exists, then the coefficient matrix, A, can be found by

$$A = X^{-1} B, \tag{64}$$

where $B = [1,1,1....,1]'$ and the dimension of B is the same as the dimension of A. The number of data points required to determine the unknown coefficients of a general polynomial of order k is one less than described by Equation 61 which also yields the row dimension of the vectors A and B.

The polynomial solution is a mathematical model of a quadric surface containing the sample data points; however, what information does this polynomial contain which would enable the determination of the basic surface shape, size, orientation and location required in many applications? A qualitative analysis of the effects of the variation of these parameters on a basic quadratic equation will provide the answers.

4.2 Determining Location and Orientation

If the resulting coefficient matrix contains no second order terms, then the surface cannot be a quadric surface but instead is a linear surface and may be described as a plane. Because a plane is not a closed surface and is infinite in size by the defining equation, location must be considered irrelevant with respect to a plane. However, if the planar surface is bounded, the location of the plane may be determined but is relative to the boundaries of the surface which are not described in the mathematical model. The orientation information may be derived from the mathematical model and may be described by the normal vector which is invariant to the surface point for a plane. The plane

$$ax + by + cz - d = 0, \tag{65}$$

would have the normal vector

$$\boldsymbol{n} = a\boldsymbol{i} + b\boldsymbol{j} + c\boldsymbol{k}, \tag{66}$$

where $(\boldsymbol{i}, \boldsymbol{j}, \boldsymbol{k})$ is the unit normal vector.

As illustrated by the planar surface, not all information can be obtained from the mathematical model if the model does not describe a closed surface. In general, only the ellipsoid and the special case of the ellipsoid, the sphere, is a closed surface and the location of all other quadric surfaces may not be represented by a single point without additional information other than the mathematical model.

If any of the second order terms of the quadratic polynomial have nonzero coefficients, then the surface is a quadric surface. All quadric surfaces may be classified by a finite set of basic shapes such as the ellipsoid, paraboloid, hyperboloid or cylinder with further subclassifications such as the elliptic hyperboloid. Each of the basic shapes may be expressed by the simpler polynomial

$$ax^2 + by^2 + cz^2 + d = 0 \tag{67}$$

where the coefficients may be any real number. If the origin of the surface were translated to a point (t,u,v) then the equation of the surface may be obtained by replacing the variables x, y, and z with $x-t$, $y-u$, $z-v$. This yields the equation

$$a(x-t)^2 + b(y-u)^2 + c(z-v)^2 + d = 0. \qquad (68)$$

By expanding this quadratic equation, the polynomial becomes

$$ax^2 + by^2 + cz^2 - 2atx - 2buy$$

$$-2cvz + at^2 + bu^2 + cv^2 + d = 0. \qquad (69)$$

It is apparent by this equation that the first order terms will appear if the origin of a quadric surface is translated to some point other than the global coordinate system origin. Equation 69 is unique in the surface it describes and the equation may, therefore, be decomposed into the more qualitatively understood form of Equation 68 from measured data. Assume that a set of data points of a surface were obtained and the mathematical model of the surface,

$$a'x^2 + b'y^2 + c'z^2 + d'x + e'y + f'z + g' = 0, \qquad (70)$$

was derived using Equation 64. By equating the terms of Equation 70 with those of Equation 69, the results $a'=a$, $b'=b$, $c'=c$, $d'=-2at$, $e'=-2bu$, $f'=2cv$, and $g'=(at^2+bu^2+cv^2+d)$ are obtained. By simultaneous solution the coefficients of Equation 68 are found to be

$$a = a', \quad t = -d'/2a',$$

$$b = b', \quad u = -d'/2b', \qquad (71)$$

$$c = c', \quad v = -f'/2c',$$

$$d = (g' - \frac{d'^2}{4a'} - \frac{e'^2}{4a'} - \frac{e'^2}{4b'} - \frac{f'^2}{4c'}).$$

By Equation 71, it is possible to determine the basic surface shape as described by Equation 67 with the origin translated to the point (t,u,v).

The most general form of the quadratic equation is

$$ax^2 + by^2 + cz^2 + 2dxy + 2exz + 2fyz + gx + hy + iz + j = 0, \qquad (72)$$

which includes all terms of second order or less. It has been shown that the translation of the coordinate axis produces first order terms in Equation 72. The "cross-product" terms, dxy, exz, and fyz, cannot, however, be attributed to translation of the axis. These terms are by-products of axes rotation. A quadratic equation of the form shown in Equation 67 will be referred to as a quadric surface in "standard position" having no rotation or translation of the coordinate system axes or origin.

If an object is rotated in the xy-plane about the x-axis by the Euler angle, α, and new $x'y'$ coordinate system is obtained where

$$x' = x \cos \alpha + y \sin \alpha$$

$$y' = -x \sin \alpha + y \cos \alpha$$

$$z' = z$$

if the rotation is in the counter clockwise direction. This may be represented in matrix form as

$$\begin{bmatrix} x' \\ y' \\ z' \end{bmatrix} = \begin{bmatrix} \cos \alpha & \sin \alpha & 0 \\ -\sin \alpha & \cos \alpha & 0 \\ 0 & 0 & 1 \end{bmatrix} \begin{bmatrix} x \\ y \\ z \end{bmatrix}. \tag{73}$$

Similarly, a rotation about the y-axis and the x-axis by the Euler angles β and γ respectively may be performed by the operations

$$\begin{bmatrix} x' \\ y' \\ z' \end{bmatrix} = \begin{bmatrix} \cos \beta & 0 & -\sin \beta \\ 0 & 1 & 0 \\ \sin \beta & 0 & \cos \beta \end{bmatrix} \begin{bmatrix} x \\ y \\ z \end{bmatrix}, \tag{74}$$

and

$$\begin{bmatrix} x' \\ y' \\ z' \end{bmatrix} = \begin{bmatrix} 1 & 0 & 0 \\ 0 & \cos \gamma & \sin \gamma \\ 0 & -\sin \gamma & \cos \gamma \end{bmatrix} \begin{bmatrix} x \\ y \\ z \end{bmatrix}. \tag{75}$$

If the three rotation transformation matrices of Equations 73, 74, and 75 are denoted by $R(\alpha)$, $R(\beta)$, $R(\gamma)$ respectively, the complete rotation transformation process may be represented as

$$[x'\, y'\, z']' = [R(\alpha)]\, [R(\beta)]\, [R(\gamma)]\, [x\, y\, z]', \tag{76}$$

or

$$[x'\, y'\, z']' = R(\alpha, \beta, \gamma)\, [x\, y\, z]', \tag{77}$$

where

$$R(\alpha, \beta, \gamma) = R(\alpha)\, R(\beta)\, R(\gamma).$$

The problem of rotating a surface described by a polynomial of the form of Equation 72 such that the resulting quadratic equation is in standard form is one of determining some $R(\alpha, \beta, \gamma)$ such that the cross product terms in the $x'\, y'\, z'$ coordinate system have zero coefficients.

The process of rotating the coordinate system may be described as a change of the basis vectors for which the coordinate system is defined. Furthermore, since the original coordinate axes are orthogonal, a rotation of the coordinate axes would also yield an orthogonal set of coordinate axes. This requires that the new set of basis vectors also be orthogonal. No "stretching" or "contraction" of any of the coordinate axes occurs during a rotation as described in

Equation 77, therefore, each of the basis vectors should have a norm or magnitude of unity. These requirements restrict the new set of basis vectors to be orthonormal.

The quadric surface described by Equation 72 may be represented in matrix form as

$$[x\ y\ z]\begin{bmatrix} a & d & e \\ d & b & f \\ e & f & c \end{bmatrix}\begin{bmatrix} x \\ y \\ z \end{bmatrix} + [g\ h\ i]\begin{bmatrix} x \\ y \\ z \end{bmatrix} + j = 0, \tag{78}$$

or

$$Y'DY + EY = j = 0. \tag{79}$$

The desired form of the quadratic equation after rotation is

$$[x'y'z']\begin{bmatrix} a' & 0 & 0 \\ 0 & b' & 0 \\ 0 & 0 & c' \end{bmatrix}\begin{bmatrix} x' \\ y' \\ z' \end{bmatrix} + [g'h'i']\begin{bmatrix} x' \\ y' \\ z' \end{bmatrix} + j' = 0, \tag{80}$$

or

$$\hat{Y}'\hat{D}\hat{Y} + \hat{E}\hat{Y} + j' = 0, \tag{81}$$

where $\hat{D}$ is a diagonal matrix. The method of matrix diagonalization is well known [35,36], and

$$\hat{D} = \begin{bmatrix} \lambda_1 & 0 & 0 \\ 0 & \lambda_2 & 0 \\ 0 & 0 & \lambda_3 \end{bmatrix}, \tag{82}$$

where λ_1, λ_2, and λ_3 are the eigenvalues of D. It is necessary however, that the diagonalization matrix P be orthonormal where $\det(P) = 1$.

Every quadric surface has at least two mutually perpendicular principal planes, and a central quadric [37] has at least three mutually perpendicular principal planes. The line of intersection between two principal planes is a principal axis or an axis of symmetry. Every quadric surface has at least one principal axis. If it has more than one, there exists at least one other principal axis which is perpendicular to each. The directions of the normals to the principal planes of a quadric, and hence the principle axes, are directed along the eigenvectors associated with matrix D [38]. Therefore, the eigenvectors may be used to represent the orientation of the quadric surface and the eigenvalues become the coefficients of the quadric surface in standard position for defining the shape of the surface.

It is desirable in many applications to obtain the location of the surface origin relative to the original global coordinate system axes. If the axes are rotated to eliminate the cross-product terms in the quadratic equation, a translation using the method described by Equation 71 would yield the location of the surface origin relative to the new coordinate axes. It is, therefore, necessary to translate the surface origin to the coordinate system origin before rotating the axes. The method of translation to eliminate the first order terms of the quadratic equation is further complicated by the presence of the cross-product terms in the equation. A surface whose origin is located at the global coordinate system origin could be described by the equation

$$ax^2 + by^2 + cz^2 + dxy + exz + fyz + j = 0. \tag{83}$$

If the surface origin is translated to a point (t, u, v) in the global coordinate system, the surface equation would become

$$a'x^2 + b'y^2 + c'z^2 + d'xy + e'xz + f'yz + g'x + h'y$$

$$+ i'z + j' = 0. \tag{84}$$

The coefficients of Equation 84 are related to the coefficients of Equation 83 by the relationship obtained by substituting $x = x' - t$, $y = y' - u$, and $z = z' - v$ into Equation 83. Through this substitution and collecting the terms of x'^2, y'^2, z'^2, $x'y'$, $x'z'$, etc., the relationship between the coefficients of 83 and 84 is found to be

$$a' = a \quad b' = b \quad c' = c$$

$$d' = d \quad e' = e \quad f' = f$$

$$g' = -(2at + du + ev) \tag{85}$$

$$h' = -(dt + 2bu + fv) \tag{86}$$

$$i' = -(et + fu + 2cv) \tag{87}$$

$$j' = j + at^2 + bu^2 + cv^2 + dtu + etv + fuv. \tag{88}$$

Given a quadratic equation of the form 84, it can be translated to the global coordinate system origin by solving the translated origin (t, u, v) through the simultaneous solution of Equations 85, 86, and 87. The term j may then be found by Equation 88 to obtain an equation of the form of Equation 83. Note that if d', e', and f' of Equation 84 are zero, the solution reduces to the form of Equation 71.

In order to decompose a quadric surface described by a quadratic equation of the form of Equation 84, the information related to shape, size, orientation and position must be found through the procedure previously discussed. First, the surface should be translated to the origin and (t, u, v) should be retained as the location information. Secondly, the surface equation must be rotated such that the quadratic equation is in standard form and the orthonormal eigenvectors should be retained to represent the orientation information. The quadratic equation in standard position represents the shape and size information and each surface may be recognized by the coefficients of the second order terms using Table 13 as a classification guide.

The classification can be based on the invariants of the original representation

$$a_{11}x^2 + a_{22}y^2 + a_{33}z^2 + 2a_{12}xy + 2a_{12}xy + 2a_{23}zy \tag{89}$$

$$2a_{14}x + 2a_{24}y + 2a_{34}z + a_{44} = 0.$$

In particular, the four quantities

$$I = a_{11} + a_{22} + a_{33} \tag{90}$$

$$J = \begin{vmatrix} a_{11} & a_{12} \\ a_{21} & a_{22} \end{vmatrix} + \begin{vmatrix} a_{22} & a_{23} \\ a_{32} & a_{33} \end{vmatrix} + \begin{vmatrix} a_{33} & a_{31} \\ a_{13} & a_{11} \end{vmatrix} \tag{91}$$

$$D = \begin{vmatrix} a_{11} & a_{12} & a_{13} \\ a_{21} & a_{22} & a_{23} \\ a_{31} & a_{32} & a_{33} \end{vmatrix} \tag{92}$$

$$A = \begin{vmatrix} a_{11} & a_{12} & a_{13} & a_{14} \\ a_{21} & a_{22} & a_{23} & a_{24} \\ a_{31} & a_{32} & a_{33} & a_{34} \\ a_{41} & a_{42} & a_{43} & a_{44} \end{vmatrix} = A_{44} \tag{93}$$

and the signs of the quantities below are invariant with respect to translation and rotation.

$$A' = A_{11} + A_{22} + A_{33} + A_{34} \tag{94}$$

$$A'' = \begin{vmatrix} a_{11} & a_{12} \\ a_{21} & a_{22} \end{vmatrix} + \begin{vmatrix} a_{11} & a_{13} \\ a_{31} & a_{33} \end{vmatrix} + \begin{vmatrix} a_{11} & a_{14} \\ a_{41} & a_{44} \end{vmatrix} + \begin{vmatrix} a_{22} & a_{23} \\ a_{32} & a_{33} \end{vmatrix} + \begin{vmatrix} a_{22} & a_{24} \\ a_{42} & a_{44} \end{vmatrix} + \begin{vmatrix} a_{33} & a_{34} \\ a_{43} & a_{44} \end{vmatrix} \tag{95}$$

and

$$A''' = a_{11} + a_{22} + a_{33} + a_{44} \tag{96}$$

where A_{ij} is the cofactor of the 4×4 determinant.

It has been shown that a mathematical model can be derived from a set of data points which lie on the surface of the object. It has also been shown that, given a general form quadratic equation, the mathematical model can be

TABLE 13

CLASSIFICATION OF QUADRIC SURFACES

		Proper quadrics $A \neq 0$			Improper (degenerate) quadrics $A = 0$				
		$A > 0$		$A < 0$	Cones and cylinders $A' \neq 0$		Pairs of planes (degenerate quadrics) $A' = 0$		
		$A'I$ and J both > 0	$A'I$ and J not both > 0		$A'I$ and J both > 0	$A'I$ and J not both > 0	$A'' > 0$	$A'' < 0$	$A'' = 0$ $A''' = 0$
Central quadrics $D = 0$	DI and J both > 0	No real focus (imaginary ellipsoid)		Real ellipsoid	Point in finite portion of space; vertex of imaginary elliptic cone; point ellipsoid				
	DI and J not both > 0		Hyperboloid of one sheet	Hyperboloid of two sheets		Elliptic cone; (degenerate hyperboloid)			

Noncentral quadrics $D=0$	$J>0$			Elliptic paraboloid (of revolution if $I^2 = 4D$)	No real locus (imaginary elliptic cylinder)	Real elliptic cylinder circular if $I^2 = 4D$)	Straight line (degenerate elliptic cylinder intersection of imaginary planes)		
	$J<0$		Hyperbolic paraboloid			Hyperbolic cylinder		Two real planes intersecting in finite portion of space (degenerate hyperbolic cylinder)	
	$J=0$ $I \neq 0$					Parabolic cylinder	No real locus (imaginary parallel planes)	Two real parallel planes (degenerate parabolic cylinder)	One real plane (coincident parallel planes)
Rank of the 4th-order square matrix A_{ik}			4		3			2	1

decomposed into shape, size, orientation and location information. It was stated previously that for a given surface model of order k, the number of data points required to determine the coefficients of the three dimensional polynomial is

$$T = \frac{(k+1)(k+2)[2(k+1)+1]}{6} + \frac{(k+1)(k+2)}{4} - 1. \qquad (97)$$

This is, however, the minimum number of required data points. Since it is often difficult to measure accurately the coordinates of surface points by any method, it is often desirable to use more than the minimum number of data points in order to reduce the error.

The method of least-squares approximation provides one method of fitting a surface to a set of data points with the minimum mean-squared error. If more than the minimum number of data points are used, then X is no longer a square matrix and cannot be inverted. A method known as the pseudo-inverse solution, however, provides a minimum mean-squared-error solution by multiplying the equation

$$XA = B \qquad (98)$$

by X' to yield

$$(X'X)A = X'B. \qquad (99)$$

Since $X'X$ is a square matrix, then, if $\det(X'X) \neq 0$, the solution may be found by

$$A = (X'X)^{-1} X'B \qquad (100)$$

to yield the minimum mean-squared error for the interpolating polynomial. Surface measurement methods can be used to obtain a number of sample data points much larger than the minimum required. In order to reduce the errors due to truncation in the sample data, it is best to use all of the data that is obtained by the process.

The sample surface points obtained by the procedure discussed in this chapter may be used to describe the surface geometry by applying the methods previously discussed. Using the sample points of the cylinder, a quadric surface is obtained having the standard form quadratic equation

$$0.489x^2 + 0.0321y^2 = 1.0111z^2 = 2.8970 \qquad (101)$$

which, from Table 13, describes an ellipsoid. However, note that the surface is extremely elongated along the y-axis. if the y^2 coefficient is assumed to be zero, the circular cylinder is seen to be elliptical, having a major axis of 1.217 inches and a minor axis of 0.846 inches. The actual radius of the clinder was measured and found to be approximately 1.375 inches. Similarly, the mug surface is described by

$$0.6805x^2 - 0.0692y^2 + 1.0010z^2 = 8.3593 \qquad (102)$$

after the surface is rotated to alignment with the axis and translated to the origin. The surface described by this equation is an elliptic hyperboloid of one sheet. In the $x-z$ plane the ellipse has a major axis of 1.75 inches and a minor axis of 1.44 inches. The actual radius at the base of the mug is 1.55 inches. Both surface representations reveal a reasonable approximation to the true surface geometry.

4.3 Summary

Numerous sample data points for a given surface cannot directly describe the surface unless some geometric or algebraic relationship between neighboring points is obtained such that areas of the surface which are not included among those sample points can be approximated. Mathematical modeling is often the best approach to surface description, since the information contained in the model includes the shape, size, orientation and location of the object or surface (where applicable). A mathematical model also provides the interpolation and extrapolation of surface points which are not included in the same data. Polynomial models are the easiest mathematical models to obtain from sample points since linear methods may be used to determine the coefficients of the polynomial. However, in cases where a surface is best described by other functions such as trigonometric, logarithmic, hyperbolic, etc., methods do exist for which the exact model can be obtained.

5. CONCLUSIONS

Curved surface representation techniques have been described whereby a model of the surface may be obtained which is applicable to surface recognition. These models may consist of mathematical models or numerical approximations. In order to obtain the necessary sample surface points to describe the surface geometry adequately, two methods of curved surface measurements were developed. One method is based upon the relationships between surface shading and surface geometry. A second method was developed which utilizes the stereo vision approach to surface measurement but requires only a single image of the surface to be analyzed.

Curved surface representation is a necessary step in curved surface recognition. Because objects comprised of curved surfaces may have no edges or vertices, surface models must be used to describe their geometry in three dimensions. These models may consist of mathematical models or numerical approximations such as splines. Mathematical modeling is often the best approach to surface description since the information contained in the model may be easily decomposed into qualitative descriptors including shape, size, orientation and location. A description of this type provides an object recognition technique which is independent of the location or orientation of the object.

In order to describe an object, some level of knowledge must exist for which a qualitative description can be derived. The level of knowledge obtained by the two techniques discussed in this paper is a set of sample surface point coordinates.

One method of obtaining sample surface points is by examining the shading of the surface under controlled lighting conditions. It has been demonstrated that the models used in image synthesis for surface shading can be applied in image analysis to obtain the surface shape. If an appropriate shading model for the surface material and texture is obtained, a measurement of the surface normal vector for the surface's points viewed by each pixel in the image can be computed. This method does not require a high resolution image in order to accurately obtain a sufficient number of sample surface points to describe the object. Obtaining an adequate shading model for a particular surface material and texture is a difficult problem requiring continued research.

A second method is described which utilizes the stereo vision principle. Vertices and edges are the features most commonly matched in a pair of images and, by triangulation, the three-dimensional location of the feature is obtained. Featureless curved surfaces may have no edges or vertices and still be closed surfaces. The solution of this problem in this paper is to impose features onto the surface through "active imaging." The system described uses both camera and projection system. Because the projection process is identical to the imaging process, with respect to the process models, this system performs stereo vision with a single image.

The automation of matching the image points and grid points is a topic requiring further research. Other research which could be considered is the total automation of the single image stereo procedure. This would require automatic calibration perhaps by computer controlled pan and tilt mounts for the camera and projector. Also, a robot arm mount for the camera and projector might be desirable. The calibration procedure should give the perspective projective matrices previously described. Further research could also be done on the automatic vertex location in the recorded images of the projected grid. Finally, the use of the measured three-dimensional points for the recognition of objects in a scene could be expanded.

REFERENCES

[1] T. B. K. Tio, J. J. Hwang, C. A. McPherson, and E. L. Hall, "Surface Location in Scene Content Analysis," Proc. SPIE Conf., August, 1981, pp. 330-337.

[2] F. I. Parke, "Measuring Three-Dimensional Surfaces with a Two-Dimensional Data Tablet," Computer and Graphics, Vol. 1, 1975, pp. 5-7.

[3] E. L. Hall, *Computer Image Processing and Recognition,* Academic Press, New York, 1979.

[4] R. C. Gonzalez and Paul Wintz, *Digital Image Processing,* Addison-Wesley, Reading, MA, 1977.

[5] J. J. Hwang, "Computer Stereo Vision for Three-Dimensional Object Location," Ph.D. Dissertation, The University of Tennessee, Knoxville, August, 1980.

[6] W. D. Renner, T. P. O'Connor, S. R. Amtey, P. R. Reddi, G. K. Bahr, and J. G. Kereiakes, "The Use of Photogrammetry in Tissue Compensation Design," Radiology, Vol. 125, November, 1977, pp. 505-510.

[7] R. Cernuschi-Frias, D. B. Cooper, and R. M. Bolle, "Estimation of Location and Orientation of 3-D Surfaces Using a Single 2-D Image," Proc. Pattern Recognition and Image Processing Conf., June, 1982, pp. 605-610.

[8] R. B. Kelley, J. R. Birk, D. Q. Duncan, R. P. Tella, and L. J. Wilson, Workpiece handling and sorting system, Patent #US4305130-D52, 1981.

[9] M. Oshima and Y. Shirai, "Representation of Curved Objects Using Three-Dimensional Information," Proc. Second USA-Japan Computer Conf., 1975, pp. 108-112.

[10] R. J. Woodham, "A Cooperative Algorithm for Determining Surface Orientation from a Single View," Proc. Fifth Int. Joint Conf. Artificial Intelligence, 1977, pp. 635-641.

[11] S. Berman, P. Parikh, and G. Lee, "Computer Recognition of Overlapping Parts Using a Single Camera," Proc. Pattern Recognition and Image Processing Conf., June, 1982, pp. 605-655.

[12] P. J. Mungaonkar, L. G. Shapiro, and R. H. Haralick, "Recognizing Three-Dimensional Objects from Single Perspective Views Using Geometric and Relational Reasoning," Proc. Pattern Recognition and Image Processing Conf., June, 1982, pp. 479-484.

[13] R. P. Kruger and W. P. Thompson, "A Technical and Economic Assessment of Computer Vision for Industrial Inspection and Robotics Assembly," Proc. IEEE, Vol. 69, No. 12, December, 1981, pp. 1524-1538.

[14] S. W. Holland, L. Rossol, and M. R. Ward, "CONSIGHT-I: A Vision Controlled Robot System for Transferring Parts from Belt Conveyors," *Computer Vision and Sensor Based Robots,* G. G. Dodd and L. Rossol, Eds., Plenum Press, New York, 1979, pp. 81-100.

[15] J. S. Albus, *Brains, Behavior, and Robotics,* Byte/McGraw-Hill, New York, 1981.

[16] P. M. Will and K. S. Pennington, "Grid Coding: A Novel Technique for Image Processing," Proc. IEEE, Vol. 60, No. 6, June, 1972, pp. 669-680.

[17] J. L. Posdamer and M. D. Altschuler, "Surface Measurement by Space-Encoded Projected Beam Systems," Computer Graphics and Image Processing, Vol. 18, 1982, pp. 1-17.

[18] R. P. Paul, *Robot Manipulators,* MIT Press, Boston, MA, 1981.

[19] B. Julesz, "Towards the Automation of Binocular Depth Perception," Proc. IFIP Congress 1962, Amsterdam, North Holland, 1962, pp. 439-443.

[20] C. A. McPherson, "Computer Vision Approach to Three-Dimensional Curved Surface Measurement and Surface Representation for Object Recognition," MS Thesis, University of Tennessee, June, 1983.

[21] B. K. P. Horn, "Image Intensity Understanding," AI Memo 335, MIT Laboratory, August, 1975.

[22] M. I. Skolnik, Radar Handbook, McGraw-Hill, New York, NY 1970.

[23] J. F. Blinn, "Models of Light Reflection for Computer Synthesized Pictures," Computer Graphics, Vol. 11, No. 2, 1977.

[24] B. T. Phong, "Illumination for Computer Generated Images," Comm. ACM 18, No. 6, June, 1975, pp. 311-317.

[25] K. E. Torrance and E. M. Sparrow, "Theory of Off-Specular Reflection from Roughened Surfaces," JOSA, Vol. 57, No. 9, September 1967, pp. 1105-1114.

[26] R. L. Cook and K. E. Torrance, "A Reflectance Model for Computer Graphics," Computer Graphics, Vol. 15, No. 3, August, 1981, pp. 307-316.

[27] T. Whitted, "An Improved Illumination Model for Shaded Display," Comm. ACM, Vol. 23, 1980, pp. 343-349.

[28] C. Sagan, *Broca's Brain: Reflections on the Romance of Science,* Random House, New York, NY, 1979, p. 17.

[29] W. K. Giloi, *Interactive Computer Graphics,* Prentice-Hall Englewood Cliffs, NY, 1978.

[30] A. K. Mackworth, "Interpreting Pictures of Polyhedral Scenes," Artificial Intelligence, Vol. 4, 1973, pp. 121-137.

[31] R. L. Burden, J. D. Faires, and A. C. Reynolds, *Numerical Analysis,* Prindie, Weber and Schmidt, Boston, MA, 1978.

[32] R. O. Duda and P. E. Hart, *Pattern Classification and Scene Analysis,* John Wiley and Sons, New York, NY 1973.

[33] M. Potmesil and I. Chakravarty, "A Lens and Aperture Camera Model for Synthetic Image Generation," Computer Graphics, Vol. 15, No. 3, August, 1981, pp. 297-315.

[34] J. B. K. Tio, C. A. McPherson, and E. L. Hall, "Curved Surface Measurement of Robot Vision," Proc. Pattern Recognition and Image Processing Conf., June, 1982, pp. 370-378.

[35] C. A. McPherson, J. B. K. Tio, E. L. Hall, and F. A. Sadjadi, "Curved Surface Representation for Image Recognition," Proc. Pattern Recognition and Image Processing Conf., June, 1982, pp. 370-378.

[36] G. B. Thomas, Jr., *Calculus and Analytical Geometry,* Addison-Wesley, Menlo Park, CA, 1972.

[37] B. Spain, *Analytical Quadrics,* Pergamon Press, New York, NY 1960.

[38] G. A. Korn and T. M. Korn, *Mathematical Handbook for Scientists and Engineers,* McGraw-Hill, New York, NY, 1961.

8

Use of Optical Reflectance Sensors

BERNARD ESPIAU

1. BASIC PRINCIPLES

1.1 Introduction

A robot may be considered as the association of four main elements:

1. A sensitive system (perception).
2. A decision-and-control system.
3. A physical device performing interactions with the environment.
4. A man-machine communication system.

Associating these functions defines a classical automatic (and "programmable") robot when the world is exactly known or given. On the other hand, working in unknown or uncertain worlds is a characteristic property of "adaptive" robots, in which perceptive devices are of most importance. It is well known that the two main fields of investigation in perception problems are based on visual and force (or compliance) informations. In the classical approach, robot vision (2D or 3D) is studied in a *global* way, often from sensors having no or few motion capabilities. *Now, it is possible to consider that in most cases, the actions and the evolution of a robot are time-and-space-local phenomena,* for which the information provided by a classical vision system are inadequate or superabundant.

Here is thus proposed a new concept, called "local environment" to which some new kind of sensors and related algorithms will be associated.

This chapter will include four main parts, following a presentation of basic principles: first, a state-of-the-art, technical review in proximity sensing will be made. Then, we shall describe a special kind of sensor, based on optical reflectance properties. Thirdly, various techniques for getting informations from these sensors will be given. Finally, algorithms for building closed-loop controls based on the outputs of optical sensors, with some examples, will be presented.

1.2 The Local Environment Concept

1.2.1 Basic Definitions

We consider a serial link, rigid manipulator with n degrees of freedom. The configuration of a manipulator is specified by n generalized coordinates q_i taken as independent joint coordinates: with respect to fixed given base R_0, the position and orientation of R_n, a frame associated to the end effector (hand for example), are specified by six operational coordinates x_j, with:

$$x = f(q) \tag{1}$$

where $q \in D_q$: $\left[q^m, q^M\right]$, D_q closed connex subspace of IR^n. We may then define some aspects of the world of a given manipulator:

Def 1 The x - subspace (or configuration space): $D_x = \{x: \forall q \in D_q\}$, $x \in IR^6$

Def 2 The reachable space: (RS): for a given end effector, it is the physical associated volume from IR^3, corresponding to D_x, referenced in R.

Def 3 The workspace (WS): related to a given task, WS is the part of RS in which, in each point, the value of x required to perform the task may be reached by at least one value of $q \in D_q$.

It is also necessary to establish some definitions related to the *interactions* between a robot and its world:

Def 4 At each instant, and for a given task, an *obstacle* is an object (on a part of an object) disturbing the execution of the task.

Def 5 In the same conditions, a *target* is an object (or a part of object) concerned with the task.

Def 6 The *global environment* (GE) is the set of all targets and all the obstacles lying in the workspace at a given time.

Def 7 The *local environment* (LE) is the intersection between the global environment and a given volume located around the *mobile* parts of the manipulator.

The last definition is here given from an intuitive point of view, but it is possible to state a mathematical model of the local environment, mainly using two approaches:

- a geometric one, by defining some simple volumetric approximations of the parts constituting the robot, and *growing* them in a given way (e.g. in [1]).
- A dynamic approach using modeling with potential functions (e.g. [2]). A convenient choice is the following (Figure 1): We define the frame R_k, related to the kth link L_k. As in the previous case, we choose a simple approximation C_k for the solid L_k, such as:
  $\forall x^k$ (point in R_k) $\in L_k : x^k \in C_k$ (cylinder and 1/2 spheres in the Figure).

An interesting choice for a potential function is then:

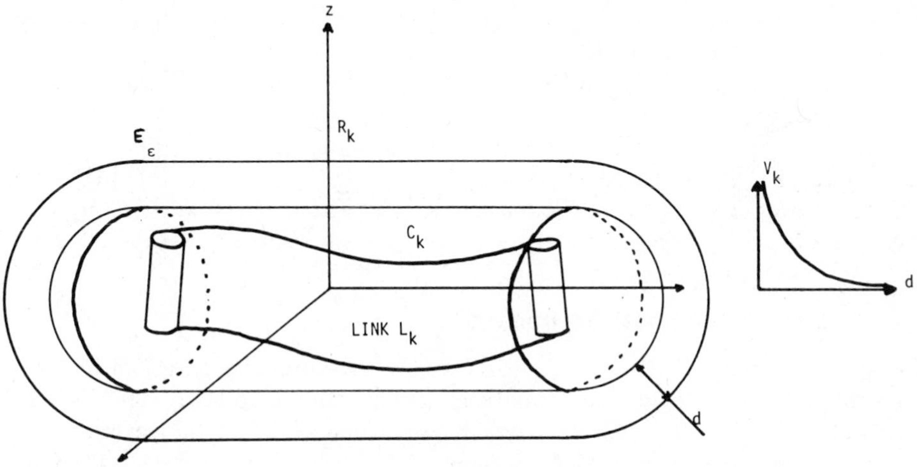

Figure 1 Potential function for a link.

$$V_k(x^k) + c = \frac{\lambda}{d} \text{ (newton)} = \begin{cases} \dfrac{\lambda}{(x^2+z^2-r^2)^{1/2}} & \text{for } y \in [y_m, y_M] \\[2ex] \dfrac{\lambda}{[x^2+(y-y_M)^2+z^2-r^2]^{1/2}} & \text{for } y > y_M \\[2ex] \dfrac{\lambda}{[x^2+(y-y_m)^2+z^2-r^2)]^{1/2}} & \text{for } y < y_m \end{cases} \tag{2}$$

with $x^k = (x,y,z)$ in R_k

($V_k(x^k)$ is only defined on the outside of C_k)

The total potential function, associated to the manipulator, is then

$$V(x) = \sum_{k=1}^{n} \tilde{V}_k(x), \tag{3}$$

where $\tilde{V}_k(x)$ is $V_k(x^k)$ *expressed in* R_0. $V(x)$ is thus dependent on the configuration (q) and we may only define *instantaneously* the local environment: Let E_ϵ be the equipotential surface: $\{x: V(x) = \epsilon\}$.

With a convenient choice of $\{V_k\}$, E_ϵ is the boundary of a finite volume, which is the instantaneous LE.

The interest of such a modeling for LE will be pointed out in the last part of the chapter.

From a practical point of view, it is obvious that using the *complete* LE for example for control purposes, is necessary only in few cases, e.g. like very complex worlds, highly mobile obstacles, multi-arms working. So, in this chapter, we shall mainly consider the case of the LE related to the end effector, as shown in Figure 2.

1.2.2 Sensing the Local Environment

Classical perception systems are concerned with the global environment, and provide a great amount of informations, which does not directly describe the *relationship* (geometric, for example) between the robot and the environment. From the opposite point of view, we may notice that a LE-based control might highly reduce processing time for local tasks, and could allow a compensation of open-loop errors, if the interactions were continuously taken into account. For such purposes, we need special sensors, and we may give some definitions, from a robotic point of view:

Def 8: A proximity sensor is a sensor providing informations related to the objects lying into the LE.

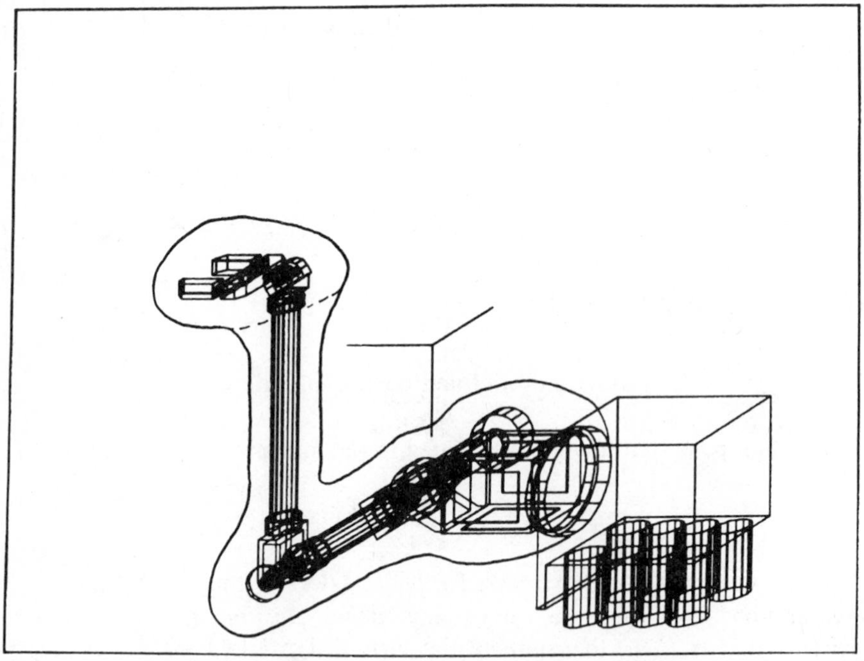

Figure 2 Local environment for a MA23 manipulator.

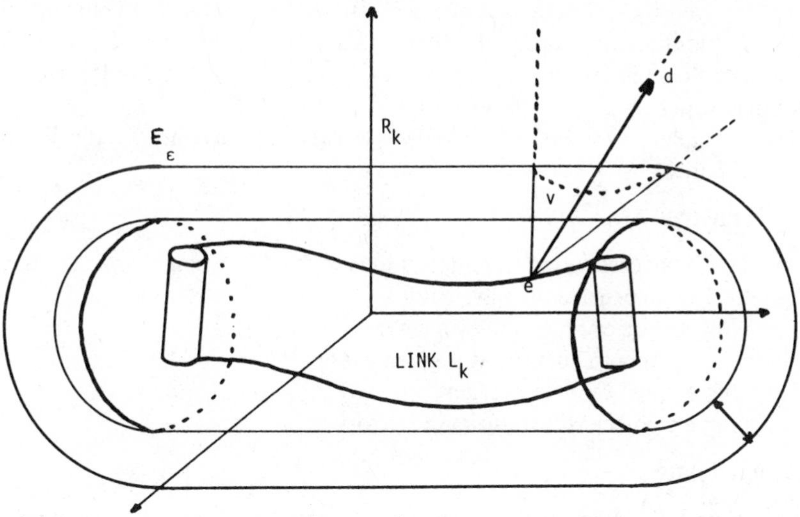

Figure 3 The observed volume from e.

More precisely, we choose, in the given frame R_k (Figure 3):

Def. 9 – a point of observation *e*

– a direction of observation d, with origin *e*

– a volume of observation, built around d, and bounded by the LE: v

From v, two kinds of informations may be extracted:

a. about the object:

a1 What are the values of some characteristics parameters of the object (geometry color,...)?

a2 Is it a target or an obstacle?

b. about relations between the object and the robot:

b1 Is it a relative motion? what are its direction and velocity?

b2 What about its instantaneous situation with regard to R_k: apparent geometry, "distances" (notion to be precised)?

b3 what is the *specific* response of the sensor to the object?

It may be noticed that a-type informations are high-level, (like pattern recognition), and that b-type informations are generally more easily processed.

Finally, using several volumes v, in a multi-sensors context, an important part of the LE may be investigated; further, if we associate the successive information due to displacements of the robot, it is easy to see that, even with elementary low-level sensors, the resulting information about the environment may be very suitable, particularly for control purposes.

1.3 Fields of Application

LE sensing is mainly useful if it is impossible to exactly know the global environment by incomplete modeling, or inadequate sensing. A second case of interest is given by the existence of uncertainties about the real behavior of the robot (flexible links for example). In both cases, a closed loop control based on LE informations is a good way to make the robot highly adaptive to its environment, generally with a low cost.

Such complex, variable, or unknown worlds may be met in the following fields of robotics:

1.3.1 Industry

LE sensing is necessary in non-deterministic cases, when classical vision is inadequate (too expensive, or physically impossible), for example:
- Grasping an object on a conveyor belt
- Following a smooth surface at a given range
- Tracking a slowly moving target
- Grasping an object with positioning errors.

1.3.2 Medical

Prosthesis and telemanipulators are used to give assistance to the physically handicapped. An aid to grasping hidden or flimsy objects may be brought about by closed-loop algorithms based on LE [3].

1.3.3 Hostile Worlds

It is the main field of applications, because of the complexity (and often unicity) of tasks and the degradation of vision in most cases:
- Nuclear plants [4]
- Space applications (local control is sometimes essential, because of transmission delays)
- Underwater working.

In whole cases of Teleoperation (master/slave) LE sensing may
- Increase the slave's autonomy
- Improve the information fed back to the operator.

2. STATE OF THE ART AND TECHNICAL REVIEW

2.1 Technological Solutions to Proximity Sensing

We consider two kinds of sensors:

The first are called "active" when they themselves emit the information which has to interact with the environment before being analyzed. In the other case, "passive" sensors are considered.

2.1.1 Passive Sensors

If we do not consider the possibility to use *active* labeling of the objects we may only point out some specific vision systems:

a. on board solid state sensors: CCD systems, which are well known, present some significant advantages e.g. small size, high sensitivity, providing high-level information. However, they are fragile and need rather complex processing algorithms.

b. Solid state sensors with transmission by optical fibers (e.g.: UNIVISION). These systems are more convenient for LE sensing because:
   - The terminal size is small
   - The number of observation-points may thus be substantially increased with regard to the previous case.
   - Each elementary sensor may have a small number of pixels, simplifying the processing software.

However, as the algorithms related to such sensors are mainly concerned with picture processing, we shall not further consider vision systems from a specific LE point of view.

2.1.2 Active Sensors

We summarize in the following main classes of active sensors; pointing out some characteristics of utilization e.g. Kind of provided information, range, limitations, sensitivity to noises and perturbations, reliability, simplicity, size and cost.

a. Tactile systems. They provide only touching information. Sensing LE is then not possible without setting some cumbersome devices on the robot (rings, needles - see [4]).

b. Pneumatic sensors. They have a small range (some millimeters), and the interpretation of information is very difficult (due to high sensitivity to external conditions, orientation, etc. - example in [4]).

c. Ultrasonic systems. They provide distance information, by measuring the flight time of some pulses (example: POLAROID system), with a piezoelectric transducer. Their size is not very small, they are sensitive to the orientation and the nature of the target, and there are some problems in measuring distances less than 20 cm. Finally they are inadequate for space applications, but work well underwater.

d. Microwaves. Several classical techniques of measurement may be used to obtain distance information, e.g. frequency modulation, phase or flight time measurements. Another possibility is to measure the back-diffused energy of a target. This principle is analogous to the measure of optical reflectance (see further). However, the working range of such systems (some meters), and the size of the sensitive device limit their use to very large scale robots e.g. for space applications [4].

e. Active vision. To obtain some information about the apparent shape of an object, it is interesting to overlighten points or planes in the scenes for example by projecting lines or grids [5,6]. Further, by using 2 or 3 CCD sensors with a sweeping laser spot, it is possible to get relative 3D coordinates of a target. Unfortunately, the complexity of such a system in not compatible with the requests of practical LE sensing.

f. Inductive sensors. These sensors are widely used for presence detection. Highly robust, with a fast response time, they have however a bad size/range ratio and are only sensitive to metallic parts. They are sometimes used in arc welding for joint tracking.

2.2 Optical Reflectance Sensors (ORS)

2.2.1 General Presentation

The ORS is one of the most interesting devices for LE sensing. The principle is given in Figure 4, and typical outputs are presented Figure 5.

In most cases, the emitter is a LED with wavelength 950 nm, and the receiver is a photodiode or a phototransistor. Focusing devices are lenses or optical fibers. The main advantages of these sensors are:

- very good size/range ratio
- low cost, good reliability
- simplicity of the associated electronics
- low sensitivity to environmental disturbances, by using synchronous modulation or pulsed emission.

However, as the output information is the *energy*, back-diffused by the target in the detection of the photoreceiver, it is clear that these sensors are sensitive to the nature and the local orientation of the object (Figure 5).

2.2.2 Examples of Sensor Embodiments

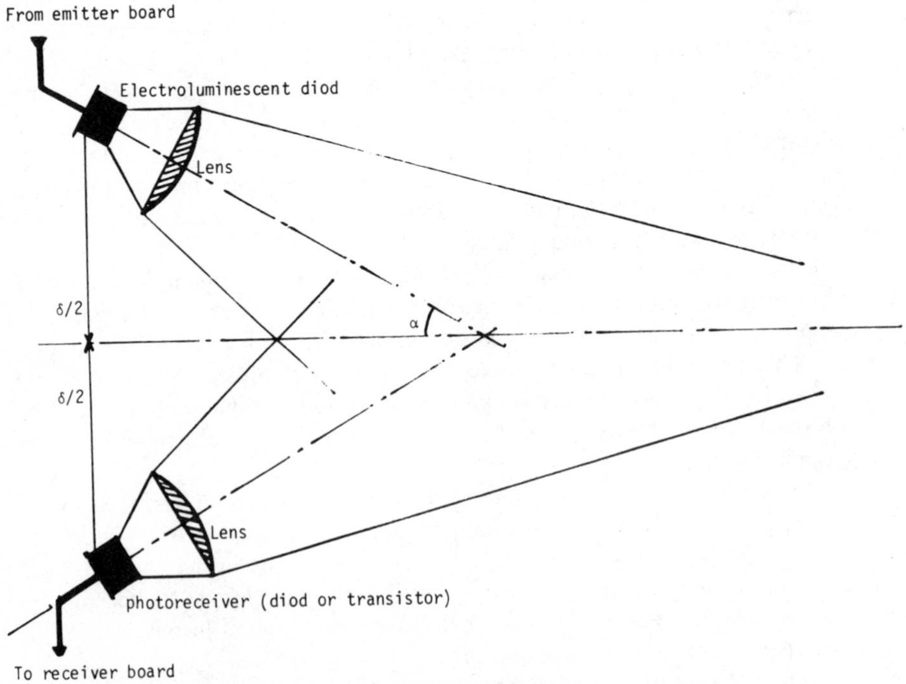

Figure 4 Principle of optical reflectance sensors.

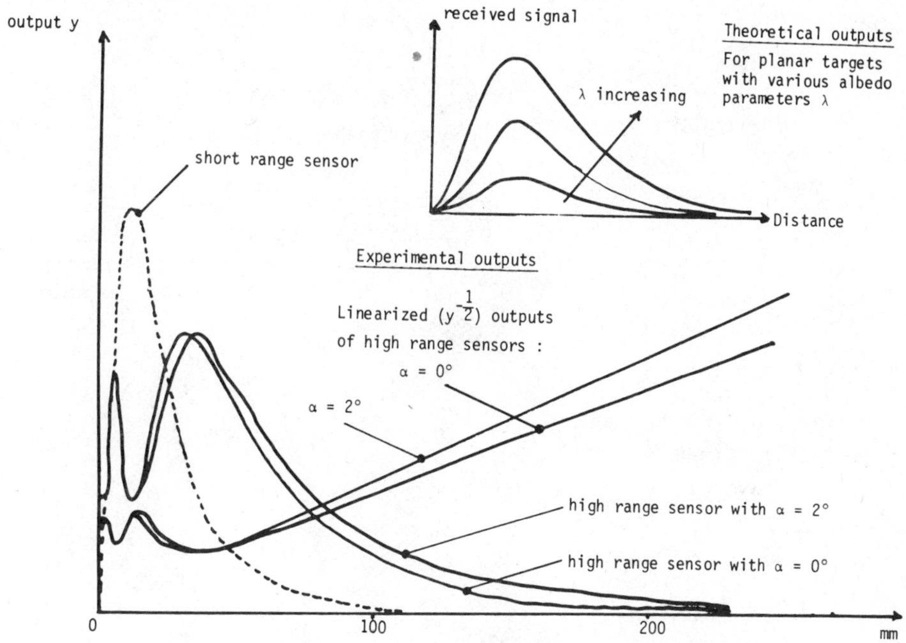

Figure 5 Typical outputs of ORS.

2.2.3 ORS with Optical Fibers

Designed by SAGEM C_y [7], the sensors presented in Figure 6 use optical fibers, with emitters and receivers located in the connectors. Optical fibers (OF) avoids the integration of any electronics component onto the gripper. However, OF do present some drawbacks: the aperture angle (without additional lenses) is fixed, and the minimal curvature radius of the fibers is not very small. Other systems exist using OF, for example with 1 mm ϕ fingers (range 10 mm).

2.2.4 Other Kinds of IR Sensors

Several optical proximity systems have been designed at JPL [8,9,10], mainly:

- The first of them, which uses prisms, with size about 3 cm^3
- A multi-sensor system, with 4 emitters and 4 receivers, defining 13 small sensitive areas in a plane, a four-claw gripper with four proximity sensors.

The IR sensors from JPL have been integrated within various end effectors, for example: a two-fingered gripper, an anthropomorphic hand, a four-fingered gripper. A multi-sensor gripper, designed by C. Wampler ([18]) is presented in Figure 7. It may be noticed that this system is used in the closed loop control of manipulators.

Some other kinds of realizations may be found in references [11], [17], [19].

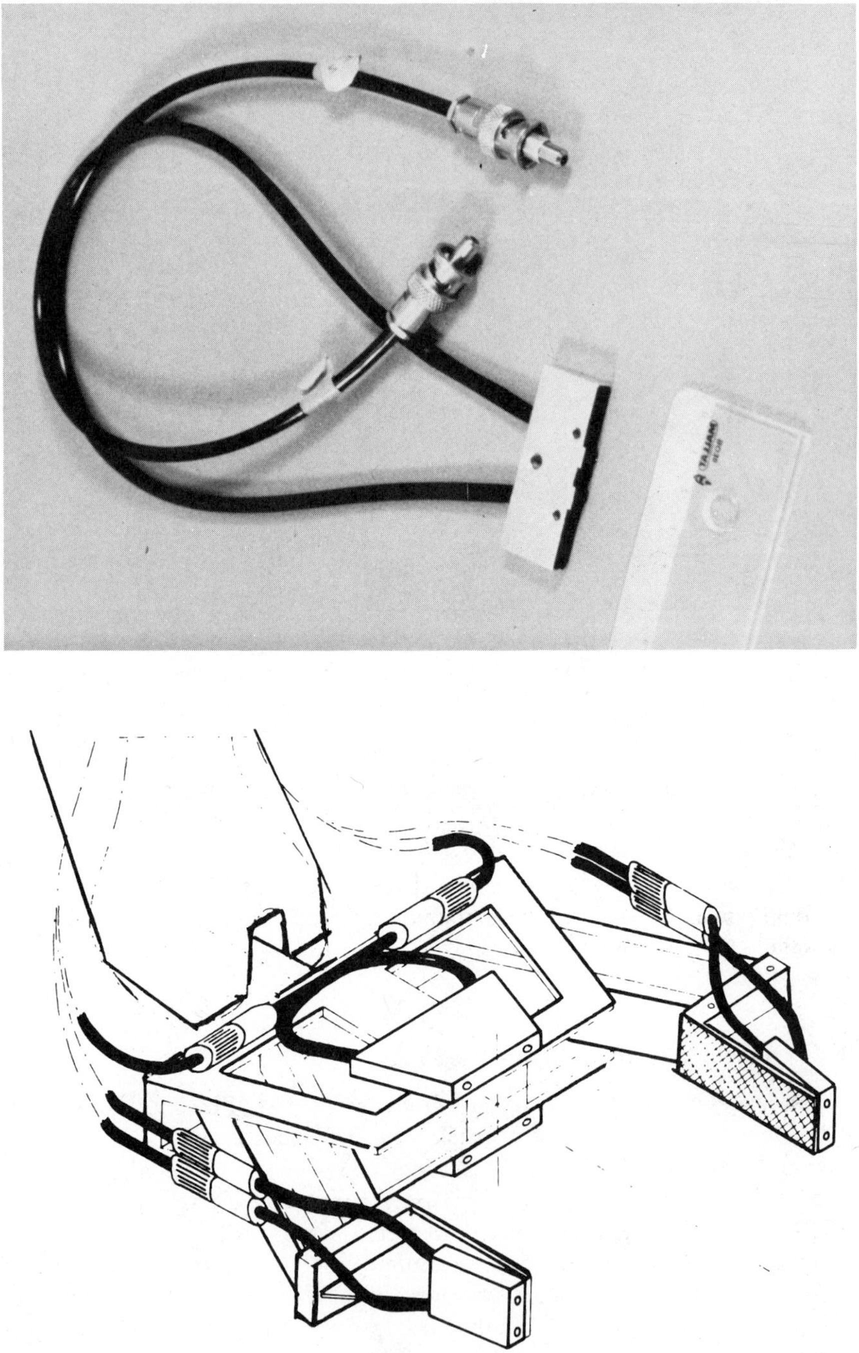

Figure 6 A proximity sensor with optical fibers, and the associated gripper from SAGEM (see [7]).

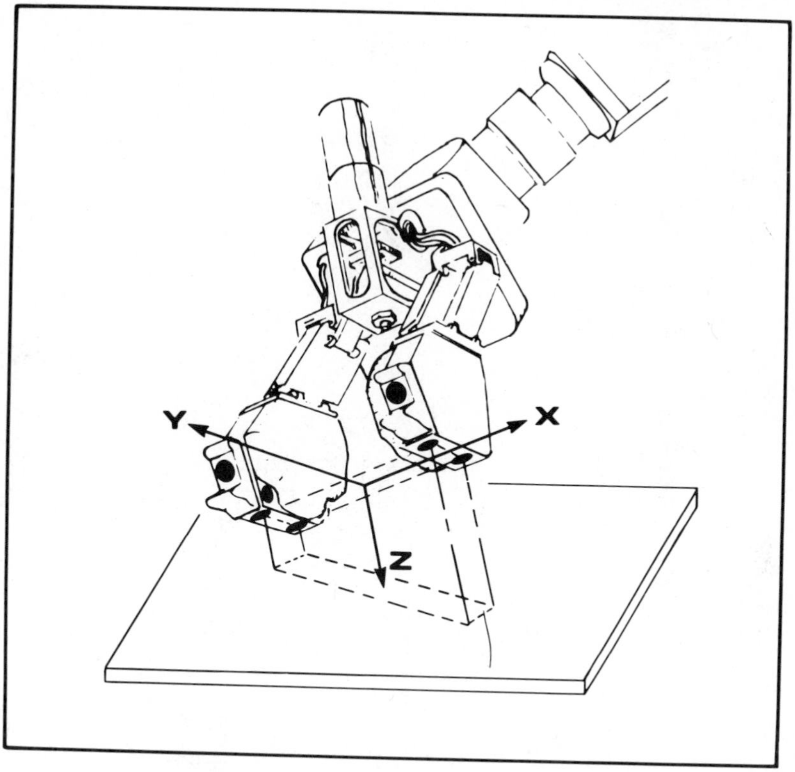

Figure 7 A 10-sensors bidigital gripper, from [18].

A new multi-sensor system. To efficiently design LE-based control algorithms it is necessary to get information about the LE:
- quickly
- with a sufficient covering of the ideal LE volume.

This requires special performances from proximity sensors:
- low response time: τ
- sufficient range: r
- very small size (volume of the sensing head: v_s)

From a study of several applications, and in relation with the main existing robots, we may draw the following specifications for a realistic LE sensor:
- *global* response time for the whole sensing system: < 10 ms
- range r: up to 250 mm, not less than 50 mm depending on the field and on the target.
- $v_s < 1 \text{ cm}^3$

A good way to make adequate sensors with low cost is to use available photocomponents with pulsed scanning control. Pulsed mode gives an high instantaneous emitting power, and scanning preserves the required resting time

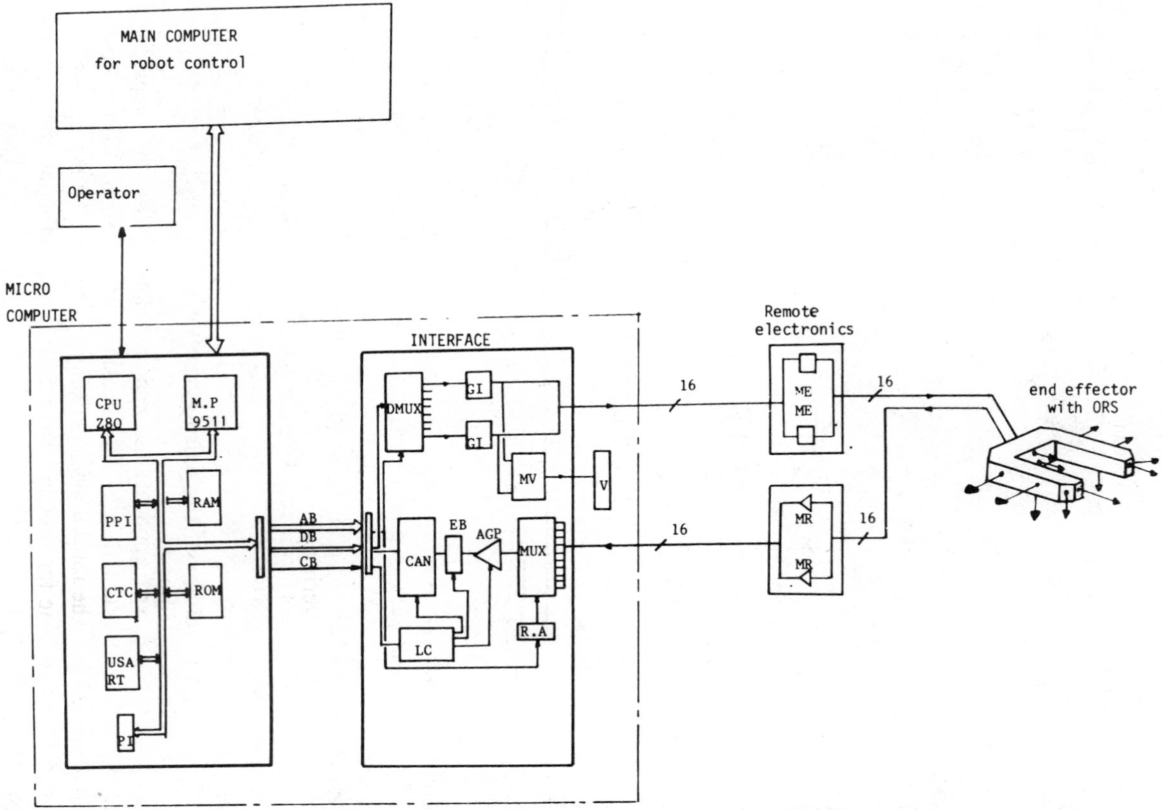

Figure 8 Principle of the IRISA's multi-sensors system.

of each sensor. Other advantages of pulsed mode lie in increasing the range, quickening the response, and decreasing the sensitivity to optical disturbances.

Based on these principles, a complete multi-sensors system has been made at IRISA [12]. Figure 8 gives the general description of the 16-channel system. It includes the following components:

A set of sensor heads, the characteristics of each one are given by:

- The kind of the emitter, E
- the kind of the receiver, R
- the values of parameters x and δ (shown in Figures 4 and 9), tilt angle and distance between axes

All these features have an effect on the shape of the sensing area and on the practical range.

From the many possible combinations, we extract four typical E/R couples given in Table 1.

The associated electronic modules:

- Emission: Generation of the power pulse to the LED. A typical scheme is given Figure 10.
- Reception: The aim of the module is to preamplify the analog output of the optical receiver. To improve signal-to-noise ratio, it is necessary to have a small pulse time τ that implies that the reception module must have a low rise time. Several schemes are useful. We give in Figure 11 the typical modules associated with a phototransistor and a photodiode.

The microcomputer and the interface. It has three main functions:

- Driving the sensors. Generation of the pulses, with programmable parameters (e.g. length τ, resting time, time delays in the sequence of successive activation of all sensors and A/D conversion of the received outputs.
- Local processing of the signals. Elimination of continuous values, filtering, detection, linearization, estimation of distance, etc. (see further the following algorithms).
- Communication with the user and with high-level control systems. Selection of active sensors, and of desired preprocessing algorithms, transmission of sensor data, visualization of the received signals.

The system described in Figure 12 uses a Zylog Z80 microcomputer with a 9511 arithmetic unit and specific interface cards. The total response time (from pulse start to data available in memory) is less than 200μs for each sensor. An industrial ORS system will be shortly available on INTEL 8086.

2.2.5 Sensing Devices

LE based control problems may be concerned with two classes of end effectors:

- The special-purposes devices.
- The general-purposes effectors.

In the first case, the integration of sensors is related to a specific application

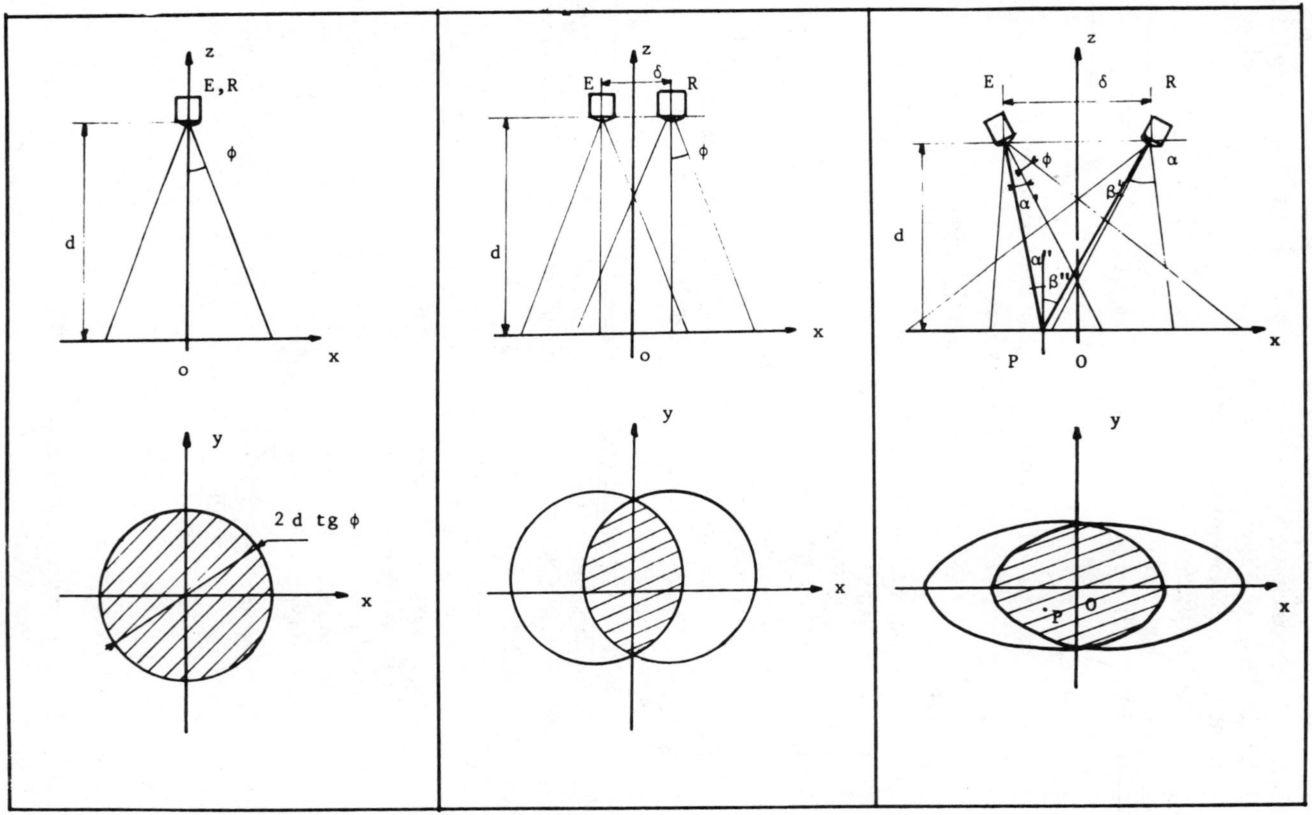

Figure 9 Different geometric dispositions of sensors.

TABLE 1

	Emitter	Receiver	α(deg)	δ(mm)	Practical maximal range (mm) with a white plane target orthogonal to sensor axis	field: ϕ(deg)
1	Siemens LD 242	Siemens BP 104 or BP 103	0	8	120	± 60
2	Siemens LD 271 or TiL 33	Texas TiL 99	4	7	200	$\phi_E = \pm 25$ $\phi_R = \pm 10$
3	Texas TiL 31	Texas TiL 81	6	9	250	$\phi_E = \pm 5$ $\phi_R = \pm 10$
4	Siemens LD 261 (Small size)	Siemens BPX 81	0	3	80	$\pm 10°$

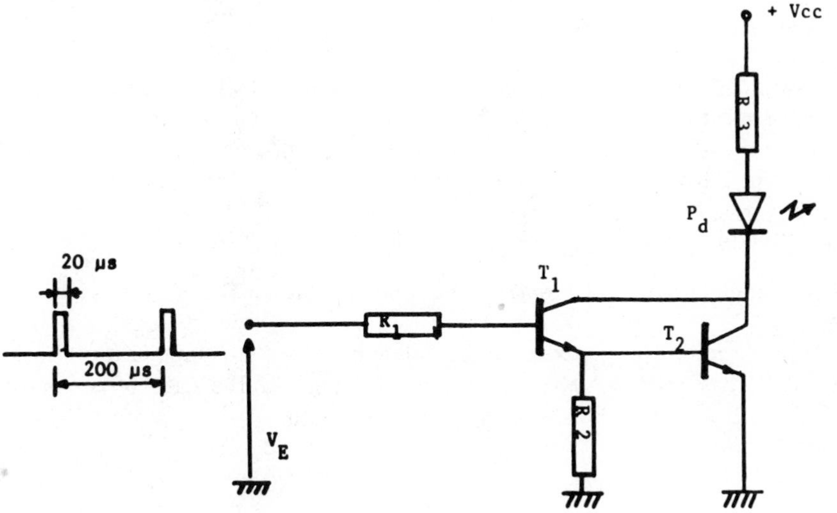

Figure 10 A typical emitting board.

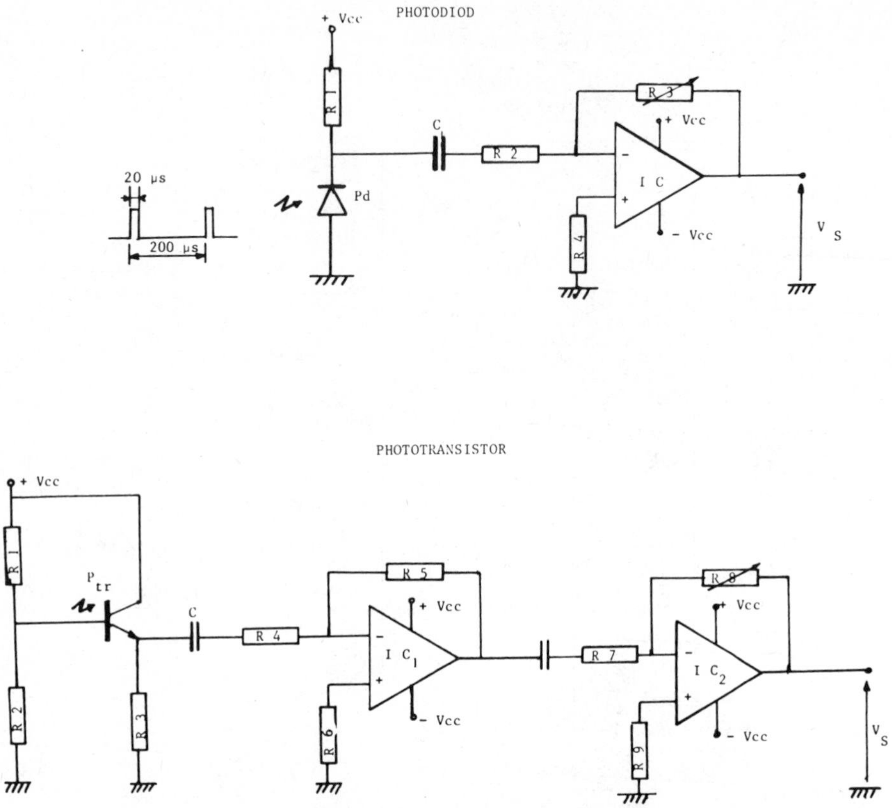

Figure 11 Two typical receiving boards.

and the sensing function may be precisely determined. In the second situation, it is necessary to define some basic general functions associated with given areas of the LE. Three classes of such functions may be considered.

a. Fine positioning of a end effector.
b. Approximate positioning.
c. Safety functions, e.g. rough obstacle avoidance.

For example, Table 2 gives the functions relating to the previously described sensors.

For designing a general-purpose effector, the first stage is to locate the sensing areas related to the previous functions: In the case of a two-finger gripper this distribution is given in Figure 13, with

Z1	:	a; b or c
Z2	:	c
Z3	:	a

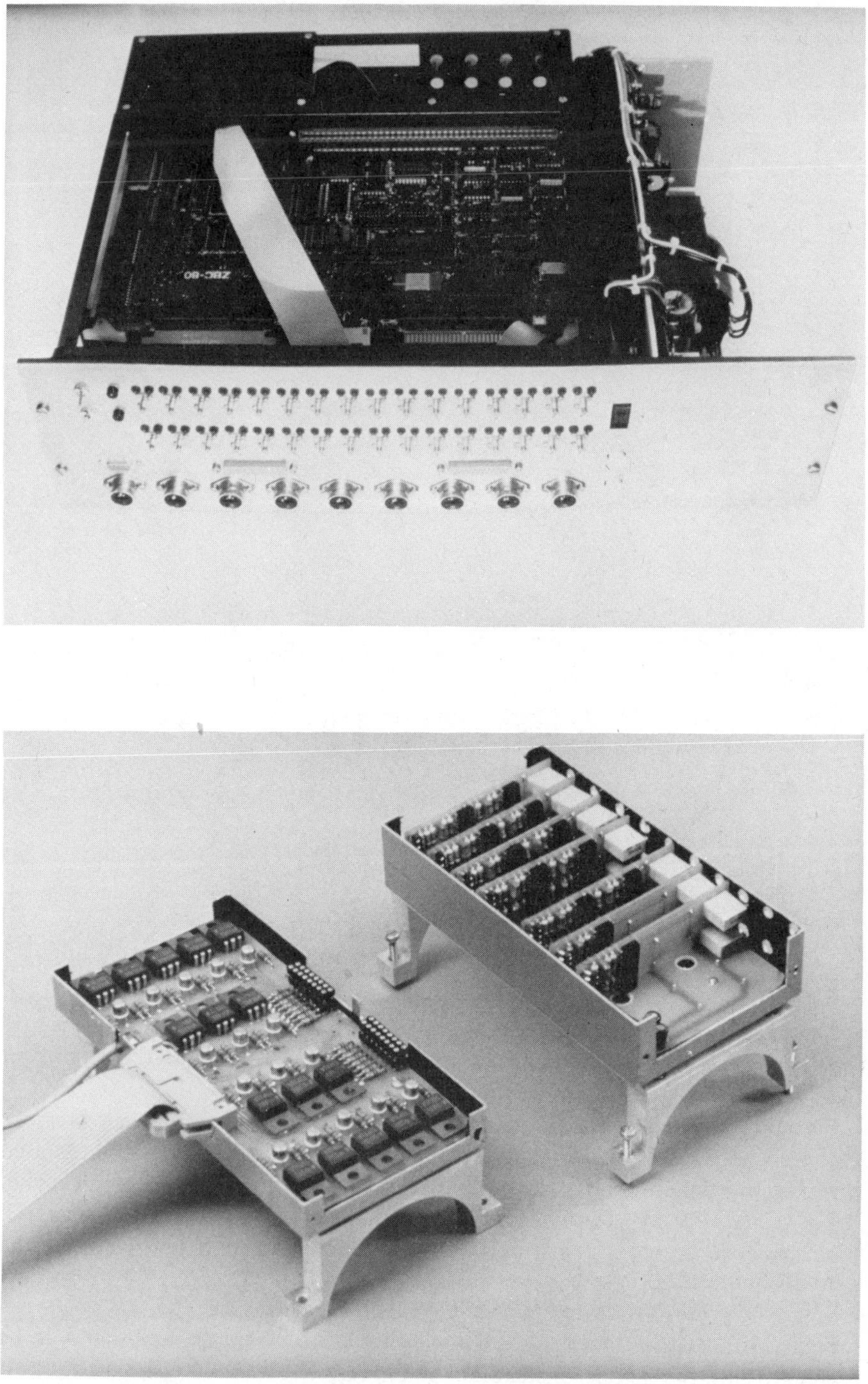

Figure 12 The proximity sensors control system.

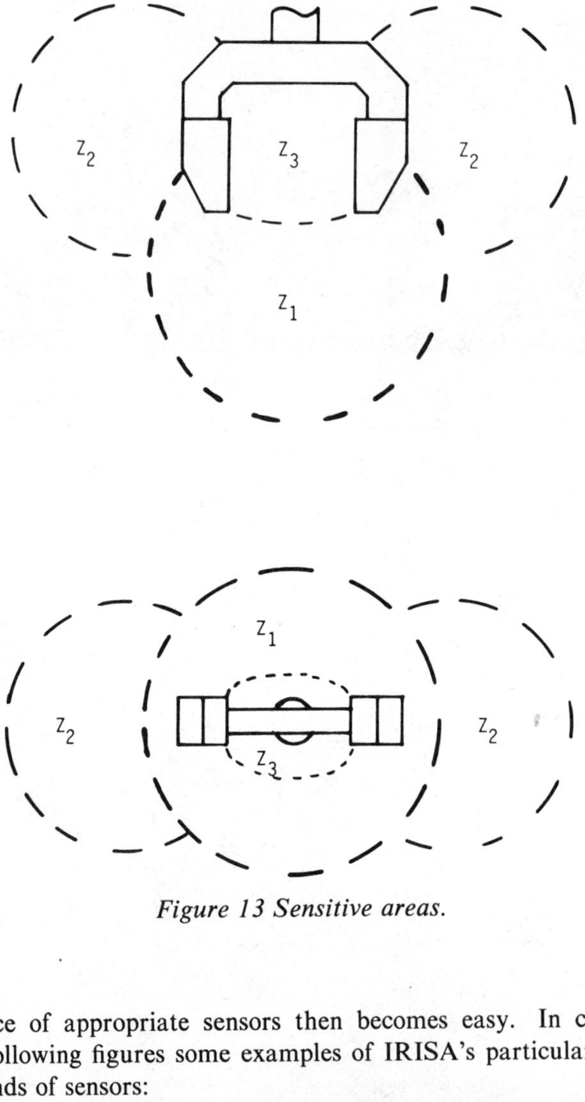

Figure 13 Sensitive areas.

The choice of appropriate sensors then becomes easy. In conclusion, we give the in following figures some examples of IRISA's particular embodiment of various kinds of sensors:

- Figure 14: A two-fingers hand designed according to the previous approach. This system includes 10 fixed sensors, 8 of which are integrated within the fingers, and a removable piece located in the back of the hand. The last device may receive an ultrasonic sensor as well as various infrared sensors; all these systems have normalized size and are driven by the microcomputer.
- Figure 15: Various devices with IR and ultrasonic sensors.
- Figure 16: A 16-sensors matrix (including local electronics), which may be plugged into the gripper instead of the fingers.
- Figure 17: A 32-sensors linear array, with inside electronics.

TABLE 2

	Functions	E/R Couples 1	2	3	4
	a_1 fine positioning			**	**
a	a_2 distance estimation			**	*
	a_3 fine orienting			**	*
b	Centering		**	*	
c	c_1: Obstacle avoiding and navigation	**	*	*	
	c_2: safety (anti-clash)	**		*	

* Good adaptation to function.
** Excellent adaptation to function.

Figure 14 A two-fingers hand with 10 sensors.

Figure 15 A 4X4-sensors matrix.

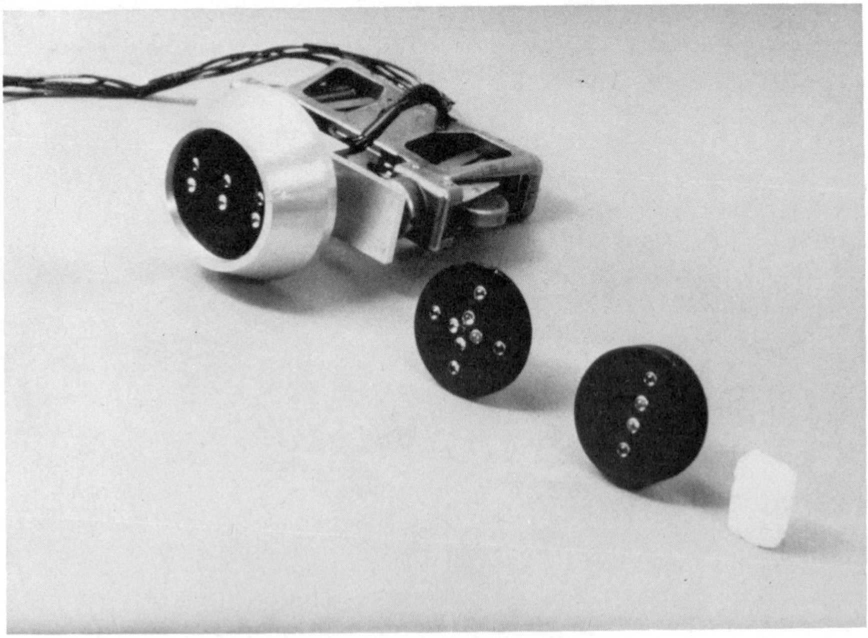

Figure 16 a Some special devices.

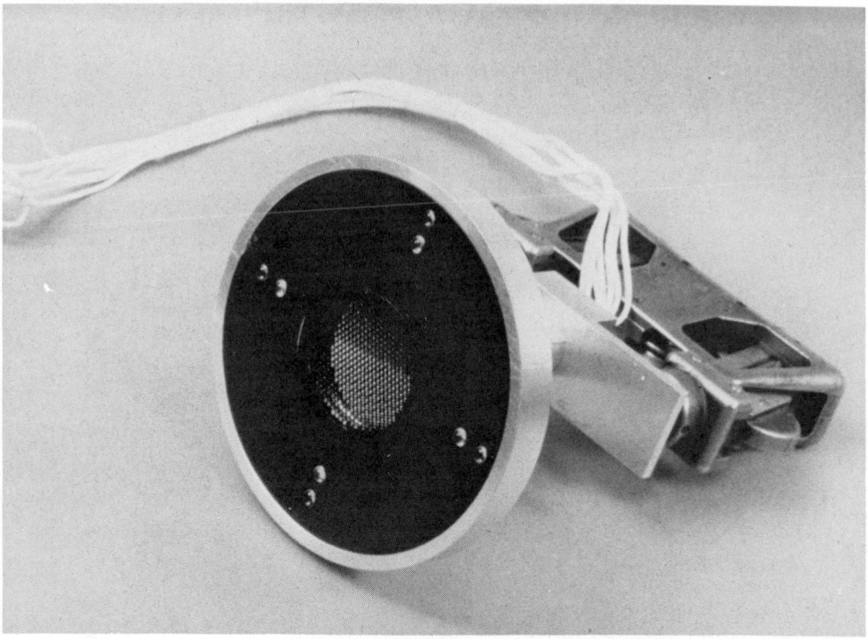

Figure 16b

Figure 17 A 32-sensors linear array.

3. MODELING OPTICAL REFLECTANCE SENSORS

A photometric study of ORS properties is necessary for a better understanding of their behavior; further, modeling ORS is useful for deriving good expressions of simulation. We present here both cases.

3.1 Hypotheses

The notations are given in Figure 9 (case 3). For modeling convenience, we assume that:

- the emitter (E) and the receiver (R) are small, with surfaces S_E and S_R
- The optical axis of E and R are in the same plane (Oz, Ox)
- the system is symmetric
- the target is a diffusing plane orthogonal to Oz, with albedo λ
- the emitted intensity has the form:

$$I = LS_E \cos \alpha' \tag{4}$$

where L is the luminance parameter.

3.2 General Model

In case 3 of Figure 9, we may write the lighting energy in P:

$$E(P) = LS_E \frac{\cos \alpha' \cos \alpha''}{r^2} \tag{5}$$

with $d = r \cos \alpha''$. The total received flux is then:

$$\phi = \iint_{\Sigma} \lambda \, S_R \, S_E \, \frac{\cos \alpha' \cos^3 \alpha'' \cos \alpha' \cos^3 \alpha''}{d^4} \, d\Sigma = \lambda f(d) \tag{6}$$

where Σ is the cross-hatched area in Figure 9.

A complete analytical model may be derived by expressing α', α'', β', β'' Σ''' as functions of x, y, d, δ. The related expressions are given in [7].

3.3 Simplified Model

Assuming that E and R (Figure 14; case 1) are merged with the same axis, eq. (6) becomes:

$$\phi = \iint_{\Sigma} \lambda \, S_R \, S_E \, L \, \frac{\cos^8 \alpha}{d^4} \, d\Sigma \tag{7}$$

where Σ is a disk with radius $d \tan \phi$.

As $\quad d\Sigma = 2\pi R dR, \quad$ with $\quad dR = \dfrac{d}{\cos^2 \alpha} d\alpha,$

eq. (7) gives

$$\phi = \frac{K}{d^2} \quad \text{with} \quad K = \frac{\pi}{4} \lambda \, S_R S_E L \, (1 - \cos^6 \phi) \tag{8}$$

This expression has to be connected to the definition of LE which uses $V_k(x^k)$ see eq. (2). If a *force* is associated with the output ϕ, it will *directly* correspond to an attractive or repulsive action derived from a Newtonian potential. That property is very interesting and important for control purposes, provided that real sensors could follow the model given in eq. (8). An other advantage of the model described in eq. (8) is to make very easy a linearization process. For both reasons, the interest of having sensors with real output like eq. (8) is obvious. Figure 5 shows the direct output y and the linearized output $(y^{-1/2})$ of the ORS with thin field. The signal presents good linear behavior for the range 50/250 mm. More precise experiments have shown that it is possible to tune α and δ to get a 10^{-3} linearity for the same range.

The simplified model may thus be considered useful for the previous sensor, for plane targets, and with $d > 50$mm.

Remark: From eq. (8) it is easy to see that changing the albedo λ of the plane leads to linear responses with different slopes: a similar result is obtained by giving a constant orientation to the plane.

4. PROCESSING OF ORS SIGNALS

4.1 The Classes of Sensor-Based Algorithms

The classical control levels of a robot are given in Figure 18.

A closed-loop control is defined when, at each instance in time, the input of a process having to stay closer to a given objective, is computed using the last measures and a "state", dynamic memory of the system. In robotics, proprioceptive servo loops are located at level 5, and we may consider that level 4 is a class of closed loop control with regard to exteroceptive measures: the actions may be directly driven by the outputs of sensors. In an other way, levels 1, 2 and 3 are *open loop controls* with respect to the previous definition.

As shown in Figure 18, LE may be taken into account in levels 3 or 4. It is thus necessary to use two classes of processing algorithms, features or parameters extraction (in a large sense) from LE in open loop cases and closed loop control algorithms in the other.

4.2 Parameters Extraction

Related to the main parameters that we may try to extract in open loop from ORS outputs, three basic cases of data handling are interesting, detection of an object, range estimation and pattern recognition. We briefly present some solutions for each problem.

4.2.1 Detection

This is the basic function of a proximity sensor. (More than 90% of industrial systems work so.) In robotics, it has been seen that the ratio range/size was an

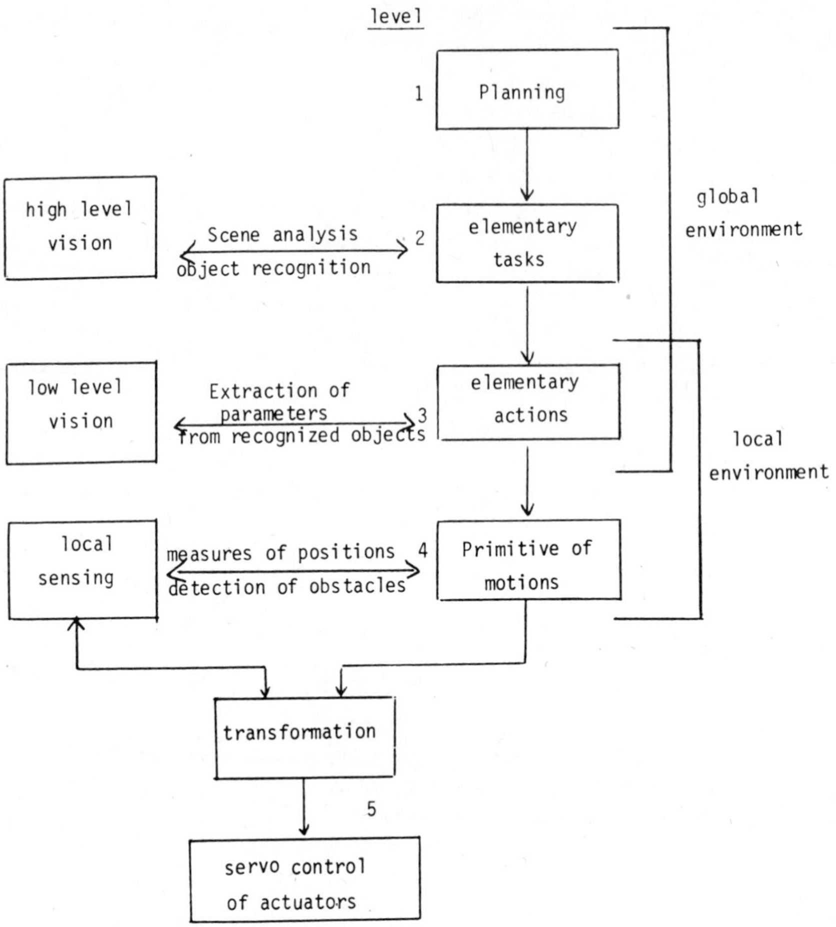

Figure 18 Classical control levels.

important parameter. In detection problem the significant parameter is not the distance but a binary value, absence or presence. Increasing the previous ratio does thus signify that *detection algorithms have to work well, even when signal-to-noise ratio is very low.* The quality criteria are then (cf. Figure 19)

- short detection delay R
- high mean time between false alarms τ

Further, realistic applications require low-cost algorithms. With all these constraints, it is obvious that a simple lowpass filter does not have sufficient performance, and in an other way that sophisticated algorithms, like matched filter may be too expensive for practical implementation.

We thus recommand to use a specific algorithm, derived from statistical techniques for detecting changes in mean value of random variables [4].

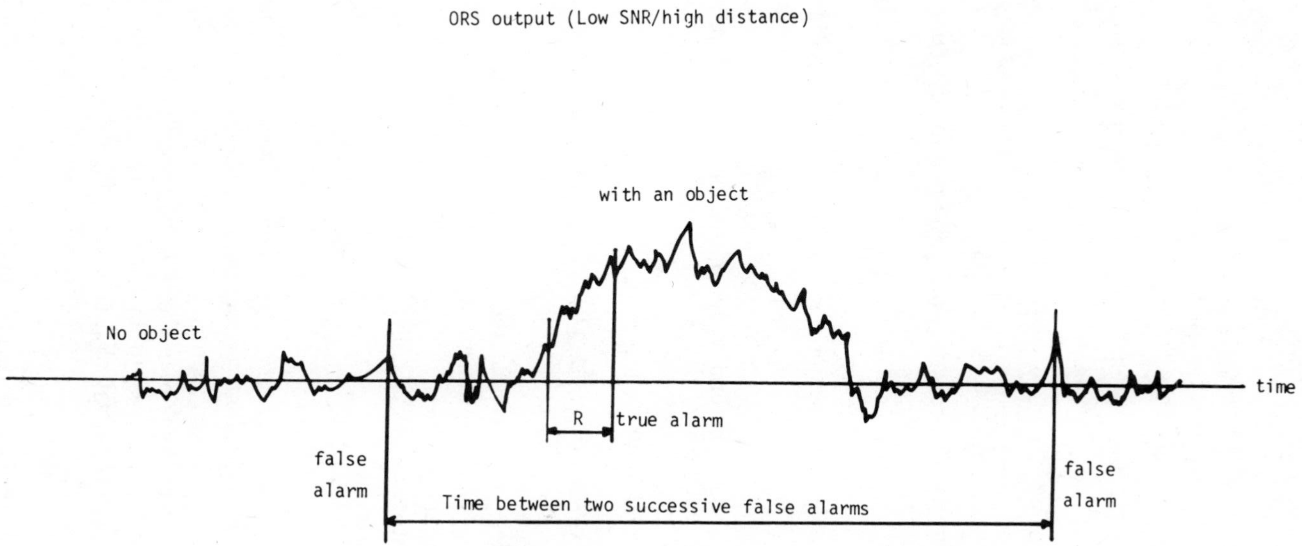

Figure 19 ORS output with low signal-to-noise ratio.

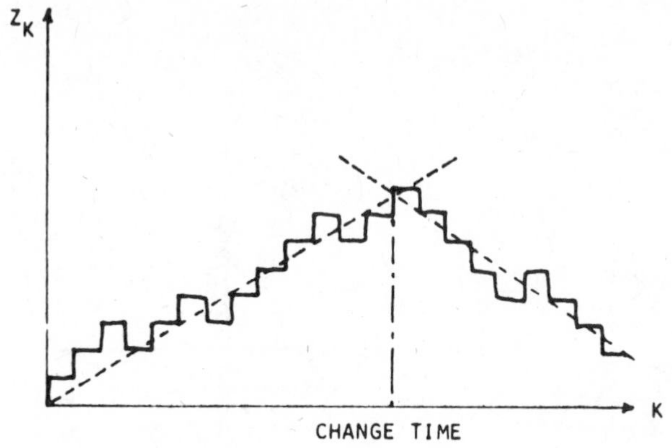

Figure 20 The random walk generated for detection.

Given at each discrete-time k:

y_k: output of the sensor, for example at each pulse in case IRISA's ORS.

μ_0: mean value of y_k when the field is free

μ_1: a given minimum value of the output level provided by any target,

then, an optimum algorithm (min R and max τ) is given by: let

$$S_n = \sum_{k=1}^{n} (y_k - \lambda) \qquad \text{with} \quad \lambda = \frac{\mu_0 + \mu_1}{2} \tag{9}$$

and $M_n = \text{Max}_{0<k<n} S_k$, then: a change has occurred if $M_n - S_n > h > 0$, h a given threshold, provided that y_k is a sequence of independent Gaussian variables.

A very low-cost version of this algorithm is:

$$S_n' = \sum_{k=1}^{n} \text{sign}(y_k - \lambda) \tag{10}$$

which is much more robust to very large instantaneous perturbations.

Using for μ_0 an estimation $\hat{\mu}_0$ provided by a low pass filter with slow response time, the only tuning parameters are $\Delta\mu$ (with $\Delta\mu = \mu_1 - \hat{\mu}_0$) and h. This algorithm, which has been shown to be only slightly less accurate than the optimal one [4] is very easy to implement with comparators and counters. The behavior of the random walk S_n' is given in Figure 20.

4.2.2 Range Estimation

From the model described in eq. (6), it is clear that ORS do not provide directly distance information. In LE sensing, only the ultrasonic system easily gives an output proportional to the range. However, optical sensors may

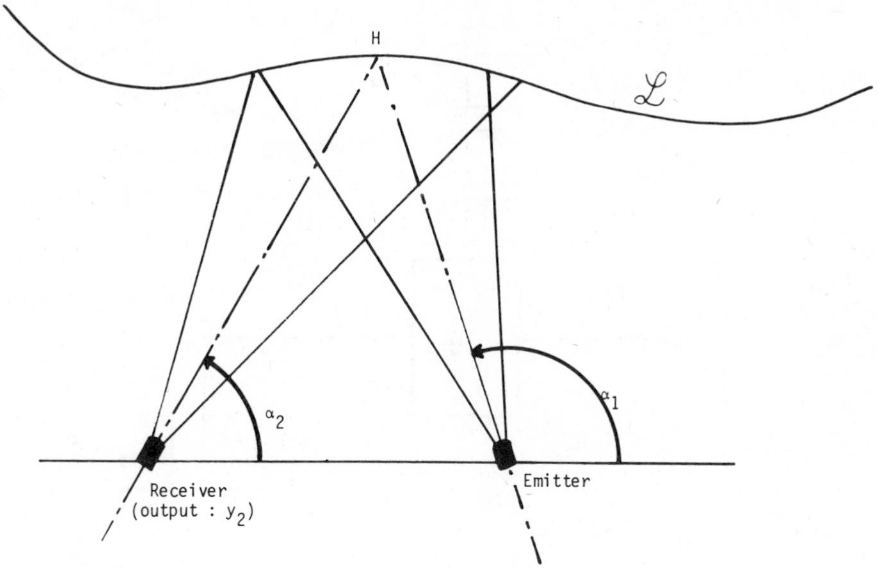

Figure 21 A sweeping ORS system.

provide range information in special conditions, by using multiple data in space or time:

a. *Mechanical sweeping systems.*

An extension of the JPL' system defining sensitive areas may be made using the principle given Figure 21. For a given angle α_1, the H point is approximately located by researching α_2 such that y_2 be maximum. A scanning procedure allows us to reconstruct the line L, parametrized by α_1. This system is slow, not very accurate, and the use of small actuators for moving R and E is delicate.

b. *Optical phase sensors.*

As the ranges concerned with LE sensing are rather small (less than 500 mm, currently 200 mm), measuring the flight time of an IR pulse is very difficult (300 mm = 1 nanosecond). We thus need a phase shift measurement, whose principle is given in Figure 22 (e.g. in [13]).

This system requires an emitter with high power, narrow optical band, and fast response time. A laser diode is necessary, and the corresponding receiver is a PIN photodiode. For practical LE ranges, the frequency modulation is about 50 MHz.

c. *Optical reflectance sensors.*

General case: If the target is a plane, the model given in eq. (6) is valid: $y = \lambda f(d)$, where $f(d)$ is the known characteristic of the sensor. When λ is unknown, it is obvious that a single measurement is not sufficient to get d. Let us now suppose that the target (or the sensor) may be linearly displaced by a given value h, such as: $y^1 = \lambda f(d+h)$.

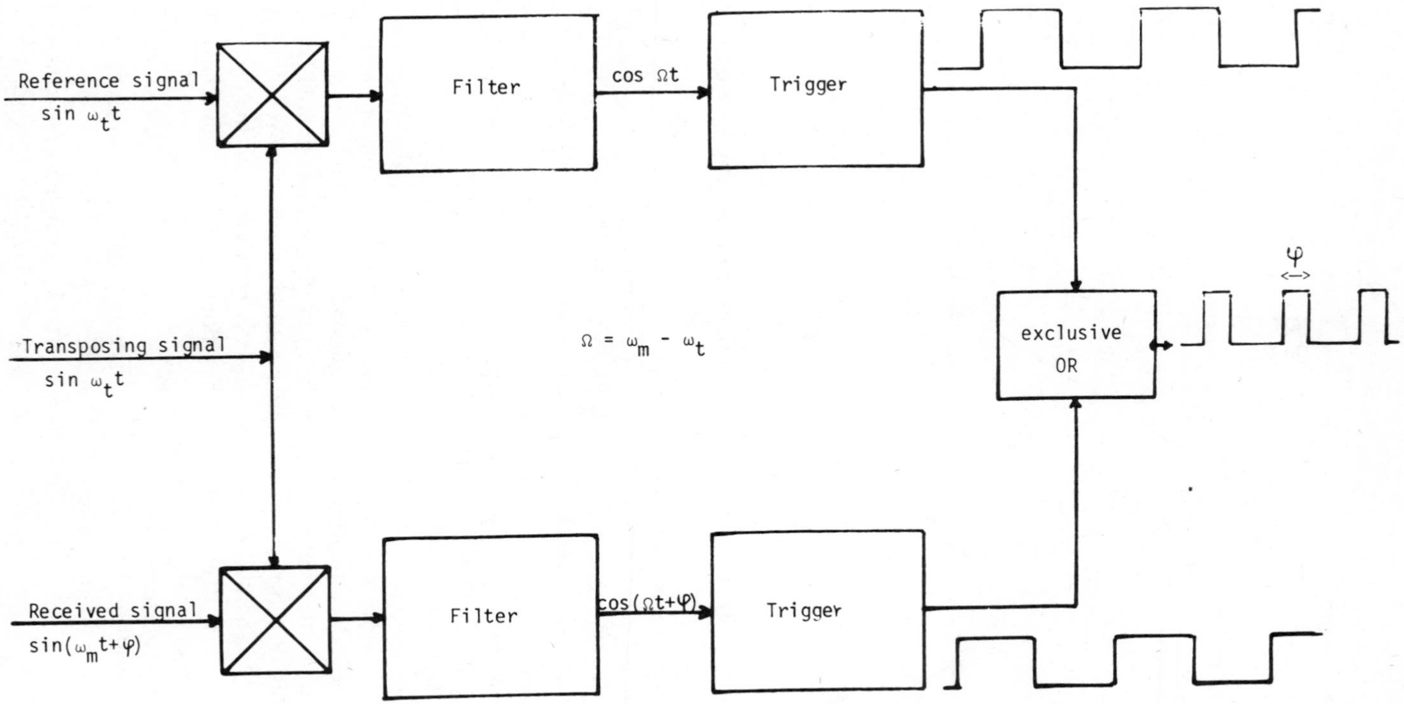

Figure 22 Principle of optical phase measurement.

Then, d is a solution of $y^1 f(d) = yf(d+h)$. However, in practical cases, the model in eq. (6) is an approximation, for:

- the targets are not always locally planar w.r.t. the field of the ORS
- the albedo and the local orientation may not be constant during the motion
- the output of the sensor is disturbed by the noise.

If we apply to the sensor a given linear motion, such as $d(k+1) = d(k) + h(k)$ at time k, a practical model is then (Figure 23).

$$y(k) = a[d(k)]f[d(k)] + \epsilon(k) \qquad (11)$$

where $a[d(k)]$ includes orientation and albedo variations at distance d_k, and $\epsilon(k)$ is the output noise.

Finding $d(k)$ from the successive outputs $y(1) \cdots y(k)$ requires a nonlinear filtering algorithm. The state and output equations may be written under the following classical form:

$$\begin{cases} d(k+1) = d(k) + h(k) + e(k) \\ a(k+1) = a(k) + e'(k) \\ y(k) = a(k)f[d(k)] + \epsilon(k) \end{cases} \qquad (12)$$

where:

$e(k)$ is a positioning error
$e'(k)$ is an artificial state noise, allowing $a(k)$ to slowly vary. e, e' and ϵ are assumed to be zero–mean white gaussian.

Some algorithms for solving eq. (12) are given in [4] and [7]. Using extended least squares techniques, they are accurate but expensive. We prefer in practice to derive simplified processing techniques using approximate filters:

Simplified case.

Assuming that:

$$\begin{cases} e(k) = e'(k) = 0 \ \forall K & H1 \\ f[d(k)] = 1/d^2(k) & H2 \\ \text{var}\epsilon(k) = R, \text{ with } R \text{ small} & H3 \\ h(k) = h < 0 & H4 \end{cases}$$

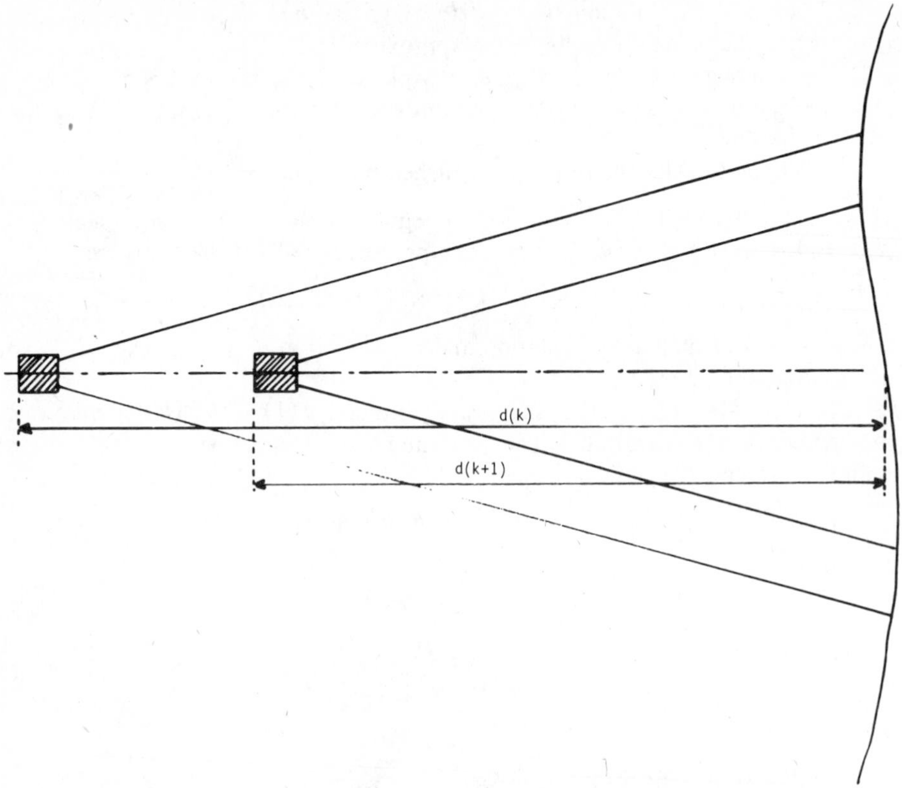

Figure 23 Linear motion of a sensor.

and putting $\boldsymbol{y}(k) = [y(k)]^{-1/2}$ the system given in eq. (12) becomes:

$$\begin{cases} d(k+1) = d(k) + h \\ a(k+1) = a(k) = a \\ \boldsymbol{y}(k+1) \simeq a(k+1)d(k+1) + \epsilon'(k+1) \end{cases} \tag{13}$$

with

$$\operatorname{var}(\epsilon') = V(k) \simeq \frac{R\, d^6(k)}{a^3}$$

by H2 and H3, which will be upper bounded in practice by Vmax.

Putting $z(k+1) = \boldsymbol{y}(k+1) - \boldsymbol{y}(k)$, we finally get the two systems:

$$\begin{cases} a(k+1) = a(k) \\ z(k+1 = ha(k+1) + \epsilon''(k+1) \end{cases} \tag{14}$$

$$\begin{cases} d(k+1) = d(k) + h \\ \\ y(k+1) = a(k+1)d(k+1) + \epsilon'(k+1) \end{cases} \tag{15}$$

where $IE[\epsilon''^2(k)] = 2V(k)$ and $IE[\epsilon''(k)\ \epsilon''(k+1)] = -V(k-1)$. IE denotes the mathematical expectation.

The solution of (14) is given by a recursive least-squares (RLS) algorithm, with an exponential forgetting factor $\alpha < 1$, allowing slow variations of $a(k)$:

$$\hat{a}(k+1) = \hat{a}(k) + \frac{1}{h}\frac{1-\alpha}{1-\alpha^{k+1}}[z(k+1)-h\ \hat{a}(k)]. \tag{16}$$

An approximate solution of (15) is then derived using the delayed estimation of $a(k)$ in a RLS filter for (15):

$$\hat{d}(k+1) = \hat{d}(k) + h + \frac{1}{(k+1)\ \hat{a}(k)}[y(k+1) - \hat{a}(k)\ \hat{d}(k)-h]. \tag{17}$$

As $d(k)$ decreases ($H4$), the signal to noise ratio increases, and $V(k)$ roughly decreases. The influence of noise $\epsilon(k)$ then becomes small and the convergence of the approximation algorithm given in eqs. (16) and (17) is very satisfying (example in Figure 24).

4.2.3 Pattern Recognition

Let us consider an isolated object lying onto a plane P. Classical "images" of the object are mainly:
- luminance (and sometimes chrominance) pictures
- 3D description in a given frame.

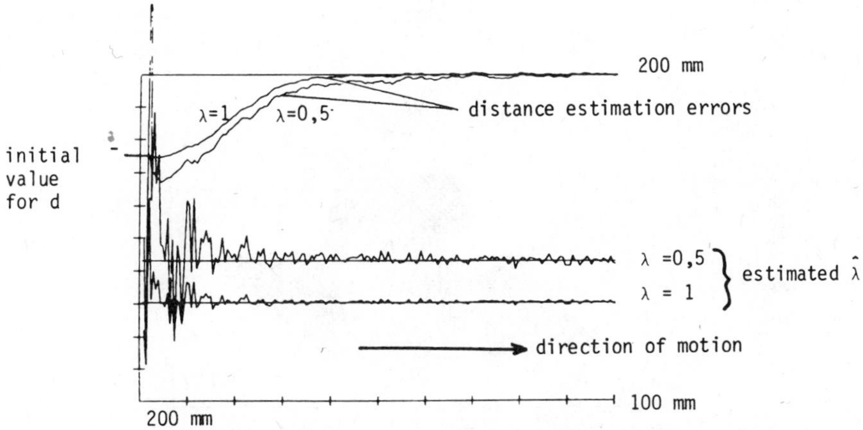

Figure 24 Convergence of the simplified distance estimation.

In the first case, the image is obtained from a TV camera, and in the second, the description is derived from a stereoscopic pair, or by scanning with a range-finder (e.g., laser sensors). ORS may provide a third kind of image, we call "Active Infrared Reflectance Image" (AIRI), in which each "pixel" is given by the output of an elementary ORS with a narrow field. If we consider a set of identical sensors, with regular spatial separation into a plane P' parallel to P, with a given constant distance (P', P), it is obvious that the obtained image depends only on the object. Each sensor output then follows the model given in eq. (6), and includes the local contributions of distance, albedo and orientation, which are object characteristics. Further, as sensors are narrow-field, model eq. (8) is valid and we may use the linearized signal $\boldsymbol{y} = y^{-1/2}$. An example of a simulated AIRI, obtained with the device presented in Figure 17 in association with a conveyor belt, is given in Figure 25.

By comparison with a gray level picture, the AIRI provides an information insensitive to lighting conditions, and characterizing the geometric features of the object: for example, edges represent transitions of slopes or heights, and

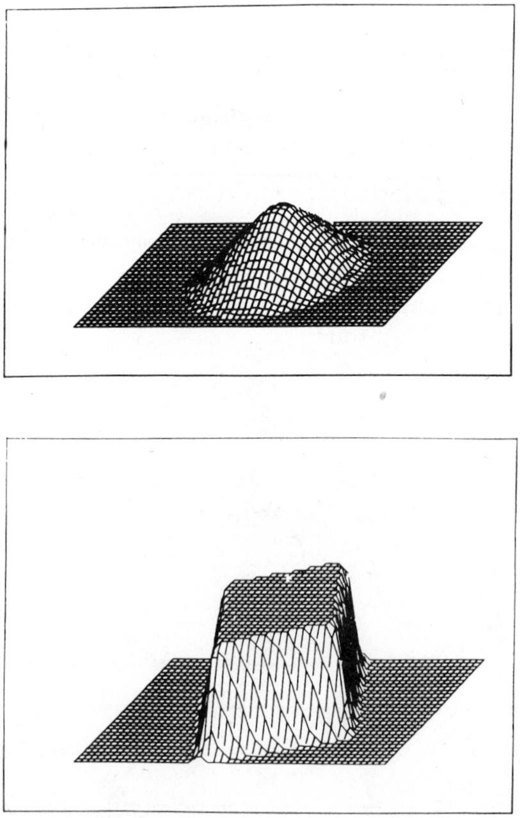

Figure 25 AIRI for a cone and a cube.

not gray level variations. We may thus use AIRI in pattern recognition problems, with the same techniques as the 3D geometric images: segmentation into regions by level or gradients, computation of shape parameters (see [12]). In conclusion, an AIRI has not the fine resolution of laser scanning, but is often a low-cost and accurate alternative to gray level pictures for certain object recognition purposes.

5. CLOSED LOOP CONTROL WITH ORS

5.1 Basic Principles

For simplicity, we only consider here the case of a six degrees-of-freedom manipulator, with ORS integrated within the end effector. The closed loop tasks will include navigation (obstacle avoidance), orienting and automatic grasping: a typical one is:

ζ: *Starting from a point A, go the given area B and grasp an object located in B, in avoiding unknown obstacles between A and B.*

Three main levels of control may occur for performing ζ:

- level 0: instantaneous ORS-based loops are sufficient (no ambiguity, no check situation).
- level 1: memoryless algorithms of level 0 are not sufficient for ζ, solving check situations (labyrinth for example) requires a little bit of intelligence (but, in that level, the navigation algorithms only use ORS informations).
- Level 2: even with intelligence, ORS outputs do not suffice to perform ζ. Additional information is required and level 2, at request, uses other kinds of sensors (vision) or works in cooperation with a human operator.

For a given task which uses ORS, it is clear that a practical purpose is to lie, on average, near the level 0. We now describe the algorithms of that level.

5.1.1 Fundamentals of Control

(a) Modeling

With the notations of § 1.2.1, it is well known that the dynamic model of the manipulator in generalized coordinates is:

$$A(q)\ddot{q} + H(q,\dot{q}) = \Gamma \tag{18}$$

where:

$$H(q) = B(q,\dot{q})\dot{q} + C(q) + D(\dot{q}) \tag{19}$$

$$B(q,\dot{q})\dot{q} = \begin{bmatrix} \dot{q}^{\mathrm{T}} V^1(q)\dot{q} \\ \vdots \\ \vdots \\ \dot{q}^{\mathrm{T}} V^6(q)\dot{q} \end{bmatrix}$$ is the centrifugal and coriolis contribution,

with $V^i(q)$ symmetric. Here, $C(q)$ is the gravity term; $D(\dot{q})$ expresses dissipative forces: strictions, viscous frictions; $A(q)$ is the inertia matrix; Γ is the control torque vector; and $\dim(q) = n = 6$.

Recalling that $x = f(q)$ eq. (1), we define the Jacobian matrix:

$$J(q) = \left\{ \frac{\partial f_i(q)}{\partial q_j} \right\}$$

Then, and when J^{-1} exists, it is easy to derive a dynamic model in operational coordinates:

$$A'(q)\ddot{x} + E(q,\dot{q}) = F \tag{20}$$

with

$$\left[\begin{array}{l} A'(q) = J^{-T}(q)\, A(q)\, J^{-1}(q) \\ \\ E(q,\dot{q}) = -A'(q)\, W(q,\dot{q}) + J^{-T}(q)\, H(q,\dot{q}) \\ \\ F = J^{-T}(q)\Gamma \end{array}\right. \tag{21}$$

where

$$W(q,\dot{q}) = \begin{bmatrix} \dot{q}^T W^1(q)\dot{q} \\ \vdots \\ \vdots \\ \vdots \\ \dot{q}^T W^6(q)\dot{q} \end{bmatrix}$$

with W^i being the derivate of the ith line of J w.r.t. q Eq. (20) is then equivalent to:

$$M(x)\ddot{x} + N(x,\dot{x}) = F \tag{22}$$

by putting

$$M(x) = A'(q)$$

and

$$N(x,\dot{x}) = E(q,\dot{q}).$$

(b) Control: Statement of the Problem

Since (18) and (22) have similar forms, the following control techniques are useful in both cases. However, we only describe here control in operational coordinates, in order to prepare the introduction of ORS-based closed loops.

Controlling eq. (22) requires two kinds of actions:
- compensating for coupling terms, inertias, and nonlinearies
- controlling the dynamic characteristics of the servo loops which track a given desired behavior.

We may distinguish the two aspects by writing:

$$\hat{F} = \hat{M}(x)u + \hat{N}(x,\dot{x}) \tag{23}$$

where u ensures the second function, and $(\hat{M},\hat{N})$ are given values for M and N.

For convenience, we introduce a reference model:

$$\ddot{x}_r = u_r \tag{24}$$

whose control is left free to the user's disposition, with a tracking error

$$\epsilon = x - x_r. \tag{25}$$

We set:

$$u = -k_p\epsilon - k_v\dot{\epsilon} + u_r \tag{26}$$

Then, eqs. (22) to (26) give the error evolution equation:

$$\begin{aligned}&\ddot{\epsilon} + M^{-1}(x)\hat{M}(x)\ [k_p\epsilon+k_V\dot{\epsilon}]\\ &= M^{-1}(x)\ [\hat{N}(x,\dot{x})-N(x,\dot{x})+(\hat{M}(x)-M(x))\ u_r].\end{aligned} \tag{27}$$

The behavior of eq. (27) depends on the choices made for $\hat{M}$, $\hat{N}$, k_p, k_V. For example, classical dynamic control uses $\hat{M} = M$ and $\hat{N} = N$, which leads to a homogenous linear form for eq. (27). However, in practice, only approximate values of M and N are known, with the extremale cases:

$\hat{M} = I$, $\hat{N} = 0$: classical regulator with constant gains: the accuracy is bad when inertia effects and speed increase.

$\hat{M}$ and $\hat{N}$ are estimations of M and N provided by adaptive algorithms using some parameter identification techniques, like recursive least-squares; results depend on the chosen parametrization for $\hat{M}$ and $\hat{N}$.

From the user's point of view, it is obvious that we need an error $(\epsilon,\dot{\epsilon})$ as small as possible: in the ideal case $(\epsilon=\dot{\epsilon}=0)$, the single eq. (24) will only have to be considered. Now, a very important result of C. Samson [14] may be used: he has shown that, for any choice of $\hat{M}$ and $\hat{N}$, under weak conditions, there exists a function $k(.)$ such as building variable gains k_p and k_V using k ensures the stability of (27); further, a method for building $k(.)$ given $\hat{M}$, $\hat{N}$ and the structural properties of the robot, is given. We may thus consider that, even in the general case, it is possible to get a behavior which is close to the one provided by an ideal dynamical control: $M = \hat{M}$, $N = \hat{N}$.

Using that theorem, *without specifying an a-priori choice for* $\hat{M}$, $\hat{N}$, we may thus "forget" eq. (27) and consider that we have only to control eq. (24), which is very interesting for ORS based control, so, it is possible without any loss of generality, to restrict the problem to the two cases:

- controlling a pure double integrator:

$$\ddot{x}_r = u_r$$

- or controlling an exact dynamical model:

$$F = M(x)\, u + N(x, \dot{x}) \tag{28}$$

5.1.2 Elementary Actions Related To ORS

We associate to the end effector, called solid(S) the frame R_6; in R_6, we give, for each sensor C_j ($j = 1 \ldots n_c$) (Figure 28)

its position e_j
its direction of observation d_j.

We define the actions as follows: each action V_i ($i = 1 \ldots n_v$) is a vector, with origin x_i, direction δ_i and the value w_i of which depends on sensors outputs y. At each action V_i is associated ($\Leftrightarrow$) a set ζ_i of sensors:

$$\left.\begin{array}{l} \forall i: \quad (x_i, \delta_i) \Leftrightarrow \left\{ C_{j_1} \cdots C_{j_{k(i)}} \right\} \\[2ex] \text{and} \\[2ex] \quad (w_i) \Leftrightarrow \left\{ f(y_{j_1}) \cdots f(y_{j_{k(i)}}) \right\} \\[2ex] \text{with: } k(i) \in [1, \ldots, n_c];\ i = 1, \ldots, n_v;\ n_v \leqslant n_c; C_{j_{\ell(i)}} \neq C_{j_{m(k)}} \ \forall i \neq k \\[2ex] \text{and } \forall \ell, m, \text{ when } x_i \neq x_k \end{array}\right\} \tag{29}$$

Parametrizing the ORS-based actions then consists in specifying x_i, δ_i, w_i, ζ_i, at a higher decision level. Finally, elementary actions are merged when the origins of some vectors are the same, and the result is the set $\{v_k, \forall k = 1, \ldots, n_v\}$ with the application points: $x_i \neq x_j \ \forall i \neq j$.

$$\forall k = 1,\ldots,n_v: \left\{ \begin{array}{l} v_k = \sum_{\ell \in e_k} w_\ell d_\ell \text{ where } e_k = \{d_i: x_i = x_k\} \\ \\ v_k \neq 0 \end{array} \right\} \tag{30}$$

Some examples of practical applications will be given in the following.

5.2 ORS-Based Control Loops

Since local environment is a concept devoted to the analysis of *relations* between robot and objects, it is clear that absolute informations are not adequates for our control purposes. In other words, in general it is not interesting to use ORS for direct absolute position control (R_6 in R_0). We thus distinguish only three classes of possible actions: force, velocity, acceleration.

5.2.1 Force Synthesis Using ORS

As seen in paragraphs 1.2.1 and 2.3, it is often possible to consider that an ORS output is the value of a gradient field associated to a newtonian potential. Each v_k is then elementary force which may represent an attractive (target) or repulsive (obstacle) action from the object to the effector. In R_6, the elements of reduction of the set of the external forces, at the origin 0_6 of R_6, are $F_e = [f\ \tau]^T$ with

$$f = \sum_{k=1}^{n_v} v_k \tag{31}$$

and

$$\tau = \sum_{k=1}^{n_v} \overrightarrow{0_6 x_k} \wedge v_k \tag{32}$$

where $\wedge$ denotes the vector product.

The set of external forces F_e is expressed in R_6 and depends on x (Figure 26). We note $F_e^o(x)$ its expression in R_o. $F_e^o(x)$ is obtained from F_e by the transformation:

$$F_e^o(x) = \begin{bmatrix} A_1 f \\ \\ A_2\ \tau \end{bmatrix} \tag{33}$$

where A_1 and A_2 are two 3 × 3 matrices depending on the rotation parameters which define the orientation of R_6 in R_o. The final control is thus

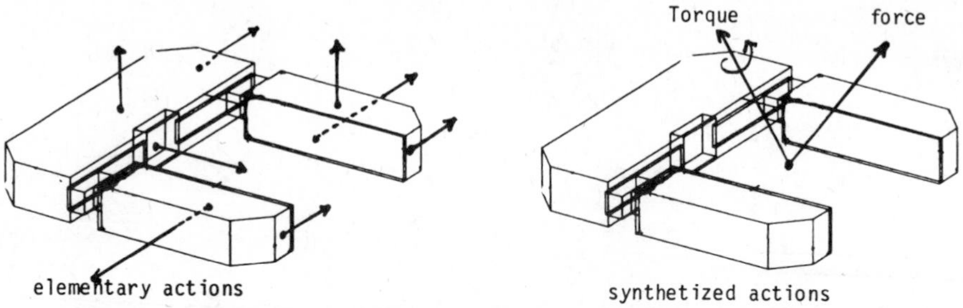

Figure 26 Elementary and synthesized actions.

given by adding F_e^o to $\hat{F}$ in eq. (23), or $J^T(q)F_e^o$ to $\hat{\Gamma}$ in the case of generalized coordinates control. In case of exact dynamic control, eq. (27) then becomes:

$$\ddot{\epsilon} + k_V \epsilon + k_p \epsilon = M^{-1}(x)\ F_e^o(x) \tag{34}$$

while the reference model is governed by, for example, a proportional/derivative loop:

$$\ddot{x}_r = u_r = -L_p(x_r - x_c) - L_V(\dot{x}_r - \dot{x}_c) \tag{35}$$

$x_c(t)$ and $\dot{x}_c(t)$ are the user's inputs to the reference model.

The behavior (stability and dynamics) of ϵ depends on the form of $F_e^o(x)$, which is related to the applications. In general case, the steady state solution of eq. (34) is obtained for a value $\epsilon_S \neq 0$: the final situation S is compromise between the desired setpoint assigned to the reference model and the action of $F_e(x)$ on the system. If a reference model is not used, the setpoints $(x_c, \dot{x}_c)$ are directly expressed in u. Eq. (26) then becomes:

$$u = -K_p(x - x_c) - K_V(\dot{x} - \dot{x}_c) \tag{36}$$

and eq (34) is valid, with $\epsilon = x - x_c$.

A very important application of such a control is the constitution of synthetic forces in master/slave teleoperation: x_c is setted by the human operator on the master arm and the artificial force/torque vector F_e may be felt by the operator as attractive or repulsive sensations which allow him to perform good manipulation tasks, even with disturbed vision. The resulting motion then depends on the resistance that the operator wants to offer to the ORS synthetic forces (cf. [15] and [20]).

5.2.2 Velocity Control

In eq. (35), the choice of x_c, $\dot{x}_c$, specify the kind of control:
$x_c = x_r$, $\dot{x}_c(r) = \dot{x}_{desired}(t)$: first order velocity control loop
$\dot{x}_c = 0$, $x_c(r) = x_{desired}(t)$: second order position control loop with damping term.

It is quite convenient to use eq. (35) as an input mode for ORS informations. As ORS position control is not interesting, we may use ORS velocity control $(\dot{x}_c)$, with or without nominal desired position trajectory $x_c(t)$, and for given L_p, L_V. In that case the behavior of ϵ does not depends on the ORS outputs, provided that the demanded motions are compatible with the performances of the control (23). If they do, ϵ is small and we will only consider eq. (35). Further, when using only velocity control, and if the dynamics of the velocity inputs are lowpass w.r.t (L_p, L_V), it is possible to state that $||\dot{x}_c - \dot{x}_r||$ is small: this is kinematic control. For ORS based loops which generally work locally in the space, that assumption is true, and such a control is very useful; in discrete-time case, it means that the desired $\dot{x}_c[t(n)]$ is reached by $\dot{x}_r$ before time $t(n+1)$.

To ensure coherence of control loops, it is better to introduce the desired (without ORS action) behavior at the velocity level than to use x_c. We choose that solution in the following by the way of a so-called "nominal trajectory".

(a) Nominal Trajectory

At each time $t(n)$, the nominal trajectory concerns two desired vectors: rotation and translation velocities of R_6 in R_0, that we describe by:

- the translation velocity vector of R_6: v_d
- the rotation velocity vector of R_6: r_d, both expressed in R_6.

In the general case, the user may define partially v_d or r_d by specifying the general form $v_d = Av_d$ and $r_d = Br_d$. For simplicity we consider only here the case: $A = I$, $B = 0$ (free rotation).

(b) Determination of Velocities Inputs (Principle)

If we consider the set $\{v_k \ \forall k, v_d\}$ as elementary desired velocities of points $\{x_k, 0_6\}$ it is obvious that this set has no reason to belong to a real velocity field of (S).

We thus define:

- A frame R linked to (S)
- the twist characteristic of (S) expressed in R, defining the velocity field $w(.)$, and the rotation vector r.
- A point A in R, where the value of the vector field $w(.)$ is written $w(A) = s$.

We may write:

$$\forall x_k = [x_k \ y_k \ z_k]^{\mathrm{T}} \text{ in } R \in (S): w(s_k) = s + r \wedge \overrightarrow{\mathbf{A}x_k} \tag{37}$$

where $\wedge$ denotes the vector product, and with:

$$A\mathbf{x}_k = [x_k - x_A, y_k - y_A, z_k - z_A]^{\mathrm{T}} = \delta_k^{\mathrm{T}} \tag{38}$$

and $r \wedge A\mathbf{x}_k = \Delta_k r$, where Δ_k is the antisymmetric matrix associated to the vector product:

$$\Delta_k = \begin{bmatrix} 0 & \delta_k^3 & -\delta_k^2 \\ -\delta_k^3 & 0 & \delta_k^1 \\ \delta_k^2 & \delta_k^1 & 0 \end{bmatrix} \tag{39}$$

Let us define $\epsilon_k = ||v_k - w(x_k)||^2$, and W a weighting matrix such as $W = [\text{diag}; \{\lambda_i > 0\}]$, $\sum_{i=1}^{n_v} \lambda_i + \lambda_d = \lambda$

Then, we search for T, defined by s and r which minimizes

$$\begin{aligned} J &= \epsilon^T W \epsilon + \lambda_d ||v_d - w(0_6)||^2 \\ &= \sum_{k=1}^{n_v} \lambda_k ||v_k - s - \Delta_k r||^2 + \lambda_d ||v_d - s - \Delta_d^r||^2 \end{aligned} \tag{40}$$

The solution is given by $\partial J/\partial r = \partial J/\partial s = 0$:

$$\hat{s} = \frac{1}{\lambda}(\ell - Kr) \tag{41a}$$

$$H\hat{r} = m \tag{41b}$$

where

$$\begin{aligned} K &= \sum_{k=1}^{n_v} \lambda_k \Delta_k + \lambda_d \Delta_d \\ H &= \sum_{k=1}^{n_V} \lambda_k \Delta_k^T \Delta_k + \lambda_d \Delta_d^T \Delta_d \\ \ell &= \sum_{k=1}^{n_v} \lambda_k v_k + \lambda_d v_d \\ m &= \sum_{k=1}^{n_v} \lambda_k \Delta_k^T v_k + \lambda_d \Delta_d^T v_d \end{aligned} \tag{42}$$

Let us now define a new solid $(\boldsymbol{S})$ constituted by pin-point masses, with weights $\lambda_1, \ldots, \lambda_{n_v}, \lambda_d$, located in $x_1, \ldots, x_{n_v}, 0_6$, and rigidly linked by zero-mass bars. By choosing A as the origin of R, and locating it in the mass center G of $(\boldsymbol{S})$, we get $K = 0$ and eq. (41a) becomes:

$$\hat{s} = \frac{1}{\lambda} \ell. \tag{43}$$

The solution of (41b) depends on the rank of H. It is possible to show that a sufficient and necessary condition for inversibility of H is that the $\{x_k, \forall_k\}$ are not located on a same line. Further, H is the central inertia operator of $(\boldsymbol{S})$. By taking R as a central inertia

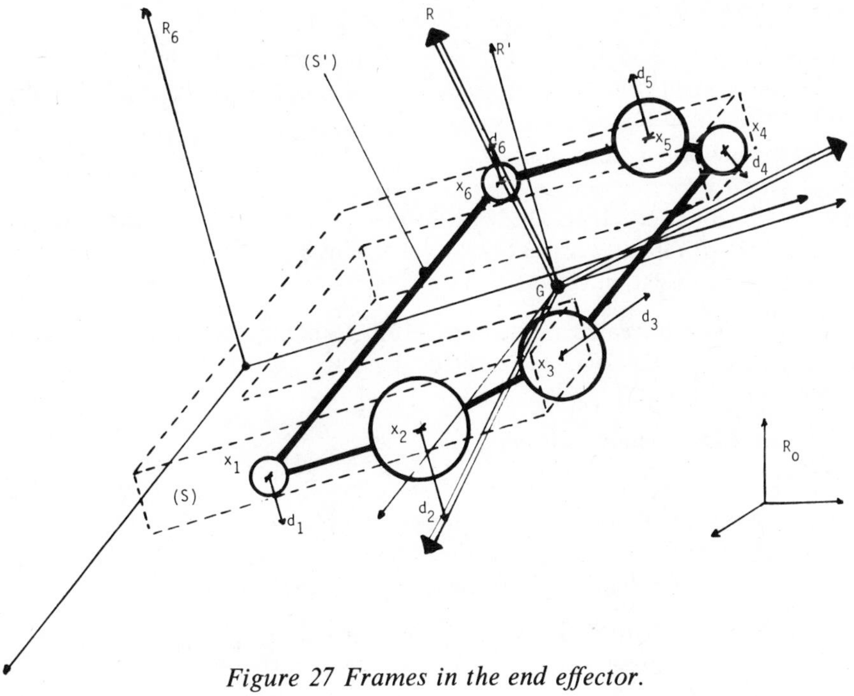

Figure 27 Frames in the end effector.

frame (Koenig's frame; Figure 27) H becomes diagonal. We thus find, *in R*, a simple expression for velocity control, with a rather similar form to eq. (31) and (32):

$$\hat{s} = \frac{1}{\lambda}\,\ell \tag{44a}$$

$$\hat{r} = (\text{diag}^{-1})m \tag{44b}$$

We thus get, expressed in R, the velocities vectors in translation and rotation minimizing eq. (40). For using these vectors in control, it is then necessary to express them in R_6, and to compute the corresponding values of operational coordinates $\dot{x}_c$, for example using Euler angles. The computed values of $\dot{x}_c$ give the ORS-based velocity inputs to be introduced in the control of the reference model (35).

Remarks:

1. Eq. (41) or (44) can valid even when $\{v_k \;\forall k, v_d\}$ really belongs to the velocity field of (S)
2. when H^{-1} does not exist, we have only two cases:
   - n_v=0: $\hat{r}$ is arbitrary
   - $n_v \neq 0$: rang(r)=2: a one-dimensional freedom exists for $\hat{r}$.

Obviously, a good choice for R_6, if possible, is: $R_6 = R$.

(c) Extensions

Extended cases may be examinated using the previous approach:

1. other specifications of the nominal trajectory: desired r, freedom of v_d . . .
2. introduction of constraints on the resulting motion: speed limitations, trajectory located onto a given subspace ...
3. specification of other kinds of elementary motions: for example it is possible to directly associate to a set of sensors a desired rotation in place of some translation vectors.

All these cases are treated in Reference [6]. Experimental results with ORS-based loops using the devices presented in Figure 14, 15 and 16, are given in reference [12] and [20].

5.2.3 Practical Considerations

(a) Examples of Elementary Functions

Figure 28 summarizes three examples of the main elementary actions useful in practice; the direct ORS signal (Kd^{-2}) is used in $y_1 y_2 y_3$ and the linearized one $(\boldsymbol{Kd})$ in $y_4 y_5 y_6 y_7$:

– *obstacle repulsion*: $\|v_1\| = \alpha y_1$ (44)

– *centering* : $\boldsymbol{v}_2 = \beta(y_2,y_3)\ (y_3-y_2)\boldsymbol{\delta}$ (45)

– *orientation* : $\begin{cases} \rho_1 = \gamma_1(y_5-y_7) \\ \rho_2 = \gamma_2(y_4-y_6) \end{cases}$ (46)

The β function is a normalizing term, such as $\beta(y_2,y_3) = [\max(y_2+y_3), y_{\min}]^{-1}$, which has the property to make v_2 insensitive to the distance [14]. In an other way, it is possible to show that using the linear outputs in orienting tasks for a plane target also leads to a behavior not distance-dependent.

(b) Stability in Case of Total ORS-Based Control

We only give here some slight intuitive notions of stability for ORS-based loops. When using force ORS feedback, it is obvious that, if it is the *only* used control, the resulting motion, for example with perfect decoupling, may have an unstable form: by setting, in (23):

$$u = F_e^o(x);\ \hat{M} \equiv M,\ \hat{N} \equiv N,$$

the result is

$$\ddot{x} = F_e^o(x). \qquad (47)$$

The existence of equilibrium points x_e, with $\dot{x}_e = \ddot{x}_e = o$ of eq. (47) first requires that $F_e^o(x_e) = o$. However, eq. (47) may only be unstable, or stable with periodic solutions; asymptotic stability, i.e., reaching and staying at an equilibrium point, requires necessarily the adjunction to F_e^o of dissipative forces, such as a damping term $\beta\dot{x}$.

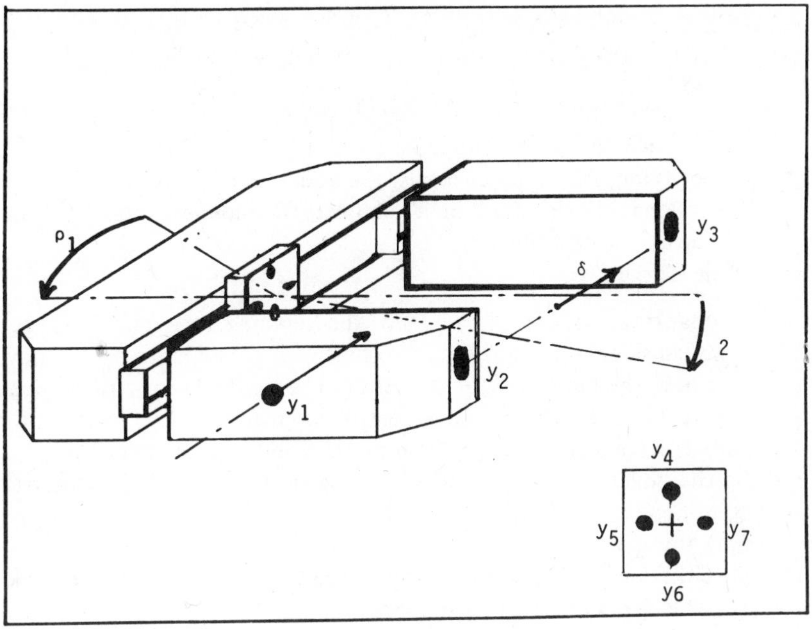

Figure 28 Example of elementary actions.

It is not a sufficient condition damping and, in addition, as it is impossible to study the stability in the general case, the minimal recommendation for using ORS feedback (with velocity or force) is to be sure that the builded ORS functions may provide a significant equilibrium point (minimal potential) where the actions are zero. For example, that implies that Eq. (44) may be completed by:

$$||v_1|| = 0 \qquad 0 < |y_1| < \epsilon$$

In case of strict velocity control, the general expression of ORS-based inputs leads to:

$$\dot{x}_c = g(x), \text{ with } x_c = x_r \text{ in the reference model.}$$

Assuming that $||\epsilon||$ and $||\dot{\epsilon}||$ are small, we have $x \simeq x_r$, and the dynamic evolution of x_r has the form:

$$\ddot{x}_r + L_V \dot{x}_r = L_V\, g(x_r) \tag{48}$$

In that case, the damping term is pre-existing and $g(x)$ has only to satisfy the requirement of being zero, with a minimal potential, at the equilibrium points.

(c) Meaning of the Weights λ_i

The λ_i parameters may have the following utility:

- tuning the a-priori relative importance of the different actions
- weighting the actions with respect to the real values of the

outputs: for example, a repulsive action eq. (44) may have a low relative weight when y_1 is weak, because the obstacle is far away. Suppression of weighting in this case would consist in demanding a very small motion for x_1, even the related action is not primordial; a possible choice for λ then is: $\lambda_1 = |y_1|$.

- Specifying, from higher levels, the active sensors related to the task. That may be inserted in a complete manipulator control system, like the one described in [10].

(d) Mobile Targets

As ORS feedbacks are really closed loop, the servoing to a mobile target is natural. If we consider, for example, the problem of grasping an object moving on a conveyor belt, the loop eq. (45) is useful. The tracking error then depends on the dynamic (gains) of the whole control loop and on the nature of the target. Improvements of the velocity loop may be obtained in several ways:

- Tracking the object with a closed loop acting on the rotation ρ_1 rather than the translation δ (Figure 28).
- Estimating the speed of the object by a Kalman filter [4].
- Using acceleration and velocity control instead of pure velocity feedback, if the noise on sensor signals is not too large.

6. CONCLUDING REMARKS

Using local environment for control purposes with ORS is an approach which deserves to be generalized. The main existing applications are in the field of teleoperation: space and nuclear interventions. In the context (hostile and complex words, disturbed vision, transmitting delays) the improvements made by ORS-based control are easily perceived. Extension to industrial applications are now necessary, and we may remark that LE sensing is useful where ever these exists uncertainties about the relations between a robot and its environment: grasping moving objects, inspection of large skew surfaces, compensation of structural elasticities of a robot.

REFERENCES

[1] N. Ahuja, R. T. Chien, Interference detection and collision avoidance among 3D objects. First National Conference on Artificial Intelligence, Vancouver, 1981.

[2] O. Khatib, Commande dynamique dans l'espace opérationnel de robots manipulateurs en présence d'obstacles. Thesis, University Paul Sabatier, ENSAE, 10, Av. Edouard Belin, BP 4032, 31055 Toulouse Cédex, France, December, 1980.

[3] J. Y. Catros and al., Le Système Spartacus: Toucher doux, approche et saisie automatique. Int. Conf. of Telemanipulators for the physically handicapped, IRIA, BP 105, 78153 Le Chesnay Cédex, France, September, 1976.

[4] B. Espiau, Prise en compte de l'environnement local dans la commande des robots manipulateurs. Thesis, University of Rennes, Campus de Beaulieu, 35042 Rennes Cédex, France, June, 1982.

[5] J. F. Albus, Proximity Vision System for protoflight manipulator arm. Report PB 291 335, NBS, Washington, January, 1979.

[6] P. M. Will and K. S. Pennington, Grid coding: a preprocessing technique for robot and machine vision. Artificial Intelligence, Vol. 2, 1971.

[7] B. Espiau and J. Y. Catros, Use of optical reflectance sensors in robotics applications. IEEE Trans. on Systems. Man and Cybernetics, Vol. SMC-10 n° 12, December, 1980, pp. 903-912.

[8] A. K. Bejczy, Effect of hand based sensors on manipulator Control performance. Journal of Mechanism and Machine Theory, Vol. 12, 1977, pp. 547-567.

[9] A. K. Bejczy, Kinesthesic and Graphic feedback for integrated operator control. Sixth Conf. Man Machine Interfaces for Industrial Control, Purdue University, West Lafayette, Indiana, April, 1980.

[10] A. K. Bejczy and M. Vuskovic, An interactive manipulator control system. Preprints of the Second Int. Symposium on Mini and Microcomputers in Control, Fort Lauderdale, Florida, December 10-11, 1979.

[11] K. Kelley, Acquiring connecting rod castings using a robot with vision and sensors. First Symposium on Robot Vision and Sensory Control, Stratford/Avon, England, April, 1981.

[12] G. Andre, Conception et modélisation de systèmes proximétriques, application à la commande de téléopérateurs. Thesis, University of Rennes, Campus de Beaulieu, 35042 Rennes Cédex, France, October, 1983.

[13] R. Masuda, K. Hasegawa, and Wei-Ting Gong, Total sensory system for robot control and its design approach. Proc. of the 11th International Symposium on Industrial Robots, Tokyo, October, 1981.

[14] C. Samson: Robust non linear control of robotic manipulators, 1983 IEEE Conference on Decision and Control, San Antonio, Texas, December, 1983.

[15] B. Espiau and G. Andre, Using Proximetry sensors in telemanipulation. Fourth Symposium on Theory and Practice of Robots and Manipulators, "ROMANSY 81", Warsaw, Poland, September, 1981, pp. 289-299.

[16] B. Espiau, Proximity sensors-based control. IRISA Report, Campus de Beaulieu, 35042 Rennes Cédex, France, February, 1984.

[17] R. Dillman - A sensor controlled gripper with tactile and non tactile sensor environment, 2nd Symposium "Robot Vision and Sensory Control", Stuttgard, West. Germany, November, 1982.

[18] C. W. Wampler, Multiprocessor Control of a Telemanipulator with Optical Proximity Sensors, International Journal of Robotics Research, Vol. 3, no. 1, 1984.

[19] J. S. Albus - Proximity vision system for protoflight manipulator arm. Technical Report PB 291-335, NBS, Washington, January, 1979.

[20] B. Espiau, G. Andre - Sensor-based control for robots and teleoperators. 4th CISM-IFTOMM Symposium "Romansy", Udine, Italia, Juin, 1984.

9

Device Organization In Advanced Robot Systems

CARL F. RUOFF

1. INTRODUCTION

A robot control system is designed to organize and coordinate the activities of system devices in order to achieve specified physical goals. The task of the robot system designer is to create a system structure which is supportable, capable of controlling a variety of manipulators, stable in the sense that user-specific and application-specific versions do not proliferate, extensible to a reasonable degree, and convenient to use in applications. This task is complicated by the fact that robots are open-ended tools. They may operate in a stand-alone manner or be part of an automated production system, receiving commands and data from hierarchically higher levels and supplying status information in turn. They may also be required to interact with other robots and automatic machines, and will almost certainly be required to control subordinate sensors and tooling which are unknown in detail at system design time. Attempts may be made as well to retrofit them with auxiliary devices which interfere with their kinematic and dynamic operation and violate system constraints.

This chapter deals with the organization of devices in advanced robotic systems, where device organization is taken to mean the way that user-device, system-device, and device-device interactions and supporting data structures are arranged. While many of the approaches described here are valid for robotic systems in general, coverage is at the level of advanced industrial robot systems that are beginning to emerge from research and development laboratories. In these systems, users are responsible for task planning, resource allocation, and collision avoidance. Tasks are programmed in a predominantly linguistic manner using explicit, ordered descriptions of actuation, sensing, and computational sequences with appropriate conditional branching. In most cases a certain degree of geometrical transformation is permitted to allow manipulators to adapt to variations in the locations of workpieces.

Internal device structure in a detailed electronic or mechanical sense and device organization from a machine intelligence standpoint are not covered in this chapter. Machine intelligence techniques including modeling, planning, perception, and task-level learning have not yet been extensively integrated into robotic systems even in the laboratory. Furthermore, a gulf currently exists between machine intelligence and control engineering as they apply to robotics. Machine intelligence seeks to generate manipulation plans automatically, while control engineering must take the resulting high-level task descriptions (such as "pick up the block") and turn them into sequences of actuator inputs that result in stable, efficient manipulator behavior.

The treatment emphasizes organizational approaches that have been found useful by the author and others in developing advanced robot systems which can be programmed to perform real tasks of fair complexity. This is not to say that a narrowly pragmatic point of view has been adopted. An attempt has been made to approach the problem in a general way which can be extended to support more sophisticated robots as system capabilities evolve. Emphasis is accordingly placed upon symbolic coordinate transformations, device exchangeability, device coordination, and parallel execution, since they form the basis for spatial adaptability, system flexibility, both device and instance independence, and temporally efficient operation. Since organizing devices is tantamount to creating at least portions of a control system architecture, much of the discussion will be from a systems point of view.

After a background description, spatial organization, device aggregation, device control structures, and task level devices control will be addressed. A discussion of limitations will follow, along with a summary. The terms "arm" and "manipulator" will be used interchangeably.

2. BACKGROUND

There is no unanimity regarding what constitutes a device. Some consider a device to be the lowest-level entity to which a scalar command can be issued [1], while others allow a device to have several degrees of freedom. In the first view, a multi-axis manipulator would be considered a collection of scalar devices, while in the second it would be considered a device itself. For purposes of this chapter, a device is a named object to which commands can be issued. It may be scalar or vector valued, may be a sensor or actuator (or both), and may receive both metrical and non-metrical commands such as control modes or object recognition requests. A device may also be physical or non-physical (the CENTER, or tool frame, described below is an example of a non-physical, or virtual device). Whether a device is considered to be an individual degree of freedom or a collection of individual degrees of freedom, it is essential that the devices and degrees of freedom participating in a task segment be controlled collectively. Mechanisms for collecting system elements into appropriate cooperative structures must therefore be provided. The individual axes of a manipulator, in particular, must be carefully coordinated if repeatable kinematic trajectories and reliable sensor-based control are to be achieved.

It is interesting to note that there is no clear distinction between sensing and actuating devices. The eye (with the visual cortex), for example, is considered a sensor, yet it will not function without its associated oculomotor apparatus. The oculomotor apparatus itself is composed of actuators and sensors (including the vestibular apparatus). For purposes of this chapter an actuator is a scalar (but perhaps nonlinear) object which can affect the environment. A detector is an object which passively provides data about the environment, where the environment includes the robot system itself. Devices, then, are composed of detectors and actuators whose behavior is organized by the control system in a manner appropriate to the task. In this view, the eye is a sensor system composed of both actuating (oculomotor) and detecting (retinal) elements. Because the control of mechanical elements is a less tractable task than the control of electronic elements, it is convenient to decompose actuators and detectors into separate control structures, associating them dynamically as the need arises. In this way the elements which must move in space can be easily coordinated for cooperative activity. Virtual devices provide a change of context or reference in some sense, and are therefore similar to detectors from a control standpoint. Position transducers attached to activators, such as axis position encoders, will not be considered as separate system devices.

In organizing a control hierarchy it is important to note that device structure is context-dependent. The internal sequencing and construction of a rivet driver, for example, are irrelevant to a robot system when the driver is being used to drive rivets. The important pieces of information are the location of relevant coordinate frames, driver operating procedures, and what to expect during operation.

It is important that devices be combinations of logically related degrees of freedom which are organizationally and functionally convenient. An arm is one such combination. A gripper is another. Because of kinematic and dynamic interactions, and because they often need to perform distinct subtasks, it is usually not convenient to collect degrees of freedom from different physical arms or from unrelated objects, such as an arm and a punch press, into a single device. On the other hand, it is imperative that different devices be able to be combined (grouped and aggregated) in a cooperative sense, as will be seen below.

2.1 Functional Requirements

As was mentioned above, this chapter is based upon approaches which the author and others have found useful in constructing robot control systems. Some of the basic functional requirements addressed by these systems are:

Manipulator and device independence. The system (with appropriate handling software and hardware) must be able to control a variety of manipulators of differing kinematic and dynamic characteristics. Furthermore, it must be straightforward to configure the system to control special-purpose devices (including sensors) defined by the user using the standard input-output mechanisms provided with the system.

Sensor-based control. The system must have the ability to acquire and organize sensory information from different types of sensing subsystems, integrating it into both low and high level control loops.

Spatial transformability. Manipulator motion sequences should be capable of being transformed in space (relocated relative to coordinate frames and changed in scale).

Parallel execution of task segments. The system must have the ability to execute task segments in parallel, both synchronously and asynchronously, as appropriate. The system should be able to support independent tasks.

Device and process coordination. The system should be able to coordinate the activity of manipulators and other devices, within their limitations, at both high (task-oriented) level and at a a servo level. Manipulators should be able to engage in cooperative tasks.

3. SPATIAL ORGANIZATION

Robotic control systems are functionally concerned with manipulating objects. This manipulation is performed relative to features of interest, which, within limits, may vary in position and orientation. The system must be able to adjust the robot motions accordingly. To do so it must be able to sense object properties, associating those properties with locations in space. While not all properties of interest are geometrical (e.g., color, temperature, and voltage), such properties are nonetheless associated with (indeed help to define) regions of space that coincide geometrically with objects of interest to the robot. Because sensors (tactile and force sensors, for example) are often attached to manipulators, kinematics plays a central role in sensing as well. Once locations are defined, the system must be able to transform the spatial coordinates into joint coordinates and control parameters for the particular manipulators being used. A consistent, convenient geometrical organization is essential to this process.

3.1 Coordinate Frames

The basic element of a consistent geometrical organization is the coordinate frame (see Figure 1). A coordinate frame is an orthonormal, right-handed set of Cartesian basis vectors with a common origin located somewhere in space. It is a six-dimensional object (three position elements and three orientation elements) defined, relative to the frame in which it is embedded, as a displacement vector and a three-dimensional rotation which determine the location of its origin and the orientation of its axes. In some cases reduced two, three, four, and five degree of freedom frames are used, but they are special cases of the general frame described here. The actual representation of a coordinate frame for computational purposes may be defined in several ways [2,3]. Frames may be defined to the system functionally using systems such as POINTY [5], which was developed at Stanford University. POINTY allows the user to define frames interactively by mounting a pointer in the hand and using the manipulator itself as a measuring device.

3.2 Workplace Organization

A workplace consists of fixtures, tools, and one or more manipulators embedded in space. Each workpiece or piece of equipment has features, such as jaws or bores, which are functionally important. It is the task of the manipulators to move the workpieces and tools in such a manner that the features of interest are brought into coincidence so manufacturing operations can be performed. Given the fact that the various features may be imprecisely located for assorted reasons, and the fact that features may be defined relative to other features (the common situation in engineering drawings), it is convenient to represent

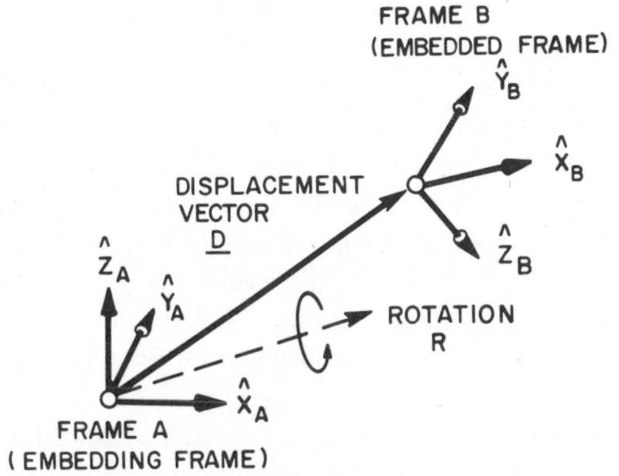

Figure 1 Coordinate frames. The location and orientation of frame B with respect to frame A are specified by a displacement vector **D** *and a spatial rotation R. The components of* **D***, which locates the origin of frame B, are referred to the axes of frame A. R is the rotation which would be required to bring the axes of frame A into alignment with the corresponding axes of frame B.*

the workspace and workpieces as hierarchically nested sets of coordinate frames, which are related in prescribed ways to the features of interest. By providing coordinate transform operators at the system level it is then possible to adapt to changing geometrical characteristics of the workplace (see below).

3.3 Universe Frames

In order to simplify the geometrical description of a workplace it is useful to define a high-level frame, the universe frame, in which all workplace features are ultimately embedded, either directly or indirectly through frame attachements (see Figure 2). Tools, manipulators and fixtures are located in space as coordinate frames relative to this universe frame. Features associated with individual pieces of equipment and workpieces, or motion excursions which must be executed relative to such features, are naturally located as coordinate frames relative to their defining frames (features). In this way a position error in the location of a fixture can be accommodated simply by updating the fixture's frame coordinates relative to the universe frame. Since the fixture's feature locations are defined relative to the fixture frame, and are not affected by motions of the fixture as a whole (assuming it is a rigid body), their absolute positions in space can be updated automatically by the system. This is a powerful way of handling relative motion. If it becomes necessary to replace a manipulator, for example, the entire task need not be retaught, nor must the new manipulator be precisely repositioned. Only the base frame locating the new manipulator in the universe must be acquired. By locating each manipulator relative to the universe, the system, which must transform the motion target frames into manipulator joint angles and other control parameters, can automatically refer the target frame in universe coordinates to manipulator base frame coordinates.

This makes it straightforward to calculate joint angles and control parameters using the kinematic and dynamic packages for the manipulator. In addition, embedding the manipulators in the universe frame means that only one set of motion targets and fixture frame descriptions is needed, those relative to the universe, rather than a separate set defined relative to each manipulator's base frame. It is essential to include such geometrical organization at the most basic level. At the risk of making manipulator replacement more difficult, tasks involving single manipulators can avoid the use of a separate universe frame by defining all motion relative to the manipulator base frame. Cooperation between manipulators, however, requires that a universe frame be defined.

The universe frame location relative to each manipulator can be defined functionally [5] by using the manipulator as a gauge to interact with physical features in space (a pattern of holes, for example), acquiring coordinate data from which the location of the universe frame relative to the manipulator base can be calculated. A procedure is executed by each manipulator relative to the same features. Since the resulting frames describe the locations of the same physical features, each manipulator has generated a description of the universe

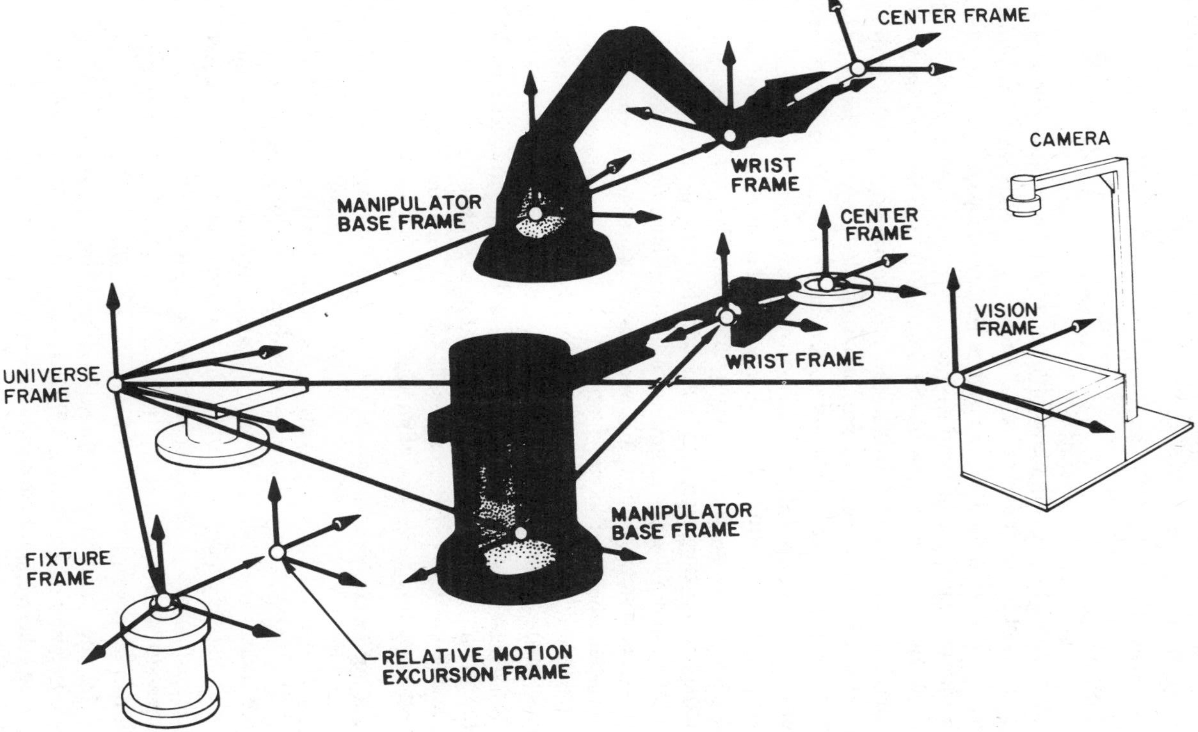

Figure 2 The workplace is organized as a hierarchical set of nested frames embedded in the local universe frame. Part and equipment relocation can be handled by updating frame descriptions.

frame relative to itself. These are easily inverted to locate the manipulators relative to the universe (see Eq. 10 below). All subsequent manipulator motion can automatically be referred to universal coordinates by the system. If the manipulators are then used to define fixture, tooling and sensor coordinate frames functionally, those frames are automatically described relative to universal space as well.

3.4 Arm Motion

As described above, robots manipulate objects, bringing various features into prescribed geometrical relationships with each other and the environment. Motions are naturally described as excursions relative to particular coordinate frames. The motions involved in acquiring a part from a feeder, for example, are naturally executed relative to the feeder. Similarly, the motions involved in inserting the part are executed relative to the subassembly of which it is to be a component. Since the part feeder and subassembly may be independently positioned, it is natural to regard the motions relative to each as being sets of relative excursions embedded in their respective frames rather than absolute motions embedded in the universe. Thus, when the subassembly moves for whatever reason, (which may be detected by a sensor) the system, using frame arithmetic, can update the necessary arm coordinates for motion segments relating to the subassembly without having to disturb those for the feeder.

Arm motion sequences, which are sequences of target frames, can be defined computationally, interactively, and in response to sensory input. In each case it is necessary to know what is meant by moving the arm to a target frame. Given a frame (tool or CENTER frame as described below) embedded in the hand (or the last link of the manipulator and rigidly attached to it), and a target frame in the environment, a manipulator is said to be located at the target frame when the CENTER and target frames coincide in space (origins coincide and corresponding axes are aligned).

3.5 CENTER or Tool Frame

As described immediately above, the tool frame [2] (which we will call the CENTER [3,4]), is intimately associated with the meaning of manipulator motion. To reiterate, the CENTER is a frame embedded in the hand. The manipulator is defined as being located at the target frame when the CENTER and target frames coincide. Thus the CENTER defines how the manipulator physically addresses the target frame. If the CENTER is regarded as a virtual device, which can be repositioned within the hand, features (defined as frames) of parts held by the hand may be brought into coincidence with the target frame simply by commanding the CENTER to coincide with the feature frame in the hand and then commanding the manipulator to move to the target frame. In this way grasping constraints can be relaxed if the sensors can provide the appropriate frame information for the CENTER. Craig et al. [4] at JPL have used such a technique to insert randomly grasped pegs into holes using force feedback. In this way manipulation sequences can be specified in terms of part motion rather than manipulator motion. The AL system [5] has taken this one step further with its workplace-wide attachment structure.

The CENTER is a frame, so it may be transformed using frame arithmetic (see below). Thus the CENTER may be made to coincide with appropriate part features once the part orientation in the hand is known, providing either that some data set contains relative feature frame locations or that a sensing strategy can provide them. If the CENTER coincides with relevant feature frames, it is also useful for transforming forces and torques [3,4], since adaptive assembly sequences are concerned with the physical behavior of the assembly process during the mating of part features.

3.6 Device Universe Frames

The discussion above regarding universe frames is valid as well for the CENTER and other devices embedded in the last link of the manipulator (a gripper or hand is not always present). A wrist mounted force/torque sensor, for example, and the virtual force torque sensor composed of individual joint torques (see below) must share a common frame description with the CENTER so sensor values can be meaningfully transformed. The hand universe frame provides such a description. The same is true of multi-fingered hands (such hands are described elsewhere in this book), which actually consist of a multiplicity of small manipulators. Just as in the case of the workplace, cooperative finger control demands a unified geometrical description. In the case of the CENTER, force-torque sensor, and related frames it is convenient to define their (common) universe frame to coincide with the wrist frame of the manipulator, or the frame attached to the last link with one base vector coinciding with the axis of rotation (see Figure 3) assuming revolute joints. Device frames in the hand are located relative to the hand universe just as in the case of the workspace except that the frame coordinates might be acquired from an engineering data set. The CENTER device origin can coincide with some feature on the hand such as its mounting flange. The extension of the universe concept to frames in the hand allows an important software and support simplification. As was the case in the workspace, movement of the various device base frames (from hand shimming or redesign, for example) can be accommodated by redefining the locations of the base frames relative to the universe origin, avoiding the need for reprogramming and reteaching. In this view the manipulator is really a mobility system for the hand universe frame. The hand devices are concatenated with the manipulator in order to calculate manipulator control parameters as we will now see.

3.7 Concatenated Devices

As we have seen, targeted arm motion is not meaningful without considering the CENTER frame. CENTER motion, on the other hand, is independent of arm motion. The arm is therefore concatenated with the CENTER, and arm motion commands in actuality are commands to move the CENTER (considered rigidly attached to the hand) in the workspace. The force-torque

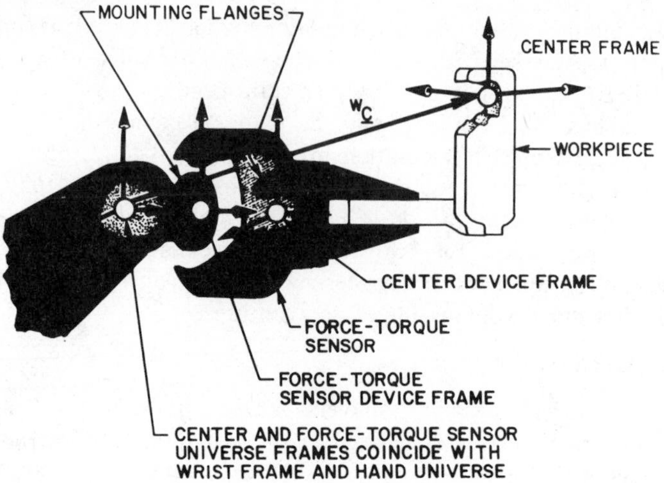

Figure 3 The CENTER frame, which is attached to the outermost link of the arm, is the frame that defines arm position and orientation in space. It can be repositioned relative to the outer link to coincide with features of grasped parts or attached tooling. It is also the frame to which force-torque readings are referred. The CENTER and force-torque sensor share a common local universe frame which coincides with the wrist frame.

sensor is another example of a concatenated device. Concatenation is an important consideration in system design because the appropriate data from the independent device must be available to the dependent one in order for the system to make sense of the actuation commands.

3.8 Kinematics

In order to operate a manipulator in space it is necessary to be able to determine both the position of the manipulator in Cartesian space given a set of axis positions and to determine axis positions corresponding to given positions in space. It is also often convenient to be able to determine the velocity of the arm in space given the axis velocities and to determine appropriate axis velocities given a desired frame velocity.

The forward kinematics of an n degree of freedom manipulator will give its position (and orientation) in Cartesian space relative to the manipulator base:

$$(\boldsymbol{X},R) = K\ (\theta_1,\theta_2,\cdots,\theta_n)\ , \tag{1}$$

where $\boldsymbol{X}$ stands for position, R stands for orientation and K is the kinematic solution corresponding to the axis positions θ_1 through θ_n. The reverse solution, on the other hand, gives the axis positions in terms of the Cartesian quantities:

$$(\theta_1,\theta_2,\cdots,\theta_n) = K^{-1}\ (\boldsymbol{X},R)\ . \tag{2}$$

If n is greater than six, difficulties arise since no unique solution exists. In that case one must resort to context-dependent rules. Even in the case of $n = 6$ not all arms are explicitly solvable.

Differentiating (1) with respect to time we obtain:

$$\frac{d}{dt}(X,R) = J(\theta_1, \ldots, \theta_n) \frac{d}{dt}(\theta_1, \ldots, \theta_n) , \tag{3}$$

where $J(\theta_1, \ldots, \theta_n)$ is the Jacobian of the position with respect to the axis positions. Formally inverting this (not possible if J is singular or non-square):

$$\frac{d}{dt}(\theta_1, \ldots, \theta_n) = J^{-1}(\theta_1, \ldots, \theta_n) \frac{d}{dt}(X,R). \tag{4}$$

See Paul [2] for a complete mathematical discussion of these topics. Note that some workers have chosen to evaluate incremental forward and reverse solutions rather than evaluating Jacobians. It is important that the system provide sufficient computational resources to meet the arm control needs. The system architecture must incorporate them in a convenient manner.

3.9 Geometrical Transformations

Nested coordinate frames have been introduced above as a convenient way of describing locations and motion sequences. To use such a description effectively, however, it is necessary to be able to perform changes of basis. That is, to change the description of coordinate frames from one coordinate frame to another. As an example, suppose a part position has been determined by a vision system, relative to its own frame (see Figure 4). Assume the position of the vision sensor is known relative to the universe, as is the position of the robot base frame. Before the robot can be positioned at the part, the part location description must be transformed from the sensor frame to the universe frame and from there to the robot base frame. Furthermore the CENTER frame must be considered. Note that in all cases the location of the same physical object (the part) is being described. The object location does not change; all that changes is the *description* of the location as coordinate frames are changed.

Coordinate transforms can be composed as sequences of two operations which here are called ORG and REL [3]. In the ORG operation (see Figure 5) the description of a frame C defined relative to frame B (which is defined in turn relative to frame A) is transformed into a description relative to frame A:

$${}^{A}C = {}^{B}C \text{ ORG } {}^{A}B , \tag{5}$$

where the left superscript indicates the frame relative to which B or C is defined.

In the REL operation (Figure 6), the description of a frame C, defined relative to A, is transformed into a description relative to B (B is also defined relative to A):

$${}^{B}C = {}^{A}C \text{ REL } {}^{A}B . \tag{6}$$

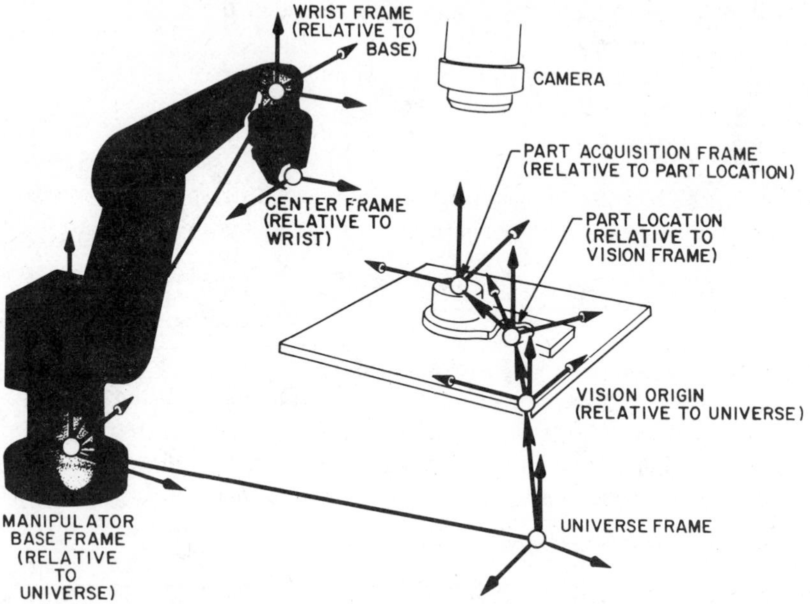

Figure 4 Relationship between frames for part acquisition. Before the gripper is closed, the manipulator must be positioned so the CENTER frame coincides with the acquisition frame. See Figure 7.

ORG and REL are inverse to one another:

$$^{A}C = (^{A}C \text{ REL } ^{A}B) \text{ ORG } ^{A}B \tag{7}$$

$$^{B}C = (^{B}C \text{ ORG } ^{A}B) \text{ REL } ^{A}B\,.$$

For a mathematical treatment of ORG and REL operators in terms of matrices and vectors see Ruoff [3]. Paul [2] deals with equivalent issues in terms of homogeneous transforms. Figure 7 illustrates the coordinate transformations

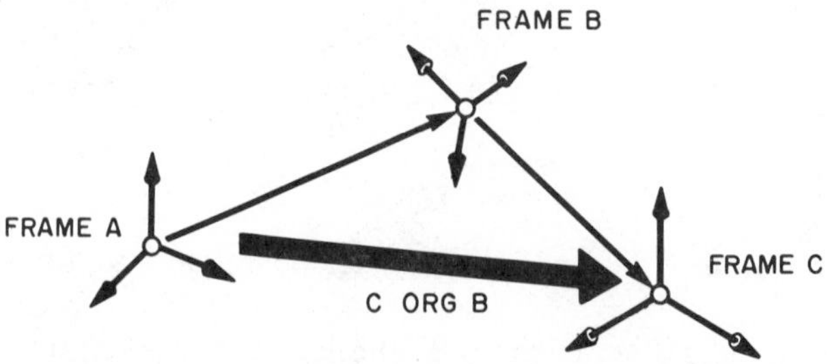

Figure 5 ORG *operation. If frame C is defined relative to frame B, and frame B is defined relative to frame A, the ORG operation defines frame C relative to frame A.*

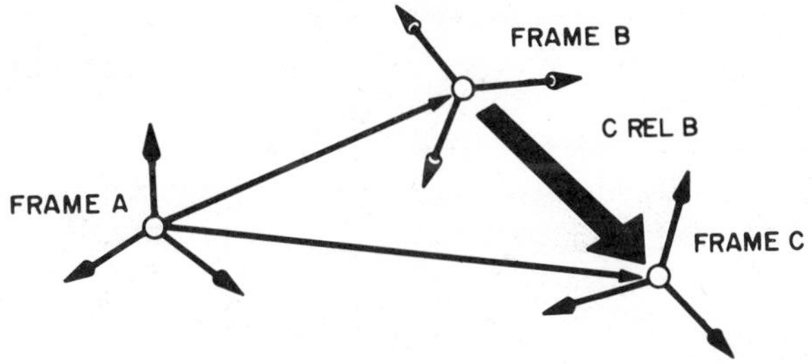

Figure 6 Rel operation. If frames C and B are both defined relative to frame A the REL operation defines frame C relative to Frame B.

involved in calculating the position and orientation of the robot wrist required for picking up the part in Figure 4. As shown in Figure 7, the acquisition frame is calculated relative to the universe frame with a string of ORG operations:

$$^{U}A = ((^{P}A \text{ ORG } ^{V}P) \text{ ORG } ^{U}V) \,, \tag{8}$$

where A is the acquisition frame, P is the part frame, U is the universe frame, V is the vision frame, and where upper left superscripts indicate the embedding frame to which the given frame is referenced. The acquisition frame is located relative to the manipulator base frame using a REL operation, where B represents the base frame:

$$^{B}A = {^{U}A} \text{ REL } ^{U}B = ((^{P}A \text{ ORG } ^{V}P) \text{ ORG } ^{U}V) \text{ REL } ^{U}B. \tag{9}$$

Before the wrist frame may be located relative to the base, its coordinates must be found relative to the universe. To do that, it is first necessary to locate the wrist relative to the acquisition frame. This can be accomplished by noting that when the part is acquired the CENTER and acquisition frames coincide by definition. Since the CENTER is located relative to the wrist, again by definition, inverting the CENTER will give the wrist relative to the acquisition frame. The mathematics involve a REL operation:

$$^{C}W = {^{A}W} = I \text{ REL } ^{W}C \,, \tag{10}$$

where C is the CENTER and I in the identity frame consisting of zero displacement and zero rotation. Finally,

$$^{B}W = {^{C}W} \text{ ORG } ^{B}A. \tag{11}$$

3.10 Force and Sensor Transformations

As mentioned above, adaptive manipulator control for part mating requires force feedback. In such cases it is useful to command the CENTER frame (in wrist coordinates) to coincide with the appropriate feature (frame) of the grasped part, and then to transform the force-torque components from the force-torque sensor to the CENTER frame.

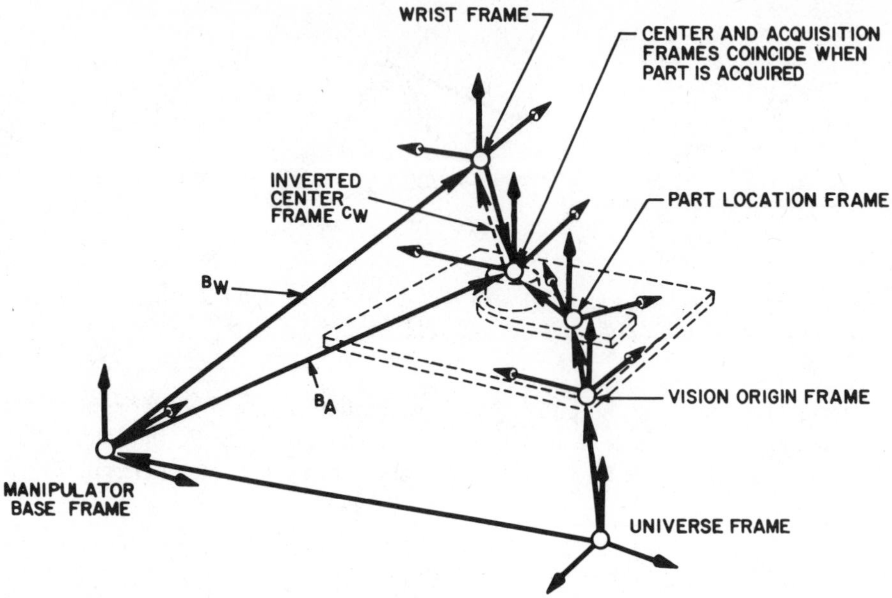

Figure 7 To acquire the part of Figure 4, the coordinates of the wrist frame which brings the CENTER and acquisition frames into coincidence must be calculated relative to the manipulator base frame. The wrist frame is then passed to the manipulator kinematics package, which calculates the corresponding joint angles. This involves inverting the CENTER frame to obtain a description of the wrist frame relative to the CENTER. The inversion is indicated with a dashed arrow.

Given a CENTER frame (see Figure 3) defined relative to the wrist by displacement vector ${}^{W}\boldsymbol{C}$ and a rotation ${}^{C}{WR}$, which rotates coordinate *axes* from the wrist frame to the CENTER frame, and given that the force-torque sensor input processes yield forces and torques relative to the wrist as well, forces $\boldsymbol{F}$ and torques $\boldsymbol{T}$ relative to the CENTER are given by

$$ {}^{C}\boldsymbol{F} = {}^{C}_{W}R \; {}^{W}\boldsymbol{F}, \tag{12} $$

$$ {}^{C}\boldsymbol{T} = {}^{C}_{W}R \left[{}^{W}T - {}^{W}\boldsymbol{C} \otimes {}^{W}\boldsymbol{F} \right], \tag{13} $$

where $\otimes$ is the vector cross product. See Paul [2] for a derivation of these relationships.

Forces and Torques at the CENTER universe (the wrist frame) can also be estimated from actuator torques, τ, by use of the Jacobian transpose $J^{+}\ (\theta_1,\ldots,\theta_n)$ [2], but in practice the results tend to be noisy and gravity loading must be subtracted. Equations (12) and (13) are still valid, but forces and torques relative to the wrist are given by:

$$\boldsymbol{\tau}_{\text{actuators}} = J^{+}\,(\theta_1, \ldots, \theta_n)\ {}^{W}(\boldsymbol{F},\boldsymbol{T}), \tag{14}$$

$${}^{W}(\boldsymbol{F},\boldsymbol{T}) = (J^{+})^{-1}\,(\theta_1, \ldots, \theta_n)\ \boldsymbol{\tau}_{\text{actuators}}\,, \tag{15}$$

where (15) is valid only if J^{+} is invertible. If not, either an invertible submatrix must be selected or an approach such as the Penrose pseudo inverse [6,7] must be used.

3.11 Reduced Degrees of Freedom

Many commercially available servo controlled manipulators have fewer than six positional degrees of freedom. In order to preserve a general system architecture it must be possible to control such manipulators. Since they have reduced mobility, it is not possible for them to position and orient objects arbitrarily in space. When a full six degree of freedom target coordinate frame is passed to the controller in such cases, it is necessary to select an appropriate set of axis values which will result in the desired position. One way of doing this is to deflate the dimensionality of the target frames by an appropriate amount, passing only the subframes which the manipulator can deal with. In four degree of freedom Cartesian manipulators, for example, the approach is usually to use x, y, z positional coordinates along with rotations in the $x-y$ plane. Another approach, used in some five degree of freedom manipulators is to solve for the full reverse kinematics as if the arm had full mobility, simply ignoring the solution component for the missing degree of freedom.

The above two approaches are simple and fast. The first approach, using reduced frame descriptions, is not as general as it requires separate frame arithmetic. It also doesn't work with all arm geometries. The second approach is general. An alternative approach would be to calculate the axis values by minimizing the errors in, say, a least squares sense, between the target frame and the wrist frame, subject to the constraint that the arm does not have full mobility. In such an approach it is assumed that position is more important than orientation in that for all cases the target and arm CENTER frame origins are made to coincide. The optimization criterion is to minimize the misalignment of the frames. That can be accomplished by maximizing the sum of the dot products of corresponding base vectors [8].

3.12 Solution Degeneracy

Reverse kinematic solutions for manipulators are often degenerate (unless joint limits prohibit alternative solutions). That is, for a particular manipulator more than one set of axis values exists which corresponds to a given position and orientation of the manipulator in space. This is always true for redundant arms (arms with more degrees of freedom than the spaces in which they work) and is true for many simpler arms. A two degree of freedom linkage, for example, can reach a given point (not the maximum extension) with the elbow either up or down. Such an arm is two-fold degenerate (see Figure 8). Many

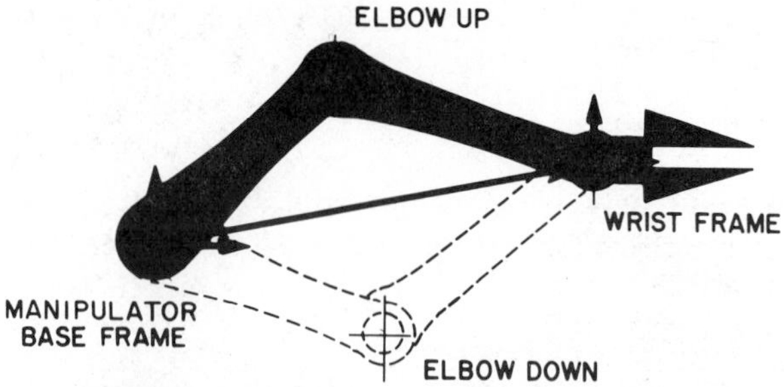

Figure 8 The two degree-of-freedom linkage shown is two-fold degenerate. Two distinct sets of joint angles give the same position of the wrist frame in space.

manipulators employ a gimbal-like device as their final three degrees of freedom. Gimbals, which are mathematically described by Euler angles [2,3], are mathematically degenerate in the sense that if Euler angles (α, β, γ) define the orientation of a frame, the angles $(\alpha+\pi, -\beta, \gamma+\pi)$ described it as well [3].

The point is that reverse kinematic solutions are not single valued. The commonly used Puma manipulator from Unimation, for example, is 8-fold degenerate. Since the axis angles describe link positions, the alternative solutions may place the arm in a radically different configuration than was expected, creating collision problems. Being able to select the appropriate arm solution dynamically is thus an important problem from a system point of view, especially when joint positions are being obtained from transformed coordinate frames rather than from a table of pre-taught values.

A couple of approaches to the degeneracy problem have been tried. Neither, in the author's view, is satisfactory, but for fairly constrained situations they are at least operable. VAL [9] has the user specify whether the elbow is up or down, whether the shoulder is to the left or right, and whether joint five is positive or negative. The arm remains in the specified configuration unless attempts to violate joint constraints are encountered or unless a mode change is commanded. A system developed by the author used a mode command [3,10] which could be given various values to specify either the absolute solution configuration or to select a solution depending upon the joint excursions from the current manipulator position. Since mode values could be variable, the system could calculate the appropriate mode at run time. Unfortunately, such computations are difficult, involving search and optimization techniques.

Degeneracy also occurs when axes align. In the gimbal mentioned above, for example, solutions are degenerate when the rotational axes of the first and third joints are parallel or antiparallel (middle axis is either zero or pi radians) [2,3]. In this case an infinite number of solutions exist, subject only to summing constraints. In practice, one must make special provisions in these cases so the kinematic solutions do not cause unwanted joint motions. It is quite possible for the workpiece to execute the proper small motions while the arm axes are gyrating wildly. This, again, creates potential collisions. Such behavior is to be expected because the reverse kinematic solutions for gimbals are singular at these points.

4. DEVICE AGGREGATION

Robot systems are commonly programmed by associating commands with named physical devices. This approach is certainly serviceable, but it is limited in that command sequences must be rewritten, using the appropriate device names, whenever it is desired to execute the sequences using different devices. This makes it difficult to write generally-accessible libraries of common actuation sequences (such as peg or fastener insertions). It is more flexible to program sequences or procedures using symbolic (or logical, in computer parlance) devices, instantiating the sequences at run time with appropriate physical devices to which commands are actually issued. If the sequences or procedures are stored reentrantly, it is possible in this way to provide for the execution of multiple simultaneous instances of a given command sequence.

To implement this idea, to prevent a proliferation of user commands, and to provide a mechanism whereby dynamic equipment changes (such as having the robot replace its own hand with a special tool) become tractable, it is convenient to introduce the concept of device aggregation in which logically-related devices are associated in a hierarchical structure [3,10]. In this structure a device is composed of one or more related degrees of freedom (perhaps in both Cartesian and joint space as appropriate), and the degrees of freedom may be continuous or discrete. Non-geometrical commands such as "find the casting" (for vision systems) or "set mode" may be issued as well.

Devices, in turn, are associated with related devices in a higher-level structure called a group. A device may be a member of only one group. Groups are defined to allow related devices to be conceptually modularized for easy replacement. A hand, for example, may be defined as a group consisting, say, of a set of fingers and a set of touch sensors. A manipulator group, on the other hand, might consist of the arm, the CENTER, and a wrist mounted force-torque sensor. The hand is not part of the manipulator group here because not all robot applications use hands. The hand may be replaced by a rivet driver, for example, but the CENTER and force-torque sensor are still necessary for the effective utilization of the driver.

Related groups, in turn, are collected into aggregates. A robot aggregate might consist of a manipulator group and a hand group. Aggregates are important because they correspond to major pieces of equipment and because the use of aggregates in control structures can simplify the consistent calling of device-oriented subroutines. It is also true that aggregates are often replaced because of breakage or wear, it being more efficient to take them off-line for repair.

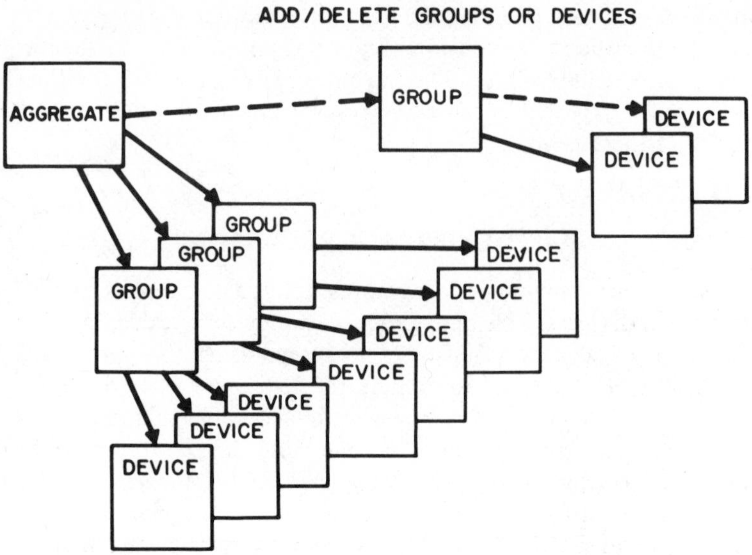

Figure 9 The aggregate-group-device structure is a linked tree. The physical description of the system may be revised by relinking as indicated by the dashed arrows.

Notice that the choice of objects to collect as devices, groups, and aggregates is quite flexible, and can be tailored to various situations. In the examples given above there is a natural functional grouping. If odd combinations are defined, programming and system functioning can become awkward. Groups and aggregates are organizational constructs, and they do not receive commands directly. They receive commands indirectly through the devices they incorporate.

The reason, again for introducing the aggregation structure is to modularize device handling for convenient program execution, dynamic device, group and aggregate replacement, and program library construction. If the aggregation structure is based upon a linked tree, device, group, and aggregate replacement are straightforward.

The utility of storing command sequences symbolically was described above. In such a scheme the devices and, consequently, the related groups and aggregates are symbolic as well. To instantiate the command sequence with given physical devices, it is necessary to associate the symbolic aggregation structure with its appropriate physical counterpart. This can be done by using an exchange array which maps the symbolic aggregate name (integer label as described below) onto the physical aggregate, providing that the two aggregate structures are functionally equivalent within the context of the command sequence. Such functional equivalence means that the symbolic groups and devices referenced by the command sequence are functionally equivalent to the corresponding groups and devices in the physical aggregate.

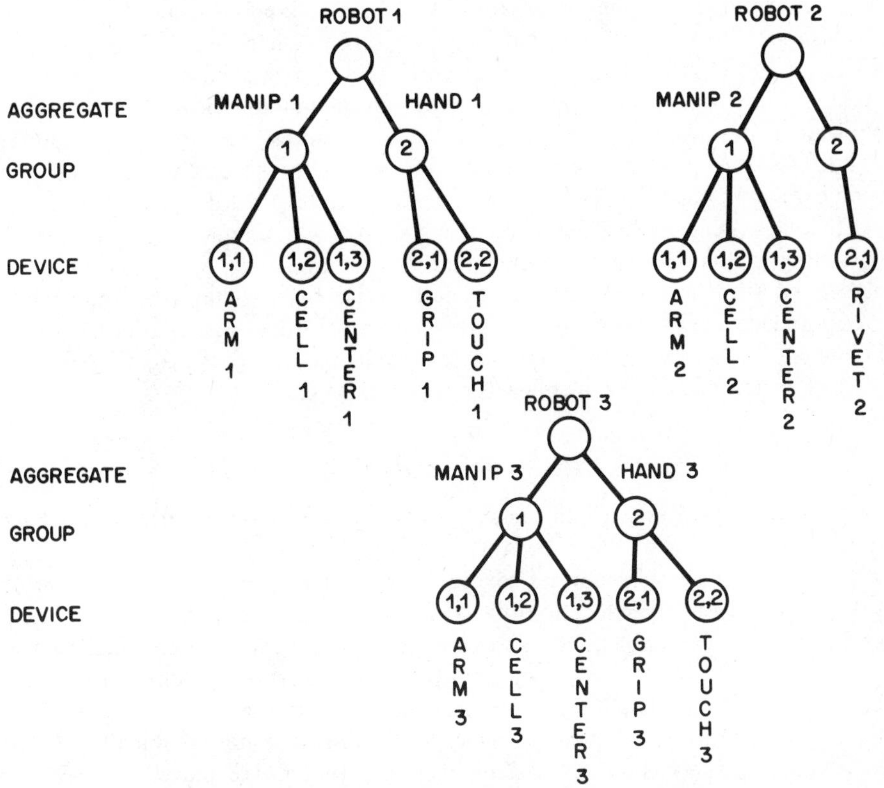

Figure 10 Aggregate trees. Node identifying integers are shown. Aggregates Robot1 and Robot3 are absolutely exchangeable, while Robot1 and Robot2 or Robot2 and Robot3 are exchangeable only for processes that reference just the manipulator group and the Arm, Cell, and CENTER devices. Cell is the name of the wrist - mounted force sensor.

A simple way of representing an aggregation structure is to use a tree (see Figure 9). The highest level node is the aggregate. Its successors are the groups. Group successors, in turn, are devices. To ensure that functional equivalence between equivalent aggregates is represented, it is obviously important to adopt a consistent labeling convention.

Aggregates which are functionally equivalent in a process are called exchangeable in that process. Aggregates (such as identical robots) which are exchangeable in all cases are called absolutely exchangeable. It is not necessary for aggregates to be physically identical to be exchangeable in some ways. For example, two robots, one with a wrist mounted rivet gun group and one with a hand group are exchangeable for pin insertion operations, providing the pin is already grasped by the hand and providing no references are made to gun group or hand group devices. This is so because the actual insertion involves only the force-torque sensor, the manipulator and the CENTER (see Figure 10). Kinematically dissimilar robots can be exchangeable as well.

Each group in an aggregate is given an integer label, distinct from that of other groups in the aggregate, but identical to functionally equivalent groups in different aggregates. Similarly, each device in a group has an integer label distinct from that of other devices in the group but identical to functionally equivalent devices in functionally equivalent groups. In this way, similar aggregates are similarly labeled, and the mapping of logical aggregates to physical aggregates described above becomes straightforward.

In this scheme, every device in the system has a unique identifying triple consisting of its aggregate, group, and device integers. A logical device is associated with its corresponding physical device by matching the appropriate group and device numbers in the corresponding physical aggregation structure indicated in the exchange array. The exchange array-aggregate structure works through arbitrarily deep levels of subroutine calls.

5. DEVICE CONTROL STRUCTURE

Geometrical organization and logical/physical device assignment schemes have been discussed above. We now turn our attention to the software structures which describe physical devices to the system, thereby allowing the operating system to call for the execution of the various device functions. For a system architecture to be stable, such software structures must be both modular and generic to avoid the necessity of multiple special-purpose patches in order to make particular devices work. It is important to note that considering a device to be a collection of independent degrees of freedom is not completely tenable in a robot systems context. The degrees of freedom are indeed independent mathematically, but the underlying task of a control system is to orchestrate the activity of the degrees of freedom (from, perhaps several devices) to achieve directed, cooperative operation. Thus the device control structures, which we shall call device control blocks (DCB's) must have provisions for considering multi-axis interactions, even if the individual degrees of freedom are considered individually to be devices. In performing coordinated point-to-point moves, for example, it is required that all participating degrees of freedom arrive at their targets simultaneously, even if they are associated with different devices. (This is not always possible for non-servoed axes unless the non-servoed axis can be a control input for the servoed axes.) To achieve this simultaneity, it is necessary first to calculate the time taken by the slowest axis to complete the move, scaling all other axis velocities accordingly. Finally the position set points must be issued (with appropriate velocities) to the servos in a burst (no interrupts are allowed during the burst since the system could switch to a higher priority task). Thus it is not possible to calculate axis-by-axis servo parameters independently. Force control, velocity control and dynamics and gravitation compensation also require that the axes be considered as interacting. Salisbury and Craig [11] for example, describe a 9×9 matrix formulation for the force control of 3 interacting three degree of freedom fingers. Even the simple matter of converting between joint and Cartesian coordinates requires a collective knowledge because in general each physical degree of freedom contributes to each Cartesian degree of freedom (contribution is zero in some cases). For a system standpoint it is important to have both Cartesian and joint space commands available to the user. Such commands should adhere to common internal representation conventions, such as turns [10,12] for angular measurement, the use of right-handed coordinate frames, etc. While most task-oriented commands are in Cartesian space, the joint space commands are important for maintenance automation and for resolving degeneracies. One may, for example, wish to resolve a degeneracy through the use of a particular joint value. By storing reference joint values for particular arm motions it is also possible to resolve degeneracies even under small coordinate transformations by looking for the solution closest to the reference solution. (Note that this makes system architecture more difficult in an aggregate exchange sense because arms with different kinematics are not kinematically exchangeable in all situations).

Device control blocks (DCB's) as alluded to above are software structures which describe devices to the system. That is, they contain or point to various routines which provide input/output, condition and transform data, etc. Not all systems call their device description structures DCB's, but the idea of table driven systems is universal. Recognizing that physical inputs and outputs to system devices may vary widely in format, it is convenient nonetheless to adopt a standard internal system representation for angles, lengths, logical variables and other quantities. This makes the job of writing system software much more straightforward. Translation between internal and external format is then handled by the routines which actually do the physical I/O. Since these routines are often customized anyway because of physical addresses, etc., this is a natural approach.

It has also proven useful to the author to organize device functions around routines. Functions associated with the device are performed by calling a particular routine, passing to the routine a pointer to a data structure which instantiates the routine in the case in question. A DCB is this organization consists largely of a header, describing to the system the type of device (degrees of freedom, etc.), which implies the format of the DCB, and a list of handling routines for the device functions. Typical functions are kinematics, coordinate transformations, reset, dynamics. Recall here that many device functions, such as servo parameter calculations, arm force control, etc., must be done collectively. This means, for example, that coordinating two arms will require separate subroutine calls for each arm to set up the servo parameters and then separate calls to the arm output routines (in a burst) to send the set points to each servo, as we saw above. Each involved device would be marked busy in this case until its axes had finished their motions. It is also convenient for the DCB to provide descriptions to the most recent device commands, busy/free state, etc.

5.1 Device Control and Coordination

Since the DCB is intended to be a modular structure intelligible to the system, it contains (or points to) the routine for detailed device control such as servos and dynamics (if they are done in the host). These are naturally associated with tables of the appropriate servo parameters. Servo routines are conveniently integrated into the system by linking them into the clock loop (servos form a linked list) at system generation time. Servos which are executed in subordinate processors may be handled by sending updated set points from the host every clock tick. The subordinate processors in that case are responsible for physical I/O to and from the servo hardware. Servos must also be coordinated in such a way that a servo difficulty in one axis is reflected in the control of associated axes in order to preserve path kinematics. Multiport memories are a convenient communication medium for master/slave processors.

In practice many devices other than servoed devices are encountered. Devices such as air cylinders have no position feedback or cycle completion sensors unless these are provided by the tooling designer. Such devices may be handled by using a timer which is linked into the clock loop when cycle is initiated. When the timer is timed out the cycle is considered complete. Such an approach is risky because a loss in air pressure is not sensed. This may have possibly catastrophic consequences.

5.2 User Configured Devices

A device may be fabricated by the user from a set of disparate elements, all of which must act in concert with definite timing requirements. It is important to be able to sequence such devices under system control because robot systems are often confronted with them. They can be controlled by writing sequencing routines which reference standard I/O channels, and attaching the I/O terminals to air valves, limit switches, etc. The clock loop is used to monitor switch closures and actuation times, and to issue appropriate commands to actuators over the output channels. These routines are then incorporated into the DCB for the device in the normal way. When an actuation command is issued the routines are linked into the clock loop. It is especially useful if devices can be defined to the system by the user, composing device functions by using normal features of the robot programming environment.

5.3 Calibration

Device calibration is an extremely important area in robotics because it is through calibration that the system is able to compensate for the (unavoidable) deviations of devices and other system elements from their idealized properties. Geometric calibration, for example, involves defining coordinate frames such as those locating vision systems and fixtures, thus compensating for the fact that workspaces cannot easily be set up to great precision without great expense. Device calibration, on the other hand, may involve data sets which condition (linearize, offset, scale) physical device parameters, transforming them to and from an internal representation suitable for a unified, system-wide processing approach. Arm kinematics, as an example, are often far from ideal because of manufacturing variations, wear, or damage. For similar arms to be kinematically exchangeable, and for arms to be accurate (as opposed to merely repeatable), these deviations must be controlled either through tight manufacturing control, which is expensive, or through the use of calibration tables. Calibration of this type involves a procedure of some sort which generates data values that are stored in the DCB's for reference by the arm kinematic handlers. The kinematic handlers may use perturbation approaches to correct the kinematics because of the intractability of exact solutions.

Geometric calibration, as we have seen, also involves procedures. Since geometric calibration data define relationships between objects, they are useful at both the user and system levels. They should be stored as named frames using, in some cases, reserved keywords, so they are accessible both to the user and to the system. These data then allow the automatic transformation from relative to universe coordinates.

5.4 Sensor Based Control

We have seen above that device control in robotic systems is a collective phenomenon. In the case of sensor-based control the problem becomes more acute because the activity of multiple devices such as manipulators, force-torque sensors, vision systems, and camera aiming platforms, must be integrated. The integration is critical because control parameters must be updated often enough that smooth, stable behavior results. In tracking and acquiring swinging objects this may involve parameter prediction in order to compensate for phase shifts due to processing time lags. The problem is further complicated by the fact that control most probably must be accomplished in task coordinates associated with objects in the workspace rather than with simple joint coordinates. Since task degrees of freedom do not necessarily align with individual actuator axes and since different task degrees of freedom may be required to obey different control laws ([4,13,14] and elsewhere in this volume), real-time transformation between joint and world coordinates is a necessity. In the final stages of peg insertion using the hybrid position/force control mode [4,13,14] for example, directions orthogonal to the peg axis are servoed to null forces and torques while directions parallel to the peg axis are controlled in position. Controlling tasks in this manner means, in general, that each manipulator axis will have an input control signal consisting of a superposition of force and position control terms.

Coordinating all this requires that the system be capable of setting up the data structures and data channels dynamically at run time because the actual control decomposition will be instance dependent. Data sets of significant size (perhaps hundreds of words) are required to specify tolerances, forces, gains, stiffnesses, and to specify the sources and destinations of the various signals. To achieve smooth, rapid control transitions, it should be possible to preallocate the data sets and quickly switch between them by relinking or flagging as task events dictate (switching from a guarded approach to an assembly sequence when the subassembly is contacted is an example). DCB structures must therefore reflect coordination requirements by having well defined routines which perform these functions.

The device control structures must further be organized so that requisite control outputs, say tracking data from vision, or forces from the force-torque sensor, are available in appropriate format at sufficient rates. Since a single processor cannot cope with all the required computations, multiprocessor architectures are required.

6. TASK LEVEL DEVICE CONTROL

Previous sections have described aspects of organizing device functions in robots. We now turn our attention briefly to the system, describing ways of organizing its interactions with the devices to support parallel execution of task segments, device coordination, subroutines, etc.

A robot task at the lowest control level is a bewildering barrage of sensor input and control output commands. Taken collectively, however, these commands result from the execution of higher level (and therefore less specific) user commands such as MOVE, GRASP, TRACK, etc. These commands, in turn, result from still higher level commands (which may be program specifications in the mind of the robot programmer) like "acquire the hub," which is less specific still. There is a clear hierarchy here. It is useful, therefore, to decompose a robot control system in such a way that different controllers in the hierarchy are responsible for controlling tasks at the appropriate level of abstraction [15].

Two control levels in wide use today are the task level, in which the user specifies task sequencing by issuing linguistic commands such as those above to devices, and device level, which deals with the details of device coordination, stability and control. Device-level control is usually accessible only to systems programmers. Axis-level control is a subset of device-level control. Device and axis control are in fact parallel processes because the computations either run in separate processors or are logically modularized as elements of a clock loop, running in the background.

Tasks are described to the robot systems considered here as sequences of macroscopic actuation commands to devices and degrees of freedom. An example of a typical command (taken from the TEACH language [10]) is

50 AFTER 10 MOVE ARM1 TO ABOVE ORG PART

The job of the task controller in this case is

1. to interpret the command;
2. to verify that the command pre-conditions have been satisfied;
3. to calculate the control parameters (joint targets velocities, etc.), for the physical device associated with the symbolic name ARM1;

4. to verify that the device is ready;
5. to issue the control parameters to the axis controllers;
6. to monitor status in order to determine when the command is complete.

If ARM1 were a non-servoed device different control parameters and controllers would be involved. Multiple devices can be actuated simultaneously by issuing the appropriate commands to their controllers. The task controller must verify that the parallelism is permissible. Again, parallel device operation can occur if the device controllers run in separate processors or if the device controller can be a modularized portion of a loop running in the background. Parallel device operation places constraints upon the system, because it implies that command sequence processing is no longer serial. Thus, one cannot issue a command, wait until it is complete and issue another. Because command time durations may vary, commands may not be completed in the order in which they were issued. Thus a mechanism must exist - the natural place is in the background process - which continually checks for events (a command completion is an event), signaling the task controller that commands which were blocked because of device busyness or because preconditions were not satisfied may possibly be executable. We will call such preconditions contingencies [10]. The phrase AFTER 10 in the above example is a contingency specifying that command 50 may be executed only after command 10 has been completed. Since the commands may be executed out of sequence it becomes somewhat more difficult to keep track of command completion status. One cannot simply assume that all commands previous to the current one are complete. It is necessary to provide a flag of some sort defining their completion status. Since the robot command sequences must be reentrant to allow multiple simultaneous instances, it is not permissible to set flags within the commands themselves. One approach is to allocate a command identity buffer for each current instance of a sequence into which all command numbers are placed as they are encountered (unless the buffer is full). When a command is initiated, it is so noted in the buffer. When it has been completed, the command identifier is purged. If the contents of the identity buffer are always kept monotonically increasing (and command sequence identifiers are monotonic as well) the absence of a command with smaller identifier than the smallest identifier in the buffer implies that the command has been executed (or skipped). Such a structure, called a cursor, has been used by the author in controlling parallel device operation [10]. Branch commands and loops can be handled straightforwardly by using the branch commands as temporary cursor loading stops and by inhibiting branching until all commands already in the cursor are complete. Branching is effected by filling the cursor beginning from the branch target.

6.1 Parallel Command Sequences

Once a cursor structure exists, the parallel execution of separate command sequences can be implemented by generalizing the cursor structure to include an additional cursor for each sequence and a mechanism for determining into which cursor(s) each command identifier should be loaded. A choice which has proven useful in practice is to allocate one cursor per aggregate. This has limitations described below, but it is a reasonable tradeoff between programming complexity on one hand and system flexibility on the other. To incorporate this idea it is necessary to associate each command (branch, arithmetic, etc.), with one or more aggregates. This is implicit in the case of device activation commands, but it must be explicitly stated (or set up by convention) for computational, device interrogation, and control commands. The system scans the command sequence, associating each command with its proper cursor(s). Note that a command may be associated with several aggregates if task sequencing demands it. The effect is to require the command identifier to be present in all associated cursors before its execution can be considered.

Coordination between sequences can be handled by contingencies or by setting flags. Cooperation can be handled by combining device commands. AL [5] at Stanford has incorporated a similar type of parallelism with the COBEGIN COEND constructs and block labeling.

6.2 Subroutines

Subroutines can be implemented by providing a new cursor structure for each aggregate passed to the subroutine, loading the new cursors in the normal manner from the command sequences which comprise the subroutine. Commands are initiated and monitored in the normal way. The cursors from the calling routine are maintained, of course, because they store the state of the calling routine. They are simply inactive (with the subroutine call marked busy) until control returns to the calling level, at which time they are reactivated and the subroutine cursor(s) destroyed. The aggregate-based calling structure is quite flexible.

Aggregates on the calling level may be asynchronously passed to separate subroutines or they may be passed to the same subroutine. It is acceptable for ARM1 to execute a pin insertion routine, for example, while ARM2 executes a riveting routine, and ARM3 continues to hold the subassembly on the calling level.

More complication is involved, of course in setting up subroutines. Data sets for local variables must be allocated (remember that routines must be reentrant) and call variables passed. Aggregates on the calling level must be associated with the proper aggregates on the level to which control will be passed. This is important because arms in multi-arm routines may not have interchangeable roles. Aggregate association, as we have seen, can be handled by an exchange array. An exchange array derived from information in the subroutine call and from the calling-level exchange array exists at each subroutine level. Data sets and exchange arrays for subroutines reside in routine control blocks, which are randomly allocated when required. A separate routine control block exists for each active instance of a routine.

Since it is not efficient to keep an aggregate busy in a routine when its role is complete, even though other aggregates are not yet finished, it is convenient to allow early returns for finished aggregates. Execution of an early return reactivates the aggregate's cursor on the calling level, marking the subroutine call done for that aggregate and deactivating the aggregate's cursor on the subroutine level. Other aggregate cursors on the subroutine level remain active. Similarly, it is efficient to allow an aggregate to enter a subroutine after it has already been initiated by other aggregates. This provides significant flexibility, but naturally the other aggregates' activities may become blocked if they depend upon the late comer.

Notice that the activation/deactivation of cursors and their associated routines is completely asynchronous. It is appropriate for such control structures to be created as linked lists from randomly allocated storage.

6.3 Programs

Given that a structure for calling subroutines exists, it is a simple matter to label the highest level routine a program, passing control to it from the console device after assigning physical aggregates, arguments, and associated data sets. These are contained in a structure called a program execution block. The program execution block is conveniently a structure composed, again, of linked list elements. Parallel programs can be instantiated in a completely straightforward way. Naturally they must use distinct physical aggregates, though they may use different instances of the same subroutines and data sets.

We have outlined a task control structure which generalizes from serial command execution to allow the simultaneous, asynchronous execution of multiple programs with multiple subroutine calls to arbitrarily deep levels, depending; only upon computational resources. Figure 11 illustrates this structure.

7. LIMITATIONS

The organizational concepts described in this chapter have proven to be very useful in programming robots at the command level to perform complex tasks. Many problems remain, however. Among them are:

1. Sensor based control techniques are not yet really well developed from either a control or a system standpoint.
2. Restricting routines to aggregates is sometimes inconvenient. It is inefficient to tie up an entire aggregate, such as a robot, on the calling level when a simple device routine is all that is required, especially if the arm can be moving as the device executes the routine. One can, of course, explicitly code the device commands on the calling level but that is additional programming effort.
3. Cooperative manipulation remains crude. A control level - called articulation control above - exists between the task level and device levels

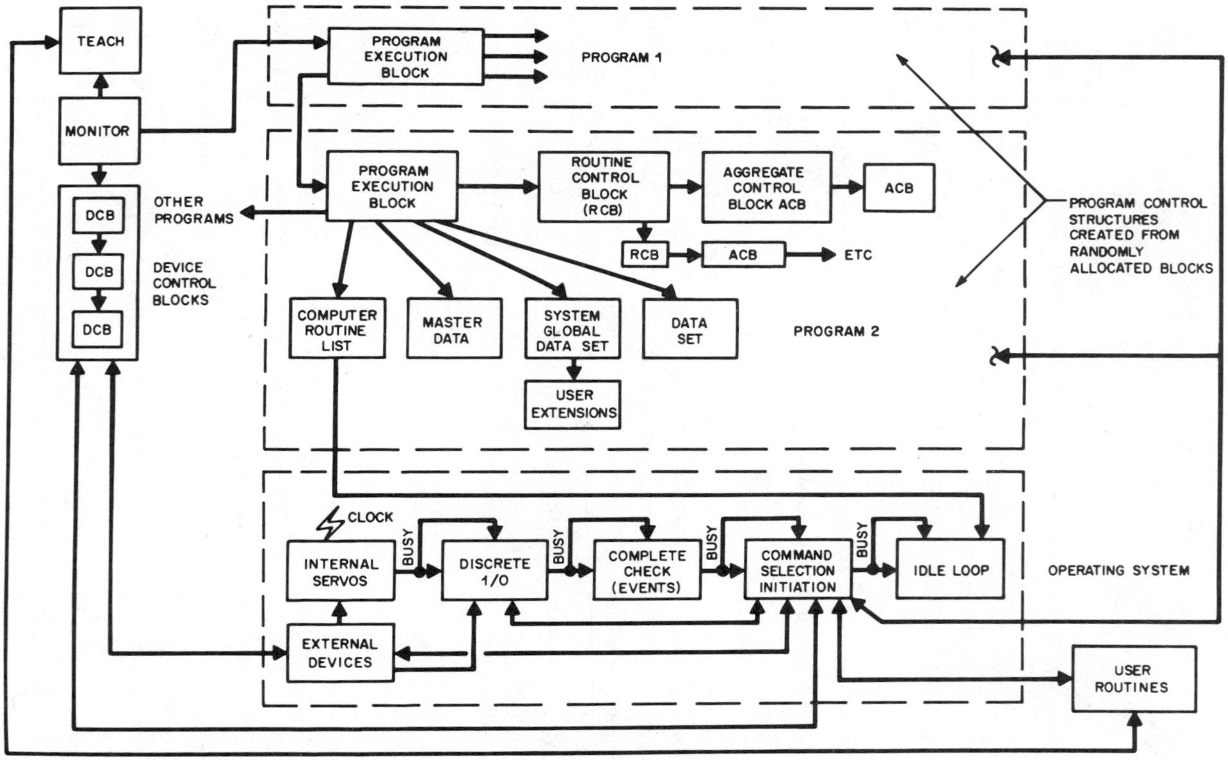

Figure 11 Task control architecture. The system relies upon randomly-allocated routine control structures and data sets to handle asynchronous, parallel command execution. The cursors mentioned in the text reside in the aggregate control blocks (ACB's).

considered here. An articulation controller would be able to coordinate all controllable axes in the system as commanded and mediated by the task controller. An articulation controller would be an ideal starting point for adaptive, or learning, sensor/motor control.

4. Device controllers do not yet integrate dynamic and inertial models to an appreciable extent, though interesting work is being done in the area by several groups.

More fundamental than the above is the fact that robots are extremely unintelligent. To program a robot to perform even a moderately complex task with error detection and recovery is a difficult task requiring significant planning, many keystrokes, and fairly strict environmental organization.

8. SUMMARY

This chapter has outlined ways of approaching the organization of devices in advanced robot systems, stressing the importance of a systematic approach. The concept of coordinate frames and transformations has been introduced and their importance in organizing both the workspace and sensory input has been emphasized. Aggregation structures were described as a mechanism for providing more generalized device descriptions, dynamic device replacement, and a basis for parallel command sequence execution. Device control structures, including organizing approaches for device control blocks, were discussed at length, followed by a description of a system organization which supports parallel processing of multiple subroutine instances.

We have seen that even though the ideas described here are useful in the sense that robots incorporating them can be programmed to do interesting tasks, robotics in actuality remains primitive. That is to be expected because robotics is in its infancy. The problems it faces are difficult, having been solved to varying degrees in animals only over the course of millions of years. Fortunately, robotic evolution is quite goal directed, and its economic niches have a special character. With the current interest in machine intelligence, microelectronics, and controls, significant improvements in robotic technology should be forthcoming.

9. ACKNOWLEDGEMENTS

I am indebted to Russell Taylor of IBM, the architect of AML and Bruce Shimano (formerly) of Unimation and the architect of VAL, for their valuable and protracted discussions with me on the subject of robot control system organization. Those discussions were very useful in helping me to organize ideas for this work.

REFERENCES

[1] R. Taylor, "An Integrated Robot System Architecture," IBM Research Report RC9824 (#42454), IBM T. J. Waton Res. Ctr., Yorktown Heights, NY, February, 1983. Also to appear in *The Proceedings of the IEEE,* 1983.

[2] R. P. Paul, *Robot Manipulators,* The MIT Press, Cambridge, Massachusetts, 1981.

[3] C. Ruoff, "The Spatial and Logical Organization of Devices in an Advanced Industrial Robot System," *Advances in Computer Technology-1980,* Vol. I pp. 239-249, American Society of Mechanical Engineers, New York.

[4] J. Craig, R. Cunningham, and C. Ruoff, "Hand-Eye Control in a Robotic Assembly Task," Proc. Autofact West Conference, Vol. II, Anaheim, CA, November, 1980, Society of Manufacturing Engineers, Dearborn, MI.

[5] S. Mujtaba and R. Goldman *AL User's Manual,* Stanford Artificial Intelligence Laboratory, Stanford University, CA, November, 1977.

[6] G. Goodwin and R. Payne, *Dynamic System Identification: Experiment Design and Data Analysis,* Academic Press, New York, 1977.

[7] N. Dorny, *A Vector-Space Approach to Models and Optimization,* John Wiley, 1975.

[8] C. Ruoff, "Minimum Error Grasps with Reduced Degrees of Freedom," JPL Memorandum #347-83-199, Jet Propulsion Laboratory, Pasadena, CA.

[9] "Users Guide to VAL," Unimation, Inc. Shelter Rock Lane, Danbury, CT.

[10] C. Ruoff, "PACS - An Advanced Multitasking Robot System," *The Industrial Robot,* Vol. 7, No. 2, pp. 87-97, June, 1980.

[11] J. Salisbury and J. Craig, "Articulated Hands: Force Control and Kinematic Issues," Proc. 1981 Joint Automatic Control Conference, Vol. 1, Charlottesville, VA, June 17-19, 1981.

[12] C. Ruoff, "Fast Trig Functions for Robot Control," Robotics Age, November/December, 1981, pp. 12-20.

[13] J. Craig and M. Raibert, "A Systematic Method of Hybrid Position/Force Control of a Manipulator," Proc. IEEE Computer Software and Applications Conference, p. 446, Chicago, IL, November, 1979.

[14] M. Raibert and J. Craig "Hybrid Position/Force Control of Manipulators," *J. Dynamic Systems, Measurement and Control,* Vol. 103, No. 3, June, 1981.

[15] J. Albus, et al., "Control Concepts for Industrial Robots in an Automatic Factory," Paper #MS 77-745, Society of Manufacturing Engineers, Dearborn, MI.

10

Tactile Sensing for Robots

LEON D. HARMON

1. INTRODUCTION

1.1 Purpose and Scope:

The purpose of this survey is to review and comment on the sense of touch in robots. This includes a presentation of the history of this relatively new technology, examination of the present state-of-the-art, and discussion of future challenges and opportunities in tactile sensing for automata. Special emphasis is placed on considerations relevant to industrial assembly robots.

Tactile sensing is only a portion of a complex web of interrelated functions, both in robots and in living systems. It is useful and to some extent necessary to introduce several related topics rather than to consider tactile sensing in isolation. Consequently, besides addressing the sense of touch in robots and in humans, we shall introduce some material on actuators, grippers, pattern recognition, and data processing. And we shall touch briefly on economic and sociological issues in robotics.

Following general introductory remarks, we first consider some aspects of the human touch sense which would seem to bear usefully on problems on machine sensing. This begins with noting relationships of touch to vision, then considering problems of coping with variables of shape, pose, object identification, and texture. Some physiological and psychophysical considerations follow, including remarks on the need for a grammar of touch and discussion of some of the active aspects of tactile search.

The next section covers the history and present state-of-the-art in automated tactile sensing. A number of aspects of automated taction are outlined, notably technical requirements like force, torque, compliance, slip, and pattern recognition. Relationships to manipulation are introduced, followed by a discussion of materials and transducers, information processing, and pattern recognition.

The final section examines outstanding problems and near-future predictions. The material is divided into several distinct but related topic areas which go beyond tactile sensing *per se:* materials and transduction, data handling and pattern recognition, hierarchical control, grippers manipulators, and revised manufacturing technology.

Owing to the wide range of topics considered here, coverage is extensive rather than intensive. The material is intended to serve as a general introduction to a relatively new discipline which couples basic understanding with practical application in many exciting and challenging ways. Ultimately one may predict that as dextrous, sensing, intelligent robots evolve, human society will be profoundly and irreversibly affected.

1.2 General Remarks:

Touch-sensing technology for robotics and prosthetics is presently very primitive. The need for robust multiple-sensor, gray-scale-responsive robotic grippers and "hands" has been widely perceived and extensively documented (32,33,86). Somewhat less often discussed, but of considerable importance, is the need to provide tactile "pressure" to a human operator for telemanipulation. Related to these technological needs are the relatively unknown operating principles of tactile perception in living systems (90) and the closely related domain of replacement touch sensing for limb orthosis and prosthesis.

The pressing requirements for sensory feedback and control in automata have spurred a great deal of research in *visual* pattern recognition in recent years. Though there have been some noteworthy and useful achievements, the state-of-the-art really is not far advanced. Device and system capabilities are modest and evolve slowly, owing mostly to many difficult problems in information processing and pattern recognition.

Unlike vision, tactile sensing for automata has been relatively neglected. Concentrated study of fundamental issues in touch began only in the last few years (cf. 15,32,33,36,74,76,86). Part of the reason for this may have been the unavailability of useful tactile sensors.

Achieving improved knowledge and technological capability in tactile sensing is a problem of many dimensions. This includes need for fundamental research and development in such matters as transduction, three-dimensional object representation of shape, orientation, location, texture, compliance, and relation of information derived from touch to system-control and manipulation operations.

In both basic understanding and in real-world application of manipulative acts employing the sense of touch, our present knowledge is virtually non-existent. This extends even to such deceptively "simple" and prosaic phenomena as delicate handling of fragile objects without slip, detection of screw-thread engagement without jamming, recognition of a paper clip (and using it), etc., etc. How does a human or can a machine competently grasp, identify what it touches (by active tactile search), sense orientation, and manipulate appropriately for a given task assignment?

In automated manipulation using touch sensing, there are two complementary classes of information to consider; one relates to properties intrinsic to objects (e.g., shape, texture, compliance), while the other relates to object response to its environment (e.g., position, orientation, deformation). The technological needs range from skin-like sensing materials to parallel processing of huge real-time continuous-variable data arrays for hierarchical control.

The emphasis throughout this essay is on *tactile sensing,* defined here to be the graded sensing of contact forces in an array of points. Consequently, a distinction is made between this and simple contact or resultant force sensing at a single point. Tactile sensing thus defined is meant to relate to skin-like properties where areas of force-sensitive and displacement-sensitive surfaces are capable of reporting graded signals and parallel patterns of touching. When reference is made to simpler contact sensing (at one or just a few points), whether binary or continuously variable, the term *simple touch* will be used.

2. THE SENSE OF TOUCH IN HUMANS AND MACHINES

2.1 Relationships of Touch to Vision:

Tactile sensing is in some ways closely related to visual sensing. Visual and tactile "pictures" have many data-acquisition, processing, and pattern-recognition characteristics in common. In both, continuous signals are sampled in two-dimensional (2-D) or 3-D space. Both modalities share a common model of the world about them. Their analogous (or even similar) spatially sampled pattern representations suggest that tactile information processing might benefit from the relatively more technically advanced art in visual scene analysis. However, tactile data processing may in some ways be easier than its visual counterpart; in touch we do not encounter the difficult problems of optical shadows environmental lighting variables, or color.

On the other hand, tactile sensing presents some totally new problems. Consider, for example, that in vision the 3-D reconstruction from a pair of 2-D representations permits the (derived) depth computation to be the burden of the central processor. That processor is usually quite separate from the *passive* sensor (which themselves are remote from the object sensed). In touch, however, not only must the transduction be at the site of the object, but real-time sensing in three dimension by *active* (manipulated) elements is mandatory. (By active I mean implementing touch by deliberate strategies of collecting information by sensor movement.) Further, tactile sensing requires handling the variables of sensor/object force and displacement, both static and dynamic. Touch, being a contact phenomenon, implies many static and dynamic attributes aside from shape (e.g., surface texture as it affects slip, conditions of load and moment, etc.). Thus, touch is, in some ways, considerably more complex than is sight.

A somatotopic (skin-surface) map representation is desirable (similar to a retinal map in vision); some "central" representation of the patterns impinging on the sensory surface is essential to intelligent processing. However, the basic features of this map (e.g., object position, shape, texture, hardness, slip, etc.) probably ought to be extracted at the low-level data processing in the hand insofar as is feasible. This implies, among other things, extraction of high-gradient pressure signals to inform about edges, holes, and the like.

Note, too, that for eye-hand coordinated action, both in robots and in humans, the 2- or 3-D touch-sensing representation needs to be overlaid or mapped onto the visual representation in some equivalent 1:1 manner. Clearly there are visual-, tactile-, and motor-map equivalences in humans. How to achieve that sort of multi-representational overlay (e.g., for touch and vision) in robotic sensing and internal representation is a most important question -- one that presently has not even the beginning of a satisfactory answer.

It is often believed that vision serves to organize spatial representation and that vision and touch alone are central to spatial perception. A completely different, important, (and probably correct) view was put forth by Sherrington in 1947. It is experimentally defended in (42), and the view is summarized in that publication as follows:

> ... because there are specific anatomical linkages between the sensory areas of the cortex and the motor cortex, the motor cortex may therefore be a pathway between the senses ... This account implies that space perception depends upon a fusion of visual, auditory, cutaneous, and proprioceptive inputs, but vision is only one element in a mutually supportive system rather than the primary spatial reference. Since the demonstration of spatial abilities requires at bottom the organization of movements in space, patterns of movement may serve to integrate spatial information.

Additional illumination of these matters is found in (19,45,57,58,110), and compelling evidence for the central role of voluntary motor control in the development of normal perception is exemplified by the studies reported in (35).

2.2 Some Relevant Aspects of Human Tactile Sensing:

It is tempting to try to make machines in man's image because the competent natural systems provide existence proof of engineering success. Further, since man has studied his own systems extensively, it could be argued (or hoped) that some of the findings will lead to useful insights in machine design.

In the past, in other fields, this sort of reasoning has failed to pay off -- usually because the rate of acquisition of information about living systems is so slow and the progress in engineering design, independently, is so rapid. Still, one supposes that consideration of some of what we know about ourselves might illuminate or at least stimulate our machine design. The following observations are offered in that spirit.

First, it is instructive to consider the transduction capability of the fingers when considering similar input systems for robots.

There are three principal types of touch receptors in human skin:

1. Pacinian corpuscles, which are buried deeply and contribute little to spatial imagery. There are about 120 per phalanx, are binary only, have very low threshold and possess a large receptive field.
2. Merkel cells, which are slowly adapting and have a density of $\geqslant 50/mm^2$; and
3. Meissner cells, which adapt rapidly and have a similar spatial density, $\approx 50/mm^2$, though they are 1/20 the size of Pacinian corpuscles and have a many-to-many mapping.

The Merkel- and Meissner-cell counts decline rapidly with age, reaching $10/mm^2$ each by age 50 or 60. Assuming a fingertip area of $20mm \times 30mm$, we estimate the initial spatial-receptor count to be $50/mm^2 \times 60mm^2$, or 30,000 for each of the two cell types. However, the different fibers subserving these receptors branch profusely as they near the skin surface, with the result that one fiber subserves $\approx 1mm^2$. Consequently, there are only 600 sensitive points to consider; i.e., a 20×30 array. Thus, one "receptive field" contains 30,000/600 = 50 each of the receptors which drops down to 10 by age 60.

A rough overall estimate of independent data lines which must subserve the human hand is: 5 fingers × 600 points (plus palm and balance of hand surface -- notably sides and backs of fingers, say a couple of hundred more). A generous upper-bound estimate therefore is 3,500 per hand. And the dynamic range of each is on the order of 20 dB minimum.

This is to be contrasted with today's experimental robotic manipulators which range from a few sensitive points (e.g., 3×3 arrays) with modest gray scale, up to a total 400 sensitive points having binary response only. However, some laboratory research models are planned to have spatial density which exceeds that of human fingertips.

It is instructive to consider the *dynamics* of the human sensing apparatus. Hearing is passive, or nearly so. Head and ears move, but beyond that and the directed attention which is common to all sensory modalitities, the act of audition is predominantly passive. Vision, in sharp contrast, notably departs from such sensory passivity by its constant, active search and exploration. Eye movement, both voluntary and involuntary, adds to the motion of its mounting system to become truly active sensing.

However, touch handsomely edges out both audition and vision as an activist. We grope, feel, fondle, slide, finger, brush, stoke, rub, manipulate, palpate, handle, roll, fold, spindle, and mutilate. All of the other senses -- like pain, heat, cold, taste, smell, time, acceleration, and muscular proprioception -- are such passive modalities that hearing seems hyperactive in contrast. Tactile sensing is the true dynamicist. See (26,28,50) for extensive accounts of this topic.

Some related and important observations in (27,50) are well worth noting. Each joint is a sense organ. The joints yield *geometric* information, the skin yields *contact* information. Object shape cannot, in general, be detected by moving an object past a fingertip -- only *vice versa*. Shape can be detected by a stick held in the hand, eliminating tactile perception as we commonly define it. Feeling can be both objective (perception of object) and *subjective* (perception of skin displacement). Also, the human equipment for *feeling* is identical to the equipment for *doing*, unlike as in any other sense modality.

Human tactile perception consists of two quite separate components -- cutaneous and kinesthetic. The cutaneous portion, mostly conveyed by the pressure-sensitive arrays on fingertips, is usually thought of when considering robotic taction. However, the kinesthetic portion plays a considerable role in the sense of touch. This component includes both *afferent* (incoming) signals developed at muscles and joints, and *efferent* (outgoing) signals which are motor (muscle) action commands. The combined cutaneous and kinesthetic senses are sometimes referred to as "haptic." For extensive and valuable treatment of these and related issues, see (45,58). A crisp summary of the view (57) is that "... the hand should properly be considered the sense organ for touch rather than the mechanoreceptors, and ... emphasis ought to be on the active seeking of information by the exploring hand." This is another way of stating the ideas presented earlier on visual-tactile-motor mapping equivalences.

The human hand achieves its stunning versatility by having 20 degrees of freedom* (though not all completely independent). Together with wrist, forearm, elbow, and shoulder which permit delivery of this manipulator with at least 11 more degrees of freedom (depending on how one tallies them), the entire system -- fingertip to thumb -- sports no less than 31 degrees of freedom. This is to be contrasted with the typical six degrees of freedom for a robotic manipulator.

When one attempts to describe or analyze the intricate movements of a hand, either human or robotic, it quickly becomes evident that an adequate language does not exist. To explore this uncharted area, one could construct a grammer of touch -- ideally a specialized human/computer language for taction and manipulation. A very interesting effort in this direction is reported in (99). Algorithmic recognition of complex shapes is attempted using psychophysically determined discriminates such as concave, convex, sharp, straight, etc., and relationships among other geometric variables. Categorical similarity measures were defined and tested for random 2-D shapes.

One way to begin thinking about a comprehensive tactile syntax is to consider three categories of description and action -- 1. Elements, 2. Object Condition, 3. Operations. These may be understood as follows:

1. *ELEMENTS* are the "elemental" or "primitive" features of touch information. "Element" and "primitive" will be used interchangeably in

---

* Four fingers have 2 joints each with one degree of freedom (DOF) an one joint with 2 DOF, one thumb with 1 joint having 1 DOF and one having 2 DOF; palm with 1 DOF.

the following discussion. By element or primitive we mean those physical properties of an object and its condition which are *necessary* to characterize it. A *sufficient* set of primitives and their relationships will permit the object's recognition.

The extraction of primitives such as line, corner, edge, etc. is standard procedure in vision systems. In touch, we need first to develop a coherent nomenclature and classification system for object description. For example, two classes of primitives in such a taxonomy might be:

a. *Intrinsic object properties*

   Shape: edges, corners, faces, radii of curvature ("sharp...dull"), contour/outline, dimensions...

   Texture: smooth...rough, pattern, ridges...

   Hardness: soft...hard,, viscoelastic, rigid...limp...

   Note, however, that tactile object primitives and related visual or geometrical primitives are not necessarily equivalent. For example, consider "edge." A tactile edge *may* be perceived by noting that two planes intersect, but it is much more common for a human to detect or describe an edge as a "sharp line" without any reference to planes which may generate that line. Further, the edge may be perceived by stroking an exploring finger either along it or else across it. This illustrates an important component of tactile-sensing object-primitive definition and set construction which must be considered. I thus hypothesize that there exists a set of primitives which is unique to the tactile world.

   Besides geometric characteristics like edge, which may be peculiar to tactile sensing, there are others like hard, flexible, limp, elastic, heavy, and so on.

b. *Extrinsic factors*

   Force, *Moment*, *Displacement* } and their time derivatives

   These are the actively derived primitives in the object's external world, together with the passive intrinsic object properties, which are required to describe the object's *condition.* Knowledge of that condition is a prerequisite to a given object *operation* (manipulation).

   With such a basic vocabulary in hand, one will then have a rational foundation on which to categorize and analyze tactile-sensing object/information relationships. Also, such elements may be useful in further steps by facilitating some grammars to be constructed from the primitive features.

2. *OBJECT CONDITION* is a statement, constructed from primitives, which describes an object and its status at a given time. The generic way in which object status is expressed constitutes a grammar of primitives. This then yields a description of the object, its orientation, position, velocity, and acceleration in the context of a 3-D force picture.

   For example, a small, smooth, heavy rectilinear object with center of mass at a particular place may be upside down and slipping from grasp, etc.

   It is such description which enables continual evaluation of progress toward completion of a specified operation, and interpretation of the feedback requisite to updating control commands.

3. *OPERATIONS* implement algorithms for active sensing, pattern recognition, and hierarchical control in order to accomplish a given task. This is made possible by the use of sequences of object-condition statements. The central problem in the study of operations is the construction of effective algorithms which lead to appropriate task performance.

There are three classes of tactile-pattern-recognition scale to consider for both humans and automata. In one, the object or quality to be recognized is smaller than the (fingertip) sensor array. In this case, the fine-grained resolution of a finger may be used to identify an object by texture. Second, an object's size may be commensurate with hand size and so be enveloped by the fingers for identification. In this case, recognition stems primarily from fingertip positions and joint angles and less from pad resolution. In the third case, the object to be identified is large compared to the hand, and there must be active, dynamic search exploration by fingers. Here the time sequences of manipulator joints also give important information. These three cases are exemplified by considering identification of a dime, a tennis ball, and a desk top.

It should be noted that most of the early robotic-touch research projects sought recognition of objects placed onto sensing-array pads. The silhouette pattern recognition of objects small compared to sensor pad (class 1, above) was directed to identification of simple geometric shapes and small parts. Information acquisition was essentially passive and static, in contrast to most human taction.

Even for fingertip recognition of small parts (e.g., our class 1 dime example above), touch sensing rarely is passive and virtually never is static. There is active stroking, rolling, squeezing, and tapping. It would thus seem that passive, static taction in automata would offer little potency and, in fact, probably draw attention away from the important problems.

So far in this discussion, tactile sensing was considered principally as it applied to object recognition. This is pattern recognition taken in its typical sense -- the identification of particular members of pre-assigned classes independent of the usual variables of size, position, and orientation.

However, another tactile operation, quite different (and generally simpler), can offer useful object discrimination. This is the process of determining only a few of the "primitives" (mentioned earlier) that make up the larger process of pattern recognition.

Very likely the most important primitives of pad sensing for robotic tactile sensing will be items like flat, concave, convex, point, edge, and the geometric relationships among them. Note that the detection of these kinds of "features" does not require pattern recognition of objects by their shapes. Clearly such detection alone may have considerable utility in many tactile-sensing tasks.

How may one construct geometric properties of large objects by touch alone? Consider, for example, feeling a desk top. The following scenario is derived from a real-time, closed-eye exercise. Note that the actions described include both cutaneous and kinesthetic components.

First, the finger(s)/hand encounters the desk top. After encounter, the hand is oriented (wrist, elbow, shoulder movement) so that an exploring fingertip moves quasi-randomly over the surface by sliding and application of approximately constant normal force. This establishes a plane regarding the feeler's coordinate system.

The encounter drops off at an "edge." A second planar surface is perceived at 90° to the first. The exploring finger then returns to the edge and traverses it, maintaining the edge roughly in the middle of the finger pad. This establishes edge location and direction.

How, analytically, is this done? One might test by a Euclidean measure, feeling whether two intersection straight lines were imbedded in the surface being explored. (A tactile straight-line could be defined by perceiving via joint proprioception that a fingertip moved from point A to point B in a constant direction in 3-D space.)

A simple first-order question is how might one determine and represent, say, horizontal planarity by moving a fingertip (or all 5 fingertips) over the desk surface? When I do this, I note that the up-down motion at the wrist joint is 0, while elbow and shoulder joint angles change.

Next, when the front edge of the desk is reached, one of two tacks is taken: 1) the wrist joint remains fixed, and the contact area moves from the fingertip (an edge scraping across it), or 2) the wrist angle is changed so that the fingertip contact position stays constant.

A right-angled meeting of planes can be defined by insisting that the same fingertip-contact area follows the angle and noting that either the wrist rotates 90° on the forearm axis, or else that the plane of the back signals of the hand does. Alternately, one could assign a point locus to the exploring fingertip, measure all joint-angles changes and thus compute a series of points in Cartesian 3-D space, noting that they form a right angle.

When a finger encounters a plane, a rough estimate of planar orientation is derived (without search) from the knowledge of the contact (depression) locus on the finger -- coupled with *a priori* knowledge of the finger's orientation in 3-space [via *dynamic* limb-joint sensory feedback (proprioception)]. As soon as contact is made, the finger can be made to move (accurately) over the surface of the arbitrarily oriented plane. This may use feedback derived from maintaining a constant pressure.

To summarize: some determinations such as planarity and recognition of "simple" objects may be made by "observation"* of a few wrist/arm/hand/fingertip angular-displacement measures. Note, too, that the fingertip-pad requirement in all cases (as speculated here) calls for maintaining a constant pad pressure and placement for all transactions; i.e., pad resolution need is modest, and the essential subsequent computations reside in the manipulator control signals required to maintain the pad-position and force signals constant.

Most of the early elucidation of human taction concentrated on receptor anatomy and physiology and on very simple perceptual responses such as spatial and temporal discrimination. Vision research took a parallel course and then, within the past twenty years, expanded rapidly into a broad spectrum of much more complex and sophisticated psychophysical exploration.

Now, research in human tactile perception is beginning to take on a similar aspect. Interest in the sense of touch from high-level perceptual aspects is growing rapidly and may be expected to parallel in many ways the intensifying interest in *automated* taction. For an important introduction to these interests and activities, see (12,39,42,50,55,57,58,90). Close interactions between perceptual psychologists and robotics engineers are in considerable evidence and are increasing. Valuable cross-fertilization undoubtedly will occur as both fields mature.

An extensive introduction to many aspects of industrial uses of human hands and their manipulative actions is found in (5). Included are relationships to tools, space and trajectory variables, and precision grip. The 192-item reference list provides a fine assortment of topics pertaining to pick-and-place and assembly manipulation and thus serves as a useful starting point for study of anthropomorphic robotic hands in relation to such tasks.

3. HISTORY AND PRESENT STATUS OF ROBOTICS TACTILE SENSING

3.1 Overview:

Manipulation by industrial robots presently includes acts of recognition, acquisition, and handling. The real-world uses to which these systems have been put so far are many. Just a partial listing of major examples would include: assembling (from bicycles to automobiles), casting and molding, foregoing, grinding and polishing, heat treating, locomotion, master-slave manipulation (usually for hazardous environments), machining, painting, pouring, sorting, stacking, transporting, and welding.

Although a large variety of tasks is already handled by these robots, many applications areas remain largely untouched. Emerging technology in touch,

* Actually, most joint angles and their changes cannot be easily or accurately contemplated consciously. This is in great contrast to skin surface signals which one can observe deliberately and in great detail.

vision, and system competence will undoubtedly open many new market opportunities. New systems will be developed for undersea salvage and research prospecting [see (89) for a complete survey], space station operations, mining, and hazardous factory and rescue operations. And, according to some optimists, even micro-surgery and agricultural harvesting by robots may not be too far off. [However, there is some reason to believe that farming operations may not adopt robotic high technology; see (32).]

The kinds of robotic tasks now executed, while useful, are quite primitive. Most are open loop, and the small proportion which uses sensing feedback does so with very simple vision- and touch-feedback information. It is clear that with more sophisticated and competent touch sensing, for example, a large increase of automated manipulation and small-parts assembly would become possible.

Such advances surely must ultimately transform our society. For example, it is already clear that the impact of industrial automation on human blue-collar employment will be enormous. Possibly as many as 20 million U.S. and Western European jobs in assembly, machining, machine operation, welding, and painting will be usurped by automata within 20 years. This undoubtedly will raise profound new issues in the socio-economic structures of society. World-wide growth of the robotics industry now approaches 40% a year. The economics alone of this new technology dictate ultimate replacement of people by machines wherever possible. [See (33) for a review of these considerations.]

Teleoperation represents another important applications area. Though potentially of considerable importance in a number of applications (space, underwater, hazardous chemical and radiation environments, mining, construction, etc.), this technology actually has received little significant attention. Manipulation tends to be gross since large objects are generally involved; however, touch sensing is extremely important since visual systems often are not useful. Remote location calls for greatest possible autonomy.

Prosthetic and *orthotic* needs for the handicapped comprise still another important use for manipulators and tactile sensors. This includes artificial limbs for the amputee and sensing and assist devices for the paralyzed. There are an estimated 7,500,000 disabled persons in the U.S. (excluding rheumatoid arthritis), of which about half probably could be helped with presently available technology. However, even as in industry, the economics do not permit as much application as one might hope. Many of the prosthetic and orthotic devices and systems having robot-like configurations, are still in prototype development, though some of these are "in the field" for realistic testing and evaluation. Typical developments in artificial hands are seen in (24,51,77,92,93).

In the *evolution* of automated tactile sensing, there has been strong and continual technological interplay between industrial robotic and prosthetic/orthotic developers. Problems of force feedback, slip, stable grasping, object recognition, position sensing, light touch, etc., are common to both. Interaction with researchers of tactile sensing in living systems has been meaningful and is growing rapidly.

It is clear that sensory feedback in robots is in need of considerable fundamental research and development, is of potentially great use to society, and presents problems which relate to industrial production, human rehabilitation, and living-systems analysis.

Despite the obviously considerable list of needs for effective touch sensing, the present real-world state-of-the-art remains surprisingly primitive. Automated tactile sensing is at a very early stage of investigation, comprehension, and competence. Up until the last few years, touch feedback systems for robots and manipulators were very simple and relatively crude. The industrial systems still employ extremely simple devices; virtually all of the more sophisticated, complex and potentially useful tactile sensors are in laboratory development, largely in academic or governmental settings.

Commercially available touch sensing has not yet been advanced past the gross-force-sensing options which have been provided by the manipulator manufacturers. However, there is a growing trend for users to purchase conventional robots and then modify them for specialized applications -- some for immediate production use and others for experimentation for future in-house special configurations. Some of these technological add-ons already are borrowing advanced prototype tactile (and visual) devices and systems from research laboratories.

It seems apparent that tactile sensing for industrial robotics is presently beginning to take on realistic and useful forms. The transition from very simple contact sensing to full manipulator taction is under way. In what follows we shall trace some of the major steps in this evolution and examine some of the problems and promises in providing robots with a sense of touch.

Sensory capability of any kind in industrial robots has had a slow development. The evolution from open-loop numerically-controlled machine technology produced surprisingly few sensing organs to provide useful feedback for control systems in commercial applications. Of course, there have been numerous laboratory research explorations, but few so far have had impact on the real world. Vision systems, after more than two decades of development, are just beginning to be useful and to find acceptance. Tactile sensing, having tremendous potential also, is only now gaining momentum in early exploratory phases.

A large assortment of devices and systems exists for robot sensing of the environment. They fall into two generic classes, contact and non-contact. Contact, of course, is the domain of tactile sensing with which we are primarily concerned here. Non-contact sensing principally includes optical, sonic, and magnetic ranging; some examples in these categories will be given in passing, where appropriate.

In robotics, and particularly in industrial automation, the matter of manipulator control is central. Closed-loop control is desirable in most cases, and that, of course, implies adequate sensing. At present the principal sensing needs in such systems are *proximity, touch/slip* and *force/torque*. A useful and illuminating tutorial overview of these and related topics appears in (9).

The major robotics manufacturers have so far felt little need to incorporate more than primitive touch sensing (resultant forces on grippers and at joints) in their products. Market capture has been accomplished with a technology of surprising simplicity. Touch sensing is accomplished in three ways: on-off switching for contact, poke-probe travel for force feedback and, just recently, simple strain-gauge force sensing in gripper jaws or fingers.

Although U.S. industrial manipulator manufacturers presently offer no complex visual or tactile feedback, they acknowledge that there are many areas where such features obviously could be offered to advantage. But, for today's applications, it is usually believed to be unnecessary to get so fancy or expensive. Indeed, many useful industrial robotic tasks are now accomplished with simple technology.

Other present off-the-staff industrial robotics sensors are used to signal impending contact. These are usually inductive proximity switches and photocell detectors. Extremely simple, they need little description. The inductive proximity sensors are used to register contact or near contact with ferrous materials. The photocell arrangements consist of beam-interrupt sensing and straightforward angle-of-reflectance proximity sensing.

A significant glimpse of the future is seen in one of the Japanese industrial robotics developments, the Hitachi HI-T-HAND (29). This seven-degrees-of-freedom articulator can feel for scattered objects, recognize forms, and position objects for firm grip. This is done using only metal contact detectors and pressure sensors of conductive rubber where the force sensing is limited to thresholded binary signals.

As recently as six years ago there was no documented evidence for real-world applications (excluding master-slave systems) of force/torque, proximity, or touch sensors as terminal devices for manipulator control (8). Force- and torque sensing developments date back only to about 1972, while touch and slip art began around 1966. Proximity sensing was then only some 3 years old.

Representative literature on force and torque sensing for manipulator control can be found in (1,11,23,82,105,106). There are four distinct approachs: 1) resultant (reaction) force sensing at manipulator joints, 2) reaction force sensing at a work table or pedestal, 3) force sensing at a wrist, and 4) passive (open-loop) compliance, such as remote center compliance (see 106-108). Owing to extremely complex problems in structuring suitable computer-control algorithms, the use of closed-loop force feedback has so far been restricted to very simple force detection and neutralization techniques. Still, a notable achievement using such techniques (actually a combination of binary wrist sensing and passive compliance) was seen in the Hitachi HI-T-HAND (29).

The state-of-the-art in resultant-force sensing is represented by the work at the Draper Labs (106-108). Two similar kinds of force-sensor systems have been developed, each having six degrees of freedom. One is a pedestal (work table) sensor system; the other is a wrist-sensor system. The wrist-sensor transducer, for example, consists of four cantilever spring bars configured as a

Maltese Cross. Each bar has bonded to it four silicon strain-gauges, one on each face (106). The work-encountered forces (coupled via sliding ball joints) are translated into assembly (effector) forces via computer control. This arrangement can insert 1/2 inch diameter pegs into close tolerance (0.0005 inch) holes.

Compliance is an important component of manipulation. There are two principal considerations. One has to do with the unwelcome fact that force cannot be measured directly; one can measure only displacement resulting from force, and thus the measurement is derivative, subject to error, hysteresis, etc. The other consideration relates to gripper-applied pressure distributions; fingers and hands must be compliant in order to ensure that handling does not insult objects handled.

Industrial robot designers are extremely sensitive to the issue that compliance, particularly in force sensing, is not an ideal fact of life. Ideally, zero compliance is desired -- for two reasons 1) compliance means uncertainty of sensor position and hence position of sensed surface (position must be computed from the measured force), and 2) compliance means reduced frequency response and, often, hysteresis.

Compliance could be put into software; i.e., "stop on touch" with preassigned force variables. However, the cost is a concern; high-bandwidth sensors imply high-bandwidth hardware, which in turn calls for high-bandwidth software -- all expensive. Some investigation of the possibilities of software compliance have been undertaken, but the developments to date are rudimentary.

Some approaches to gripper compliance are simplistic but do the job adequately. An ingenious example is Cincinnati-Milacron's solution to the raw-egg handling problem. Overall gripper force is sensed and (binary) limited by gripper spring-loaded limit switches, while local pressure distributions are controlled by compliant, shaped, sponge-rubber gripped pads in self-centering jaws. Pad texture and friction characteristics control slip. As with most industrial robot technology to use so far, this empirical solution is hardly elegant but works well.

A simple but very functional solution to the now common mechanized handling of raw eggs is obtained by a vacuum-operated accordioned rubber snout (59). Another relatively novel concept uses multiple vacuum "fingers" to adapt to complex object surfaces for gripping (96). This work appears to have been stimulated by earlier 3-D object-shape-recognition research (47).

A cleverly designed mechanism to control applied force consists of a segmented structure which, snake-like, wraps itself around a grasped object (37). Control is obtained by monitoring tension in the internal tensioning cable system. This gripper conforms to objects of almost any shape and over a wide range of sizes while applying nearly uniform pressure all over. The interesting claim is made that the design lends itself to harvesting fruit, handling eggs, capturing animals, and transporting people.

A large variety of specialized mechanisms has been developed for industrial robots. This variety often reflects the need to introduce compliance as well as to satisfy particular object-handling requirements. The principal types are finger- and tong-type grippers, vacuum grippers, and magnetic grippers. Open-loop compliance is introduced via flexible fingers, self-aligning still fingers, padding, and spring loading.

One of the desired physical characteristics of manipulator fingers for some applications is that they be viscoelastic, much like human flesh. Compliance such as that provided by elastometric technology (78) seems to be a start in the right direction. One study specifically directed to a pilot comparative assessment of high polymers developed both models and measuring equipment to evaluate appropriate viscoelastic parameters (44).

Just as human flesh compliance has been modeled to better implement viscoelastic gripper technology, so too has the prehension of the human hand been studied for automated multi-fingered design considerations (83,84). Similar anthropomorphic reference several years earlier led to improvements in prosthetic hands for amputees, including such refinements as a thumbnail which was very advantageous in picking up objects (20).

Two parallel streams of activity in gripper-design practice are apparent. In one, human-hand-like considerations enter strongly. In the other, design is based primarily on engineering principals which seem useful, and *no* attempt is made to replicate nature. Examples are found in the vacuum grippers mentioned above and in other mechanical designs such as those exemplified in (66) and (101). A brief review of some of these grippers is found in (25); overviews of several recent Japanese approaches appear in (32,95); and (59), illustrating many techniques, surveys early designs. A quite thorough and lucid survey of gripper mechanical-design is found in (17).

Compliance to limit forces in non-gripping operations (like pin insertion) generally takes the form of series-inserted elastic members. In other approaches, servo interaction is used; this is found in some limb prosthesis applications where a "soft touch" is desirable for a mechanical hand (e.g., 51). Feedback loops are arranged where sensed upward forces on a prosthetic wrist reduce downward shoulder torque, thus resulting in a softer touch. (This can be viewed as a small step in the direction of developing active compliance.)

Slip sensing is essential to proper performance in many manipulation tasks unless considerable gripper overpressure can be tolerated. Information about contact areas and pressure distributions and their changes in time are needed in order to achieve the most complete and useful tactile sensing.

In contacting, grasping, and manipulating objects, adjustments to gripping forces are required in order to avoid slip and to avoid possibly dangerous forces both to the hand and to the workpiece. Besides the need for slip-sensing transducers, there is the requirement for the robot's being able to determine at each instant the necessary minimum new force adjustments to prevent slip.

One of the earliest attempts to detect and correct slip used a piezo-electric crystal which sensed object slide relative to it (5). In the same year (1968) a device was reported which used strain gauges to report shear (slip) forces for prosthetic limbs (79). During the period 1972-1975 there was extensive device development of slip-sensing. Techniques included the use of piezo-electric, semiconductor, and electromagnetic transducers where the motion of a workpiece-contacting stylus was sensed, and photoelectric and magnetic rolling-all devices where discrete markings on the balls (optical or magnetic) produced $A \rightarrow D$ converted signals (98).

In both the early stylus-motion slip detectors and the early ball rotating types (where the balls rolled on an axle), only unidirectional slip could be detected for each transducer. This difficulty was overcome by a later development using a freely mounted (cup capture) optically marked ball (97). In this device, photo sensing of ball rotation in any direction yields all slip information in one transducer. Further, only tangential motion is detected, and thus noise rejection is superior.

All of the slip sensors cited so far are qualitative devices. The first report of a quantitative transducer is found in a development by Hitachi (61). Single-direction slip is optically measured by a rolling cylinder coupled to a radially slitted disc.

Until recently, there had been no specific effort reported to assess quantitatively the direction of slip. A modification of the rolling ball scheme is the use of a surface-dimpled ball in contact with a sensing needle. The needle is fastened to a circular plate in such a way that slip-generated vibrations in a particular direction cause the circular plate to contact 2 or 3 pairs of surrounding electrodes, thus signaling slip direction in "any" direction (10). Since there are 16 parts of ring-surrounding electrodes, omnidirectionality is quantized into 16 directions.

One potentially useful prosthetics program, whose technology could be relevant to robotics, is developing pressure and slip sensors for denervated hands (A. Schoenberg, University of Utah, personal communication). The lab-prototype pressure sensors use PVF_2 to pulse ultrasound and detect bone reflection (time-of-flight), thus measuring fingertip compression. Similar techniques, using ring transducers, measure finger-joint angle. Additionally, PVF_2 microphonics will be tried for slip detection.

Proximity Sensing, the detection of approach to a workpiece or obstacle prior to touching, is required for really competent general-purpose manipulators. Even in a highly structured environment where object location is presumably known, accidental placements and orientations may occur, and foreign objects could intrude. Avoidance of damaging collision is imperative in such "accidental" encounters by a blind blundering, groping hand.

However, even if the environment is structured as planned, it is often important to slow a working manipulator from a high slew rate to a slow approach just prior to touch. Since workpiece positional accuracy always has some tolerance, proximity sensing still is useful.

In the more general and interesting case of a relatively unstructured environment, a robot arm/hand *must* have advance warning of impending contact at all times.

Vision is only partially satisfactory as a proximity-sensing modality -- at least insofar as present technology provides. Blockage in a complicated 3-D scene presents a basic difficulty; automatic optical depth perception is primitive; realistic 3-D scene analysis is pre-stone age; and cameras tend to be large and obtrusive. Further, in proximity reporting which does not use scene analysis (optical triangulation, for instance), formidable problems are introduced by variables such as ambient light and target position, orientation, and reflectance.

Other proximity-sensing techniques presently available (e.g., sonar, EM, magnetic fields, intergeometry) appear to offer acceptable performance in some specialized, constrained situations. Still, general-purpose, multi-use, pre-contact sensing remains an important technological problem to solve.

Response time presents one non-trivial difficulty. There must be fast response in order to slow down, change course, or stop before damaging collision can occur. A deceleration-time problem then arises: Assume that a typical arm assembly moves 2 M/sec and that the structure and the driving forces can accommodate a 2-G deceleration. This implies a halting lead time of about 100 msec. If a modest 10% positional resolution is required, then a 10-msec sensing and data processing cycle time is needed. Information flow may be a bottleneck here, and system design demands will be rather stringent.

3.2 Transduction:

Since tactile sensing implies contact, this section is devoted principally to the technology of contact transduction. However, for the purposes of industrial robotics and remote manipulation, object position, shape, and even surface detail -- properties for which touch sensors are configured -- are at times sensed without contact. Several types of remote "touching" are discussed at the end of this section.

Although present touch sensors are primitive, it is of interest to note the considerable variety. As of about 1971, the only devices available for tactile sensing were microswitches, pneumatic jets, and binary pressure-sensitive pads (e.g., 53). Such sensors serve principally as limit switches and provide little or no means for detecting shape, texture, or compliance.

An obvious methodology for obtaining a continuous measurement of force is potentiometer response to a linear (e.g., spring-loaded rod) displacement. Early sensors in many laboratories used such devices, and they still are in limited use today. Specialized types of potentiometer arrangements have been developed, such as that in the "Belgrade Hand" where resistive-paint electrode arrays compress compliant material to achieve continuous conductance change (77).

In the early seventies the search was already underway for shape detection (52,93) and for "artificial skin" which could yield tactile information of complexity comparable to that detectable by the cutaneous portion of the human sense of touch. Two fields had primary interest: robotics and prosthetics.

It should be noted in passing that essentially all previous and present robotic tactile sensing research is analogous to human *cutaneous* sensing only. That is, the afferent and efferent *kinesthetic* components mentioned earlier which contribute to human *haptic* perception are not yet part of robotic space perception and object manipulation in any integrated sense. When actuator control signals and manipulator joint sensors are intimately coupled functionally to tactile sensing surfaces, then robots will have acquired a true haptic sense. It is important to keep in mind that a complete sense of touch relies on everything from fingertip to shoulder.

One of the earliest attempts to develop transducers that were analog and which had more spatial extent than a single contact point used arrays of conductive-sponge elements; progressive cell-collapse caused progressive resistance-change (53,54). Another approach by the same workers used graphite filaments in fiberglass mats. These were deliberate attempts to circumvent the well-known hysteretic difficulties with conductive rubbers. [Ref. (53) contains an interesting set of simple sketches for stress or pressure patterns for 23 classes of shapes.] Other early (and primitive) efforts to achieve spatial and, to some extent, quantized sensing used spring-adjusted switch arrays (54).

One of the most widely used force-transduction elements is the strain gauge. Many configurations have been used, both in laboratory exploration and in commercial application (e.g., 10,29.102,105). The 3-orthogonal-axes resultant-force strain gauges in each of the two gripper jaws of the current IBM RS-1 typifies the off-the-shelf state-of-the-art.

Several exotic types of sensors are described in (102), including a pneumatically delicate whisker which is a simple but effective non-intrusive poke-probe switch. Another unusual device mentioned there is a bi-stable metal dome that snaps to contact under pressure and which can be assembled into matrix arrays.

A load-profile analyzer, consisting of an array of pressure sensitive transducers, was designed to indicate distributions of force over large areas (8). Sensing elements included variable inductance and piezo-electric elements. Variable-inductance proximity switches, used in some industrial robots, are rugged and reliable. Arrays of piezo-electric transducers have been used to (simplistically) model the tactile sense (48).

A novel transduction scheme for wrist sensors employed three component wrist motions transduced by LED/phototransistor pairs which sense relative shadow area cast by a pin in the light path (81). The proportional pin intrusion is a function of the position of the wrist element to which it is attached. A similar use of proportional light sensing is used in the jaw touch-sensors of this manipulator; force applied to a compliant pad progressively shutters a light path. Both systems are simple, elegant, and linear.

Let us consider now in some detail specific classes of transducers which have been developed for simple touch and tactile sensing.

1. Conductive materials and arrays.

   Resistive/conductive materials have several advantages. They are simple, cheap, heat resistant, and they are early fabricated. However, the choice of materials to be used for tactile sensing via conductive elements is sharply limited. Developmental models using graphite embedded in an elastic medium (54), conductive rubber (40,92), and conductive foam (54,91) have not been very satisfactory. Noise, non-linearity, hysteresis, fatigue, long time-constants, low sensitivity, and drift, in various combinations have called for improved materials technology.

Some improvements have been obtained by more sophisticated fabrication techniques. A novel approach is described in (74) where a matrix of crossed silicone-rubber cords provides force-sensing conductance changes at the intersections. Besides being small, inexpensive, and rugged, the described system provides for minimal crosstalk among sensing sites.

A related conductive silicone-rubber transducer is reported which, owing to deposition techniques, is anisotropic (36). A single mating layer of etched conductive lines suffices to obtain X-Y pressure-sensing resolution. It is of additional interest to note that the proposed system approach considers *active* touch. Future plans call for a tendon-actuated mechanical finger to probe and press objects for tactile exploration.

A variety of conductive and semiconductor material designs for elastic carriers is described in (65); and the early use of electrode arrays in conductive elastomers is outlined in (92).

There has been considerable laboratory experimentation with various small arrays to statically and passively detect simple-object outlines. Tongs which could recognize cubes, spheres, cylinders, cones, and pyramids used a pair of 3 × 3 arrays of conductive polymer buttons (30). The buttons were said to yield proportional signals, but no details were given. In another 3 × 3 array using conductive plastic, small spike electrodes were pressed directly onto the plastic surface (104). A simple force distribution pattern analysis was employed to orient workpieces automatically.

In a prototype 4 × 8 array using crossed electrode strips sandwiching a thin sheet of conductive elastomer (10), pressure distributions were studied on a visual display after microprocessing. In another research project, a 4 × 4 electrode conductive elastomer array was used to study detection of small ensemble of surface discontinuities -- touch, point, pit, spur, ridge, crack, edge (91).

A long-term and widely-known program to develop artificial hands for amputees has been undertaken in Yugoslavia. This is one of the oldest projects concerned with artificial grasping and touch sensing for prosthesis. It was initiated in 1964. The "Belgrade Hand" project started collaborative experiments on so-called "artificial skin" with workers in France in 1975 (92). The aim was to achieve perception of roughness, hardness, pressure, and slip with a material which, like skin, was soft, elastic, and tough. A metalized foil surface on a conductive rubber sheet is used; sensing electrode arrays on the other side of the sheet provide distributions of force signals (18,93).

2. Semiconductor sensors.

Semiconductor approaches have considerable potential. Small and sensitive, they readily permit preprocessing at the transducer level. However, present semiconductor technology is a fragile one, and the materials tend to have friction characteristics which are not well suited for sensitive "skin". Also, semiconductors are readily distracted by their environment (e.g., by heat, EM noise, ambient fields).

Recent technology has brought about semiconductor strain gauges sensitivity, reliability, linearity, and versatility. These devices have typically been used in robot manipulator force sensing. However, their costs are on-trivial, usually several hundred dollars for a fully signal-conditioned unit. Consequently, their use has been rather limited. See (9) for a representative review of experimental approaches, and (106-108) for typical application descriptions.

The discrete-element semiconductor force sensors are now being seriously challenged by two simpler and far less expensive devices, bonded doped-silicon bars and silicon-doped resistive elements formed by IC techniques. These piezo-resistive transducers bring the costs down by an order of magnitude, signal-conditioned units being available for just a few tens of dollars.

Piezo-resistive sensors have considerable advantage over the elastomeric sensors described earlier with respect to sensitivity, linearity, noise characteristics, stability, speed, and hysteresis. However, they do not compete well with respect to physical flexibility, 3-D spatial distribution, and low cost. The latter characteristics are readily achieved with elastomers in the search for "artificial skin".

3. Other types of sensors.

The types of contact transduction discussed so far were principally resistive, semiconductive, and piezo-resistive elements. Other applicable techniques include piezo-electric, electromagnetic, hydraulic, optical, and capacitative sensors, though these are relatively rarely seen in robotics touch sensing.

Piezoelectric transduction has high potential and can be quite useful. A major disadvantage is that there is no D.C. response. This technology, too, is so far a fragile one. *Piezoresistivity,* as mentioned above, is considered to have promise, and it is compatible with silicon technology, but sensors tend to be stiff, flat, and slippery. The proper piezo technologies have yet to be developed.

Capacitative sensing is prone to influence by external fields. Additionally, it, like *magnetics* sensing, is too strongly materials dependent to be of general interest.

Photoelectric approaches are easy to implement, but the conventional uses so far appear to be limited to proximity sensing. Even in that application there are problems of erratic behavior and unwelcome sensitivity to uncontrolled variables such as reflectance, ambient illumination, and smoke. Multi-element transducer structure, using flexure of diaphragms or bending of waveguides or lightpipes, might offer a useful new approach.

Fiber-optic sensing techniques using deformable reflective elastic surfaces can offer very fine-grained sensing and freedom from EM noise effects. The novel design reported in (87) permitted a spatial resolution of 6200 sensitive spots per square inch. Present intrinsic problems with robustness, temporal response, and dynamic range are cause for concern, but further materials development could be expected to make the approach attractive for a number of applications.

A new departure in tactile sensing, designed to avoid many of the shortcomings of conventional technologies, is seen in the discrete-magnet sensor described in (31). Fine-grained arrays of magnetic dipoles are embedded in an elastic medium. The dipoles, which can be readily packed with 1-2 mm spacing, are deflected upon applied pressure. Their position and orientation are reported by underlying magnetoresistive elements. Both normal and tangential forces thus are detected, yielding novel performance for single elements. Real-world problems with ruggedness, hysteresis, and susceptibility to magnetic targets and fields may present problems; however, the method is sufficiently novel and potentially useful that its development bears watching.

In a unique transduction system employing VLSI technology (76), a conductive-polymer sheet is overlaid on an IC electrode array. Conductance changes with applied forces provide input for processing arrays in the adjoining layer. Matrix analysis to compute touched shape will be done by a sophisticated processor using filtering, shifting, and template matching. Hence the system is a self-contained "smart finger". Prototype small arrays having 1-mm spacing have been produced.

VLSI approaches like this may present robustness problems in harsh industrial environments. However, the peripheral signal conditioning, information pre-processing, data reduction, and multiplexing easily provided by this technology have valuable promise.

In a recent and innovative use of polymers, mechanical transduction is separated from electrical transduction (J. Rebman, LORD Corporation, Cary, NC, personal communication). A surface "skin" layer of elastomer is separately tailored for task-specific load-deflection characteristics. Mechanical coupling to discrete transducers (e.g., opto-electronic) in a "subdermal" layer permits tailoring and optimization of signal-response characteristics. These multi-point arrays are sensitive, fast, non-hysteretic, and robust. They have both normal-force and shear-sensing capability.

This development is the only tactile sensor presently in evidence which is specifically intended for industrially competent robotic use. A first-generation moderate-resolution sensing pad (8 × 8 sites, 0.3" spacing) is presently available as off-the-shelf technology. Finer-resolution (8 × 12,

0.1") will appear soon, and a considerably denser fine-array technique is under development.

A comprehensive review of comparative transducer technologies and manufacturers may be found in (62), and a brief look at solid-state microsensors appears in (64). An overview of sensor requirements and types is available in (102). Some general remarks on robotic sensors are given in (67) which contains a useful 37-item reference list.

4. Non-contact (remote) "touching".

Two types of remote sensing are useful to robots. One, of course, is vision in the sense of scene portrayal. The other is ranging, usually accomplished optically or by ultrasound, although radio-frequency and magnetic techniques also have been employed.

Both ranging and visual analysis are implicitly related to tactile sensing. They are useful precursors of touch for finding objects and adjusting and positioning the attitude of a gripper prior to contact.

Scene analysis has several aspects which closely parallel tactile analysis. Most fundamentally, both use sampled-space representations for pattern recognition 3-D objects. Although vision systems lie outside the purview of this report, a very brief summary of the state of the art is given below to indicate that advances in visual sensing, although useful, are also primitive.

One of the most advanced of the vision robots appears to be the Hitachi system (56). Its seven TV cameras and pressure-sensing pads in one hand (the other is larger and has no touch feedback) enable it to bin-pick jumbled parts, recognize and orient them, then fit parts together. So far it appears to be only a demonstration device, but it has been successful in assembling vacuum cleaners.

The CONSIGHT system of General Motors is one representative example of current vision technology in actual production-line use (103). Manipulators pick up unoriented (but isolated) parts from a moving conveyor belt using a starkly simple linear-array camera. Sighting along the belt and sensing structured-light-beam-interruption profiles, the camera builds up a silhouette image of the part. This is an intermediate step between beam-interrupt sensing and full scene analysis.

In a visual developed by Autoplace, seat-latch assemblies are inspected for completeness. Many parts are present, and the optical images (silhouettes) are often of low contrast; consequently, the acceptable real-world performance is impressive. Perhaps the most complex of the Autoplace vision robotics applications employs 10 cameras looking at an automobile chassis to detect and measure nonplanarity prior to body installation. An incredible accuracy of 0.005" at working distances of greater than 5' is used to specify shims which then are automatically installed.

Other vendors currently offer silhouette recognition technology which principally grows out of early SRI research in vision systems (68).

Machine Intelligence, Inc. is representative. Their view of the field is given in an informative survey of the relatively recent state-of-the-art of vision systems (80).

Early optical-triangulation ranging systems (ca. 1969) worked over ranges greater than several feet. The much closer distances required for robot manipulator sensing are accommodated by systems such as the one described in (10,41) where near-IR LED light sources work with silicon detectors. The arrangement is straightforward; a fixed optical geometry defines a volume in space in which a reflecting object is located by triangulation. Typical sensing distances are a few centimeters; accuracy is on the order of a few tenths of a centimeter. (Precise ranging information in such systems is not generally achieved easily since received light depends on target reflectance.) More complex arrays were developed to provide 2-D sensing, and thus bounded object regions became detectable. Experiments with similar technology are described in (69,98). Ring arrays providing a conical light "probe" (43), are interesting and useful. Another approach, using infra-red fiber-optic proximity sensing (16,22), is intended for grasping fingers -- both for proximity estimation and for simple-touch object-recognition using multiple sensors.

Emergent electronics technology has made possible a number of sophisticated, miniaturized systems that would not have been feasible much earlier.

Ultrasonic (time-of-flight) sensors are less commonly encountered but are useful (102). One unusual approach uses a matrix of transducers/sensors for primitive shape recognition, and another employs resonance phenomena in a compliant blanket to assess surface details of an object (18). This kind of system may be used both for range information and for object shape/texture determination.

One experimental model of robot proximity sensing using pulse-ranging ultrasound used up to 16 transducers operated by a microprocessor (54). Complex sonar-mapping strategies were explored.

Magnetic proximity sensors work with reluctance, Hall-effect, and eddy currents. Outside of the Cincinnati/Milacron use of inductive proximity switches, there seems so far to be little commercial robot manipulator use of magnetic techniques. Surveys of proximity and touch-sensing technology are found in (82) and (102).

The evolution of tactile sensing devices and systems had been largely *ad hoc* and seldom guided by rigorous specifications. In an effort to obtain a more precise set of numbers for desirable transducer characteristics, a detailed questionnaire survey obtained the opinions of 55 robotics researchers and manufacturers (33).

A summary of those responses provides reasonable guidelines for near-future tactile-sensing transducer research and development. The main sensor properties that seem desirable are listed below.

Space resolution should be about 1-2 mm. Arrays with much finer mesh (down to thousandths of an inch) might be useful for some ultra-demanding tasks, but one or a few mm (say 1/10") would very likely be suitable for most industrial uses. Since finger-like arrays were seen as reasonable first approaches, this implies a transduction matrix of between 5 × 10 elements and 10 × 20 elements for each "finger tip".

It should be noted that in human tactile perception there is a difference between *two-point resolution* and *pattern discrimination.* In the former, the ability to discriminate *two* simultaneous indentations on the finger tip from *one* is lost when the interpoint distance becomes less than about 2 millimeters.* This (simultaneous) two-point discrimination capability thus implies a resolution matrix for a typical fingertip of about 10 × 15 points. However, the underlying transducer array is implicitly much finer since such a discriminable array can be shifted and/or rotated minutely and continuously over the surface.

The questionnaire mean response of about 5 × 10 to 10 × 20 elements just brackets the human fingertip matrix characteristic of 10 × 15 points.

The range of 50 to 200 elements per fingertip which spans most responses to the questionnaire very likely would suffice for most tactile requirements. Of course, for extremely special applications (say micro-surgery or mechanical-watch assembly), resolution finer than human may be important. However, in the majority of applications, from assembly to welding and from tooling to harvesting, there appears to be little need to sense spatial gradients (e.g., radii of curvature or burrs) with a resolution capability far different from a millimeter or two.

Another approach to specification of spatial resolution used a top-down estimate. It was suggested that spatial resolution at a fingertip might be between 10X and 100X the repositioning accuracy of a robot arm. Since positioning repeatability typically is 0.1 mm, this would imply a fingertip resolution of 1-10 mm which is consistent with the above guesstimates.

Sensitivity is obviously also strongly applications determined. Interrelationships among variables are important considerations; velocity, acceleration, response time, and strength of materials are mutually dependent design parameters which relate to sensitivity requirements as well as to stability criteria. Among other considerations, a touch-sensing transducer and its appendage must not move or damage a touched (or collided with) object. This depends primarily on the touched object's mass and thus is applications bound at the start.

Most light-assembly tasks were seen to require force sensitivity (for each matrix element) of at least one or a few grams. A few respondents saw 0.5 gm as desirable, while others would be content with 5-10 gms.

Dynamic range was observed to be more important than linearity. A logarithmic response would be satisfactory, and a range of 1000:1 would be

* If the two points are stimulated in time sequence, interpoint discrimination is finer.

reasonable. Several ranges might appropriately be built into the same structure.

Linearity and *hysteresis* are often mentioned together. Non-linearity may be tolerated since, if the device is stable, inverse compensation is easy. Hysteresis, on the other hand, is intolerable. Touch sensing devices should be stable, monotonic, and repeatable. (An interesting reflection on our engineering system concepts and capabilities is that human touch, obviously potent and successful, is fairly hysteretic. We somehow manage to live very well with it.)

Time resolution is particularly important in terms of overall control loop response. The touch-transducer response time needs to be small compared to that for the loop cycle. Even with extremely fast computer technology, multi-element sensor array processing and pattern-recognition computation may be expected to be relatively time consuming. Consequently, there is a strong need to ensure that transducers have high bandwidth.

Reaction time of 1-10 msec is desirable for the entire automaton according to some, but ranged up to 300 msec for other respondents. Despite the fact that transducer response was the point in question, there was a mix of considerations which included thoughts about multiplexing, data processing, loop stability, etc. Some emphasis was placed on a perceived need to update data at 1 msec intervals. Some estimates require transducer bandwidth to cover from D.C. to 100 Hz; however, if the fastest response time thus is on the order of 10 msec, then the previous data-updating requirement cannot be met by an order of magnitude.

It should be noted that in human touch the vibration bandwidth *reportable* at the fingertips is about 20 Hz for separate touches and several hundred for sensing vibration *per se*. However, the reaction time (whole-loop) of the human to, say, push a lever as fast as possible after *any* stimulus, averages about 150-200 msec.

The probable time requirements for industrial manipulator touch sensing indicate that a useful transducer high-frequency limit should be at least 100 Hz, but 1 KHz would be more desirable. It should not be forgotten, however, that in *simple* touch sensing one can do a lot with much more sluggish response. So, too, the more potent matrix taction may be able to provide *some* utility with quite modest transducer bandwidths.

In summary, perceived ideal characteristics for tactile sensing transduction are:

1. A typical array will be 10 × 10 force-sensing elements on a one-square-inch flexible surface, much like a human fingertip. Finer resolution may be desirable but is not essential for many tasks.
2. Each element should have a response time of 1-10 msec, preferably 1.
3. Threshold sensitivity for the element ought to be 1 gram, the upper limit of the force range being 1,000 grams.
4. The elements need not be linear, but they *must* have low hysteresis.

5. This skin-like sensing material has to be robust, surviving harsh industrial environments.

To follow up on the foregoing complication of perceived tactile-sensing requirements and estimates (33), a supplementary study has undertaken to acquire similar information derived from the real world of industrial assembly lines (34). Concentration was on estimates of requisite gripper and touch-sensor design parameters derived from close observation of 41 selected (generic) assembly tasks.

This was done in the context of noted time-and-motion variables, object configuration, weight, and other physical characteristics, and any perceived requirements having to do with speed, accuracy, and delicacy. Small- and medium-sized parts assembly were used to develop numbers for tactile pad-array size, resolution, and sensitivity. For each task studied, estimates were also made for gripper configuration and for overall difficulty of achieving the requisite technology.

The estimated tactile array spatial resolution needs were surprisingly small for a large variety of these selected industrial assembly tasks. [And interestingly, they match well the armchair guesstimates of (33).] Arrays of no more that 8 × 8 pressure-sensitiver points with interpoint spacing of 0.1" and having slip sensing appear to promise considerable capability. Such arrays placed on 2-jawed grippers and 3-fingered anthropomorphic hands were concluded to be able to permit robotic replacement of humans in more than 80% of the particular types of assembly tasks studied.

As an aside, it is trivial more than interest to note what can be beautifully assembled with no more manipulator articulation than a 2-jawed gripper. Consider, for instance, a robin's nest.

There are, of course, many small, fine, delicate tasks requiring tactile sensing which undoubtedly will not be challenged by automata any time in the foreseeable future. Also, hyper-fine sensitivity and resolution is called for in many of the human inspection-by-feeling operations that often accompany manipulation, pick-and-place, and assembly.

3.3 Information Processing and Pattern Recognition:

Automatic recognition of patterns, whether geometric shape, handwriting, music, or stock-market trends, is no easy matter. Most of our accomplishments to date are rudimentary and *ad hoc*. This is very obviously true for robot tactile sensing. However, as research in artificial intelligence proceeds briskly and interest in applied problems remains high, continuing progress may be anticipated -- but slowly, because the problems are extremely non-trivial.

One of the first forays into tactile shape and force-distribution recognition consisted of an interactive system; processed signals were displayed to a human operator for recognition (88). A flexible mirror, pressed against a surface, reflected to the operator (via fiber optics, videcon and TV monitor) the image of a regular checkerboard grid. Surface irregularities and outline shape distorted the grid field. Distortions representing a small number of geometric shapes were easily learned. The optical grid design was adopted after trying photoelastic stress pattern generators and moire pattern generators.

Pattern-recognition studies have been conducted with 5 × 10 and 10 × 10 arrays (14). Conductive polymer sensing was used, but in binary mode only owing to poor transduction characteristics. Objects such as circles, rectangles, and "L" shapes were successfully recognized; however, as the authors discovered, the great magnitudes of the downstream problems were just coming into focus.

In one early use of 5-fingered robot hand, discrimination by groping was studied (47,48). Twenty-one pressure-sensitive on-off switches arrayed on fingers and palm provided signals as grasped objects were manipulated to determine shape. Discrimination between a cylinder and a square pillar was high.

This last study began with a consideration of human characteristics. While such anthropomorphic reflections are intriguing, the artificial constructs generally continue to be *ad hoc*, chiefly because we really know so little about living system design. Still, at the transducer level (for touch) there may be some payoff by considering skin and finger characteristics. This approach is seen in several of the references cited here (e.g., 36,46,48,49,53,79,85,97,100).

A speculative essay on tactile shape recognition by parallel (2-D array) poke-probing is rather novel (72,73). The probe array obtains a number of 3-D object samples simultaneously. Contour-following algorithms represent the object by a series of cross-section slices.

A probabilistic approach to tactile shape recognition stemming from research with the well-known "Belgrade Hand" is outlined in (15). Four-fingered grippers are used to grasp objects having simple geometric shapes, and articulator angles yield shape information. This study is interesting in its demonstration of serious and sophisticated attention directed to analytical touch sensing. Particularly noteworthy is the start toward achieving kinesthesis in addition to cutaneous sensing.

Recent study of gripper tactile-pad sensing (6,7,109) touches on a number of issues relevant to robotic taction -- 3-D form perception, software design, and manipulator control. The material both surveys the field and reports on specific experiments. The research centers on texture assessment using an 8 × 8 opto-electronic LORD Corporation pad sensor and a 133-site conductive-foam French finger. Emphasis is on dynamic image-processing. Time-sequenced "snapshots" of simple geometric figures are analyzed. With 4 bits of dynamic range for each element and active sensing of objects larger than the "fingertip", here is another early start toward true tactile capability.

4. OUTSTANDING PROBLEMS AND NEAR-FUTURE PROJECTIONS

In order to achieve more advanced, useful, and economically visible tactile sensing for robots, two classes of problems need to be addressed intensively.

For one, improved tactile sensors requires imaginative materials and device research and development. In the other, there are the considerably less tractable problems of data management, pattern recognition, and control. What is required is sophisticated analysis, interpretation, and use of a welter of sensory feedback signals. These systems research and development requirements include not only intelligent interpretation of tactile information, but also the transformation of that information into effective actuator and manipulator control signals in response to task commands.

The problems of how to *process* and *use* information for control loom largest. Although the development of sensor devices and manipulators poses many challenges, those problems are relatively easy compared to the system design achievements needed to use the feedback signals to greatest advantage. This includes the need to develop much more complex hierarchical control structures than are presently available. Algorithmic control procedures are needed in robot environments which may have a large number of variable couplings, shapes, inertias, surfaces, etc. Optimal control in real time (at the servo level) is the big problem. Operating systems for process control, where either visual or tactile inputs of great complexity serve as inputs, are presently quite primitive. Computer control, of course, is the critical core of these matters, and despite some excellent advances in programming languages, multi-sensory feedback, and combined sensing/manipulating, the level reached so far, even in well-controlled laboratory conditions, remains rather low. An appreciation for the complexity and magnitude of this task will be obtained by consulting (11,52,63,70,71,86,89,94).

The range of outstanding problems in robotic tactile sensing can conveniently be summarized in several categories, as follows.

4.1 Materials and Transducers:

The clear need for small, linear, non-hysteretic, industrically robust, flexible, compliant, high resolution, inexpensive, finger-like arrays has not yet been met by current technology. Though these are evident advantages in each technology, none presently satisfies the multi-characteristic needs that industrial real-world requirements impose. However, as has been seen, several recent designs are promising contenders for providing most or all of those characteristics.

Conductive elastomers provide no satisfactory solution. Piezo-electric materials seem promising, but despite some years of work, no really satisfactory solutions have surfaced. Piezo-resistive compounds have not demonstrated sufficiently flexible and non-hysteretic characteristics. Semi-conductive "intelligent" skin may offer reasonable properties, but local compliance, conformable flexibility, and robustness are not easily projected. Lumped-component transduction shows considerable promise, though it is so far relatively expensive barely-sufficient spatial resolution.

4.2 Data Handling and Pattern Analysis:

Competent and general serial digital algorithms for dealing with complex 2- or 3-D shapes, either visual or tactile, are nonexistent. A few restricted and essentially trivial special cases exist. However, they add little, if anything, to the fundamental understanding needed to handle the general cases encountered in robotics. While a great deal of work has been done on object modeling and spatial reasoning, the art has barely begun.

The deceptively simple but elegant discussion of 3-D shape representation in (60) provides a good starting point for consideration of shape-recognition and manipulator-control problems (both visual and tactile). The discussion of coordinate systems as they relate to spatial models elucidates a central and mainly unsolved problem in robotics. As desired degrees of freedom (i.e., robotic manipulator joints) increase, so, too, do difficulties in computation. With increasing coordinate position-specifications and attendant coordinate-axes computations and transformations, analysis, data management, storage, and processing times expand to unmanageable proportions.

A completely untouched territory is that of true parallel processing for pattern recognition. Any reasonable development of theory and practice here should be expected to be of immense importance. Lack of parallel information-processing capability very likely accounts in large part for the sharply limited successes in multiple-object recognition capability in natural, real-world environments. There seems to be little evidence of concentrated, continuous effort on this topic.

4.3 Hierarchical Control:

A major unsolved and centrally important problem facing the next generation or sensing, smart robots is how to achieve maximum flexibility and autonomy with minimum rigid programming. Ideally, one would wish to enter a task description in high-level language and then leave it to the automaton to figure out all required steps to execute that task in some efficient way.

The linked intermediate hierarchical steps required to execute the global command implies language understanding, pattern recognition, spatial reasoning, optimal multi-axes actuator trajectory planning, recognition of completion, etc., etc. Where touch sensing is involved, this also includes the need for the software to determine orientation, have soft touch and gentle load transfer, sense slip, and provide 2-handed cooperation -- all using active (exploratory) tactile sensing.

An extremely useful, high-payoff R&D effort in this area would be an attempt to integrate know-how for all steps from task description and command to execution completion. The development of overall system-design concepts, including continuous real-time sensory feedback, is of prime importance to future robotic technology. At present, no one knows how to do it.

It may well be that our present serial digital techniques and relatively simple, linear mathematics may not be the way to go. Complete formal analysis (e.g., equations of motion) for just a six-degrees-of-freedom manipulator appears to be intractable.

One possible way out of the predicament is to avoid the explicit spatial coordinate specifications of the computer and try to model and replicate the relatively imprecise but manifestly effective ways of nervous systems. Such is the hope that the so-called cerebellar models which rely on large numbers of "computational elements" (and consequent large memory requirements) whose control-sequence potency (and accuracy) resides in network properties; e.g., (2). Alternatively, other interesting techniques have been explored which avoid both very large memory and extremely complex dynamic-feedback control analysis by use of parametric equations using table lookup where state-variable information has been acquired from practice motion-sequences (e.g, 38,75).

Distributed processing in some form appears to hold considerable promise and is currently receiving concentrated attention. Parallel and perhaps even analog or hybrid techniques to interface sensor information to effector operation via job description seem attractive, although little progress has been made. Memory is needed in fingers; as in humans, there may be considerable virtue in partial peripheral processing which puts less demand on central processor design. All-in-all, we need to find new ways to make these systems flexible, adaptive, and heuristic.

4.4 Grippers and Manipulators:

It may well be that the anthropomorphic approach using hands with three or more fingers (e.g., 3,46,70,71,86) is the way to go. But there are many basic open questions pertaining to ideal or ultimate actuator design. To what extent are human-like actuation needs important to the near- and far-future requirements of industrial automation? To what extent is there virtue in early exploration or alternative assembly and manipulation technologies which might be configured to match robotic capabilities instead of human capabilities? Answers to such questions could very well shape important new design thinking for actuators.

Perhaps the single most important question presently perceived in this domain relates to actuator power sources. It is widely acknowledged that electrical pneumatic and hydraulic systems each have one or more serious shortcomings (power, accuracy, robustness, efficiency, etc.). Consideration of alternative kinds of actuating technology presently is only at early stages.

There are a number of innovative possibilities to consider: systems using change of state, contractile polymers (muscle-like), etc. It is instructive to note that present robotic technology, in replacing a human hand/arm wielding a paint spray can, requires a 400-lb arm!

4.5 Revised Manufacturing Techniques:

The foregoing observations and analysis obviously pertain to assembly processes as they *presently* exist and are designed for *humans*. Industry in many quarters is already in transition toward meeting machines halfway, or more, by designing operations for the convenience of *machines*. Bowl feeding, palletizing, resistor strip-feeding, etc., are examples of familiar and now well-established beginnings.

The next steps, already in progress, are aimed to make life easier for assembly robots. This includes maintaining parts orientation as far upstream as possible, configuring parts for automatic identification and easy grasping, aiming for new sophistication in the design of parts presenters, and reconfiguring mechanical design -- from parts to fasteners -- to suit robots instead of people [see, for example, (13,21)].

As these revolutions in assembly are incorporated in automated lines, the heavy demands on robotic performance implicit in many of the operations described in this study should be considerably lightened. This will be especially and importantly true for the complex tasks of achieving recognition and control algorithms for the robot's brain, such as it is.

5. CONCLUSION

Despite a number of ingenious new approaches in laboratories around the world, sensitive, dextrous, smart hands are not around the very next corner. Outstanding problems in materials, transduction, signal processing, pattern recognition, control, and actuation will yield but slowly since they are complex and difficult. Besides technical obstacles, there are non-trivial economic considerations. However, as the robotics age responds increasingly to industrial and economic demands, progress should accelerate.

Robots are undoubtedly beginning to come of age, both technically and economically. The nearly 40% annual growth rate of today's still relatively simple automata indicates health in a potent new industry.

Industrial automation clearly is the target of opportunity for tactile-sensing robotics. Such operations as assembly, pick-and-place, grinding, and inspection will become increasingly more automated.

Touch sensing, now extremely primitive, may be expected to evolve into a high technology in the near future, closely paralleling automated vision capability. There are many rough problems ranging from theoretical information processing to practical materials fabrication. The promise and growing demand for tactile feedback in industrial robots will increasingly attract serious and extensive research and development programs.

Intelligent, dextrous, competent manipulating robots will surely make their appearance and pay their way in society within the coming decade.

6. ACKNOWLEDGMENTS

I am grateful for extremely helpful corrective feedback from Karen Harmon, Jack Loomis, Bill McMillan, and Marc Raibert. Their collective criticisms of an early draft were most valuable. And without the dedicated, eagle-eye conversion of scrawl to typescript by Kim Davis and Paula Delaney, there would have been nothing to criticize. I thank all, profusely. Finally, I am eternally grateful to the inventor of the word processor.

REFERENCES

[1] Abraham, R. G. et al., *State-of-the-art in adaptable-programmable assmely systems.* SME Technical Paper, MS77-757, 1977, pp. 12.

[2] Albus, James S., *Brains, Behavior, and Robotics.* Peterborough, NH: McGraw-Hill/Cyte Books, 1981.

[3] Asada, Haruhiko and Hideo Hanafusa, Prehension and handling of objects by robot hands with redundant fingers. *Trans. Soc. Instrum. & Ctrl. Eng., 15(5),* September, 1979, 666-671.

[4] Baits, J. C. et al., The feasibility of an adaptive control scheme for artificial prehension. In: *Proc. Instn. Mech. Engrs. Vol. 183, Pt. 3F,* 1968-69, 54-59.

[5] Armstrong, T. J., *Manual performance and industrial safety.* Report prepared for Dir. Safety Research of Nat. Industry for Occup. Safety and Health, Morgantown, W.V., Task Order No. 78-10433, 1978 (NTIS PB-287 330), pp. 83.

[6] Bajczy, Ruzena, Shape from touch. Unpublished manuscript, January, 1983, pp. 55 and 28 figures.

[7] Bajczy, Ruzena et al., What can we learn from one-finger experiments? Philadelphia: University of Pennsylvania, Comp. and Info. Sci. Department, Memorandum, 1982, pp. 24 and figures.

[8] Barron, E. R. et al., Load profile analyzer in the attached specification. U.S. Patent 3,818,756, June 25, 1974.

[9] Bejczy, A. K., Smart sensors for smart hands. In: *AIAA/NASA Conf. on "Smart" Sensors,* Hampton, VA, November 14-16, 1978, pp. 17.

[10] Bejczy, A. K., Effect of hand-based sensors on manipulator control performance. *Mechanism and Machine Theory, Vol. 12,* 1977, 547-567.

[11] Bejczy, A. K. and R. L. Zawacki, Computer-aided manipulator control. In: *Proc. of the First Int. Symp. on Mini- and Microcomputers in Control,* San Diego, CA, January 8-9, 1979.

[12] Birk, John, et al., General methods of enable robots with vision to acquire, orient, and transport workpieces. 7th report, University of Rhode Island to National Science Foundation, December, 1981.

[13] Boothroyd, G., Design for producibility - the road to higher productivity. *Assembly eng., 25(3),* March, 1982, 42-45.

[14] Briot, M., The utilization of an "artifical skin" sensor for the identification of solid objects. In: *9th Int. Symp. on Industrial Robots,* Wash., D.C., March, 13-15, 1979, 529-547.

[15] Briot, M., M. Renaud, and Z. Stojiljkovic. An approach to spatial pattern recognition of solid objects. *IEEE Trans. Syst. Man., and Cybernetics, SMC-8 (9),* September, 1978, 690-694.

[16] Catros, J. Y. et al., Automatic grasping using infrared sensors. In: *8th Int. Symp. on Industrial Robots,* Stuttgart, W. Germany, May 30 - June 1, 1978, 132-142.

[17] Chen, Fan Yu, Gripping mechanisms for industrial robots - an overview. *Mech. and Mach. Theory, 17(5),* 1982, 299-311.

[18] Clot, J. et al., Project pilote - Spartacus. In: *Convention de recherche,* Toulouse, France, July, 1977, pp. 21.

[19] Critchley, Macdonald, Tactile thought, with special reference to the blind *Brain, 76,* 1953, 19-35.

[20] Crossley, F. R. Erskine and F. G. Umholtz, Design for a three-fingered hand. *Mechanism and Machine Theory, Vol. 12,* 1977, 85-93.

[21] Dewhurst, P. and G. Boothroyd, Computer-aided design for assembly. *Assembly Eng., 26(3),* February, 1983, 18-22.

[22] Espiau, B. and J. Y. Catros, Use of optical reflectance sensors in robotics applications. *IEEE Trans. Syst., Man, and Cybernetics., SMC 10 (12),* December, 1980, 903-912.

[23] Flatau, C. R., Force sensing in robots and manipulators. In: *Second CISM-IFToMM Symp. on Theory and Practice of Robots and Manipulators,* Warsaw, Poland, September 14-17, 1976, 294-306.

[24] Fletcher, J. C. et al., Tactile sensing menas for prosthetic limbs. U.S. Patent 3,751,733, August 14, 1973.

[25] Frohlich, W., Grippers and clamping elements-modular elements for the automation of production technology. *The Industrial Robot, 6(4),* December, 1979, 195-199.

[26] Gibson, J. J., Observations on active touch. *Psychol. Rev., 69(6),* November, 1962, 477-491.

[27] Gibson, J. J., *The Senses Considered as Perceptual Systems.* Boston: Houghton-Mifflin, 1966.

[28] Gordon, G., ed., *Active Touch - The Mechansims of Recognition of Objects by Manipulation: A Multideisciplinary Approach.* Oxford: Pergamon Press, 1978.

[29] Goto, T. et al., Precise insert operation by tactile controlled robot. *The Industrial Robot, September, 1974, 225-228.*

[30] Gurfinkel, V. A., et. at., Tactile sensitizing of manipulators. Engineering Cybernetics, *12(6), 1974, 47-56.*

[31] Hackwood, S., G. Beni, L. A. Hornak, R. Wolfe, and T. J. Nelson, Torque-sensitive tactile array for robotics (to be published).

[32] Harmon, Leon D., Touch-sensing technology: A review. Society of Manufacturing Engineers Technical Report No. MSR80-03, Dearborn, MI, 1980, pp. 57.

[33] Harmon, Leon D., Automated tactile sensing. *Int. J. Robotics Research 1(2),* May, 1982, 3-32.

[34] Harmon, Leon D., Robotic taction for industrial assembly. *Int. J. Robotics. Res.* (in press).

[35] Held, Richard, Plasticity in sensory-motor systms. *Sci. Am., 213(5),* November, 1965, 84-94.

[36] Hillis, W. D., Active touch sensing. MIT Artificial Intelligence Laboratory Memo 629, April, 1981, pp. 37.

[37] Hirose, S. and Y. Umetani, the development of soft gripper for the versatile robot hand. *Mechanism and Machine Theory, Vol. 13,* 1978, 351-359.

[38] Horn, B. K. P. and Marc H. Raibert, Configuration space conrol. Mass. Inst. Tech. A.I. Lab. Memo. #458, December, 1977, pp. 25 and figures.

[39] Johnson, K. O., Neural mechanisms of tactual form and texture discrimination. *Fed. Proc.* (to appear June, 1983), pp. 15 and 5 figures.

[40] Johnson, J. B. and T. R. Huffman, A tactile prosthetic device for the denervated hand. In: *11th Annual Rocky Mountain Bioengineering Symp. and Int. ISA Biomedical Science Instrum. Symp.,* April 11, 1974, 9-10.

[41] Johnston, A. R., Proximity sensor technology for manipulator end effectors. *Mechanism and Machine Theory, Vol. 12,* 1977, 95-109, Pergamon Press.

[42] Jones, Bill, Spatial perception in the blind. *Br. J. Psychol., 66(4),* 1975, 461-472.

[43] Kanade, Takeo and Haruhiko Asada, Noncontact visual three-dimensional ranging devices. In: *3-D Machine Perception. Proc. Soc. Photo-Optical Instr. Engnrs., Vol. 283,* 1981, 48-53.

[44] Kato, I. et al., Artifical softness sensing - an automatic apparatus for measuring viscoelasticity. *Mechanism and Machine Theory, Vol. 12,* 1977, 11-26.

[45] Kennedy, J. M. Haptics, In: *Handbook of Perception, Vol. 8.* E. C. Carterette and M. Friedman, Eds., New York: Academic Press, 1978, 289-318.

[46] Kinoshita, G.-I., Classification of grasped object's shape by an artifical hand with multi-element tactile sensors. *Information-Control Problems in Manufacturing Technology,* Pergamon Press, 1977, 111-118.

[47] Kinoshita, G. et al., Pattern classification of the grasped object by the artificial hand. In: *3rd International Joint Conf. on Artificial Intelligence,* Stanford, CA, August 20-23, 1973, 665-669.

[48] Kinoshita, G.-I. et al., A pattern classification by dynamic tactile sense information processing. *Pattern Recognition, Vol. 7,* 1975, 243-251.

[49] Kinoshita, G. and Y. Uegusa, A study of an artificial hand with tactile sensors. *Bull. of the Faculty of Science and Engineerings, Chuo, Univ., Vol. 20,* 1978, 231-245.

[50] Kruger, L. E., David Katz's Der Aufban der Tastwelt (The World of Touch): A synopsis. *Percept. and Psychophys., 7(6),* 1970, 337-341.

[51] Kwee, H. H., The spartacus telethesis: "soft touch". In: *Int. Conf. on Telemanipulators for the Physically Handicapped,* Rocquencourt, France, September 4-6, 305-308.

[52] Larcombe, M. H. E., Tactile perception for robot devices. In: *1st Conf. on Industrial Robot Technology,* Univ. of Nottingham, UK, March 27-29, 1973, 191-196.

[53] Larcombe, M. H. E., Tactile sensing using digital logic. In: *Proc. Shop Floor Automation Conf.,* Paper #9, Birniekill Inst., National Engineering Lab, East Kilbride, December 11-13, 1973, pp. 7.

[54] Larcombe, M. H. E., Tactile sensors, sonar sensors and parallax sensors for robot applications. In: *3rd Conf. on Industrial Robot Tech. and 6th Int. Symp. on Industrial Robots,* Univ. of Nottingham, UK, March 24-26, 1976, 25-32.

[55] Lederman, Susan J., The perception of texture by touch. Ch. 4 in: *Tactual Perception: A Sourcebook,* W. Schiff and E. Foulke, Eds., Cambridge: Cambridge University Press, 1982, 130-167.

[56] Lerner, Eric J., Robots to crash assembly lines. *High Technology,* February, 1980, 44-52.

[57] Loomis, Jack M., Tactile pattern perception. *Perception, (10),* 1981, 5-27.

[58] Loomis, Jack M. and Susan J. Lederman. Tactual Pattern Perception. Chapter in *Handbook of Human Perception and Performance,* K. Boff, L. Kaufman, and J. Thomas Eds., New York: Wiley (to be published).

[59] Lundstrom, G., Industrial robot grippers, *The Industrial Robot,* December, 1973, 72-82.

[60] Marr, David, and H. K. Nishihara, *Representation and recognition of the spatial organization of three dimensional shapes.* MIT Artificial Intelligence Laboratory Report No. AIM 416, May, 1977, pp. 33.

[61] Masuda, R. et al., Slip sensor of industrial robot and its application. *Electrical Eng. in Japan, Vol. 96, No. 5,* 1976, 129-136.

[62] Mcdermott, Jim, Sensors and transducers. *Electronic Data News,* March 20, 1980, 122-142.

[63] Meyer, Jeanine, An emulation system for programmable sensory robots. *IBM J. Res. Develop., 25(6),* November, 1981, 955-962.

[64] Middlehoek, Sinom et al., Microprocessors get integrated sensors. *IEEE Spectrum, 17(2),* February, 1980, 42-46.

[65] Mitchell, R. J., Pressure resopnsive resistive material. U.S. Patent 3,806,471, April 23, 1974.

[66] nerozzi, Andrea and Gabriele Vessura, Study and experimentation of a multi-finger gripper. In: *Proc. 10th International Symposium on Industrial Robots and 5th International Conf. on Industrial Robot Technology,* Milan, Italy, March 5-7, 1980, 215-223.

[67] Nitzan, David, Assessment of robotic sensors. *NSF Robotics Research Workshop,* Newport, RI, April 15-17, 1980, pp. 13.

[68] Nitzan, D., et al., *Machine intelligence research applied to industrial automation.* SRI International, Menlo Park, CA, August, 1979, pp. 209.

[69] Okada, T. A short-range finding sensor for manipulators. Electro-technical Lab., Tokyo, Japan, May 4, 1978, 28-40.

[70] Okada, T., Object-handling system for manual industry. *IEEE Trans. on Systems, Man, and Cybernetics, Vol. SMC-9, No. 2,* February, 1979, 79-89.

[71] Okaha, T., Computer control of multikjointed finger system for precise object handling. *IEEE Trans. Syst., Man, and Cyber., SMC-12 (3),* June, 1982, 289-299.

[72] Page, C. J. et al., New techniques for tactile imaging. *The Radio and Electronic Engineer, Vol. 46, No. 11,* November, 1976, 519-526.

[73] Pugh, A et al., Novel techniques for tactile sensing in a three dimensional environment. *The Industrial Robot,* March, 1977, 18-26.

[74] Purbrick, John A., A force transducer employing conductive silicone rubber. In: *Proc. 1st Robot Vision and Sensors Conf.,* IFS Pubs., Letd.: Kempston, Bedford, England, 1980, 73-80.

[75] Raibert, Marc H., A model for sensorimotor control and learning. *Biol. Cyber., 29,* 1978, 29-36.

[76] Raibert, Marc H., and John E. Tanner, Design and implementation of a VLSI tactile sensing computer. *Int. J. Robotics Res., 1(3),* 1982, 3-18.

[77] Rakic, M., The "Belgrade hand prosthesis". In: *Proc. Instn. Mech. Engrs., Vol. 183, Pt 3F,* 1968-69, 60-67.

[78] Rebman, Jack, Compliance: The forgiving factor. *Robotics Today,* Fall, 1979, 29-34.

[79] Ring, N. D. and D. B. Welbourn, A self-adaptive gripping device: its design and performance. In: *Proc. Instn. Mech. Engrs., Vol. 183, Pt. 3F,* 1968-69, 45-49.

[80] Rosen, C. A., Machine vision and robotics: Industrial requirements. In: *Computer Vision and Sensor-Based Robots,* G. C. Dodd and L. Rossol, eds., New York: Plenum, 1979, 3-22.

[81] Rosen, Charles, et al., *Exploratory research in advanced automation.* First Semi-Annual Report. Stanford Research Institute, Menlo Park, CA, April 1, 1973 to September 30, 1973.

[82] Rosen, C. A. and D. Nitzan, Use of sensors in programmable automation. *IEEE Computer Society Magazine,* December, 1977, 12-23.

[83] Rovetta, Alberto, On biomechanics of human hand motion in grasping: A mechanical model. *Mechanism and Machine Theory, Vol. 14,* 1979, 25-29.

[84] Rovetta, Alberto, On the prehension of the human hand. *Mechanism and Machine Theory, Vol. 14,* 1979, 385-388.

[85] Sakai, I. et al., Approach and plan: most suitable control of grasping in industrial robot. In: *5th Int. Symp. on Industrial Robots,* IIT Research Institute, Chicago, IL, September 22-24, 1975, 525-532.

[86] Salisbury, J. K. and J. J. Craig, Articulated hands: force control and kinematic issues. *Int. J. Robotics Research 1(1), 1982, 4-17.*

[87] Schneiter, J. L. An optical tactile sensor for robots. M. S. Thesis, Department Mechanical Engineering, Mass. Inst. Tech., August, 1982, pp. 104.

[88] Sheridan, T. B. and W. R. Ferrell, *Measurement and display of control informantion.* Progress Report, Engineering Projects Lab., Mass. Inst. Tech., October 1, 1966 - March 31, 1967, 13-22.

[89] Sheridan, T. B. and W. L. Verplank, *Human and computer control of undersea teleoperators.* Man-Machine Systems Lab., Mass. Inst. Tech., July 15, 1978.

[90] Sherrick, C. E., and J. C. Craig, The psychophysics of touch. Ch. 2 in: *Tactual Perception: A Sourcebook,* W. Schiff and E. Foulke, eds. Cambridge: Cambridge Univ. Press 1982, 55-81.

[91] Snyder, W. E. and J. St. Clair, Conductive elastomers as sensor for industrial parts handling equipment. *IEEE Trans. on Instrum. and Measure., Vol. IM-27, No. 1,* March, 1978, 94-99.

[92] Stojilskovic, Z. and J. Clot, Integrated behavior of artificial skin. *IEEE Trans. on Biomed. Eng., Vol. BME-24, No. 4,* July, 1977, 396-399.

[93] Stojiljkovic, Z. and D. Saletic, Learning to recognize patterns by Belgrade hand prosthesis. In: *5th Int. Symp. on Industrial Robots,* Chicago, IL, 1975, 407-413.

[94] Takases, K., Task-oriented variable control of manipulator and its software servoing system. In: *Proc. of the IFAC Int. Symp. on Information-Control Problems in Manufacturing Technology,* Tokyo, Japan, October, 1977, 139-145.

[95] Takase, Kunikatsu, Robotics research in Japan. *Robotics Age, 1(2).* Winter, 1979, 30-39.

[96] Tella, R. et al., A contour adapting vacuum gripper. In: *Proc. 10th International Symposium on Industrial Robots and 5th International Conf. on Industrial Robot Technology,* Milan, Itlay, March 5-7, 1980, 175-189.

[97] Tomovic, R. and Z. Stojiljkovic, Multifunctional terminal device with adaptive grasping force. *Automatica, Vol. 11,* 1975, 567-570.

[98] Ueda, M. et al., Sensors and systems of an industrial robot. *Memoirs of the Faculty of Engineering, Nagoya Univ., Vol. 27, No. 2,* 1975, 163-207.

[99] Umetani, Yoji and Kan Taguchi, Discrimination of general shapes by psychological feature properties. In: *Proc. 1st Int. Conf. on Robot Vision and Sensory Controls,* Bedford, England: IFS Pubs., 1981, 135-148.

[100] Vainshtein, G. G. et al., Some principals of functional organization of sensory systems and their technical counterparts. In: *2nd CISM-IFToMM Symposium,* Warsaw, Poland, September 14-17, 1976, 326-338.

[101] Van der Loos, H. F. M., Design of a three fingerd robot gripper. *Industrial Robot, 5(4),* December, 1978, 179-182.

[102] Wang, S. S. M. and P. M. Will, Sensors for computer controlled mechanical assembly. *The Industrial Robot,* March, 1978, 9-18.

[103] Ward, Mitchel R. et al., Consight: an adaptive robot with vision. *Robotics Today,* Summer, 1979, 26-32.

[104] Warnecke, H. H. et al., Pilot work site with industrial robots. In: *5th Int. Symp. on Industrial Robots,* Chicago, IL, September 22-24, 1975, 71-86.

[105] Watson, P. C. and S. H. Drake, Pedestal and wrist force sensors for automatic assembly. In: *5th Int. Symp. on Industrial Robots,* Chicago, IL, September 22-24, 1975, 501-511.

[106] Whitney, D. E. et al., Robot and manipulator control by exteroceptive sensors. In: *Proc. 1977 Joint Automatic Control Conf.,* 1977, 155-163.

[107] Whitney, D. E. et al., *Part mating theory for compliant parts.* First report, R-1407, Cambridge, MA: C. S. Draper Lab., Inc., 1980, pp. 123.

[108] Whitney, D. E., et al., *Part mating theory for compliant parts.* Second report, R-1537, Cambridge, MA: C. S. Draper Lab., Inc., 1982, pp. 170.

[109] Wolfeld, J. A., *Real time control of a robot tactile sensor.* M. S. Thesis, Univ. of Penn., The Moore School of Electrical Eng., August, 1981, pp. 62.

[110] Zinchenko, V. P. and B. F. Lomov, The functions of hand and eye movements in the process of perception. *Problems of Psychology., 1,* 1960, 12-26.

NOTE: With the permission of John Wiley and Sons, Inc., this material will also appear in the *Proceedings of the NATO Advanced Study Institute on Robotics and Artificial Intelligence,* Lucca, Italy, June 26, 1983 - July 8, 1983.

Index